MINIATURIZED AND INTEGRATED FILTERS

MINIATURIZED AND INTEGRATED FILTERS

Edited by

Sanjit K. Mitra
Department of Electrical
and Computer Engineering
University of California
Santa Barbara, California

Carl F. Kurth
AT&T Bell Laboratories
North Andover, Massachusetts

A Wiley-Interscience Publication

JOHN WILEY & SONS

New York • Chichester • Brisbane • Toronto • Singapore

***Library of Congress Cataloging in Publication Data*:**

Miniaturized and integrated filters / edited by Sanjit K. Mitra, Carl F. Kurth.
p. cm.
"A Wiley-Interscience publication."
Includes bibliographies and index.
1. Electric filters. I. Mitra, Sanjit Kumar. II. Kurth, Carl F.
TK7872.F5M55 1989 88–28219
621.3815'324--dc 19 CIP
ISBN 0-471-84496-9

Printed in the United States of America

10 9 8 7 6 5 4 3 2 1

To our wives
Nandita Mitra and Ursula Kurth

Contributors

DR. ERICH CHRISTIAN, 5704 Edgedale Drive, Raleigh, NC 27612

DR. JALIL FADAVI-ARDEKANI, AT&T Bell Laboratories, 1247 South Cedar Crest Blvd, Allentown, PA 18103

DR. RUBIK GREGORIAN, Sierra Semiconductor Corporation, 2075 North Capitol Avenue, San Jose, CA 95132

MR. A. GUILABERT, Standard Electrica S.A., Una Asociada Espanola a ITT, Ramirez de Prado 5, Madrid 7, Apartado 50702, Spain

MR. ROBERT JOHNSON, CTSD Filter Products, Rockwell International Corporation, 2990 Airway Avenue, Costa Mesa, CA 92626

MR. D. KLEIN, Standard Elektrik Lorenz AG, Postfach 400749, 7000 Stuttgart 40, West Germany

MR. CARL F. KURTH, AT&T Bell Laboratories, 1600 Osgood Street, North Andover, MA 01845

PROFESSOR JOHN MAVOR, Department of Electrical Engineering, University of Edinburgh, Edinburgh, EH9 3JL, United Kingdom

PROFESSOR SANJIT K. MITRA, Department of Electrical & Computer Engineering, University of California, Santa Barbara, CA 93106

DR. KALYAN MONDAL, AT&T Bell Laboratories, 1247 South Cedar Crest Blvd, Allentown, PA 18103

DR. R. L. ROSENBERG, AT&T Bell Laboratories, 1600 Osgood Street, North Andover, MA 01845

PROFESSOR ADEL SEDRA, Department of Electrical Engineering, University of Toronto, Toronto, Ontario M5S 1A4, Canada

DR. R. C. SMYTHE, Vice President, Piezo-Technology Inc., 2525 Shader Road, Orlando, FL 32854-7859

PROFESSOR GABOR C. TEMES, Electrical Engineering Department, University of California, Los Angeles, CA 90024

Contents

Preface

The origin of electronic filters dates back to 1915 when K. W. Wagner in Germany and G. R. Campbell in the United States independently introduced passive electric wave filters to meet the needs of the young communications industry. Since then, there have been significant and continuous advances in filter theory and technology along with the broadening of the term *filter*. Presently, an electronic circuit having a prescribed frequency-domain or time-domain response to a given excitation is called a filter. During the first half of the last 73-year period, the dominant technology was the passive lumped filters which consequently received the most attention with regard to the development of both theory and practice. However, during the last 40 years, as the passive lumped filter technology reached a plateau with respect to performance, size, and cost reductions, a number of developments precipitated the search for different filter technologies. The rapid growth of the telecommunication industry along with the use of filters in various other systems, increased the market for filters; many applications required filters with more exacting performance characteristics; and furthermore, high volume production at low cost of some types of filters required technologies offering higher precision in tuning and stability over time and required temperature ranges. This book provides a review of some of these newer filter technologies that have led to the automated fabrication, testing and tuning of miniaturized and integrated filters. A number of technologies investigated during this period turned out to be not sufficiently economical to justify their introduction into practice, and as a result they are not included here. Our emphasis is on technologies that are currently in use or have a promising future. This book has evolved from a special issue on Miniaturized Filters of the Proceedings of the IEEE published in January 1979 for which one of the editors of this book (SKM) served as a guest editor. The special issue was well received by the readers, and thus it was felt that a hardcover book covering the important filter technologies would be of interest to practicing engineers and researchers in the field.

This book consists of nine chapters. The first chapter provides an introduction to the basic concepts, definitions, and terminologies. It also describes the evolution of filter technologies and the rationale behind miniaturization along with some typical applications. Chapter 2 focusses on the miniaturization of passive lumped LC filters.

For low frequency applications, inductors tend to be bulky and expensive. As a result, considerable effort has been devoted during the last 40 years to the development of inductorless filters. One of the early approaches was based on the use of active elements in conjunction with resistors and capacitors. This approach became economically competitive with the availability of inexpensive monolithic operational amplifiers along with advances in thin- and thick-film resistor and capacitor fabrication technologies. Chapter 3 reviews this approach in the current framework. An alternate approach to the design of active inductorless filters that has become highly popular in recent years is based on the use of periodically operated electronic switches, amplifiers, and capacitors. This approach, known as the switched-capacitor filter technology, has led to the fabrication of all monolithic active inductorless filters and is the subject of Chapter 5.

In many applications mechanical and crystal filters with very sharp attenuation characteristics have replaced passive lumped LC filters since they offer a number of advantages with regard to performance, lower size and cost, and manufacturability. These filters make use of mechanical resonances to perform the filtering of electrical signals. Mechanical filter technology uses wire-coupled discrete mechanical resonators and is described in Chapter 4. Crystal filter technology, on the other hand, makes use of deposited electrodes on quartz crystal substrates. The trapped-energy regions between the electrodes serve as the resonators, while the coupling is provided by the nonplated regions between the electrode pairs. Chapter 7 is concerned with the crystal filter technology of the bulk-resonator types. Another alternate *crystal* filter technology, described in Chapter 8, makes use of acoustic waves propagating over and/or being reflected by grated crystal surfaces. These filters are known as surface acoustic wave (SAW) resonator filters.

In the late 1960s, analog delay lines based on charge-transfer devices were introduced. Here, analog samples are stored individually in capacitors formed within the devices, and filtering is accomplished by the introduction of delays, as well as summation and scaling of the individual charge packets. Such filters can be fabricated in all monolithic integrated circuit forms. This novel approach appears to be promising for the implementation of miniaturized transversal filters with programmable coefficients and is the subject of discussion in Chapter 6.

Unlike all other filters discussed so far, the signals in digital filters (Chapter 9) are totally digital in form; that is, they are binary coded versions of samples obtained by uniformly and linearly sampling an analog signal. This radically new approach to filtering has been gaining popularity rapidly since the introduction of inexpensive single-chip digital signal processors in the early 1980s.

The authors of this work are recognized experts in their respective filter design technologies. Together they have over 250 years of industrial experience and over 75 years of teaching experience.

The editors thank each of the authors for their timely contribution of high-quality texts. We also thank Dr. Renato Gadenz of AT&T Bell Laboratories and Professor Edgar Sanchez-Sinencio of Texas A & M

University for carefully reading the complete original manuscript and making numerous comments that have improved the style and content of the materials presented. We appreciate the assistance of Ms. Renee Leach of the University of California, Santa Barbara in proofreading the galleys and page proofs.

SANJIT K. MITRA

Santa Barbara, California

CARL F. KURTH

North Andover, Massachusetts

MINIATURIZED AND INTEGRATED FILTERS

1 Introduction

CARL F. KURTH
AT&T Bell Laboratories, North Andover, Massachusetts

SANJIT K. MITRA
University of California, Santa Barbara, California

Electronics have played a key role in the development of modern societies and are involved directly or indirectly in every facet of our daily activities. Applications are wide and diverse, ranging from communication, business, defense, education, entertainment, health care, space exploration, transportation, and various other areas. The use of electronics began in the early twentieth century and almost from the beginning, the electrical filter has been an important part of most electronics systems.

An *electrical filter* processes an electrical signal applied to its input such that the signal at the output has desirable properties according to the specifications in a particular application. In some applications, the filter is an electrical network exhibiting frequency selective properties. Such a network may be used to *pass* certain frequency components and *stop* other frequency components in the input signal. In this case, the processed signal appearing at the filter's output contains essentially the frequency components that are allowed to pass through without much degradation. A frequency selective filter may be used to compensate for the dispersion caused by a transmission medium such as the wideband cable and its inevitable reflection due to mismatching so that the overall system appears *close* to an ideal transmission channel with a constant delay. There are applications where the filter is designed to exhibit specific time-domain characteristics. In fact, presently, any electrical network designed to develop a specific response, whether in the frequency-domain or in the time-domain, for a given excitation is called a filter.

Generally speaking, filters can be classified into two broad groups: *continuous* and *sampled-data*. This classification is primarily based on the type of electrical signal being processed by the filter. In continuous filters the signal is a continuous function of time with its amplitude at each instant of time being permitted in principle to take any value. In a sampled-data filter, on the other hand, the signal is sampled and processed at discrete instants of time that usually occur at equally spaced intervals. This latter type of filter can again be one of two types. In one type, called *analog sampled-data filters*, the sampled signal in principle can take any value, whereas, in the other type, called *digital filters*, the sampled signal is further discretized and can take one of a fixed number of quantized values represented by binary numbers. This last type of signal is commonly called a digital signal in contrast to the non-digitized ones which are referred to as analog signals.

Each technology can be identified with a specific type of filter. Presently, technologies used to construct continuous analog filters are one of the following: passive lumped LC filters, mechanical filters, crystal filters, surface acoustic wave (SAW) filters, and active resistance-capacitor (RC) filters. Likewise, technologies in use for implementing sampled-data filters are the following: charge-transfer device (CTD) filters, switched-capacitor (SC) filters and digital filters, where the first two are strictly speaking analog sampled-data filters. This book consists of chapters on these filter technologies with an emphasis on practical aspects that are relevant in miniaturized and integrated filter implementations.

1.1. EVOLUTION OF FILTER TECHNOLOGY [19]

The passive lumped LC filter technology was introduced first and is still most widely used. It is based on the interconnection of discrete elementary components consisting of inductors, capacitors, and resistors. Here electrical resonances are provided by tuned circuits composed of inductors and capacitors. Since its introduction in 1915, the passive LC filter technology has dominated in most applications until about 1965. One of the major applications of the passive lumped LC filter has been in the design and implementation of channel bank filters in frequency-division multiplex (FDM) telephone systems. Here voice signals of multiple subscribers are multiplexed into adjacent frequency bands of the transmission channel with the individual signals being demultiplexed at the receiving end with the aid of a bank of bandpass filters. The FDM channel filters were initially designed with discrete components. Chapter 2 (**Miniaturized LC Filters**) describes some more recent techniques used to improve these traditional designs and to reduce their size to compete with newer technologies.

As the telecommunication industry began expanding rapidly and numerous other applications of filters started proliferating, the cost, size, and performance requirements have shifted the design of filters to other technologies that are more amenable to automated manufacturing methods. One of the successful alternative approaches made use of mechanical resonators to achieve filtering. The two filter categories based on this approach are the mechanical filters and the crystal filters. With the introduction of each of these technologies the implementation of filters with very sharp attenuation characteristics became possible and resulted in improving the economics of

FDM telecommunication systems. These two newer approaches are discussed in Chapter 4 (**Mechanical Filters**) and Chapter 7 (**Crystal Filters**). Another similar technology is based on acoustic waves propagating or resonating on grated crystal surfaces. They are known as SAW filters. This new technology is considered in Chapter 8 (**SAW Filters**). It should be noted that the crystal filters described in Chapter 7 are of the bulk-resonator type. Most of the commercial applications of SAW filters have been associated with their use as intermediate frequency (IF) filters in television sets, correlation filters in radar systems and time recovery circuits for high precision signals.

The development of silicon integrated circuit technology in the 1960s led to an extensive search for the design of inductorless filters for very low-frequency applications where inductors tend to be bulky and expensive. With the availability of monolithic operational amplifiers in the mid-1960s, filters were designed using these amplifiers together with discrete resistors and capacitors. They are more commonly called active RC filters and became an attractive alternative. The introduction of computer-controlled laser trimming of thin-film and thick-film resistors provided a convenient way of adjusting the performances of active RC filters and thus led to an increased use of these filters in low-frequency, high-volume applications. For example, in the pulse code modulation (PCM) telecommunication system, a low-pass filter is used to limit the frequency spectrum of the voice signal to about 4 kHz. The output of this filter is then converted into a PCM digital signal before it enters into the digital transmission network. Because of the large numbers required for interfacing subscribers into a time-division multiplex (TDM) digital system, this filter has attracted considerable attention during the last 20 years. The **active RC filter** technology is the subject of Chapter 3.

Due to the difficulty in making fully integrated resistors, the active RC filters were not amenable to fabrication in monolithic form on one silicon chip. The search thus continued to develop active circuits without inductors and without resistors. Around 1960, it was found that some configurations of circuits containing capacitors and periodically operated switches can be made to look like resistors under certain conditions. This new development towards resistor replacement became an attractive alternative to miniaturized active RC technologies, thus finally fulfilling the dream of a complete filter on a single chip of silicon. Since then, the production of a monolithic low-pass filter with an A/D converter for PCM telephony applications has indeed become a reality. Chapter 5 (**Switched-Capacitor Filters**) describes this new filter design approach.

With the introduction of analog delay lines using charge-transfer devices in the late 1960s, it became logical to investigate the development of transversal filters, based on the tapped delay line concept. The CTD approach appears attractive for implementing very high order filters with programmable coefficients and is amenable for fabrication in fully-integrated circuit form. Chapter 6 (**Charge-Transfer Device Filters**) provides a review of this filter design approach.

With the dramatic advances in digital integrated circuit (IC) technology in recent years, the emphasis of electronics has shifted towards digital techniques. A rapid growth in the use of digital techniques for frequency-selective filtering took place due to the availability of commercial low-cost fully-integrated programmable digital signal processing (DSP) devices in the 1980s. Their use is expected to increase quite rapidly as more and more engineers opt for all digital designs for a great variety of systems. The last chapter of this book, Chapter 9 (**Digital Filters**), examines this radically different approach.

1.2. FILTER CHARACTERIZATION

In this book we restrict our attention to single-input, single-output filters. Moreover, the filter is assumed to be linear, time-invariant, causal and stable. Because of the linearity assumption, the output variable (usually called the *response*) is linearly related to the input variable (called the *excitation*). The precise nature of this relation depends on the type of filter being considered.

1.2.1. Continuous-Time Filters

In the case of filters described in Chapters 2, 3, 4, 7, and 8, all signal variables are continuous functions of time. A block diagram representation of a single-input, single-output continuous-time filter is shown in Figure 1.1 where $u(t)$ and $y(t)$, respectively, denote the input and the output variables. In the case of a linear, time-invariant, causal filter, the input and output variables are related through the convolution integral given by

$$y(t) = \int_0^t h(t-\tau)u(\tau)\,d\tau \tag{1.1}$$

where $h(t)$ represents the *impulse response* of the filter, that is, the response of the filter to a unit impulse $\delta(t)$ applied at its input.

An equivalent relation in the frequency-domain is obtained by taking the Laplace transform of both sides of Eq. (1.1) resulting in

$$Y(s) = H(s)U(s) \tag{1.2}$$

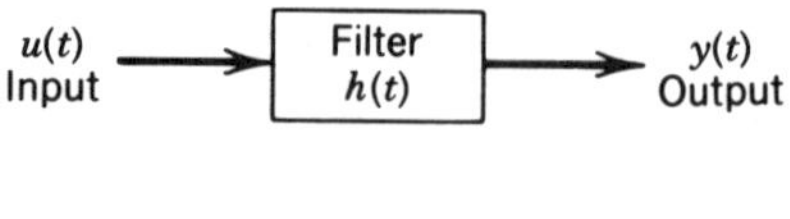

Figure 1.1 A single-input, single-output continuous-time filter.

where $Y(s)$, $U(s)$, and $H(s)$ are, respectively, the Laplace transforms of $y(t), u(t)$, and $h(t)$. The quantity $H(s)$ is more commonly known as the *transfer function* of the filter and is thus given by the ratio of the Laplace transform of the response to the Laplace transform of the excitation.

In the case of continuous-time filters composed of lumped elements, the transfer function $H(s)$ is a real rational function of s; it is a ratio of two polynomials in s with real coefficients

$$H(s) = \frac{B(s)}{A(s)} = \frac{b_0 + b_1 s + b_2 s^2 + \cdots + b_N s^N}{1 + a_1 s + a_2 s^2 + \cdots + a_N s^N} \tag{1.3}$$

The transfer function can be alternately written in the form

$$H(s) = \frac{b_N}{a_N} \prod_{i=1}^{N} \left(\frac{s - \xi_i}{s - \lambda_i} \right) \tag{1.4}$$

by factoring the numerator and denominator polynomials. The zeroes of the polynomial $A(s)$ given by $s = \lambda_i$ are called the *poles* of $H(s)$ and are the *natural frequencies* of the filter. For the stability of the filter, all poles of $H(s)$ must lie in the left half of the complex s-plane excluding the $j\omega$-axis; that is, $\text{Re}\ \lambda_i < 0$. The zeroes of $H(s)$ or, equivalently, the zeroes of $B(s)$, given by $s = \xi_i$ are known as the *transmission zeroes* of the filter as, at these complex frequencies, the filter produces no output for any finite input.

On the $j\omega$-axis, for a stable filter the transfer function reduces to $H(j\omega)$, which is the continuous-time Fourier transform of $h(t)$ and can be written as

$$H(j\omega) = |H(j\omega)|\, e^{j\theta(\omega)} \tag{1.5}$$

where $|H(j\omega)|$ is called the *magnitude function* and $\theta(\omega) = \arg H(j\omega)$ is called the *phase function*. The *gain function* of the filter is given as

$$\alpha(\omega) = 20 \log_{10} |H(j\omega)|\ dB \tag{1.6}$$

The negative of the gain function is called the *loss function*. The filter's specification is often given in terms of its loss or the phase as a function of the frequency ω. The *group delay* of the filter is defined as

$$\tau(\omega) = -\frac{d\theta(\omega)}{d\omega} \tag{1.7}$$

1.2.2. Discrete-Time Filters

In this book we shall also consider filters in which all signal variables are discrete functions of time. These filters are discussed in Chapters 5, 6, and 9. Here each signal can be considered as a sequence of numbers with each element being a function of an integer variable n. A block diagram representation of a single-input, single-output discrete-time filter is sketched in Figure 1.2 where $u[n]$ and $y[n]$, respectively, denote the input and the output signals.

The input and output variables, in the case of a linear, time-invariant, causal discrete-time filter, are related through the convolution sum given by

$$y[n] = \sum_{k=0}^{\infty} h[k]u[n-k] \tag{1.8}$$

where $\{h[n]\}$ represents the *impulse response sequence* of the filter, that is, the response of the filter to a unit impulse sequence $\{\delta[n]\}$ applied at its input.

An equivalent relation in the frequency-domain is obtained by taking the discrete-time Fourier transform of both sides of Eq. (1.8) resulting in

$$Y(e^{j\omega}) = H(e^{j\omega})U(e^{j\omega}) \tag{1.9}$$

where $Y(e^{j\omega})$, $U(e^{j\omega})$ and $H(e^{j\omega})$ are, respectively, the discrete-time Fourier transforms of $y[n]$, $u[n]$ and $h[n]$. The quantity $H(e^{j\omega})$ is more commonly known as the *frequency response function* of the filter and is thus given by the ratio of the discrete-time Fourier transform of the response to the discrete-time Fourier transform of the excitation.

In the case of discrete-time filters it is often convenient to express the input-output relation through the use of z-transforms. Thus, a z-transform of both sides of Eq. (1.8) yields

$$Y(z) = H(z)U(z) \tag{1.10}$$

where $Y(z)$, $U(z)$, and $H(z)$ are, respectively, the z-transforms of $y[n]$, $u[n]$, and $h[n]$. The function $H(z)$ is usually called the *transfer function* of the discrete-time filter and is thus given by

$$H(z) = \sum_{n=0}^{\infty} h[n]z^{-n} \tag{1.11}$$

Unlike the continuous-time case, there are basically two types of discrete-time filters. If the impulse response sequence $h[n]$ is defined for values of n within a finite range $0 \leq n < N$, the filter is called a *finite-impulse response* (FIR) filter, otherwise, it is called an *infinite-impulse response* (IIR) filter. The transfer function of an FIR filter of length N is thus a polynomial in z^{-1}

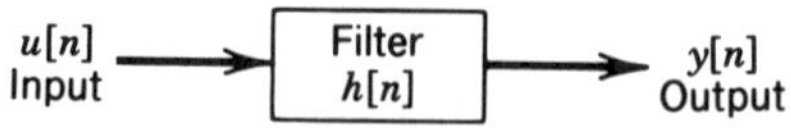

Figure 1.2 A single-input, single-output discrete-time filter.

$$H(z) = \sum_{n=0}^{N-1} b_n z^{-n} \tag{1.12}$$

where $b_n = h[n]$, $n = 0, 1, \ldots, N-1$, are the impulse response samples.

A class of IIR filters is characterized in the time-domain by a constant coefficient difference equation of the form

$$y[n] = \sum_{k=0}^{N} b_k u[n-k] - \sum_{k=1}^{N} a_k y[n-k] \tag{1.13}$$

where N is the *order* of the difference equation. Such an IIR filter is described by a transfer function that is a rational function of z^{-1}, that is, a ratio of two polynomials in z^{-1}

$$H(z) = \frac{B(z)}{A(z)} = \frac{b_0 + b_1 z^{-1} + b_2 z^{-2} + \cdots + b_N z^{-N}}{1 + a_1 z^{-1} + a_2 z^{-2} + \cdots + a_N z^{-N}} \tag{1.14}$$

The above transfer function can be alternately written in the form

$$H(z) = b_0 \prod_{k=1}^{N} \left(\frac{z - \xi_k}{z - \lambda_k} \right) \tag{1.15}$$

by factoring the numerator and denominator polynomials. The zeroes of the polynomial $A(z)$ given by $z = \lambda_k$ are called the *poles* of $H(z)$ and are the natural frequencies of the filter. For the stability of the filter, all poles of $H(z)$ must lie strictly inside the unit circle ($|z| = 1$) in the complex z-plane, that is, $|\lambda_k| < 1$. The zeroes of $H(z)$ or, equivalently, the zeroes of $B(z)$, given by $z = \xi_k$ are known as the *transmission zeroes* of the filter as, at these complex frequencies, the filter produces no output for any finite input.

On the unit circle, for a stable filter the transfer function reduces to the frequency response function $H(e^{j\omega})$, the discrete-time Fourier transform of $h[n]$, that can be written as

$$H(e^{j\omega}) = |H(e^{j\omega})| e^{j\theta(\omega)} \tag{1.16}$$

where $|H(e^{j\omega})|$ is called the *magnitude function* and $\theta(\omega) = \arg H(e^{j\omega})$ is called the *phase function*. The *gain function* of the filter is given as

$$\alpha(\omega) = 20 \log_{10} |H(e^{j\omega})| \; dB \tag{1.17}$$

The negative of the gain function is called the *loss function*. The discrete-time filter's specification is usually given in terms of the loss or the phase as a function of the frequency ω. The *group delay* of the filter is defined as

$$\tau(\omega) = -\frac{d\theta(\omega)}{d\omega} \tag{1.18}$$

Unlike the IIR transfer function, an FIR transfer function can be designed with a constant group delay. If the FIR filter is characterized by a symmetric impulse response $h[n]$ given by

$$h[n] = h[N-1-n] , \quad \text{for } 0 \le n \le N-1 \tag{1.19}$$

then the filter has a linear phase response given by

$$\theta(\omega) = -\left(\frac{N-1}{2}\right)\omega , \quad -\pi \le \omega \le \pi \tag{1.20}$$

and as a result, has a constant group delay. For an FIR filter with an anti-symmetric impulse response $h[n]$ of the form

$$h[n] = -h[N-1-n] , \quad \text{for } 0 \le n \le N-1 \tag{1.21}$$

the phase response is linear and is given by

$$\theta(\omega) = -\left(\frac{N-1}{2}\right)\omega \pm \frac{\pi}{2} , \quad \text{for } -\pi \le \omega \le \pi \tag{1.22}$$

and the corresponding group delay is again a constant.

1.3. THE FILTER DESIGN PROCESS

The design and implementation of a filter consists roughly of seven steps as indicated by the flowchart of Figure 1.3 [28]. The first step (Box 1) is concerned with the development of the frequency-domain or the time-domain performance specifications that the filter is required to meet. Such specifications may be in the form of desirable loss and phase (delay) responses usually given with some tolerances as shown in Figure 1.4 for a typical bandpass filter loss characteristic. In addition, the specifications may include desired signal levels, size, weight, cost, and other equally important practical requirements.

The second step (Box 2) involves the selection of an appropriate type of filter that can be designed to meet the specifications of the previous step. The selection process also must take into account other practical considerations such as the frequency range of operation, degree of frequency selectivity desired, signal and impedance levels, size and cost. As the later chapters of this book indicate, a number of different types of filters are at the disposal of the engineer designing the filter.

The next step, shown in Box 3, is usually known as the *approximation problem* in which a stable filter transfer function ($H(s)$ for the continuous-time filter or $H(z)$ for the discrete-time filter) is determined meeting the specifications developed in the first step (Box 1) while at the same time

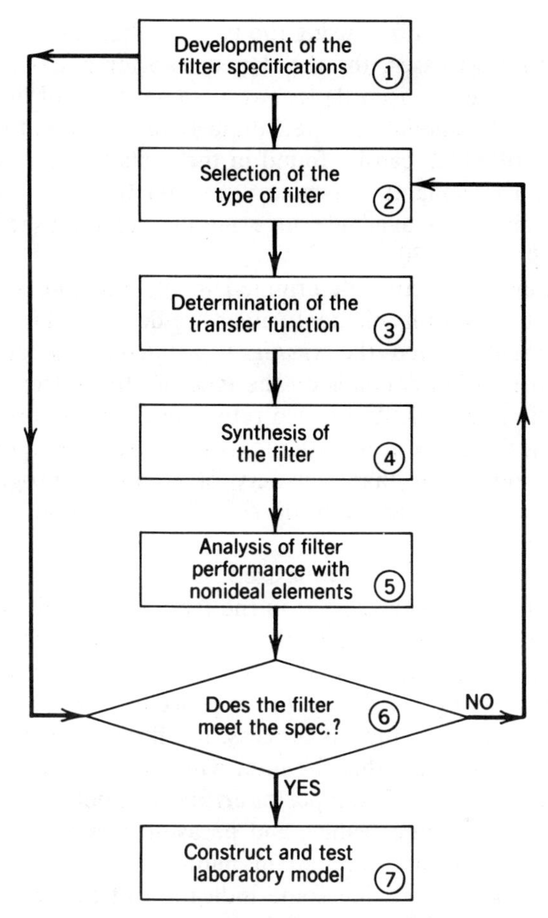

Figure 1.3 The steps in the filter design process [28].

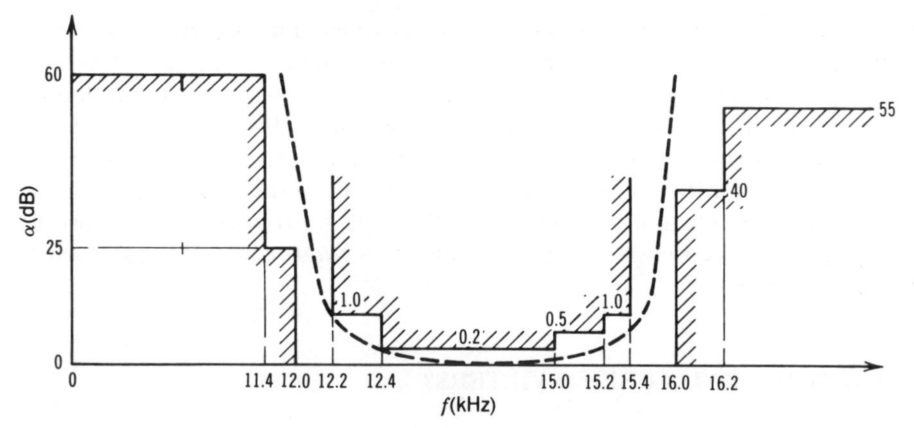

Figure 1.4 Typical loss specifications for a bandpass filter. Acceptable loss response represented by the broken lines [28].

ensuring that it can be realized using the type of filter selected in the second step (Box 2). In most cases, the developed transfer function *approximates* the desired response in accordance with some type of minimum error criteria. A variety of standardized techniques are available to carry out this step, the details of which can be found in the texts listed at the end of this chapter. Often, a reasonably satisfactory transfer function can be found by searching through tables available in a number of published filter design handbooks [5, 20, 23, 27, 30].

From the transfer function determined in the previous step, a suitable filter structure along with its element values are developed in the fourth step (Box 4), commonly called the *synthesis problem*. The actual synthesis process to be carried out depends on the type of filter selected and the final filter configuration desired. Again, we refer the reader interested in details of various synthesis procedures to the texts listed at the end of this chapter. It should be pointed out that a variety of computer programs are now commercially available to carry out this step either on personal computers or main-frame computers.

Once the filter has been synthesized, it is important to analyze its performance before it is constructed in the laboratory. This is performed in the fifth step (Box 5). Usually, in this step, first the synthesized filter is analyzed with the designed nominal element values to ensure that no error has occurred in carrying out the synthesis procedure. After the verification of the correctness of the synthesis process, the filter is analyzed with errors in the components to ensure that the filter will work satisfactorily when built with practical components. The types of errors considered here may include tolerances in the component values and parasitic effects. In most practical situations, these errors are random in nature and often a Monte Carlo analysis is performed to provide some indication of the statistical distribution in the filter's performance. This step is always carried out on a computer.

The filter's performances under a variety of conditions carried out in step five are then compared with the specifications of Box 1 in the sixth step (Box 6). If the specifications meet all the conditions, a final test under laboratory conditions may be carried out in the seventh step (Box 7) by constructing the filter in hardware form with practical components and analyzing its performance with a signal generator. If, on the other hand, the original specifications are not met in the sixth step, steps two to five must be repeated with different parameters and configurations until a satisfactory filter is designed.

1.4. WHY MINIATURIZED FILTERS?

The miniaturization of filters has been largely driven by competitive forces that have influenced the evolution of communication equipment and associated technologies. While in the early years of single-sideband equipment each toroid coil of a channel filter was individually packaged and adjusted to its proper value, modern filter technologies present a stark contrast in size, packaging, and adjustment procedures. An almost exponential increase in the number of channels to be transmitted over one cable required the design of multiplexers (analog or digital) with an ever-decreasing space available for the filters and circuitry involved. This came together with higher demands on performance of these filters, which in turn required technologies that offered higher precision in tuning and stability over time and required temperature ranges. Last, but not least, the cost per channel had to decrease in order to be competitive with the technology of the previous vintage of equipment on the market. The evolution of new equipment came approximately at a rate of one new generation every five years.

This evolution had its impact on filter technologies in various ways, depending on applications and availability of materials and techniques at different times in different countries. Basically, the evolution of filter technologies can be broken up into four different categories as shown in Figure 1.5:

A.1 Continuous Data Filters, Passive.
A.2 Continuous Data Filters, Active.
B.1 Sampled Data Filters, Analog.
B.2 Sampled Data Filters, Digital.

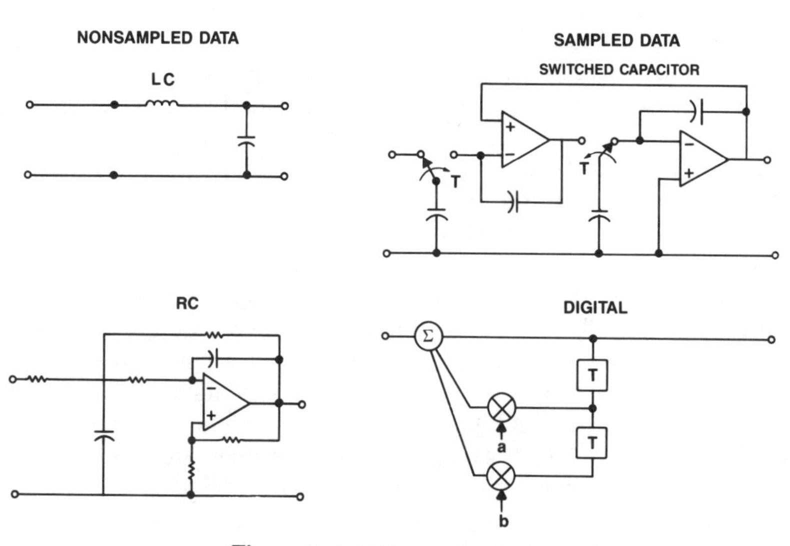

Figure 1.5 Filter technologies.

Category A.1 comprises the basic LC filters, which can be implemented by using discrete components like inductors and capacitors, or crystal and mechanical filters that are designed using LC equivalent circuits. This category is also referred to as *passive filters*, since no external power is necessary to operate them. Crystal and mechanical filters contributed greatly to size reductions of analog single-sideband channel banks in FDM systems.

Category A.2 comprises all *active filters*, which use active devices with continuous feedback techniques to generate complex pole pairs without using inductors. The only storage element left is the capacitor at the expense of operational amplifiers and resistors. Active filters can be based on LC equivalent circuits; however, many designs evolved without using LC techniques. The most important breakthrough was the introduction of thin-film and thick-film technologies, which allow the use of automation in the manufacturing and tuning process. These technologies also had a great impact on cost and size reduction. Active filters cannot universally replace passive filters, yet, their application revolutionized the design of PCM channel banks.

Category B.1, *Analog sampled data filters*, led to the first use of fully integrated complementary metal oxide semiconductor (CMOS) technology for filters. The fact that field-effect transistors (FETs) and capacitors together with operational amplifiers can be mass produced on silicon led to a second breakthrough in active filter design. Again the only storage element left is the capacitor, however, resistors have been replaced by combinations of FET switches and capacitors. This led to *SC filters* and *CTD filters* which operate on the principle of transferring analog signal samples (represented as electrical charges on capacitors) from one storage element to another. Manufacturing takes place without tuning by using precision photolithography as developed for silicon devices on a large scale. Cost reductions are related to the size of the silicon chip, and size becomes less of an issue.

Category B.2, *Digital filters*, is based on the representation and processing of signal samples in binary form, which means that conversion of the signal from analog to digital and vice versa has to be done before and after the filter, respectively. However, the digital filter is used in a digital system where the conversion for the entire system is performed on a composite signal. This makes the digital filter more attractive than every analog filter being replaced by a digital filter with its own A/D and D/A conversion. Thus, digital filters are today widely used in digital processors. Their performance is virtually not affected by analog tolerances and variations of components, as long as the entire digital circuitry is running properly at the desired speed. Digital filters are based on three fundamental signal processing functions: addition, multiplication, and storing of signal samples in registers or memory. All linear operations on signals are based on these functions. Consequently the linear operations invite integrating filtering with other operations, such as modulation and demodulation, as will be shown later. Digital filters have been implemented in all available silicon tech-

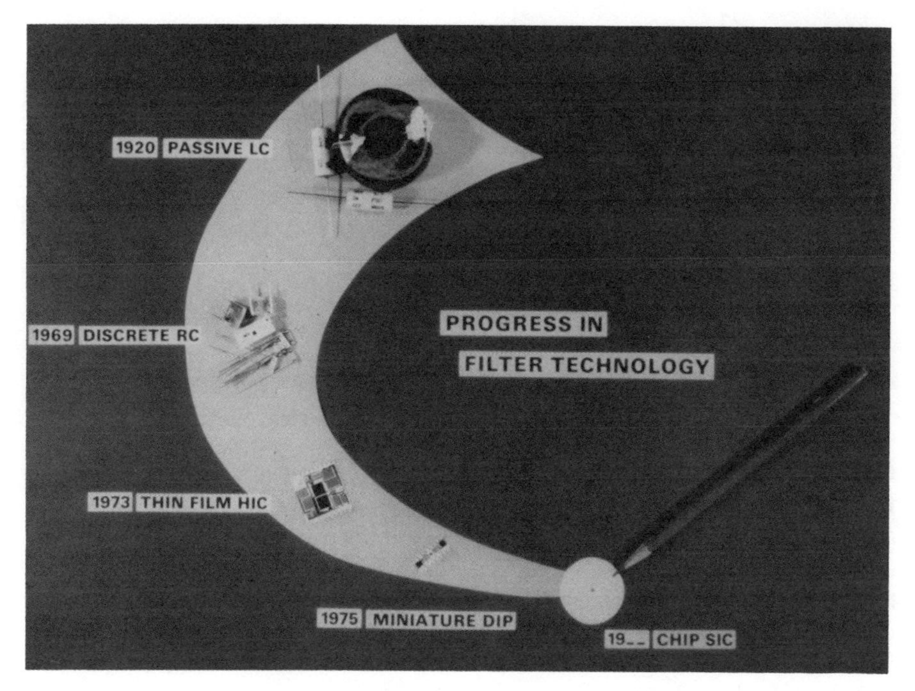

Figure 1.6 Progress in filter technologies.

nologies. They benefit greatly from any further evolution of very large scale integration (VLSI) technologies, including cost and size reductions.

Looking back into the history of filter technologies, one can see enormous improvements in performance as well as cost and size reductions, a process that still is going on as indicated in Figure 1.6.

1.5. SOME TYPICAL APPLICATIONS

Channel separation of single-sideband signals in FDM systems presented one of the greatest challenges in history for the filter designer. According to world standards, voice signals in the baseband range of 0.2–3.4 kHz must be transmitted in a 4 kHz frequency band, tightly stacked into a group of 12 channels in a frequency range of 60–108 kHz. This does not allow much guard band for the suppression of adjacent channels, which typically requires a 65–70 dB attenuation in the stop-band of each filter. To meet these requirements a strictly numerical approximation typically results in a fourteenth order transfer function, which corresponds to $N = 14$ in Eq. (1.3). The choice of technology decisively determines the frequency range in which such filters can be implemented and successfully manufactured. This in turn commits the system designer to a modulation scheme that accommodates

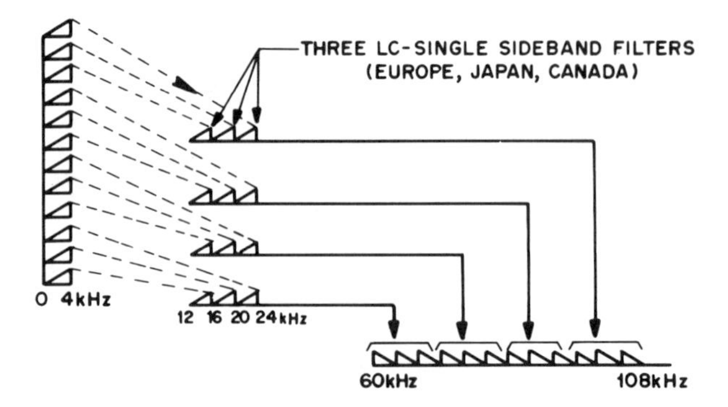

Figure 1.7 Multiple pregroup FDM modulation scheme.

the filters designed in the chosen technology. In most cases the choice of technology was determined by economics and availability of particular materials and manufacturing techniques in industries of various countries at a given time. This explains the different modulation schemes that have been used throughout the world, as indicated in Figures 1.7–1.10 [13, 14]. In the pregroup modulation scheme in Figure 1.7, the filters were implemented in ferrite coil technology and discrete capacitors. It was found that this technology allowed the realization of these three filters with best performance in the range of 12–24 kHz. Moving the pregroup further down would increase the relative bandwidths of the lowest filter, which results in large differences of element values between filter branches; this is to be avoided for technological reasons. By moving the pregroup above 24 kHz, dissipative losses in the inductors begin to impair the performance of the highest filter. The direct modulation scheme in Figure 1.8 (most convenient for the system designer), requires very high Q inductors for discrete LC components. Most

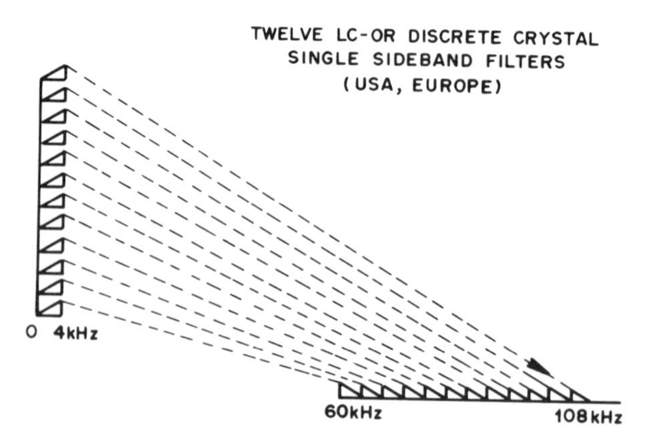

Figure 1.8 Direct FDM modulation scheme.

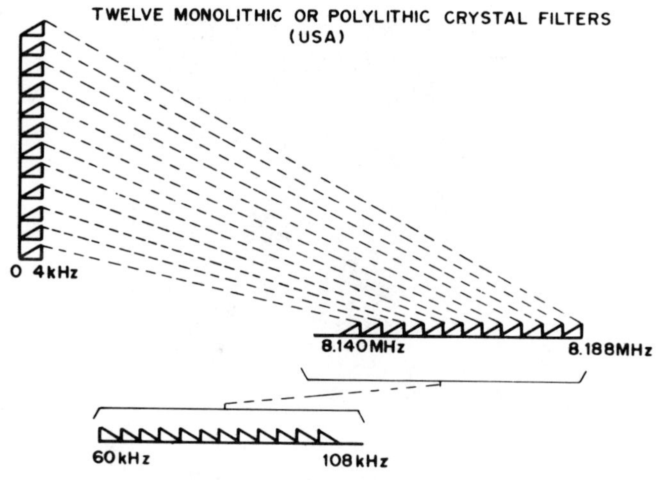

Figure 1.9 FDM premodulation scheme with 12 different filters.

realizations used crystal technology with bulk resonators that more easily met the requirements on resonator Qs and temperature stability [26].

The main advancement with respect to miniaturization occurred with the introduction of the monolithic or polylithic crystal filter. These filters where based on multiresonator designs on one crystal plate, implemented in the lower MHz range as indicated in Figure 1.9. Photolithography and laser tuning allowed automated production and led to cost reductions [1, 2, 22, 25]. Figure 1.11 represents a typical monolithic design. Figure 1.12 shows the drastic size reduction in crystal filter designs by converting from bulk resonators to the monolithic design. The active filter in thin-film technology is shown as a contrast to crystal technologies.

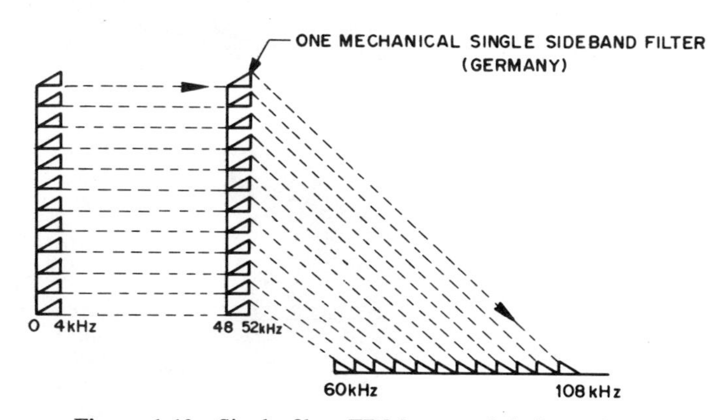

Figure 1.10 Single filter FDM premodulation scheme.

Figure 1.11 FDM-SSB monolithic crystal filter.

Figure 1.12 Miniaturization of FDM-SSB crystal filters.

Another breakthrough in miniaturization occurred with the introduction of mechanical filters. Mechanical channel filters were implemented in frequency ranges slightly below the range of 60–80 kHz or above 108 kHz. One modulation scheme with a uniform filter design at one frequency for all channels is shown in Figure 1.10. Later designs in Europe, Japan, and the US preferred single frequency designs above 108 kHz. Mechanical filters require mechanical precision techniques for manufacturing and assembling metallic resonators; yet, the technique led to cost effective designs using automation and laser tuning. Design examples are shown in Chapter 4.

The development of active filter technologies has been highly connected to the evolution of pulse code modulation, which is the basic technique underlying all digital voice transmission systems. In order to avoid aliasing in the sampling process, any A/D and D/A converter has to be preceded or followed by a band-limiting filter. In case of voice bands of 0.2–3.4 kHz sampled at 8 kHz, band-limiting is required with about 30 dB attenuation above 4 kHz. Theoretically this leads to a low-pass filter with a transfer function of fifth or sixth order, which is from a numerical point of view not a very difficult design. However, the large quantities required (one transmit and one receive filter for each channel) attracted great attention to the technological implementation of these filters. As indicated in Figure 1.13 each incoming channel is filtered and subsequently sampled by a commutating sample-and-hold device that is connected to the A/D converter. The commutating sample-and-hold device generates time-division multiplexed analog samples of several channels (in most systems 24 or 32 channels) which are sequentially converted to digital samples with a word length of

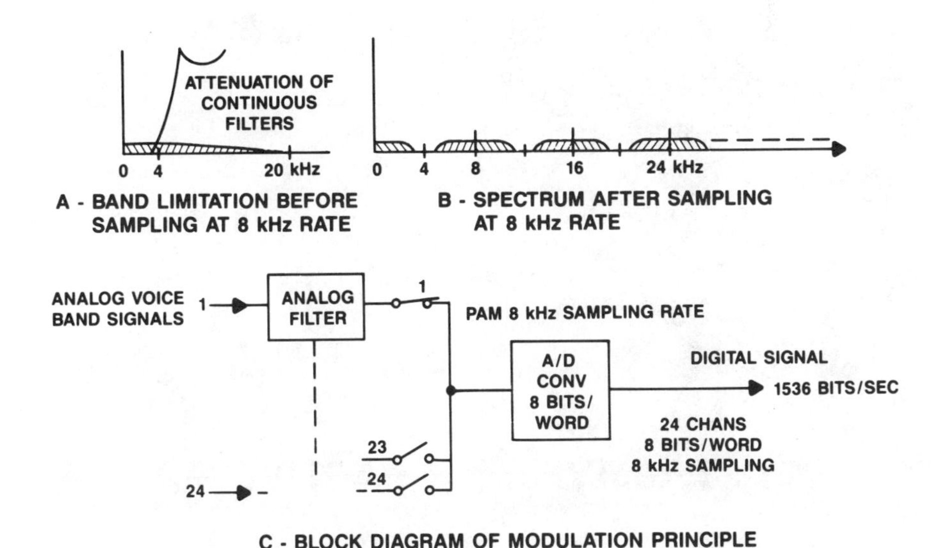

Figure 1.13 PCM A/D conversion with continuous filtering.

typically 8 bits. Thus, the A/D converter is time-shared by all channels. While the electronics for the sample-and-hold device and the A/D conversion were rapidly reduced in size and complexity, the analog filters constituted a major physical part of the front-end of a system. This changed drastically by introducing active filters in thin-film technology. This new technology offered simpler packaging, automation in assembling, but most of all, size and weight reduction. In addition, long term stability and temperature sensitivity improved due to component tracking on one substrate. The drastic difference between the two technologies is shown in Figure 1.14. The active filter itself was produced by using tantalum and gold film techniques with laser trimming of resistors to adjust the filter's frequency response. The operational amplifier chips were bonded to the substrate, and in later designs an FET chip for the sample-and-hold function was added into the filter design. Figure 1.15 shows a filter with two operational amplifiers and the FET switch bonded to the substrate. Next to it, a size reduced subsequent design is shown [7, 9, 16]. Figure 1.16 shows the evolution of the active RC thin-film filter at Bell Laboratories from a laboratory model to a manufacturable substrate within two years.

More recent designs of PCM systems employ SC technology for the filters and the associated A/D or D/A converters. Figure 1.17 shows the general diagram for a codec using a SC filter. Since the SC filter can operate at a relatively high sampling rate, in this case 128 kHz, a simple first-order RC section suffices as an analog pre-anti-aliasing filter. At the output of this filter, the signal is sampled at 128 kHz to provide the SC filter with analog samples of that rate. The SC filter performs the sharp band-limiting function

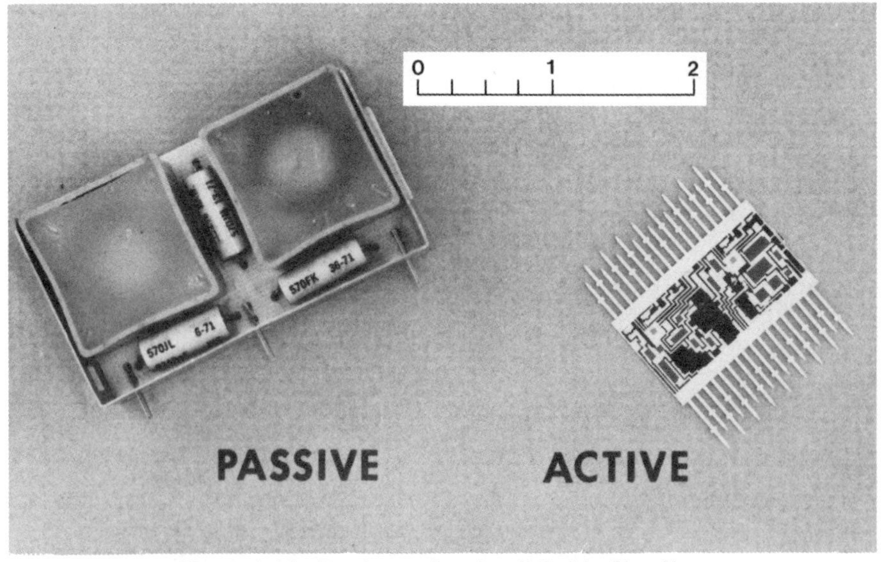

Figure 1.14 Passive and active RC thin-film filter.

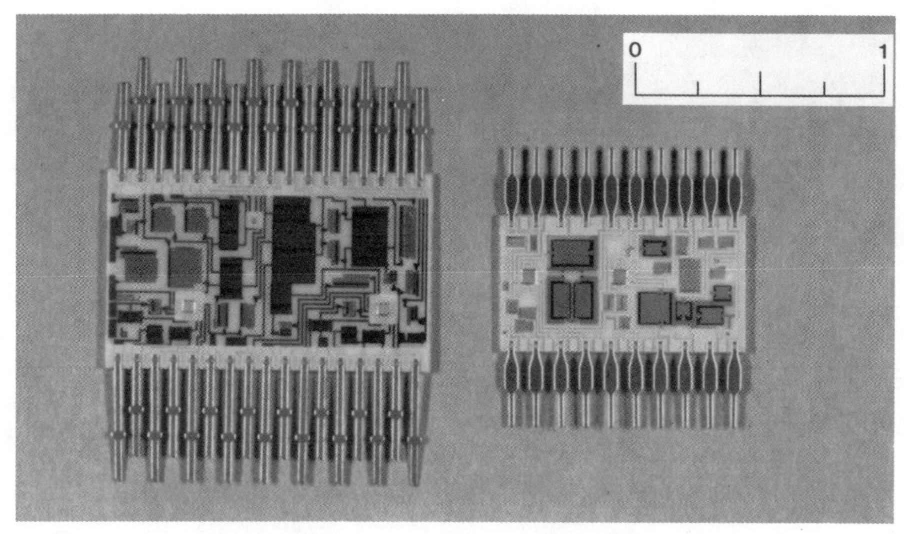

Figure 1.15 Active RC thin-film filters.

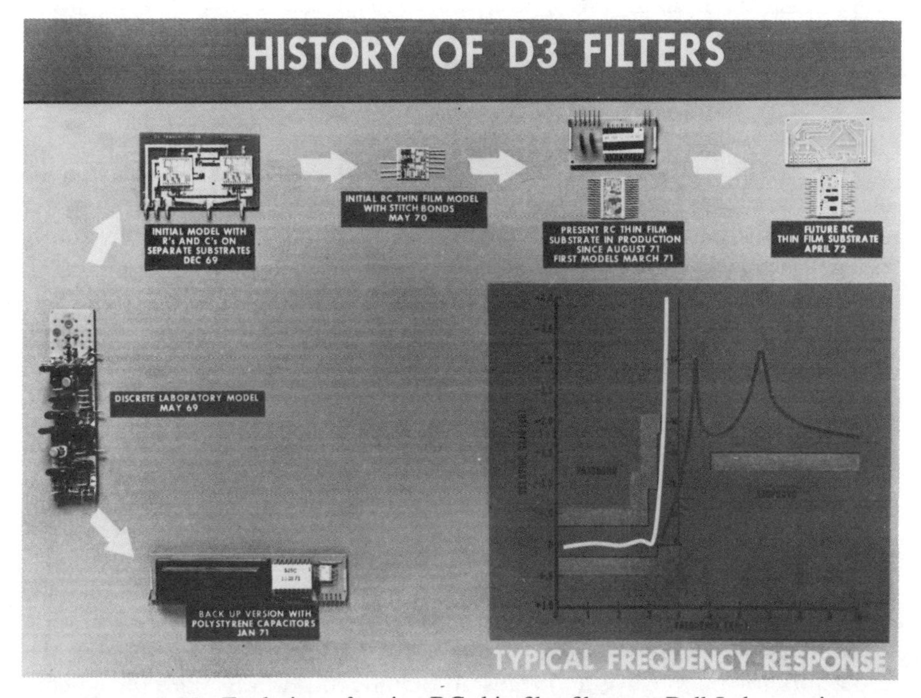

Figure 1.16 Evolution of active RC thin-film filters at Bell Laboratories.

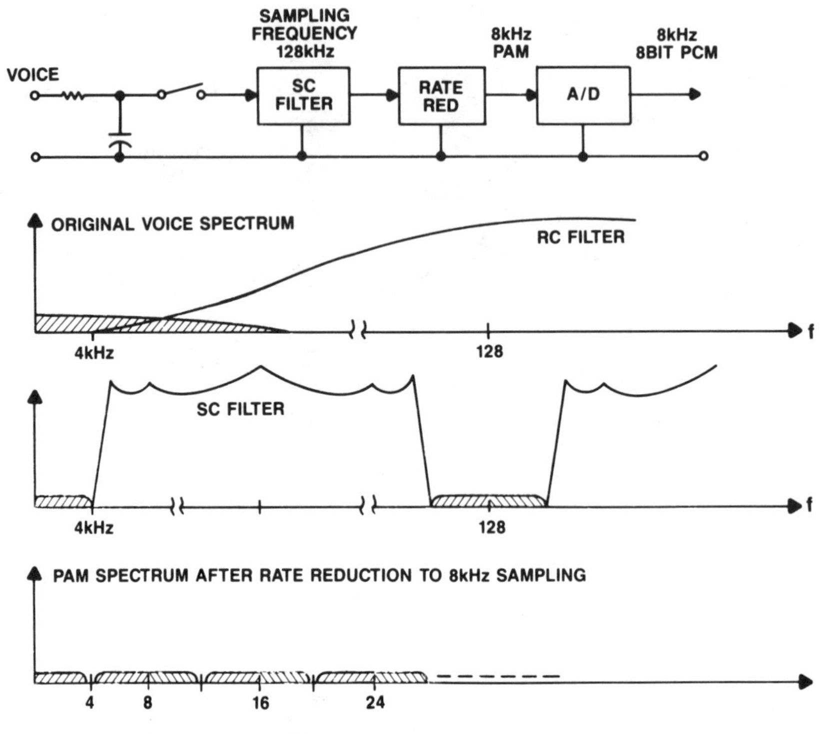

Figure 1.17 PCM A/D conversion with SC-filtering.

for the 4 kHz baseband. It is usually more advantageous to implement the A/D or D/A converters at the lowest rate possible, hence, after the SC filter a rate reduction takes place. This corresponding circuit synchronously resamples the output of the SC filter at 8 kHz, which corresponds to dropping 15 out of every 16 samples. As indicated in Figure 1.17, this rate reduction, or decimation process, shrinks the spectrum of the oversampled signal to a spectrum corresponding to an 8 kHz sampling rate. The implementation of the SC filter together with the sampling processes in CMOS technology invites the use of a charge redistribution technique for the A/D and D/A converters. Charge redistribution uses an array of capacitors as the weighting network in the A/D or D/A conversation process. As a result, the entire system shown in Figure 1.17 was implemented on one chip. An even more efficient design, shown in a microphotograph in Figure 1.18, combines both the coder and decoder for one channel on one chip. The two larger, uniform-looking areas comprise the capacitor arrays for the A/D and D/A converter. The two rows of less regular capacitor arrays in the center are the capacitors for the receive and transmit SC filters. At both sides of the chip the logic control circuitry can be seen [17]. This application of SC technology demonstrates how drastically the miniaturization influenced the design of

Figure 1.18 PCM CODEC chip with SC-filter in CMOS technology.

PCM systems. The encoding and decoding of the analog signal is now performed on a per-channel basis. The economics are favorable enough, such that a time-sharing of the A/D and D/A converter is not necessary. In addition, it offers performance advantages, since the time-division multiplexing can now be done with the digitized samples, which eliminates any potential crosstalk problems related to charge leakage in a sequentially used analog sample-and-hold device. Cost reductions are directly related to the fully automated VLSI technology.

The evolution of digital filters is closely related to the evolution of VLSI technology. While many software implementations were studied in the 1960s, the first hardware implementations of digital filters started at the end of that decade and rapidly improved during the 1970s [10]. Most early implementations were based on cascading second-order sections as shown in Figure 1.19. This had two fundamental reasons. One, which still is very valid, is the observation that higher order feedback loops in a direct implementation lead more easily to overflow situations in adders and to limit cycle oscillations. They may also require more bits for representing the signal. The other reason was a technological one. Since these earlier designs were based on serial processing, it was prudent to develop a family of building blocks that could be easily used as modules for second-order sections. A second-order section requires four single adders, four serial multipliers and two shift registers for storing signal words. Figure 1.20 shows a family of such building blocks developed by Bell Laboratories in 1974. The

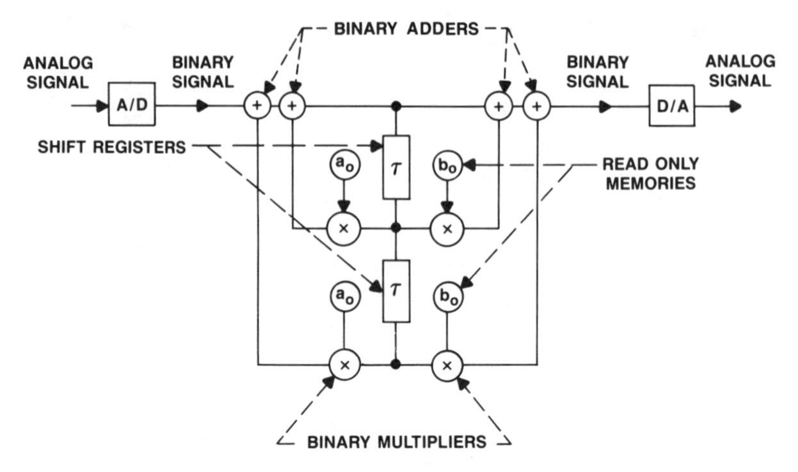

Figure 1.19 Second-order digital filter.

256-bit shift register with various taps was used as the delay component for the signal. The 256-bit length allows time-sharing of the filter over several channels by serially processing channels through the same filter. The bit rate at which the filter operates is proportional to the number of channels. The quad adder is used to perform the four single-bit additions at the input and output of the section. It is also used to form a 4-bit pipeline multiplier together with one of the dual 8-bit shift registers, which provides the delayed clock signals required in the serial process of multiplying a 4-bit

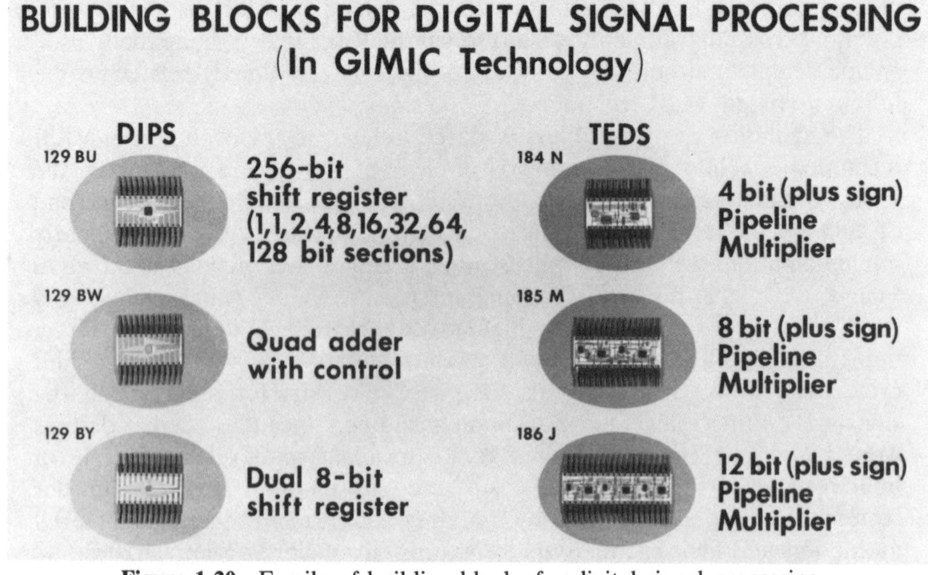

Figure 1.20 Family of building blocks for digital signal processing.

coefficient into a serial bit stream of a signal. Tandemizing the 4-bit multiplier allows multipliers with longer coefficients in multiples of four bits to be built. The building blocks were realized in guard ring isolated monolithic IC technology (GIMIC) which was designed for high speed circuits with low power consumption. The 12-bit multiplier performed about one million multiplications per second, while dissipating less than 1 W [11]. The rapid advancements in VLSI technology finally led to the implementation of digital filters in parallel processing. This in turn stimulated the development of digital signal processing chips, which are based on a computer architecture as indicated in Figure 1.21. The arithmetic unit is performing multiplications and additions on incoming data stored in RAM according to a program which is stored in ROM. The controller essentially operates the chip according to the stored program. Functions are broken down to sequences of fetch data and coefficients, perform the arithmetic and store the result, very much like in a specialized mini-computer. Figure 1.22 shows some of the earlier versions developed at Bell Laboratories [3, 6]. The processing power of these chips is considerable, which allowed the physical implementation of very complex signal processors. One outstanding example in the communication industry is the *Transmultiplexer*, which converts a 12-channel single sideband (SSB) FDM group into 12 channels of TDM-PCM channels. The processor is based on forming a group of frequency division multiplexed SSB channels using a digital filter bank, where each digital filter is fed by a linearly encoded PCM channel [8, 15]. In Figure 1.23 the principle is indicated by showing the spectra of four PCM channels before filtering. The four digital filters, offset in their passband frequencies

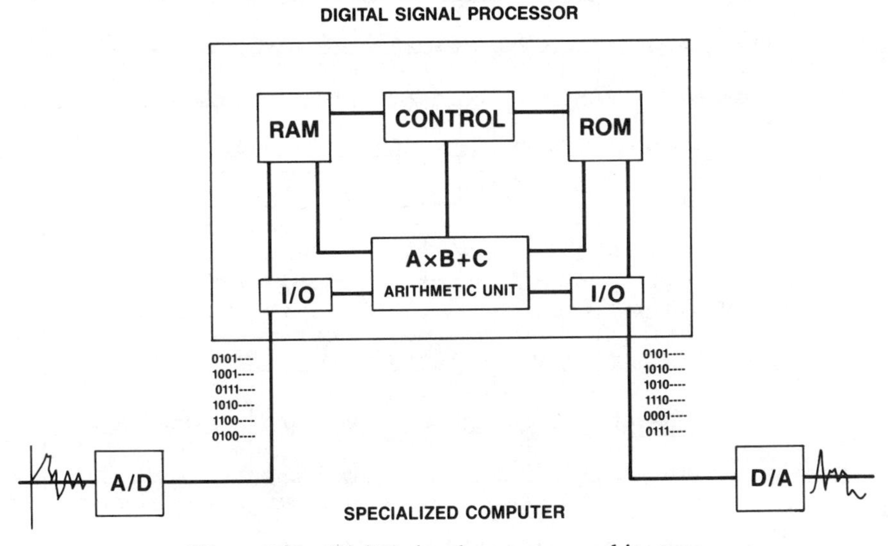

Figure 1.21 Digital signal processor architecture.

Figure 1.22 Digital signal processor chips.

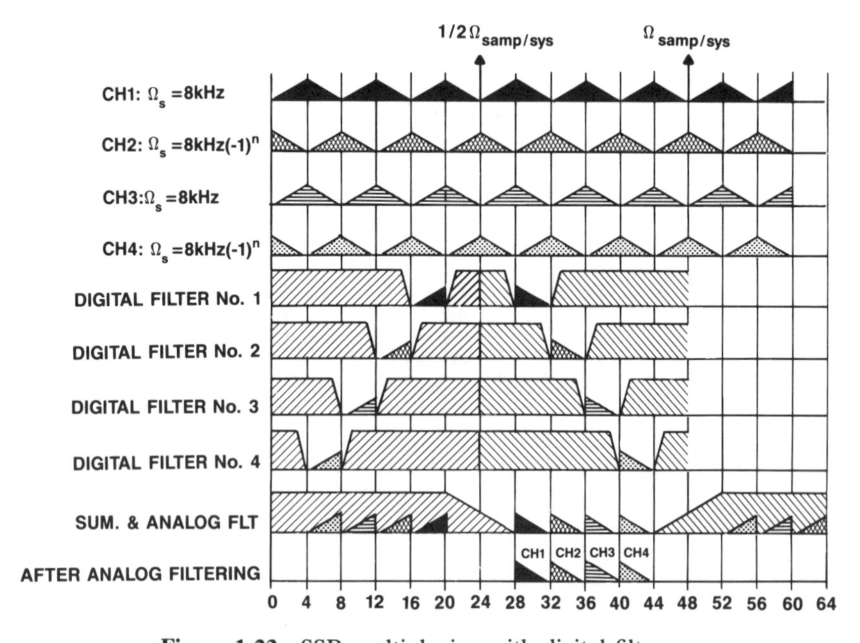

Figure 1.23 SSB multiplexing with digital filters.

by 4 kHz, automatically perform the filtering and the SSB modulation, since the desired frequency component is present in the original spectrum of each PCM channel. Periodically alternating the sign of the incoming PCM channel in every other channel assures that the SSB signals have all the same frequency orientation at the output of the filters. Digital summation of all four output signals and D/A conversion generates an analog group of four SSB-FDM channels. This is a typical example where modulation and filtering functions are integrated and designed into one algorithm that can be programmed on a DSP chip. By generically writing the algorithms in mathematical form the number of arithmetic operations can be minimized. This, in many cases, may lead to an overall block diagram of the signal flow that eliminates the recognition of each individual filter function. This was a subject of intensive research in industry and at universities for more than a decade [24].

Finally it may be mentioned that digital filters have been used for band-limitation in PCM A/D and D/A converters. As indicated in Figure 1.24, the signal has to be prefiltered with an analog filter. Subsequently a linear A/D converter running at a sampling rate higher than the minimum Nyquist rate for the filtered signal, generates a digital bit-stream that is filtered by a digital filter. The digital filter provides the sharper band-limiting function. A

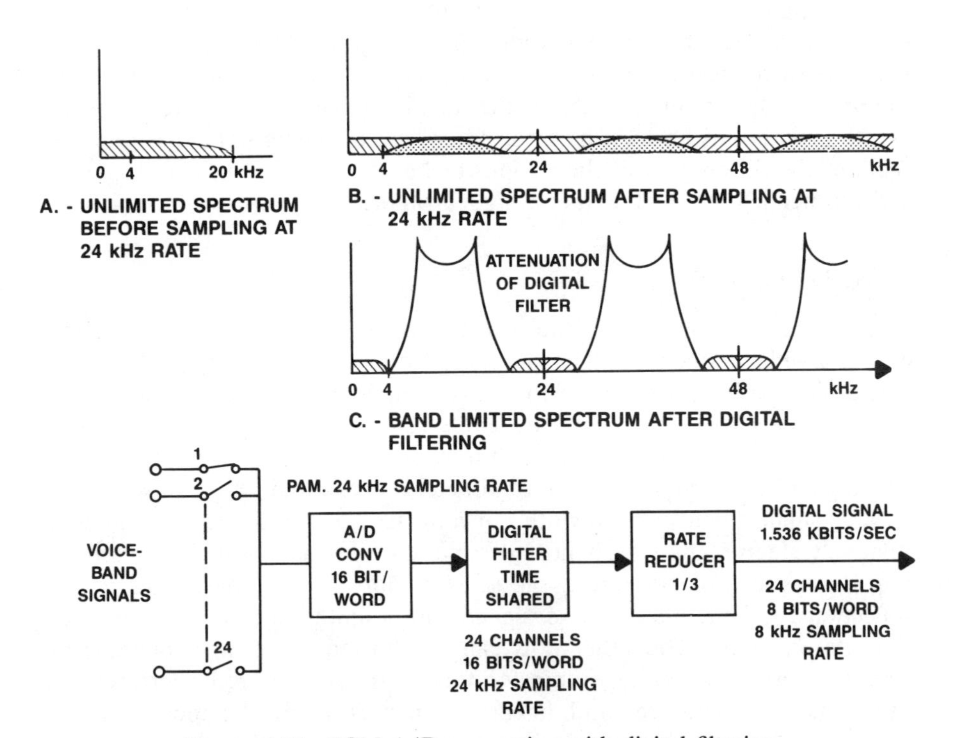

Figure 1.24 PCM A/D conversion with digital filtering.

subsequent rate reduction brings the digital signal back to the minimum Nyquist rate of the filtered signal. For economic reasons the digital filter may be time-shared over several channels. One economic disadvantage is the high speed A/D converter, which becomes more expensive the higher the speed. This has been addressed by using sigma delta modulators, which inherently run at very high speeds. A subsequent digital processor converts the delta modulated signal to a desired digital PCM signal. An ultimate custom design may result in a very cost competitive all-digital implementation of the PCM converter [4, 18, 29].

As indicated by these examples, digital filters have stimulated the development of digital signal processing chips that perform basic linear arithmetic operations as required for filtering. Yet, this evolution also opens the way for using these DSP chips for more complex and integrated functions. This in turn has stimulated a whole new industry that addresses the use of signal processors in an almost infinite variety of applications in consumer electronics as well as sophisticated commercial and military equipment.

1.6. PERFORMANCE COMPARISON

As one might expect, performance advantages and limitations vary with the technology chosen for implementing a filter. Figure 1.25 represents a typical set of performance requirements. Frequency or time response are usually the predominant requirements for the function of the filter, while the others are the conditions under which the filter has to function. In the following sections the differences of the various technologies are pointed out and are set into perspective with respect to each other.

1.6.1. LC Filters

The implementation of LC filters with discrete inductors and capacitors covers a wide frequency range, probably from several kHz up to 200 MHz. The limitation in functional performance is mainly related to the dissipative losses in coils and, at higher frequencies, also in the capacitors. Parasitic components can severely influence the choice of the filter circuit diagram. The higher the frequency the more difficult it is to build LC bandpass filters with a small relative bandwidth, or low-pass and high-pass filters with relatively steep rising stop bands. Passive filters require proper matching of impedances at both ports to avoid selective losses in the passbands due to reflections. LC circuits can be designed with a finite impedance termination only on one port. The other port can be shorted or open, depending on whether current or voltage transfer functions are required. Performance over time, temperature, and linearity range is solely dependent on the choice of components. Most designs require an opposite match of tempera-

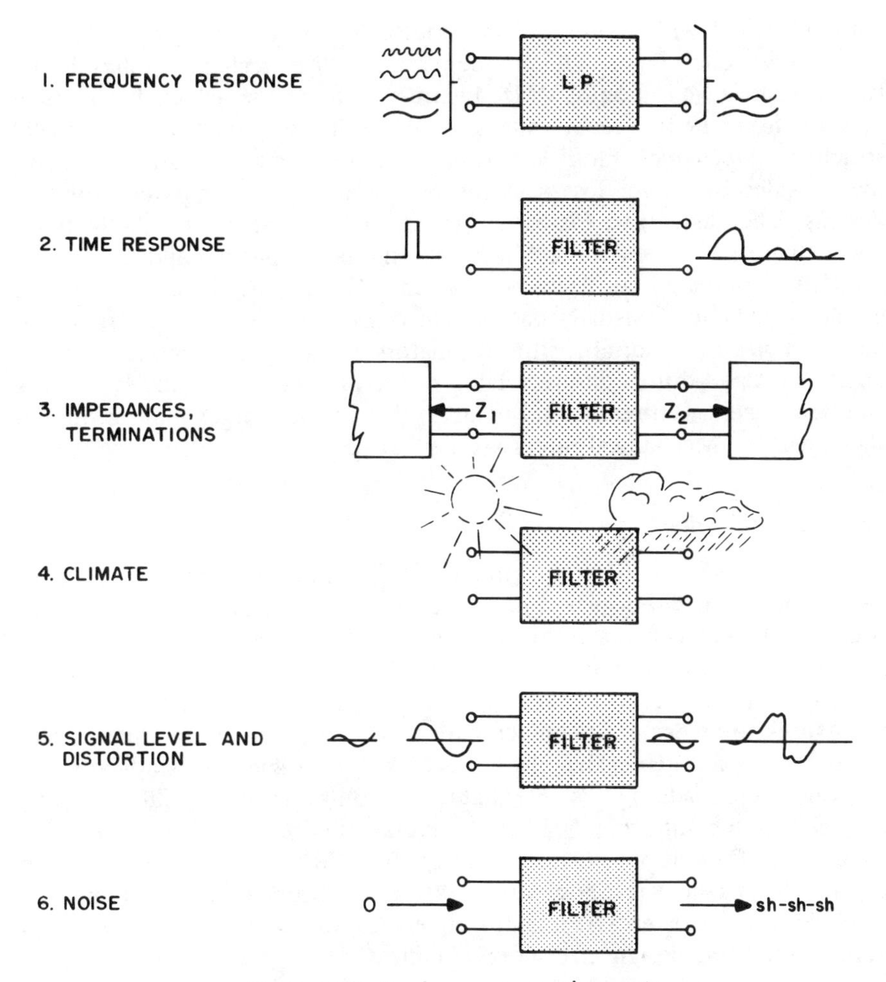

Figure 1.25 Performance requirements.

ture coefficients between inductors and capacitors to stabilize the frequency in a certain temperature range. In the case of filters with ferrite coils, the physical volume of the ferrite and the applied airgaps in the coil design determine the applicable dynamic range of the signal level for a tolerable nonlinear distortion. In filters with air coils the breakthrough voltage of the capacitors determines the maximum signal level. LC filters are virtually noise-free systems.

1.6.2. Mechanical Filters

Mechanical filters lead inherently to bandpass filter designs. This is due to the principle of mechanically coupling metallic resonators and exciting the entire structure with bilateral electro-mechanical transducers in a relatively narrow frequency band. The frequency range for mechanical filter is between 0.5 kHz and about 1 MHz. Figure 4.5 shows the achievable relative bandwidth versus frequency ranges for three different types of mechanical structures. Mechanical filters can achieve relative narrow bandwidths due to the considerably lower losses in the resonators when compared with LC designs. The dissipative losses in mechanical resonators are 10–30 times lower than in LC resonators. Mechanical filters are passive and have to be properly terminated at each port as in LC designs. Temperature and humidity behavior is usually better than in LC designs, but are dependent on the choice of material for the resonators and the packaging techniques. Signal level and distortion are related to the physical size of the resonators and transducers. Within limits, the larger the size, the larger the transmitted signal can be. Mechanical filters are noise free.

1.6.3. Crystal Filters

Crystal filters are famous for the design of narrow-band, bandpass or band-elimination filters. The piezo-electrical coupling factor causes a light coupling of the mechanically resonating crystal to electrodes, which in turn transform the resonator to an equivalent LC-circuit exhibiting two closely spaced resonances (one serial and one parallel). The Q factors of these resonances are very high and can reach 100,000 and more, hence very narrow filters are achievable. In bulk resonator designs the bandwidth can be widened by adding discrete inductors. Bulk resonator filters can be designed in the range of 50 kHz to several hundred kHz with a relative bandwidth of about 5%. Overtone designs, as described in Chapter 7, can go to higher frequencies; however, parasitic capacitors are likely to limit the bandwidth considerably. At higher frequencies, from several MHz to tens of MHz, monolithic designs are more suitable.

Impedance matching of the ports is vital to obtain good frequency responses. Temperature stability is the best of all technologies, due to the inherent temperature stability of quartz resonators. At moderate signal levels, nonlinear distortions are small and noise is virtually absent.

1.6.4. Active Filters

Active filters can be designed from several hundred Hz to 200 kHz. Their main constraints are related to the ultimate gain roll-off of the active devices, like operational amplifiers, used in the design. The roll-off limits the gain and increases the phase in the feedback path which is necessary to generate complex pole pairs. This may not only limit the applications of active filters at higher frequencies, it can also render the feedback loop unstable, if too high Qs are required for steeper filters. The temperature instability of active devices usually sets a limit to these designs. Otherwise, the RC components on thin-film or thick-film substrates can be designed with compensating temperature coefficients. Active filters do not have to be matched, since the output of the operational amplifier can be used as a voltage source. The input port often has a high impedance and is driven by a voltage source. This makes the tandem connection of second-order sections relatively easy. Transfer functions can be realized as a product of second- or first-order sections, without interference of the sections. The choice of pole-zero pairing, however, has an impact on the dynamic range the entire filter can handle. The active devices introduce distortions at higher signal levels, and lower signal levels may suffer from a low signal-to-noise ratio, due to the noise introduced by the active devices. Hence, noise floor and nonlinear distortions determine the practical signal levels for the filter. Packaging techniques of the substrates determine the resistance to humidity.

1.6.5. Switched-Capacitor Filters

SC filter designs are, in general, subjected to the same constraints as active filters. Additional limitations are related to the MOS technology used for their implementations. When the opening and closing times of the FET switches reach the time constant of the capacitors with the leakage resistance of the FET switch, severe performance deterioration sets in. This limits the sampling rate to several hundred kHz for standard CMOS VLSI designs. The accuracy of the frequency response is mainly dependent on the precision of the photolithography of the VLSI technology. This also determines a limit on the capacitor ratios in each section, which in turn may have an influence on the achievable pole-zero configuration of a section. The dynamic range of the signal level is again determined by linear distortions and noise of the CMOS amplifiers. Clock stability for the sampling rate determines the stability of the frequency range of the entire filter.

1.6.6. Charge-Coupled Device Filters

Charge-coupled device (CCD) filters, based on charge transfer in CMOS structures, are well suited for the implementation of FIR filters that do not require sharp separations between stop- and pass-band regions. They have found many applications in correlators where charge leakage and nonlinear effects in CMOS devices are less critical. These filters are active devices and require well-defined clocking of their propagation cycles.

1.6.7. Surface-Acoustic Wave Filters

SAW filters are passive, noiseless devices which in principle resemble FIR filter designs. Via the piezo-electro effect, a surface wave is induced on the surface of a crystal that propagates through the filter structure being filtered and finally reconverts into electrical quantities. SAW filters enjoy the same stability as the crystal filters; however, as natural FIR filters, they require a large number of taps for narrow as well as wide-band filters. Fortunately, photolithographic technology allows the implementation of several hundred taps for filters in the range of 50 MHz to several GHz. For modest signal levels nonlinear effects are negligible. As passive devices, these filters have to be properly terminated at each of their points.

1.6.8. Digital Filters

As indicated in §1.5, digital filters are actually a class by themselves. Since they process already digitized samples of signals, their performance depends only on the number of bits representing the signal and the coefficients. The more bits one uses for signal representation, the larger the dynamic range of the signal can be. In recent devices, floating point arithmetic offers virtually unlimited range. The number of bits in the coefficients determines the precision with which a desired frequency response can be met. A sharp and narrow filter requires more coefficient bits than a filter with a modest rise between pass-band and stop-band. In summary, digital filters have virtually no limitations in their functional performance as long as one is willing to spend the logic circuitry for implementing the necessary binary arithmetic. The conditions under which the digital filter can work is solely dependent on the choice of the technology for the logic circuitry. For technological reasons a natural limit for the number of bits in the A/D and D/A conversion is given. Depending on sampling rates, below 100 kHz, 16 bits may not be unreasonable, where at hundreds of MHz, 4–6 bits is presently the limit [12, 21].

REFERENCES

1 G. W. Bleisch and G. W. P. Michaud (1976) The A-6 channel bank: putting new technology to work. *Bell Labs Record* **49**, 251–254.

2 G. W. Bleisch (1972) The A-6 channel bank. *I.E.E.E. Trans. Commun.* **20**, 48–52.

3 Special issue on DSP (1981) *Bell System Tech. J.* **60**, 1449–1462.

4 J. C. Candy et al. (1981) A voice band coder with digital filtering. *I.E.E.E. Trans. Commun.* **29**, 815–830.

5 E. Christian and E. Eisenmann (1966) *Filter Design Tables and Graphs*. Wiley, New York; (1977) reissued by Transmission Networks International, Knightdale, NC.

6 AT & T (1986) WE-DSP16 Digital Signal Processor Data Sheet.

7 J. Dupcek and R. H. De Groot (1974) The manufacturing of thin-film active filters. *The Western Electric Engineer* **18**, 18–25.

8 S. L. Freeny, R. B. Kieburtz, K. V. Mina and S. K. Tewksbury (1971) Design of digital filters for an all digital frequency division multiplex–time division multiplex translator. *I.E.E.E. Trans. Circuit Theory* **18**, 702–711.

9 R. A. Friedenson et al. (1975) RC active filters for the D-3 channel bank. *Bell System Tech. J.* **54**, 507–529.

10 L. B. Jackson, J. F. Kaiser and H. S. McDonald (1968) An approach to the implementation of digital filters. *I.E.E.E. Trans. Audio Electroacoustics* **16**, 413–421.

11 J. Kane (1976) A low-power, bipolar, two's complement serial pipeline multiplier chip. *I.E.E.E. J. Solid-State Circuits* **11**, 669–678.

12 C. F. Kurth (Guest Editor) (1978) Special issue on analog/digital conversion. *I.E.E.E. Trans. Circuits Systems* **25**.

13 C. F. Kurth (1974) Channel bank filtering in frequency division multiplex communication systems. *I.E.E.E. Circuits Systems Mag.* **7**, 5–13.

14 C. F. Kurth (1973) Analog and digital filtering in multiplex communication systems. *I.E.E.E. Trans. Circuits Systems* **20**, 408–415.

15 C. F. Kurth (1971) SSB-FDM utilizing TDM digital filters. *I.E.E.E. Trans. Commun. Theory* **19**, 63–71.

16 M. L. Liou et al. (1970) Computer optimization of active and passive switched low-pass filters for PCM systems. *Proc. 20th Electronic Components Conf.*, pp. 301–306.

17 D. G. Marsh et al. (1981) A single chip CMOS PCM codec with filters. *I.E.E.E. J. Solid-State Circuits* **16**, 308–315.

18 T. Misawa et al. (1981) Single-chip per channel CODEC with filters utilizing delta-sigma modulation. *I.E.E.E. J. Solid-State Circuits* **16**, 333–341.

19 S. K. Mitra and D. F. Sheahan (1979) Scanning the issue. Special issue on miniaturized filters. *Proc. I.E.E.E.* **67**, 3–4.

20 G. S. Moschytz and P. Horn (1981) *Active Filter Design Handbook*. Wiley, New York.

21 A. Mounce (1982) Computationally efficient determination of D/A and A/D noise in digital transmultiplexers. *I.E.E.E. Trans. Commun.* **30**, 1477–1482.

22 S. H. Olster et al. (1975) A-6 monolithic crystal filter design for manufacture and device quality. *Proc. 29th Ann. Symp. on Frequency Control*, Washington, DC, pp. 105–112.

23 R. Saal (1979) *Handbook of Filter Design*, AEG-Telefunken, Berlin, West Germany.

24 H. Scheuermann and H. Goeckler (1981) A comprehensive survey of digital transmultiplexing methods. *Proc. I.E.E.E.* **69**, 1419–1450.

25 D. F. Sheahan (1975) Polylithic Crystal Filters. *Proc. 29th Ann. Symp. on Frequency Control*, Washington, DC.

26 T. H. Simmonds, Jr. (1979) The evolution of discrete crystal single-sideband filters in the bell system. *Proc. I.E.E.E.* **67**, 109–115.

27 J. K. Skwirzynski (1965) *Design Theory and Data for Electrical Filters*. Van Nostrand Reinhold, New York.

28 G. C. Temes and S. K. Mitra (1973) *Modern Filter Theory & Design*. Wiley, New York.

29 B. A. Wooley and J. L. Henry (1979) An integrated per-channel PCM encoder based on interpolation. *I.E.E.E. J. Solid-State Circuits* **14**, 14–20.

30 A. I. Zverev (1967) *Handbook of Filter Synthesis*. Wiley, New York.

BIBLIOGRAPHY

A. Antoniou (1979) *Digital Filters: Analysis and Design*. McGraw-Hill, New York.

M. Bellanger (1984) *Digital Processing of Signals: Theory and Practice*. Wiley, New York.

N. K. Bose (1985) *Digital Filters: Theory and Practice*. North-Holland, New York.

L. T. Bruton (1980) *RC-Active Circuits*. Prentice-Hall, Englewood Cliffs, NJ.

M. S. Ghausi and K. R. Laker (1981) *Modern Filter Design: Active RC and Switched Capacitor*. Prentice-Hall, Englewood Cliffs, NJ.

L. P. Huelsmann and P. E. Allen (1980) *Introduction to the Theory and Design of Active Filters*. McGraw-Hill, New York.

L. B. Jackson (1986) *Digital Filtering and Signal Processing*. Klewer, Boston, MA.

D. E. Johnson (1976) *Introduction to Filter Theory*. Prentice-Hall, Englewood Cliffs, NJ.

C. F. Kurth (1977) Filters. Chapter 6, *Electronic Designer's Handbook*, 2nd edn. McGraw-Hill, New York.

L. C. Ludeman (1986) *Fundamentals of Digital Signal Processing*. Harper & Row, New York.

A. V. Oppenheim and R. W. Schafer (1975) *Digital Signal Processing*. Prentice-Hall, Englewood Cliffs, NJ.

T. W. Parks and C. S. Burrus (1987) *Digital Filter Design*. Wiley, New York.

L. R. Rabiner and B. Gold (1975) *Theory and Applications of Digital Signal Processing*. Prentice-Hall, Englewood Cliffs, NJ.

R. A. Roberts and C. T. Mullis (1987) *Digital Signal Processing*. Addison-Wesley, Reading, MA.

A. S. Sedra and P. O. Brackett (1978) *Filter Theory and Design: Active and Passive*. Matrix, Forest Grove, OR.

G. C. Temes and J. W. LaPatra (1977) Introduction to Circuit Synthesis and Design. McGraw-Hill, New York.

C. S. Williams (1986) *Designing Digital Filters*. Prentice-Hall, Englewood Cliffs, NJ.

M. E. Van Valkenburg (1982) *Analog Filter Design*. Holt-Rinehart & Winston, New York.

2 Miniaturized LC Filters

E. CHRISTIAN
ITT-Telecom., Raleigh, North Carolina

D. KLEIN
SEL, Stuttgart, West Germany

A. GUILABERT
SESA, Madrid, Spain

In the large class of frequency-selective networks, miniaturized LC filters were first used in radio and communication systems. Originally, LC filters had to be constructed with bulky inductors and capacitors as they were the only ones available until about World War II. After the war, due to an explosive growth in telecommunications, especially the national and international telephone systems, the equipment had to be reduced in size in order to remain practical and cost effective. The demand for miniaturization caused an increasing pressure on manufacturers to reduce the size of components, and on designers to find alternative technologies for transmission circuits. As seen today in the evolution of VLSI technology, the objective was, and always will be, the reduction of size and cost.

Miniaturization was a dominating factor for the success of the telephone industry since the telephone administrations proceeded to multiplex an ever-increasing number of telephone channels on a single communication link. In most of the equipment, the channel banks interfacing the voice band (VB) of conventional telephones, and particularly the filters associated with the first modulation step, are primarily responsible for size and cost. Therefore, methods to design miniaturized filters for frequency-division multiplex systems (FDM) are worth a closer study. Some of the concepts are also useful in other applications. This chapter deals with the design concepts and the selection of modern, miniaturized LC components as they have been developed by many companies throughout the world. It particularly reflects the experience with miniaturized filters at three ITT companies worldwide [3, 8, 15, 18, 20, 32, 38].

An important aspect in the successful design of filters is the engineering work of the system designer, which precedes the work of the filter designer. The system engineer has to generate the layout of the system in the form of a block diagram and, subsequently, has to specify performances, responses, levels and interfaces, such as impedances, for the individual blocks. In addition, he may want to specify physical size and environmental conditions for storage and operation. Usually a system must interface with already existing equipment and with systems that are designed somewhere else at the same time. The interface with other equipment puts some constraint on the system design. However, within these limitations the system engineer will have some freedom; and with his creativity and the consultation of the filter designer, suitable specifications can be worked out.

In systems that use predominantly frequency-selective networks, questions about the location of loss or delay equalizers, the types of combining networks, impedance levels and the allocation of pass-bands, should be raised because they may have a significant influence on the size and number of components. Special attention should be given to specifications of return loss which is a logarithmic measure of interface matching [4], and stop-band attenuation. Overspecification can drastically increase the number of components.

Over the last two or three decades considerable efforts have been made to replace LC filters in multiplex systems (particularly the channel filters) by filters implemented in other technologies. The result was that in some countries and companies designs of mechanical filters and monolithic crystal filters succeeded in replacing the LC channel bandpass filters, but only in conjunction with a modification of the system's modulation plan [12, 16, 31, 35, 39]. RC active filters replaced band-limiting low-pass filters at the voice frequency end of the systems [9, 14]. In all cases a high production volume justified the considerable development costs. With the exception of the above-mentioned channel filters and very narrow crystal filters for high frequencies, as of today, LC filters with miniature components offer the most economical implementation of frequency-selective precision networks at frequencies of 100 kHz and above. Their advantages are:

(a) good stability in the pass-band if properly designed;
(b) easily adjusted and controlled stop-bands;
(c) low noise and distortion for large dynamic signal ranges;
(d) no dc power consumption;
(e) modern computer programs, design aids and text books are available for their design.

For their construction, components of high quality are available that fit modern PC boards regarding heights and pin configuration.

2.1. DESIGN CONCEPTS FOR MINIATURIZED LC FILTERS

The general design of LC filters by modern synthesis methods is a well-established science for which an extensive literature is available. Because of similar or identical notations used in this chapter, some publications may be of special value for those who may want further details [4, 24, 29, 34].

In any communication system, electrical circuits have to perform a certain function in the signal path. A typical example is a single sideband modulator for FDM systems as shown in Figure 2.1. We may assume that the input signal, voice or data, enters from the left and passes through a low-pass filter

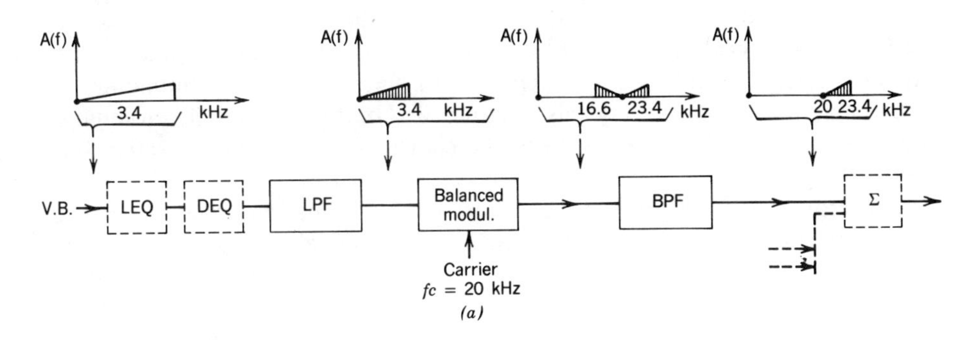

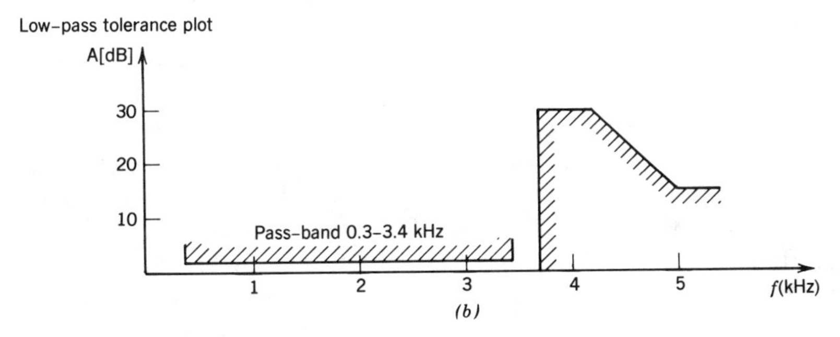

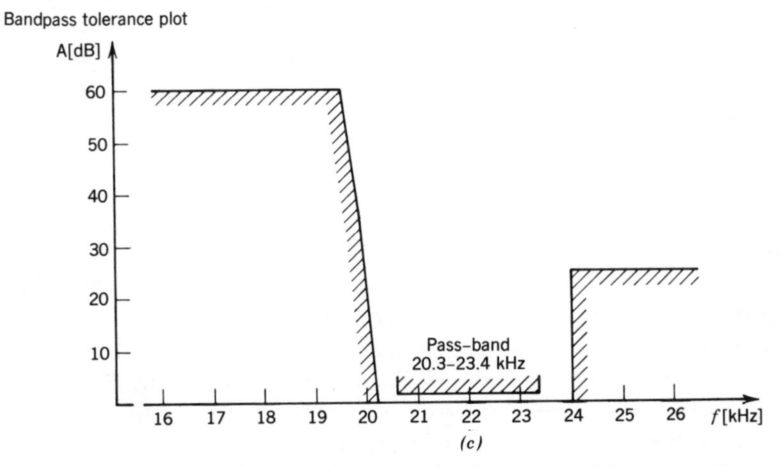

Figure 2.1 (a) Block diagram of a typical FDM system, (b) loss specifications of the LPF, and (c) loss specifications for a typical BPF.

(LPF), a modulator and a bandpass filter (BPF), and that the output signal is the upper or lower sideband generated in the modulation process. We may further assume that the input signal is restricted to the standard frequency range 300–3,400 Hz and that inband signaling is employed. In principle, the modulation process could be carried out with an almost arbitrary carrier frequency. However, the choice of this frequency most definitely restricts the technology that must be employed to construct the filter. If the carrier is 8 MHz, for instance, the bandpass can only be designed as a monolythic or polylythic crystal filter [31]. For modulation processes with carriers between 30 and 130 kHz, the bandpass filters may either be mechanical filters or circuits that employ LC components in combination with quartz crystals. Below 30 kHz, filters are conventionally designed as LC filters. For the LPF, the choices are either LC or active RC networks [9, 14].

Regardless of the frequency allocation of the modulation process, the first step in the design is to carry out a strictly mathematical process, conventionally called *approximation*, by means of which one or two rational functions of a complex variable are constructed [4, 24, 29]. The poles and zeroes of these functions must be chosen such that tolerance plots specified by a systems engineer are not violated. For the block diagram of Figure 2.1, the tolerance plots shown are typical. Obviously from this figure, the objective of the LPF is to pass the standard telephone frequencies (0.3–3.4 kHz) and to suppress frequencies in the stop-band range, 4 kHz and above. Thus, in the modulation process with a carrier of 20 kHz, for instance, we generate a lower sideband 16.6–19.7 kHz and an upper sideband 20.3–23.4 kHz. The objective of the BPF is to suppress one of the two sidebands in order to utilize a given frequency band economically, and to remove any carrier leak due to imperfections of the balanced modulator. Which one of these sidebands is selected depends on the modulation plan of the system. In the block diagram of Figure 2.1, it is assumed that the balanced modulator suppresses the baseband signal and the carrier in excess of 40 dB.

According to the tolerance plot in Figure 2.1, the upper sideband is passed and the lower sideband suppressed. Theoretically, the lower stop-band of the BPF needs only to cover the band from 16–20 kHz and an upper stop-band would not be necessary, since the LPF removes all frequency components above 4 kHz. However, distortion and intermodulation products generated in the modulation process, and baseband leaks through the modulator, require some attenuation for frequencies 4 kHz above and below the carrier. Attenuation in these bands will aid the stop-band of the LPF, and some trade-offs are possible. The trade-off can be discussed between the system and filter designer in order to meet a balance between economics and tolerable signal leaks. Also shown in the block diagram of Figure 2.1 are two networks that the filter designer may have to provide to satisfy the requirements of the overall channel translation. The first is a loss equalizer (LEQ) that compensates for loss distortions of the BPF and the LPF. The second is a delay equalizer (DEQ) which may be necessary if limits for the group delay distortion are part of the specifications. In principle, both equalizers could be inserted at any arbitrary point of the signal path. They could even be designed for the correction of the overall delay and amplitude distortion in the receiver and transmitter of the signal path. However, a correction would make it impossible for the channel translator to work with a system designed by some other vendor. For this reason, corrections are made individually. An essential part of the specifications relates to the interface conditions with other parts of the system. They are usually expressed as impedance levels, balanced or unbalanced, the quality of matching expressed by return loss specifications, and limits on the loss variations in the pass-band with respect to a reference frequency, normally 800 or 1,000 Hz. The return loss specifications can be most stringent, and can be improved at the expense of some loss by inserting attenuation pads. At the internal interfaces between LPF, modulator and BPF, impedance levels and matching conditions should be left to the discretion of the circuit designer. The availability of balanced transistor modulators has made the design considerably simpler, since they provide unilateral isolation between the LPF and BPF compared to the transparent diode modulators of the past.

For specifications similar to those discussed above, FDM systems have been designed for many decades. Historically initial filter designs were based on image parameter methods [21, 30]. To be successful, filter designers had to have a high level of experience to find the right termination of filters with frequency-dependent characteristic impedances. Eventually, with the arrival of adequate computing capacity about 30 years ago, numerical computation-intensive synthesis methods replaced the earlier methods. The synthesis will in most cases yield the same filter configuration as the old image parameter method. However, the circuit elements are different and the filter is designed for predetermined resistive terminations. The flexibility that the synthesis methods offer, can be used advantageously in the design of miniaturized LC filters or such LC circuit, which are the equivalents for filter structures in other technologies. Given an acceptable set of specifications, the following objectives should be considered.

(a) Design of a circuit with the minimum number of components, especially inductors. In principle, this will reduce size and cost [36].

(b) Apply network transformations that produce as many identical components as possible (especially capacitors), without increasing their total number. If this is not possible, the range of the capacitors should be within practical limits.

(c) Design the circuit such that expected parasitics can be compensated for by absorbing them in actual components in series or parallel. This is of special importance in filter types at higher frequencies (50 MHz and above), where the parasitics often have the same order of magnitude as the design values.

(d) As an extension to (c), tailor the design such that parasitic elements appear as circuit elements. This can often be accomplished by placing poles and zeroes of the characteristic function at suitable locations on the real axis of the complex s-plane or at some conjugate complex points.

(e) Reduce the number of circuits by combining, for instance, the specifications of two different filters into only one. As an example, in an FDM channel the LPF can be designed such that it compensates the delay distortions of the BPF, eliminating the need for an all-pass filter.

In the following, these topics will be discussed in some detail and demonstrated by examples. A major part of these examples applies to the special case of filters for FDM channels, the basic design concept, however, applies to LC filters in general. These considerations are especially valid for cases in which LC filters remain the only possible technology.

2.2. COMPONENT-EFFICIENT FILTER CONFIGURATIONS

2.2.1. On the Realization Process of Ladder Circuits

As in most fields of engineering, a design starts with the consideration of potential solutions to a given task before proceeding to quantitative evaluations. In the case of filter design, that means contemplating circuit configurations that potentially satisfy given specifications. A certain amount of experience and some familiarity with image parameter sections [21] should be a valuable help. A proposed circuit configuration can then be related to the pole-zero pattern of a characteristic function, to a transducer function, and to the two-port parameters of the circuit [4, 21, 34].

The placement of poles and zeroes in the characteristic function is of first concern, in particular, the determination of a sufficient number of attenuation poles in one or several stop-band ranges, which will basically define the degree of the filter. Reflection zeroes for equal ripple pass-bands may be directly calculated from the location of these poles or they may be placed heuristically by the designer [21, 29, 34]. To the placement of both singularities, a computer program for their optimization is the most efficient tool [6, 33]. In many filter circuits, these poles and zeroes of the characteristic function will be located in the pertinent frequency ranges of pass-band and stop-band(s) on the imaginary axis in the s-plane. However, in order to arrive at the desired configuration or element values, it is often necessary to place some of the poles and zeroes at locations in the s-plane where their influence on the response is minimal. Such is the case, for instance, with even- or odd-degree parametric filters which may be of special benefit if miniaturization of LC filters is contemplated [4, 24]. As an example, even-degree parametric bandpass filters may lead to a ladder circuit with a minimum number of inductors, and also lead to configurations where each inductor is tuned to suppress the transmission at one of the attenuation poles in the stop-band range. They are useful for the design of filters constructed by piezo-electric resonators and capacitors only, and in two other types of applications: first, when an LC circuit is intended to be implemented by an active RC circuit; and, second, for matching networks in high-power transmitters where the use of (vacuum-) capacitors is more expensive than the use of inductors.

The next step in the design is the realization of one or several basic circuits arrived at by the decomposition of the input impedance or of the chain matrix that were determined from the characteristic function [4, 21, 24, 29, 34]. Both types of decomposition are processes in which ladder sections with attenuation poles are sequentially generated. In these processes, every attenuation pole must be removed once and only once. If there are n such poles, pole pairs or pole quadruplets, there are $n!$ possible removal sequences and therefore, as many different basic circuits. Obviously, the total number must be divided by the permutation of those poles with a multiplicity higher than one. For instance, for a channel filter of twelfth degree with four pairs of attenuation poles at finite frequencies, two at zero and two at infinity, there are

$$8!/(2!)(2!) = 10{,}080$$

removal sequences and therefore as many circuit configurations. Obviously, most of these will have no practical value and the designer must restrict his search to those that are most likely to produce practical results [4].

The decomposition process yields a so-called basic circuit which consists in most practical cases of ladder sections and, theoretically, the response will be identical to numerical predictions of the transducer function by using specified terminations. It might be useful to consult Figures 2.2 and 2.3 to predict the resulting circuit for any sequence of pole removals. The table of Figure 2.2 in the left- and right-most column shows the basic removal sections that are identical to the half-sections of image parameter methods. The four columns in the center correspond to the four possible cases of the input reactance at the point of pole removal. For instance, let a typical characteristic function $K(s)$ be of eighth degree with a triple attenuation pole at zero, one pole each above and below the pass-band and a simple pole at infinity (see Figure 2.3). Then the open circuit input reactance Z_{11} is a reactance function with a seventh degree polynomial in the numerator and an eighth degree polynomial in the denominator. The degree of Z_{11} is abbreviated as 7/8 and corresponds to the column $j = 2$ in Figure 2.2. Let the designer select the removal of all poles in the following sequence

$$0 - \omega_1 - \omega_2 - 0 - 0 - \infty$$

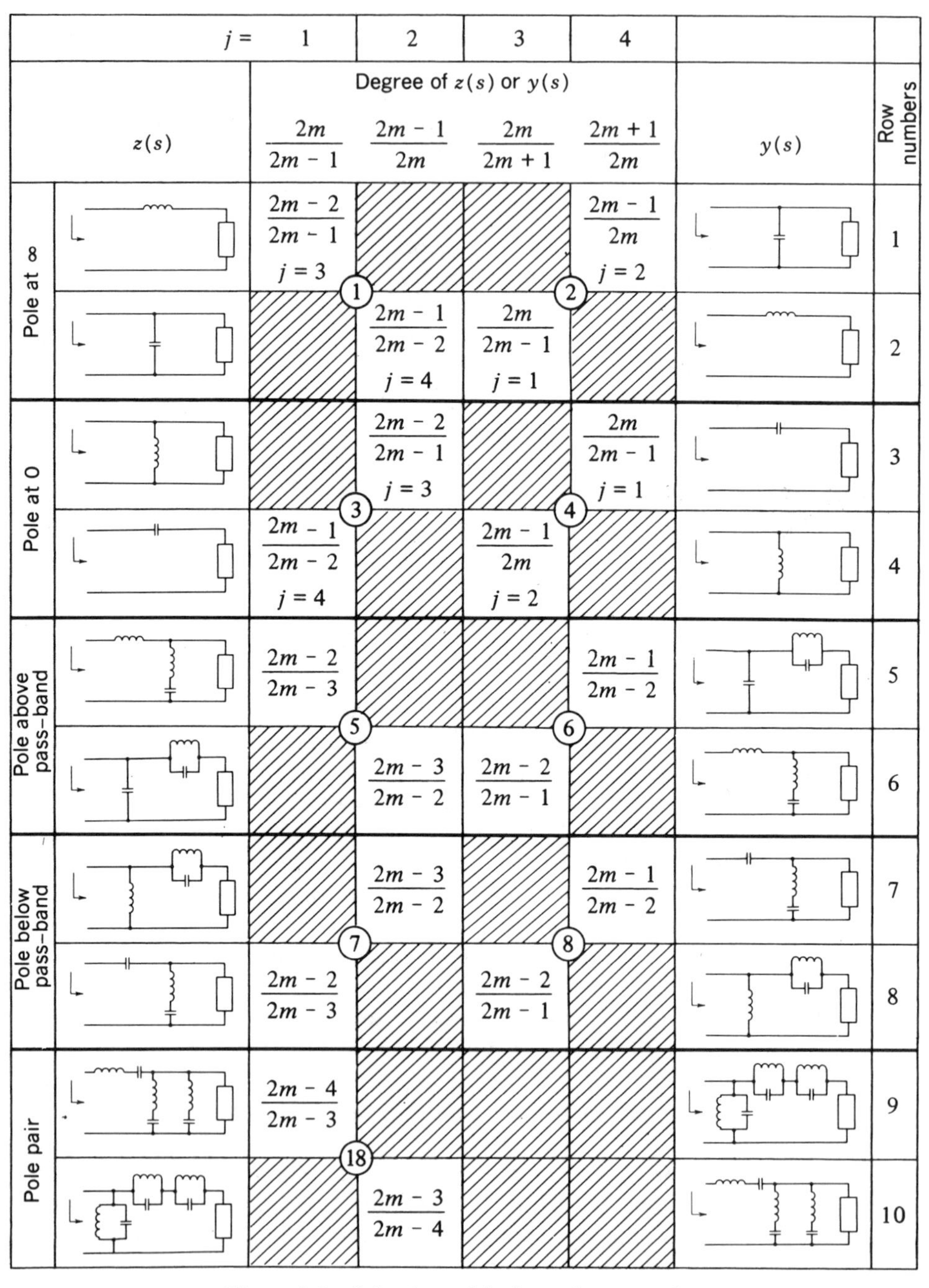

	$z(s)$	$j = 1$	$j = 2$	$j = 3$	$j = 4$	$y(s)$	Row numbers
		Degree of $z(s)$ or $y(s)$					
		$\frac{2m}{2m-1}$	$\frac{2m-1}{2m}$	$\frac{2m}{2m+1}$	$\frac{2m+1}{2m}$		
Pole at ∞		$\frac{2m-2}{2m-1}$ $j = 3$			$\frac{2m-1}{2m}$ $j = 2$		1
Pole at ∞			$\frac{2m-1}{2m-2}$ $j = 4$	$\frac{2m}{2m-1}$ $j = 1$			2
Pole at 0			$\frac{2m-2}{2m-1}$ $j = 3$		$\frac{2m}{2m-1}$ $j = 1$		3
Pole at 0		$\frac{2m-1}{2m-2}$ $j = 4$		$\frac{2m-1}{2m}$ $j = 2$			4
Pole above pass-band		$\frac{2m-2}{2m-3}$			$\frac{2m-1}{2m-2}$		5
Pole above pass-band			$\frac{2m-3}{2m-2}$	$\frac{2m-2}{2m-1}$			6
Pole below pass-band			$\frac{2m-3}{2m-2}$		$\frac{2m-1}{2m-2}$		7
Pole below pass-band		$\frac{2m-2}{2m-3}$		$\frac{2m-2}{2m-1}$			8
Pole pair		$\frac{2m-4}{2m-3}$					9
Pole pair			$\frac{2m-3}{2m-4}$				10

Figure 2.2 Selection table for pole removals.

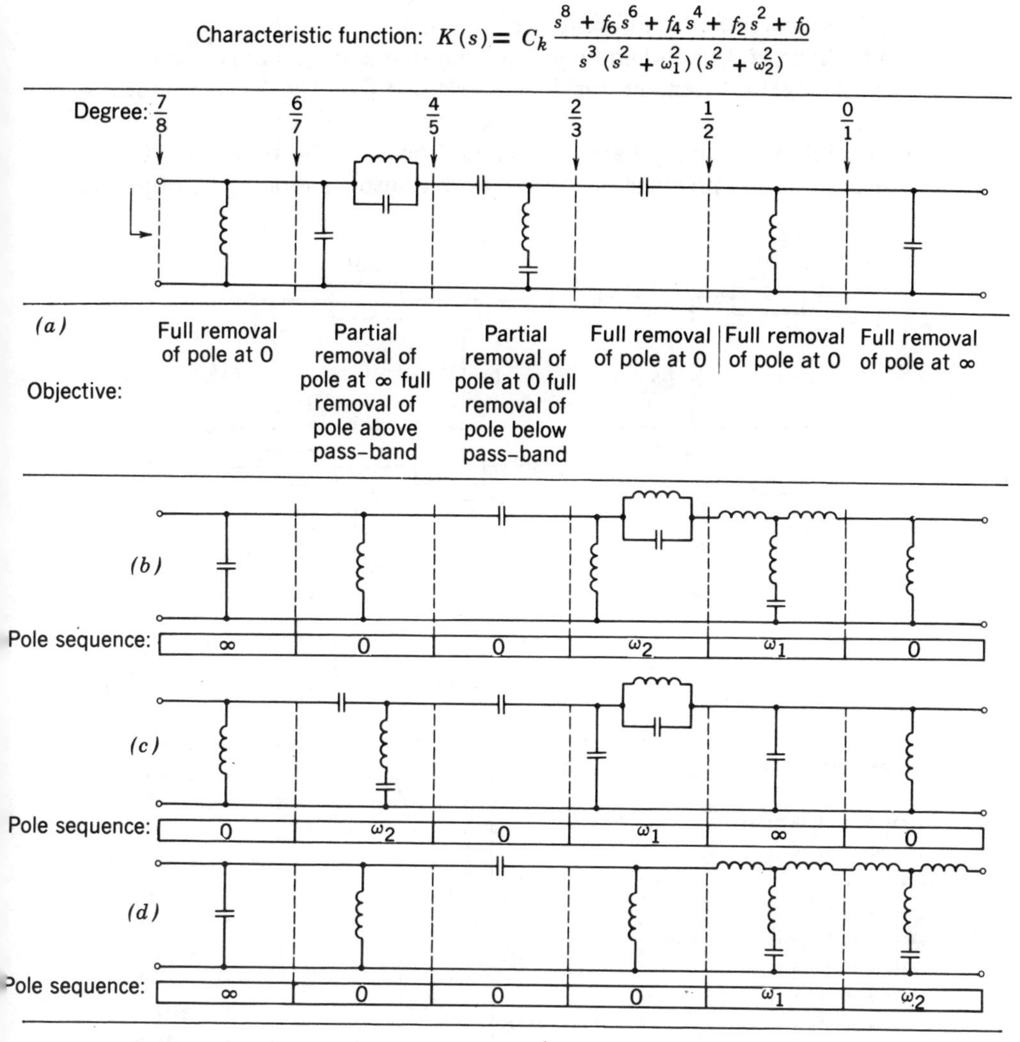

Figure 2.3 (a) The realization of an eighth order BPF for a typical pole removal sequence, (b), (c) and (d) circuits resulting from other removal sequences.

This sequence is complete since all poles are removed exactly once. It leads to a complete decomposition of the circuit. The first step is the removal of a pole at 0. As the table of Figure 2.2 shows, the removal section must be an inductor in shunt because the alternative, a capacitor in the series branch, would be incompatible with the degree of Z_{11}. After its removal, the degree of the reduced input reactance is 6/7 which corresponds to the column $j = 3$ in the selection table. In an analogous manner, the designer may predict the section types resulting from the removal of the remaining poles. Other removal sequences yield different circuit configurations, three of which are shown in Figure 2.3. It should be pointed out that certain removal sequences may yield negative elements or Brune sections [4]. The basic circuit obtained may be identical to the circuit the designer had contemplated as the desired solution. In many bandpass filters, however, the basic circuit must be converted into a practical one by network transformations. The objective

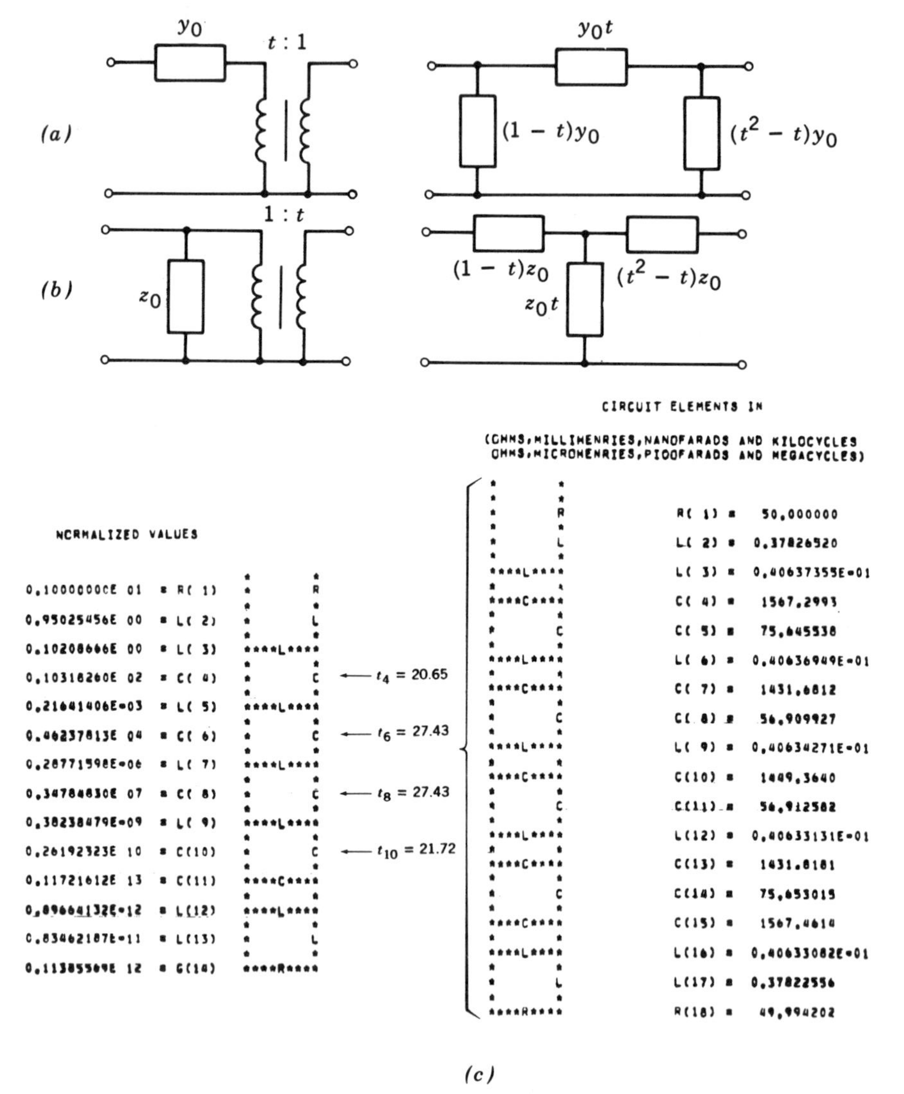

Figure 2.4 Norton transformations: (a) transformation applied to a series branch; (b) transformation applied to a shunt branch; (c) example illustrating transformation shown in (a) where the objective is to create equal valued shunt inductors in the transformed circuit.

is to arrive at a modified circuit with more favorable element values. Most often applied and useful are transformations according to Norton [23] which are essentially impedance transformations without the use of transformers as shown in Figure 2.4. Other types of transformations will be discussed in later subsections. These network transformations must be kept in mind during the removal sequence in the basic circuit, since the circuit must lend itself to such transformations.

2.2.2. Conventional Channel Bandpass Filters

Figure 2.5 shows a typical filter circuit for a channel bandpass filter before and after Norton transformations. The circuit (a) results from a removal process in the following sequence of attenuation poles: pole at infinity; poles below the pass-band; pole at zero; pole above the pass-band; pole at infinity; pole at zero. The resulting circuit must be converted by several Norton transformations applied to the series and shunt capacitors to arrive at the final configuration.

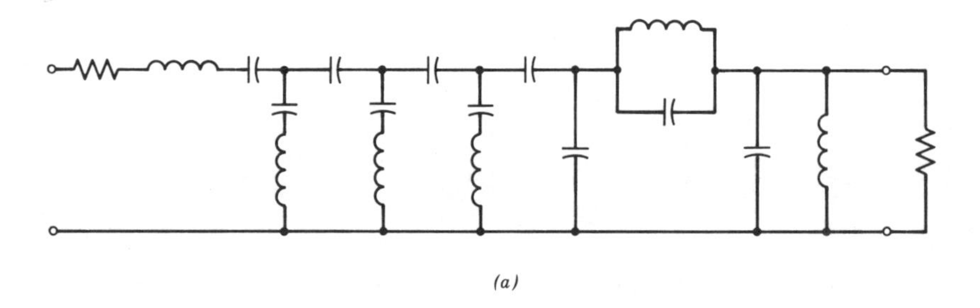

(a)

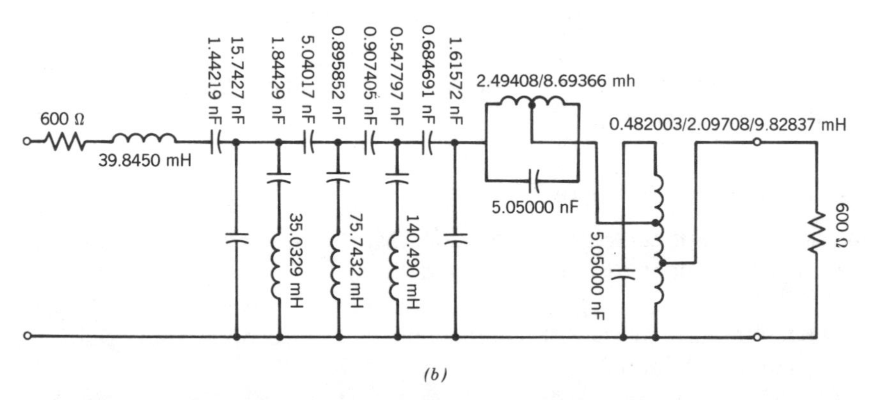

(b)

Figure 2.5 Conventional channel BPF with a 20.3–23.4 kHz pass-band. (a) Basic ladder circuit resulting from the decomposition; (b) ladder circuit obtained after transformation.

In order to compare the various resulting circuit configurations, it is useful to introduce two quantities which serve as figures of merit for size and economy. For reasonable termination impedances, let

$$S_1 = \sum_{i=1}^{n} C_i/C_0 \tag{3.1}$$

represent the sum of all normalized capacitors. In this equation

$$C_0 = \frac{1}{\omega_0 R_0} \tag{3.2}$$

where R_0 = reference impedance, normally equal to the generator impedance, and $\omega_0 = 2\pi f_0$ where f_0 is the reference frequency normally chosen within the pass-band range of the filter. For size and economy, the quantity S_1 should be small. Also, let

$$S_2 = \sum_{i=1}^{n} (\Delta C_i/C_m) \tag{3.3}$$

with

$$C_m = \frac{1}{n} \sum_{i=1}^{n} C_i \tag{3.4}$$

and

$$\Delta C_i = (C_i - C_m) \tag{3.5}$$

represent the sum of deviations from an average capacitor C_m. This quantity should also be small. For the circuit of Figure 2.5, the two quantities are

$$S_1 = 3.19 \quad \text{and} \quad S_2 = 15.9$$

These values are typical for circuits of this type and can be improved only by means of additional capacitors.

2.2.3. Alternative Bandpass Circuits

In view of miniaturization and economy of the bandpass circuit, it is reasonable to adopt the following guidelines:

(a) minimum number of components with acceptable sensitivities to element tolerances;

(b) the circuit configuration should lend itself to such network transformations by which as many equal capacitors as possible can be produced;

(c) capacitor values should be standard;
(d) inductance values should be optimum regarding stability and Q-factor.

The absolute minimum number of circuit components is obtained with canonical circuits where attenuation poles at finite frequencies are realized by Brune sections [4, 27]. In earlier decades, such sections were considered impractical since the technology of coils was restricted to coil forms that do not permit a near perfect coupling of coil windings. With the arrival of cup

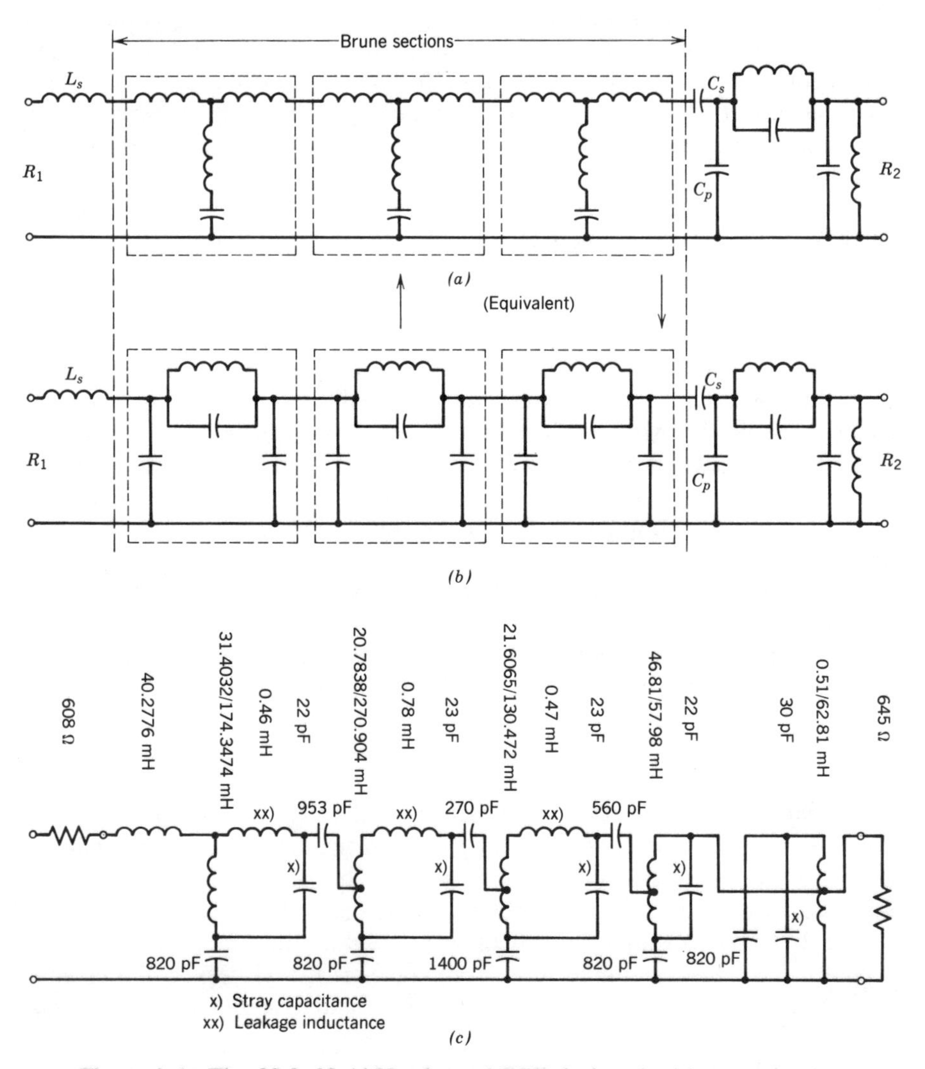

Figure 2.6 The 20.3–23.4 kHz channel BPF designed with tapped coils.

cores in the frequency range of up to 100 kHz, Brune sections can be realized and are successfully employed in LPF. Unfortunately, in BPF they often produce element values that are impractical. This situation can be overcome by means of network transformations that increase the number of capacitors. The overall circuit is then to longer canonical. However, the price to be paid for additional components is well worth it, since it leads to practical advantages in coil and capacitor sizes. Typical canonical bandpass configurations are shown in Figure 2.6(a) and (b). Both are basic circuits decomposed by the same sequence of attenuation poles as in the circuit of Figure 2.5(a). Unfavorable element values make network transformations imperative. For the transformation of these circuits, two options are available. The first is to distribute portions of the series capacitor C_s in the circuit (a) of Figure 2.6 to the two locations in between the first and second, and the second and third Brune section. To this end, at each of these two locations insert a pair of series capacitors of opposite sign. Of these, move the one with negative sign through the Brune sections by means of Norton transformations (Figure 2.4(b)) and, eventually, absorb it in C_s. In these consecutive transformations the series resonance circuits of the Brune sections serve as the branch Z_0 in the equations of Figure 2.4(b). This process will modify the element values, but the Brune sections will remain Brune sections. By a proper choice of the inserted capacitors it is in most cases possible to arrive at Brune sections with identical resonance capacitors. In a modified form, starting with the circuit (b) of Figure 2.6, the procedure described above can also produce Colin sections [5] with specified capacitance of the winding. The three degrees of freedom introduced by these transformations may be utilized to arrive at identical values for the tuning capacitors in the series resonance circuits. After these transformations, we may also compensate for leakage inductances, if necessary. To this end the transformation shown in Figure 2.7(a) may be employed for each section and the accumulated inductances may be absorbed by the series inductance L_s at the input.

The second option is to apply three so-called Colin transformations [5] as shown in Figure 2.7(b) and in Appendix B. This transformation is identical to Brune transformations between T- and π-circuits when $K = 0$. For $K > 0$, the resulting circuit has a capacitor C_p in shunt to L_p which may, for instance, be the winding capacitance. In order to apply the transformation to the canonical circuit (Figure 2.6(b)), it is necessary to move a portion of the capacitors C_p and C_s to locations between the Brune sections, again by means of Norton transformations. After this has been done, we may, in addition, carry out a compensation of leakage inductances according to Figure 2.7(a). This is theoretically incorrect because the shunt branch of the section is no longer a simple series resonance circuit. However, the error is tolerable with small leakage inductances and capacitances across the overall winding.

The resulting circuit derived from a canonical bandpass configuration is shown in Figure 2.6(c). The bandpass filter employs only eight capacitors as

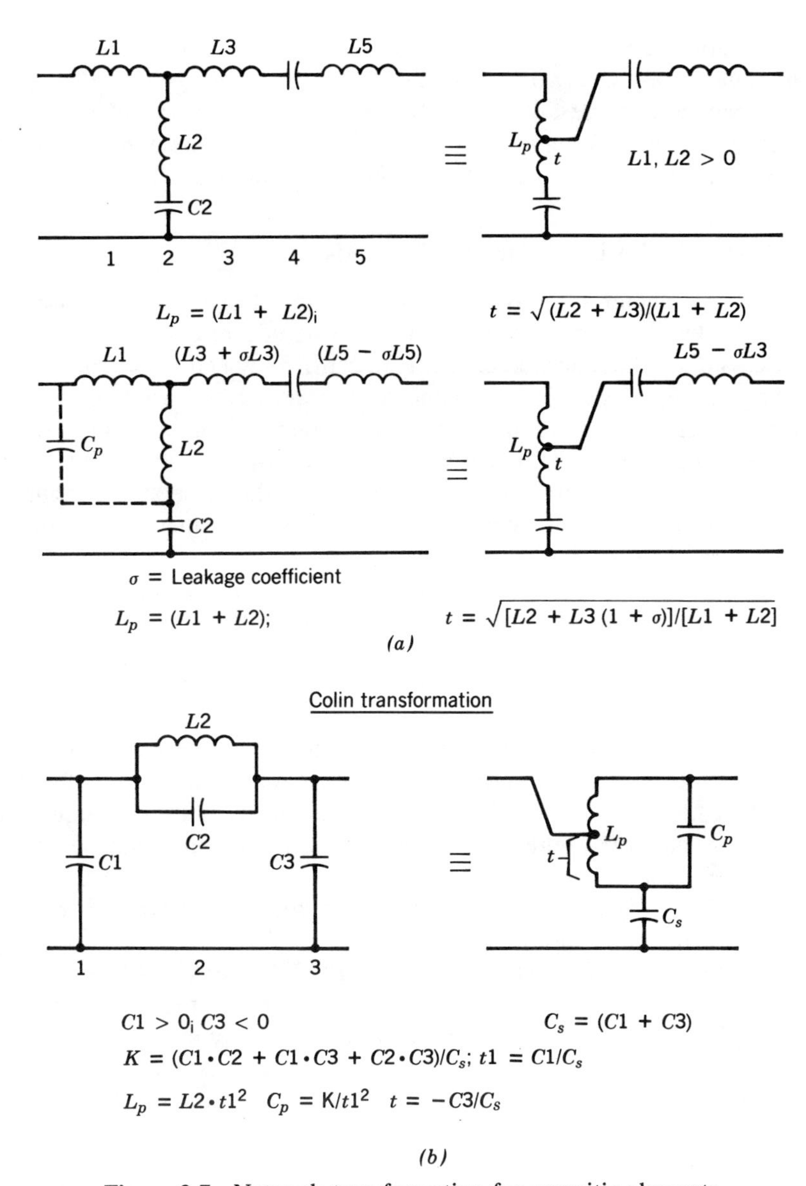

Figure 2.7 Network transformation for parasitic elements.

compared to eleven in the conventional circuit. Of these, four capacitors have identical values of 820 pF. By means of an optimization program, all capacitors were brought to standard values of the E12 and E48 series. The parameters S_1 and S_2 for the final circuit are

$$S_1 = 0.537 \quad \text{and} \quad S_2 = 1.09$$

Obviously, the new circuit permits smaller capacitor values for the same terminations as previously obtained and, furthermore, a significantly smaller range of capacitor values. It is clearly superior to the circuit of Figure 2.5 although it does have a higher sensitivity with respect to element tolerances. However, the test results indicate that these are still acceptable.

2.2.4. Design for Specified Parasitic Elements

Whenever a designer intends to employ canonical filter circuits or transformers, the leakage inductances may cause a significant deterioration of the response. Such inductances become significant, for instance, when the cup cores have large air gaps or when the turns ratio of two windings become too large. In such cases, an inductance in a series branch of a ladder circuit may be utilized to absorb the parasitic leakages. A typical example is the channel bandpass filter of Figure 2.6 where the inductance at the input has been reduced to absorb the leakage inductance of all tapped coils. Normally, after the reduction the inductance will remain positive and, therefore, a physical coil must be provided for the circuit. A complete elimination will occur if the leakage inductances reduce the input inductance to zero. If this is the case, the circuit has five rather than six coils. The filter circuit can be designed such that the series inductance is equal to the leakage inductance by selecting a suitable pole/zero pattern of the characteristic function.

Figure 2.8(a) shows a filter circuit consisting of one Brune section [4]. According to its characteristic function $K(s)$ and related transducer function $H(s)$, it will have a pair of reflection zeroes at the normalized frequencies $\pm j1.0$ and a pair of attenuation poles at $\pm j2.0$. If denormalized with a reference frequency of 12 kHz, they will fall on 12 kHz and 24 kHz, respectively. When constructed with ferrite pot cores, the response of the filter will be very close to the theoretical results.

If, however, the reference frequency is 12 MHz and the two significant frequencies become 12 MHz and 24 MHz, respectively, the pot core to be utilized will have a much larger air gap which normally causes a significant leakage inductance. This inductance will cause an attenuation pole at infinity and a shift of the reflection zero from its desired location at 12 MHz when the pole frequency is adjusted to its nominal value.

In order to arrive at a circuit with a series inductance of specified value, the characteristic function $K(s)$ of the circuit is augmented by a reflection zero at some frequency σ on the real axis. The circuit is then of third degree with one attenuation pole at infinity. The removal of this pole at the input will produce a series inductor with a normalized value of about $1/\sigma$. For a desired normalized inductance of 0.011 (corresponding to 10% leakage), $\sigma = 88.5$ was chosen initially. Finally, by trial and error, $\sigma = 83.33$ was selected. The circuit of Figure 2.8(b) shows the result. In this particular example, the changes in element values are not very significant. However, in the case of several Brune sections in cascade, considerable changes may occur, especially in the return loss.

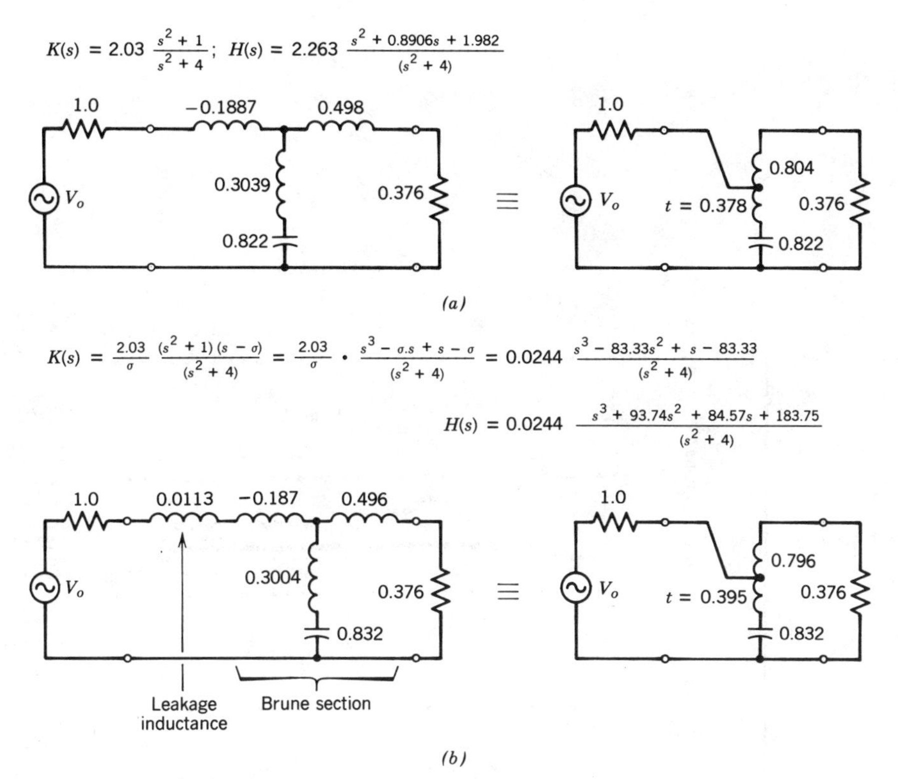

Figure 2.8 Leakage compensation in Brune sections.

An alternative method to compensate for any added parasitic elements appearing in the circuit diagram is to employ an optimization program capable of modifying the original elements of a given circuit. The location and the values of the parasitics in the circuit have to be determined. The complete circuit, including all parasitics, is used as initial circuit of a computer-aided optimization. All parasitic elements are kept at fixed values. The selection of variable elements that are to be modified is subject to the same considerations that apply to computer-aided tuning algorithms.

2.2.5. Tolerances and Tuning

An important step in the design of LC filters is the specification of tolerances and the tuning procedure before any engineering documents can be released for manufacture. Tolerances and tuning procedures are related to the sensitivity of the response to element variations [1, 2, 28]. For stop-band ranges it is usually only necessary to adjust either the inductor or the capacitor of a resonance circuit to a desired transmission zero. In most cases setting these frequencies is sufficient to assure that the response will stay above a specified lower boundary.

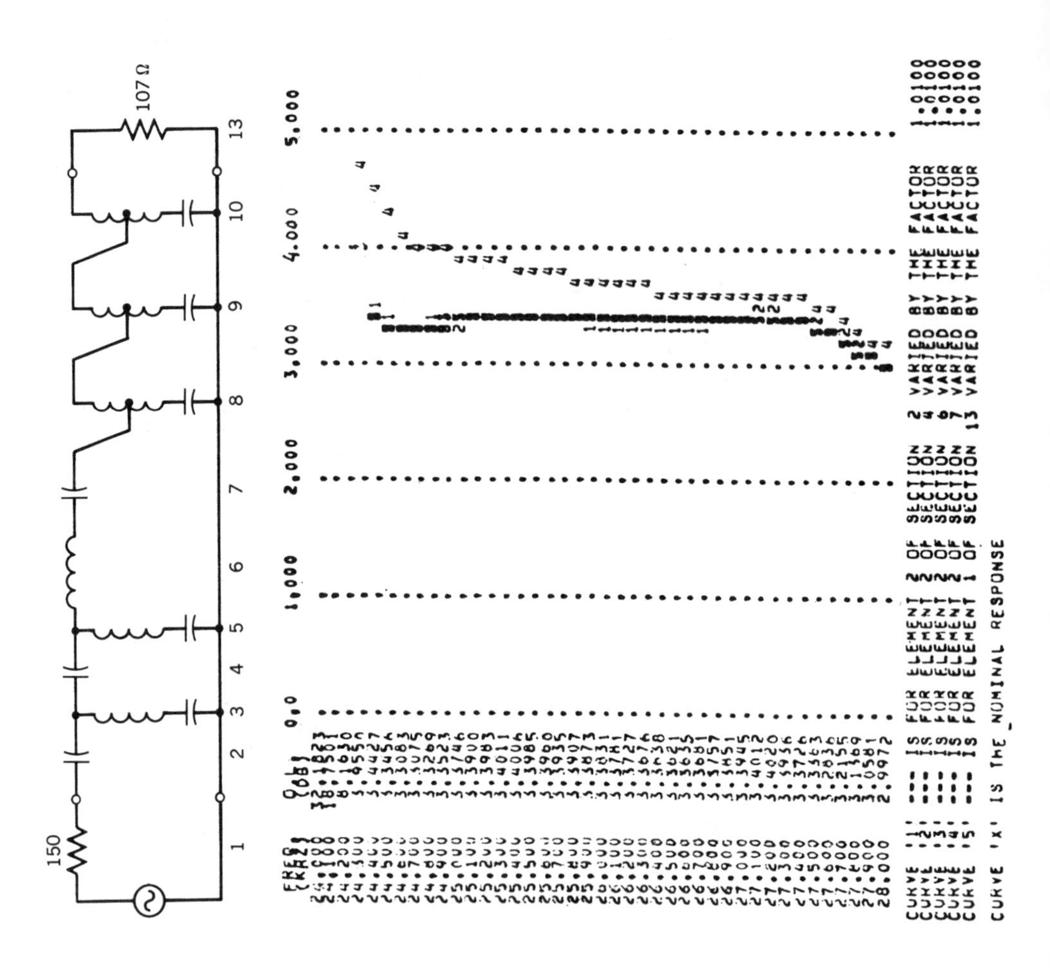
150
107 Ω
1
2
3
4
5
6
7
8
9
10
13
0.0
1.000
2.000
3.000
4.000
5.000
FREQ (KHZ)
CURVE 'X' IS THE NOMINAL RESPONSE

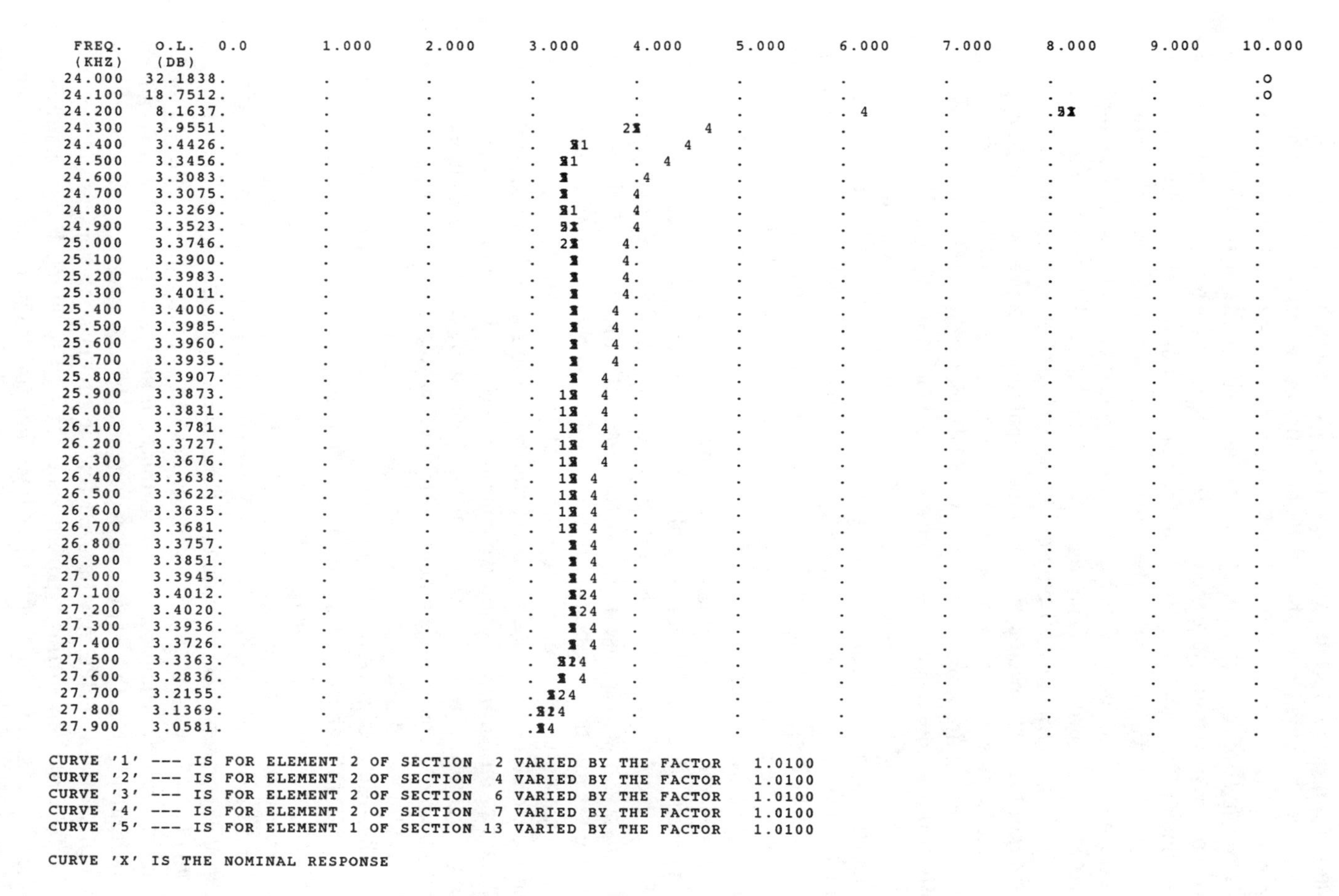

Figure 2.9 Sensitivity analysis of a bandpass circuit.

To assure the required pass-band response (transducer loss, return loss and group delay) it is necessary that the nominal tolerances of circuit elements be small enough and that a suitable set of variable elements be provided to compensate for the composite deviation due to the actual values of the fixed circuit elements. For the analysis of the sensitivities, suitable computer programs have been prepared, some of which are commercially available. Quite often, a user may be able to modify an available computer program for the purpose of determining those circuit elements that are most sensitive for the desired response and at which frequencies. An example is shown in Figure 2.9. The significant feature of the pertinent computer program is a graphical display of the sensitivities on a single page.

Figure 2.9 represents the computer analysis of a predistorted bandpass circuit as shown at the top. As is often done in the case of ladder networks, the branches are numbered consecutively underneath. For instance, branch no. 3 represents a series resonance circuit consisting of the elements nos. 1 and 2 (inductor and capacitor, respectively). In the series branch no. 4, the element no. 2 is the single capacitor. In the plot underneath the circuit, the two left-most columns contain a list of frequencies and the corresponding transducer losses for the initial condition with all elements having nominal values. The pertinent program employs chain matrix analysis in the calculation process. It is conventional in such analyses methods to calculate the response for a set of frequencies first with the nominal set of element values and then, sequentially, for several sets of toleranced elements. However, in this particular program the order of calculations is reversed. At each frequency, the response is computed first with the nominal elements and plotted with the symbol 0. Subsequently, at the same frequency, the computation is repeated several times, at each time with one and only one of the elements toleranced. The results are plotted with the print symbols 1, 2, and so on. In these repetitive computations the print line is not advanced and, therefore, overprinting will occur. Eventually, after the completion of all computations with toleranced element values the frequency is incremented and the print line advanced.

The result, displayed in Figure 2.9, is typical for this type of analysis. It shows a tolerance analysis of the bandpass with element tolerances of +1.0% applied to the capacitors in the series branches 2, 4, 7, the inductor of branch 6, and the load resistor. Obviously, from the plot with the print symbol 4 we can conclude that the capacitor in branch 7 is most critical, especially in the lower portion of the passband, that is, at frequencies about 24.3 kHz. Compared to it, the sensitivities of the other components are only minor. From information similar to this, the designer can deduct tolerance requirements for the various components. It also gives an indication of which set of elements may be used for the tuning of the circuit.

Considerations of this type are valid for all LC filters. For miniaturized filters an additional question arises regarding how to provide the tuning and adjustment in order to save space. The answer to this question depends on the type of filter and the frequency range of interest. In low-frequency applications, the range of capacitor values is usually too high to have trimmer capacitors and, therefore, tuning is normally provided by means of the trimming assemblies of most commercially available inductors, for instance, the tuning slugs of pot cores or those of ceramic coil forms. They eliminate the need for additional parts on the PC board. If in low-frequency applications a capacitor becomes very sensitive it is usually more economical to specify closer tolerances.

At higher frequencies the capacitors will be of low enough values that trimmer capacitors are feasible. Such capacitors of small physical size are commercially available and may be employed as tuning elements in addition to tunable inductors. This is especially true for LC filters with piezo-electric resonators that pass or suppress a narrow frequency band.

In practice, the tuning of an LC filter is carried out in its operating condition with proper termination. The first step is to set the transmission zeroes by tuning the respective resonance circuits for minimum transmission. Subsequently, the passband loss or return loss is adjusted by sweeping the frequency equipment over the pass-band range. If carried out manually, by observing the screen, the operator adjusts the remaining inductors or capacitors that are not responsible for transmission zeroes. In this second step of the tuning process, the operator may also adjust attenuation poles far away from the pass-band in order to bring the response within the specified boundaries. The time required for this operation depends on the complexity of the network and the experience of the operator. In order to simplify the tune and test operation, computer-controlled test methods and equipment have been introduced at many companies [1, 2, 28]. The advantage of these methods is that the computer takes into account the simultaneous changes of a set of element adjustments and gives specific instructions to the operator on what to do next. Not only will these computer-aided processes reduce the tuning drastically; they also will require less training for the operators. A detailed treatment of the theories and some practical examples with several types of networks is given by Mueller [22].

2.3. LOSS AND DELAY EQUALIZATION

Typical attenuation and group delay curves of a channel bandpass in the range of 20–24 kHz are shown in Figures 2.10 and 2.11 in combination with tolerance masks according to CCITT recommendations for voice transmission. For channels carrying data over telephone lines, the tolerance mask for group delay distortions is considerably narrower. In addition to the distortions contributed by the bandpass filter, the low-pass filter will add to both distortions. The loss distortions are due to the ohmic losses especially in the coils. They become worse when the network must be constructed with miniaturized components.

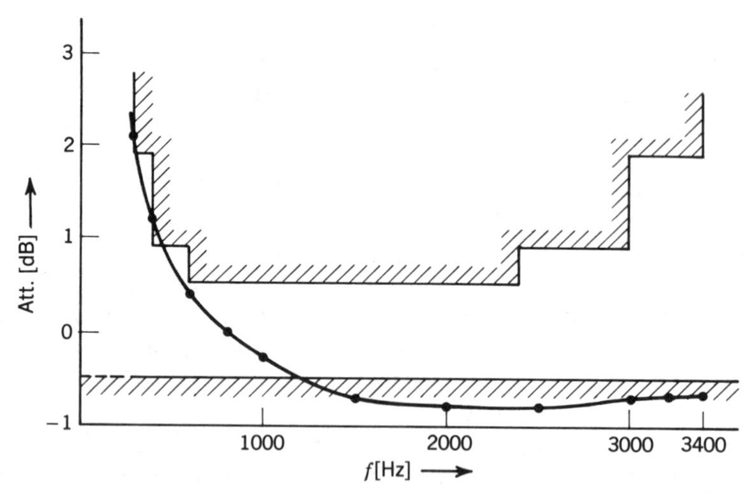

Figure 2.10 Typical loss responses of a 24–28 kHz channel BPF referenced to voice band frequencies.

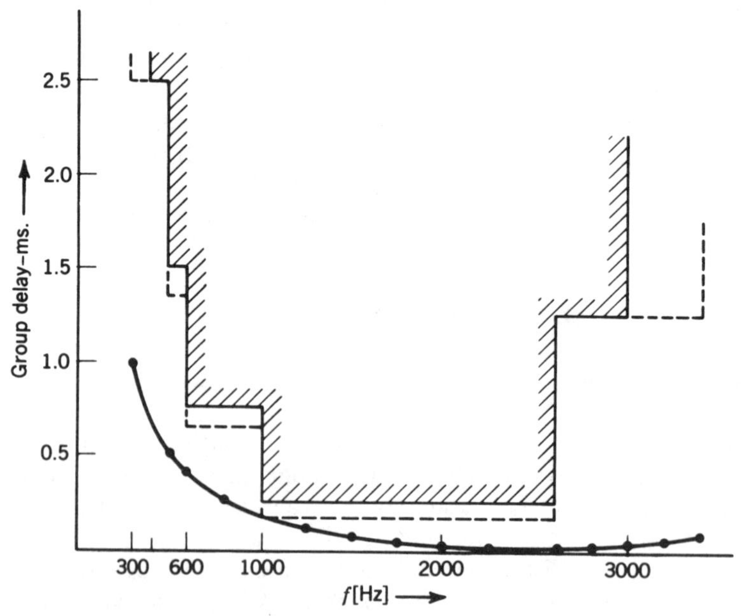

Figure 2.11 Typical group delay distortion of the BPF of Figure 2.6.

In order to satisfy the specifications with sufficient margins for deviations due to component tolerances, the designer must in some way correct for these deficiencies. The following options can be chosen from:

(a) insertion of correction networks, that is, loss and delay equalizers;
(b) predistortion of the networks to compensate for loss distortions;
(c) design of the low-pass filter that compensates for loss and/or delay distortions.

2.3.1. Loss Equalizers

Conventionally, equalizers have been designed as constant-impedance devices containing in most cases two inductors in a bridged-T configuration [4]. In essence they emphasize the low frequencies of the voice band. The

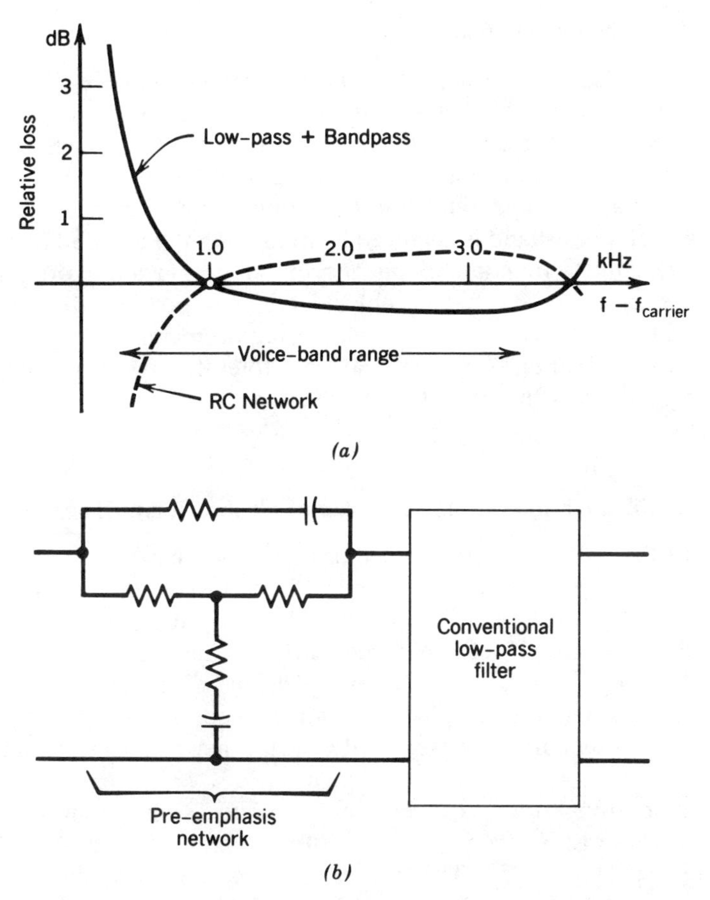

Figure 2.12 Pre-emphasis of voice-band frequencies.

advantage of such networks is an improvement of the return loss at the point of insertion, except at those frequencies at which their transfer loss is zero. Their disadvantages are added overall loss, and additional cost due to additional circuitry which is detrimental to miniaturization.

An alternative to conventional equalizers is shown in Figure 2.12. The network is a passive RC circuit in bridged-T configuration which pre-emphazises the low and the high frequencies. Obviously, its response must be the inverted response of the composite attenuation of LPF and BPF. The network is no longer of the constant-impedance type and, therefore, certain sections of the voiceband will have a decreased return loss. However, such deteriorations can be tolerated and, if necessary, corrected by an attenuation pad. The components of the RC network are determined either manually by trial and error or, more economically, by one of the available optimization programs for circuit optimization.

2.3.2. Predistortion Methods

The addition of any correcting equalizer can be avoided by employing predistortion to either of these filters. The predistortion of the BPF, however, will yield a circuit that is very sensitive to element tolerances. Therefore, the predistortion should be applied to the LPF such that its pass-band shape compensates for the attenuation response of the BPF. Shaping of the pass-band is achieved by modifying the location of the zeroes of the transducer function by means of an optimization program. This method is successful for this problem and does not lead to a significant increase in circuitry. Since the stop-band requirements of the LPF are not too severe, this filter is not critical and tolerates the worsening of the sensitivity, which is the price for the shaping process.

2.3.3. Combined Equalization of Loss and Group Delay Distortions

Channel LPFs are normally designed as minimum phase networks. For such networks, once the attenuation has been fixed, the phase and its derivative, the group delay, are uniquely defined [19]. Specifying the requirements for attenuation and group delay independently, means the conventional addition of an all-pass to control the group delay, which in turn implies the addition of a nonminimum phase network. Theoretically the all-pass does not alter the attenuation; consequently, such a network can be added upon request.

An alternative to the conventional method is the design of a non-minimum phase LPF containing a quadruplet of conjugate complex attenuation poles [4, 11, 13, 25]. The comparison of the conventional and the nonconventional solution is shown in Figure 2.13. Part (a) of this figure has

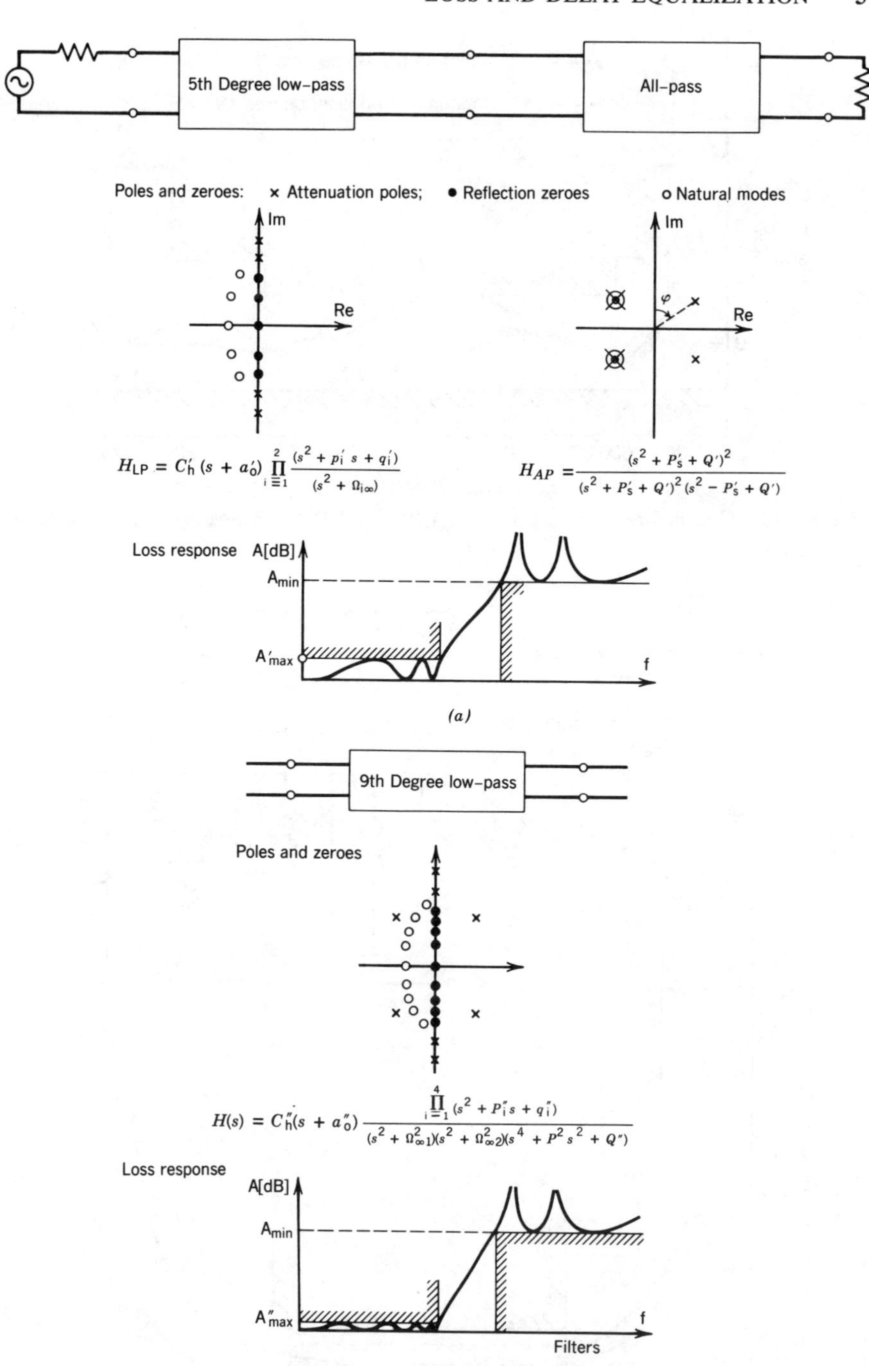

Figure 2.13 Typical pole-zero patterns: (a) conventional cascade realization of a LPF and an all-pass filter; (b) nonminimum phase LPF.

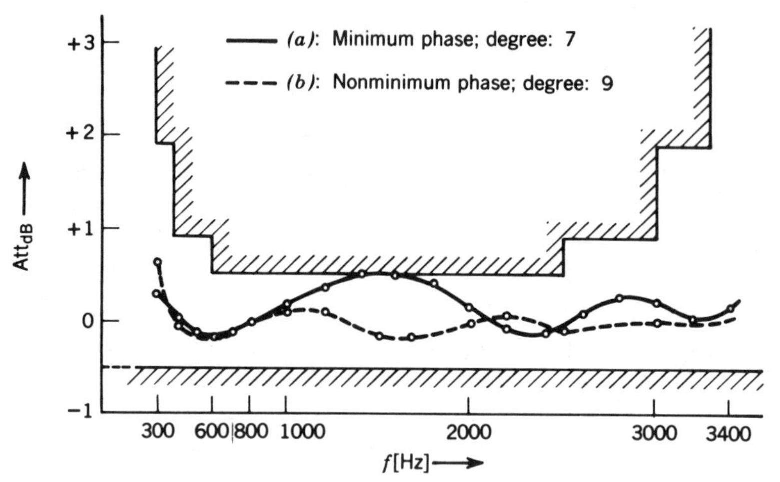

Figure 2.14 Composite loss response of LPFs and BPFs referenced to voice-band frequencies.

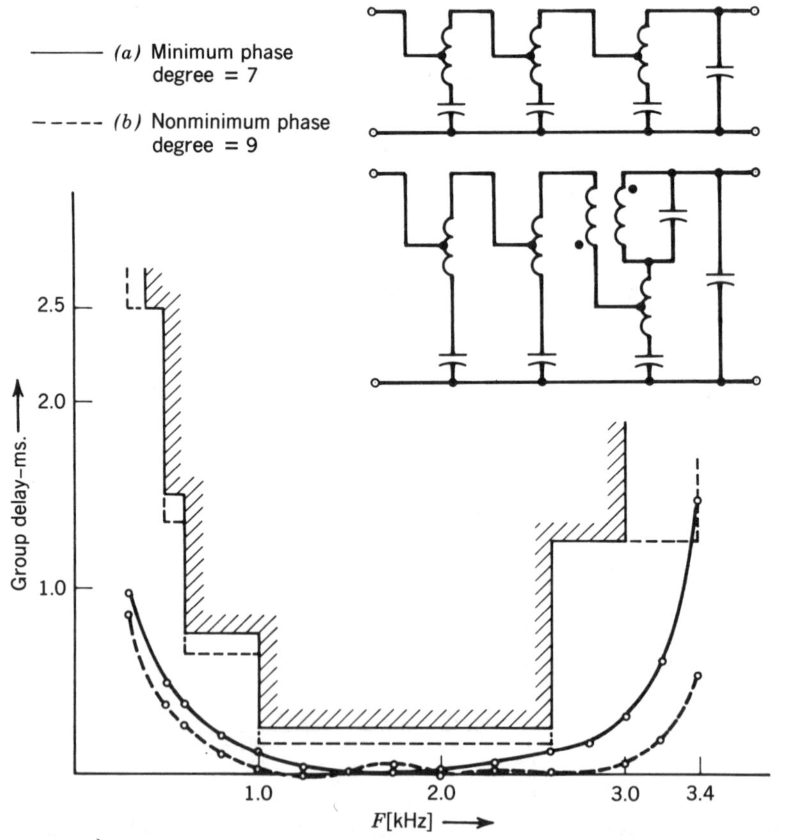

Figure 2.15 Composite group delay response of LPFs and BPFs referenced to voice-band frequencies.

an equal-ripple low-pass of fifth degree in cascade with a fourth degree all-pass. The addition of the all-pass filter will not alter the equal ripple pass-band response and its upper boundary A'_{max}. The overall network, low-pass plus all-pass filter, is a composite reactive network of ninth degree, in which only two pairs of reflection zeroes are available for the pass-band.

In contrast, the design of a ninth degree low-pass (Figure 2.13(b)), offers four pairs of reflection zeroes available for distribution in the pass-band range and, consequently, the upper boundary A''_{max} becomes considerably smaller [11, 13, 25]. More reflection zeroes offer more parameters available for the optimization of a specified response.

The results of a practical design of a compensating LPF confirm the predicted advantages. The objective is a LPF that compensates for the distortions displayed in Figures 2.10 and 2.11. In addition, stop-band requirements are moderate: about 30 db attenuation between 3,750 and 4,200 Hz with monotonic attenuation decrease down to 15 dB at 5 kHz. Two solutions are displayed in Figures 2.14 and 2.15. Case (a) pertains to a minimum phase low-pass of seventh degree. Case (b) to a nonminimum phase lowpass of ninth degree with one attenuation pole quadruplet.

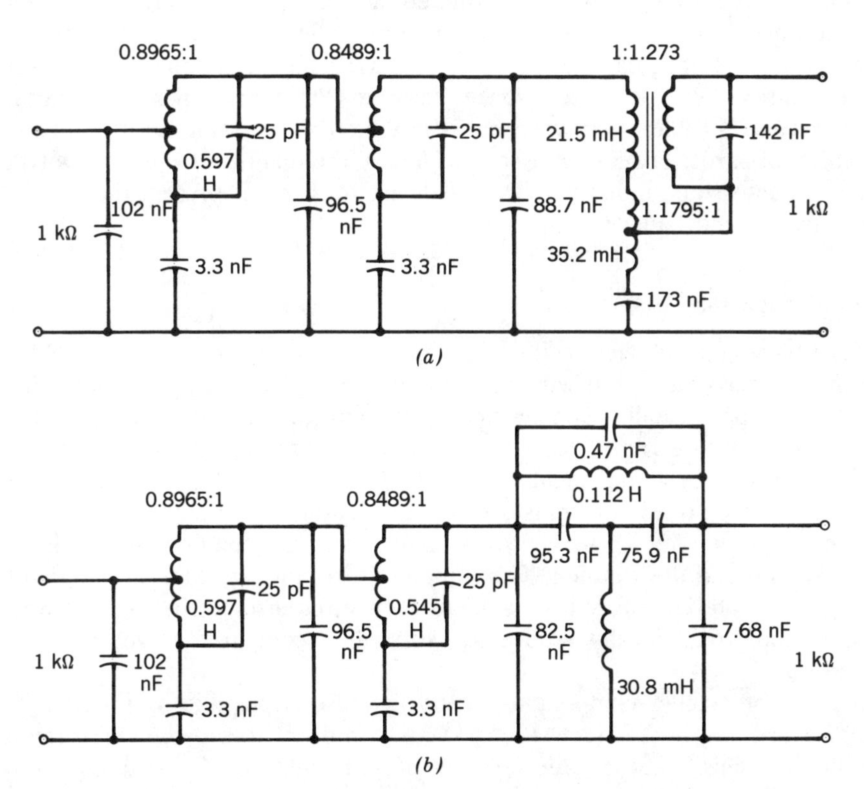

Figure 2.16 (a) Canonical and (b) noncanonical realizations of the pole quadruplet of the LPF.

Stop-band specifications are met in either case. The objective for the least-square optimization is to compensate for the composite loss and delay distortions of the low-pass plus bandpass response in the pass-band. The solution (b) is clearly superior to (a). Improved group delay response in solution (a) would require additional circuitry of an all-pass, increasing size and cost.

For the realization of the pole quadruplet, the options are canonical removal of sections that have the smallest number of elements (see Figure 2.16). At frequencies where such sections are not practical the designer may resort to noncanonical removal of sections in a bridged-T configuration. The conversion equations are shown in Appendix A.

2.4. THE COMPONENT ASPECT OF MINIATURIZED FILTER CIRCUITS

The modern design of miniaturized equipment has led to standards of PC board dimensions and their spacing adjacent to each other. In order to automate the insertion of components, the location of pins or connecting straps in the case of surface mounted devices (SMDs) have also been standardized. These trends lead to component heights of about 12.5 mm or less and to rectangular cross-section to permit high component density on the boards. The decreased space between PC boards has also caused interference problems between filters mounted at the same location of two adjacent boards. For miniaturized LC filters, the main components involved are capacitors and inductors. The following subsections present the state of the art regarding these components.

2.4.1. Inductors

The objective of efficient coil designs is to concentrate the magnetic field in a small area around the winding. This is accomplished by using materials with high permeability and by proper shaping of the magnetic path. The introduction of ferrite materials first led to pot cores and, eventually, to the presently preferred rectangular module (RM) cores [12, 17]. Unless these cores are employed for transformers, for reasons of stability and adjustment, both employ an air gap of defined size in the center post that holds the bobbin and the winding. The complete surrounding of the winding with magnetic material makes it possible to reduce the distance between components and to drastically reduce the size when compared to toroids used previously.

An even further component density increase is possible by providing a bigger opening in the sidewall of pot cores in order to provide coil formers with pins. This led to the development of RM cores, which are identified by the addition of a number indicating their size in multiples of 0.1 inch

(=2.54 mm). For instance, an RM-5 core has a square shape with a side 0.5 inch long. RM cores are available from RM-4 up, although in miniaturized filters RM-5 is about the largest that would be considered due to restrictions in height. (Note: A nonstandardized RM-3 core is available without air gap for use in miniature transformers.) A significant advantage of RM cores lies in the fact that the hardware necessary for assembly has also been standardized, except for some details of the adjustment slugs. Pot cores are identified by their dimensions: diameter × height (in millimeters). For instance, a pot core identified as 1811, which is about the largest pot core for

Figure 2.17 Typical inductors for miniaturized LC filters (courtesy of TDK Components, Ltd.).

miniature filters due to its height, has a diameter of 18 mm and a height of 11 mm. Typical ferrite cores are shown in Figure 2.17. Ninety percent of pot cores and RM cores are mainly used in filters for frequencies below 100 kHz. The reason is twofold. First, the permeability of ferrites suited for high frequencies is much lower than those for low frequencies and, second, the airgap in the center post must be larger. Consequently, such ferrites are no longer capable of containing the magnetic field as well as the materials with high permeability. Inside the core, the field will protrude into the winding, causing additional losses due to eddy currents in the winding itself. Externally, the spread of the magnetic field leads to an interference with other parts of the circuit. Because of their construction, pot cores do have an advantage over RM cores in this respect and, in some special forms, have been employed as transformers into filter networks well above 100 MHz. However, at frequencies above 10 MHz many designers prefer the open or shielded construction of ceramic coil formers or other constructions that are commercially available.

For a given task, most designers will attempt to employ inductors of the smallest possible size. There are two response criteria that limit the selection regarding smallness. The first is the limit postulated by magnetic field density in the magnetic material which sets an upper boundary for the power that the core can handle. If this power is exceeded, intermodulation and distortion may become excessive. The second limit is the available space for the winding. It sets an upper limit to the wire size that can be used for a given number of turns and, therefore, a lower limit for the DC resistance of the winding, which is one of the contributing factors for coil losses. The other contribution comes from eddy currents in the conducting material surrounding the winding, for instance, the magnetic material itself or any electrostatic shields either between windings or external. Outside shields often are a major obstacle and concern, when ceramic coil formers with open construction are employed. Both loss contributions, conventionally called *copper losses* and *core losses*, respectively, are modeled by a series and a shunt resistor and are responsible for the bell shaped Q-curves when plotted over a logarithmic frequency scale [4].

Suitable ferrite materials for frequency applications from about 1 kHz to 100 MHz are compiled in Figure 2.18. Definitions and tables of material constants and information for the design of coils with ferrite materials may be found in the pertinent vendors' catalogues.

For the channel bandpass in the 20–24 kHz range, materials of the type N28, N48, H6B or similar are conventionally used. Their core losses, which range about 20–30% of the total losses, are significantly better in comparison to other materials. In the low-pass filter, the copper losses remain dominant. Obviously, because of the positive temperature coefficient of these ferrite materials, in all resonance circuits capacitors with negative temperature coefficients of the dielectric must be used. RM cores offer significant advantages, such as volume, size, and cost, for bandpass realiza-

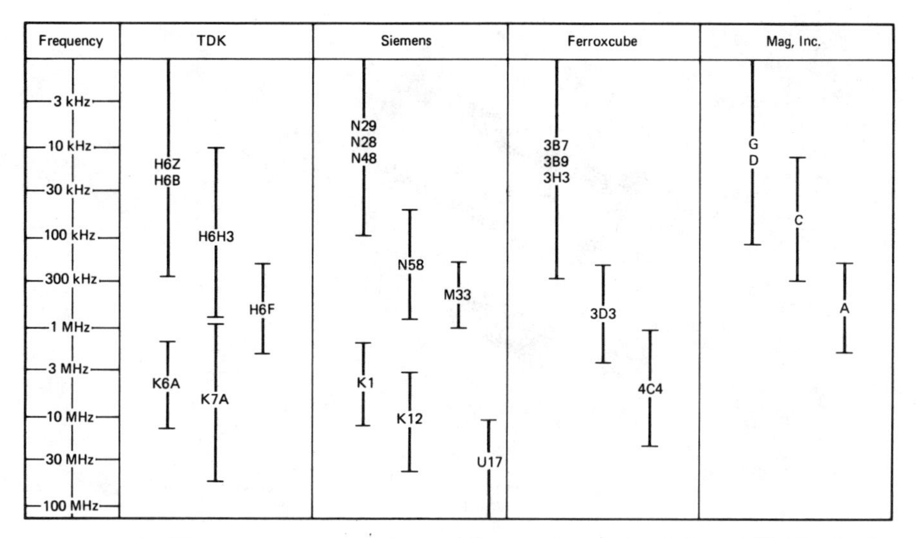

Figure 2.18 Frequency range of operations of commercially available ferrite materials.

tions. In combination with the compression technique to be described below, the core type RM-4 with dimensions $10.5 \times 9.8 \times 9.8$ mm is the best choice. For this core, an inductance factor $AL = 160$ mH per 1000 turns is an acceptable compromise regarding tuning range, stability and Q-factor. Q-values of 350 at 20 kHz can be reliably achieved.

2.4.2. The Compression Technique for Coil Windings

Over the years there has been a continuous improvement of ferrite materials. The winding technique itself did not keep in step. In the low-frequency range, copper losses are dominant. Therefore, a coil operating at low frequencies can be reduced in volume only if it is possible to improve the winding technique significantly. The following describes a technology that permits an optimal use of the available winding volume and, simultaneously, a reduction of the winding capacity.

The method is based on the published technique of bobbinless windings using an enamel copper wire with thermoplastic coating [8]. By means of thermal baking, the winding becomes mechanically stable. In addition to this, for RM-4 coils the finished and heated winding is subjected to a lateral compression of two diametrically opposite sides. The result is a winding that has an oval shape, the smaller diameter fitting the available space in the RM-4 core (see Figure 2.19). The uncompressed portion of the winding protrudes from the open flanges of the core. In the compression processes of the heated winding the thermoplastic coating fills up the normally open spaces between the turns. Obviously, the method permits the use of wire

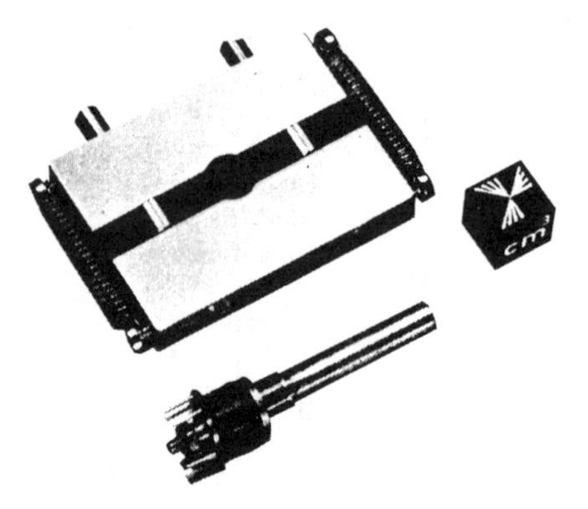

Tool for the compression of the coil winding.

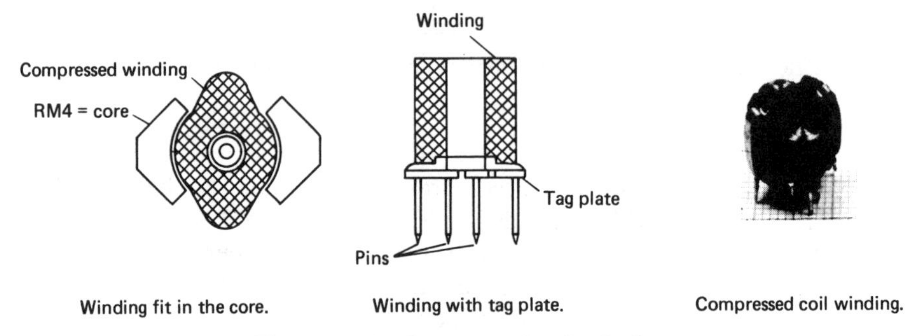

Winding fit in the core. Winding with tag plate. Compressed coil winding.

Figure 2.19 Compressed coil winding.

sizes that would be too large for conventional winding methods. It is applicable to all cores with significant openings such as the RM series. A typical RM-4 coil with compressed winding is shown in Figure 2.19.

2.4.3. Measured Results

Cross-sections of the compressed winding show that the turns are almost ideally side-by-side in the compressed portion as compared to conventional windings. Measurements also show an approximately 50% decrease of the copper and dielectric losses, a reduction of the winding capacity and, consequently, a reduction of the sensitivity to humidity. Figure 2.20 shows Q-curves and winding capacities of various RM coils. Therefore, in comparison to other methods, the partial compression method yields a better utilization of the volume and a higher reliability. Unfortunately, the compression technology does not lend itself easily to automation processes.

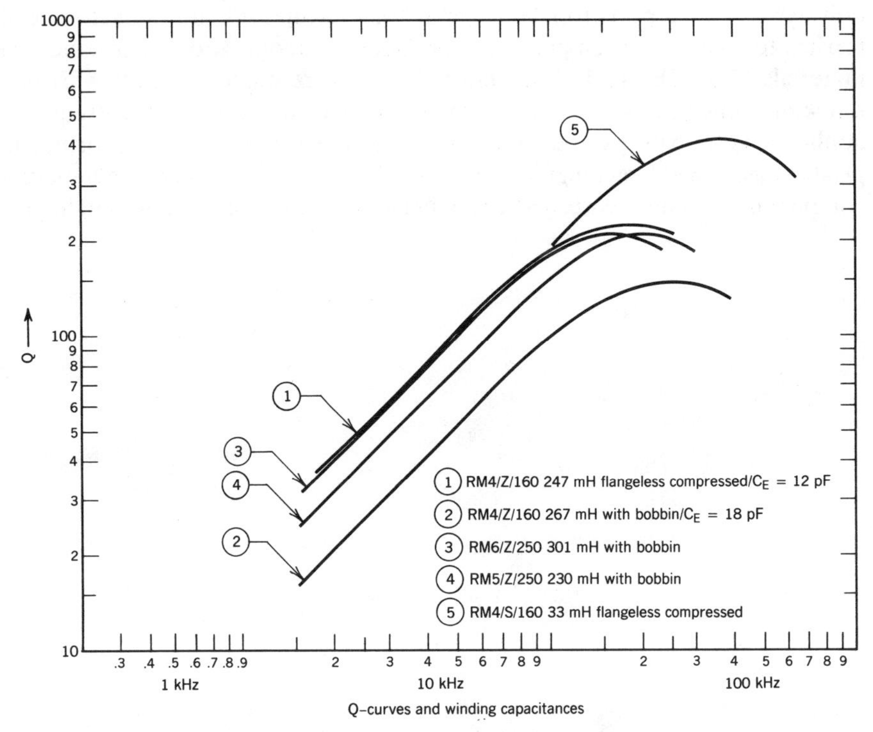

Figure 2.20 *Q*-curves and winding capacitances.

2.4.4. Capacitors

The choice of capacitors is determined by the following aspects:

(a) the dielectric is chosen according to required electrical properties; it determines the temperature behavior in operation and in storage and also its degradation due to humidity;

(b) the construction of the capacitor: capacitors with a film dielectric are smaller in size and are especially useful for miniaturization; however, the thinness of the film reduces the applicable voltage to about 63 V dc;

(c) standard or high reliability versions: the latter are usually somewhat larger in size and considerably more expensive;

(d) the temperature extremes to which the capacitor can be exposed.

Whenever a capacitor is a part of a resonance circuit, it must be chosen such that its temperature coefficient is about the opposite of that of the inductor. For low frequency applications with ferrite inductors, the choice is usually a polystyrene or a polypropylene capacitor. Exceptions are inductors with ferrites for which the temperature coefficient is about zero, for instance the materials H6Z, 3B7 and G of Figure 2.18. For resonance circuits with such inductors, mica or NPO ceramic capacitors can be used. In low-frequency applications polypropylene capacitors are most often used in protected plastic cases with rectangular cross-sections. They are ideal companion components in filters with RM cores because of their shape and height. For

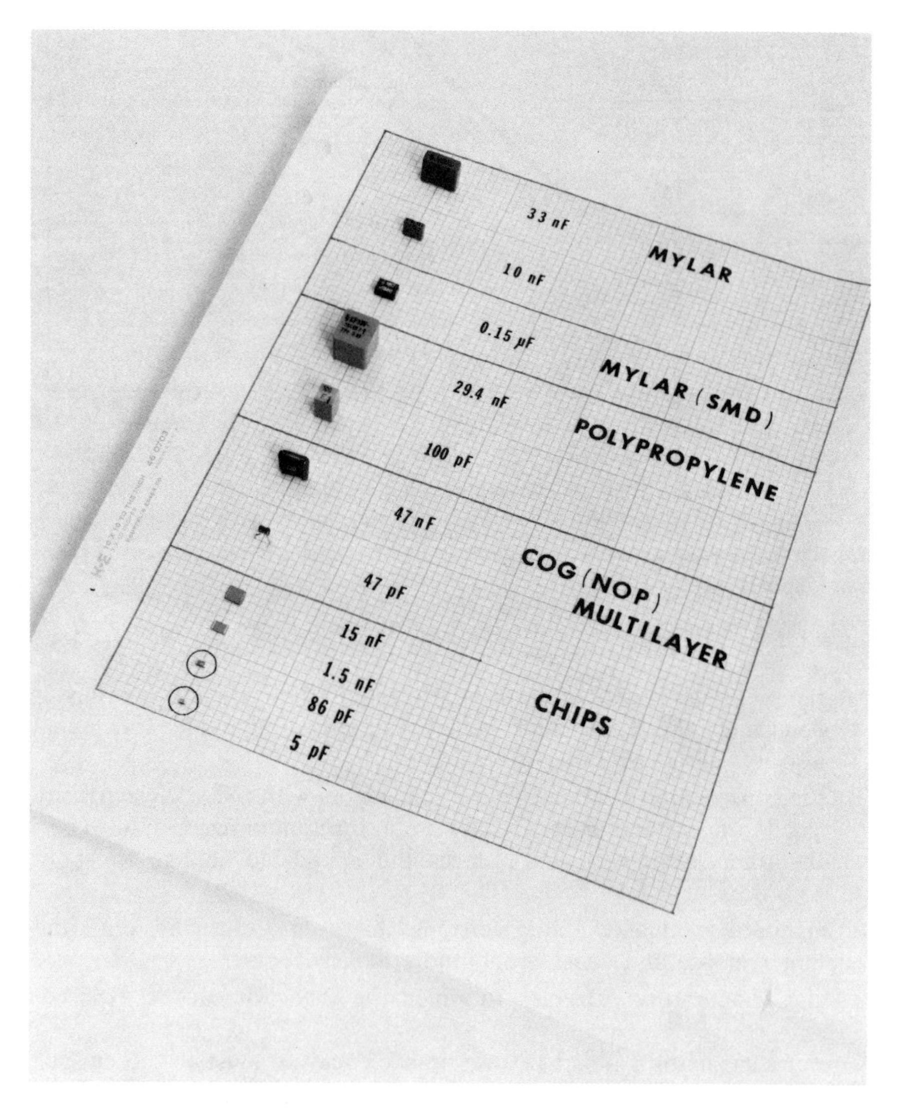

Figure 2.21 Typical capacitors for miniaturized LC filters (courtesy of Siemens Components, Inc.).

those capacitors in low-frequency applications that are not part of a resonance circuit, metalized film capacitors are normally the best choice because of their small size and their self-healing property in case of a break-down due to high voltages. Their temperature coefficients, which can be positive or negative, are considerably larger than those of polystyrene capacitors. Of all metallized film capacitors, those with polycarbonate dielectric have the lowest temperature coefficients.

For miniature filters in high-frequency applications, mica or ceramic capacitors are usually the best choice. At high frequencies, lead inductances can become a problem. To reduce these in mica, as well as in ceramic capacitors, a multilayer construction is preferred. The metal layers are connected on opposite sides of the package. The leads may be brought out by means of connection wires that permit an upright mounting or, in the case of ceramic chip capacitors, as a metal bracket for direct surface mounting. In the latter case, it is of advantage to mount the chip capacitor directly under its companion coil in the case of resonance circuits. Typical capacitor types are compiled in Figure 2.21.

2.5. TEST RESULT OF THE BANDPASS FILTER

The actual bandpass filter is displayed in Figure 2.22. It has been constructed with RM-4 coils with compressed windings and a ferrite material with zero temperature coefficient, and mica capacitors. Its volume is 8.8 cm^3 and, in combination with the low-pass and the signaling filter, the total volume occupies 20 cm^3. In this set of three filters, the signaling filter is a bandpass filter, which is necessary for the insertion or extraction of a dialing frequency of 3825 Hz by which pulsed or frequency-shifted dial tones are transmitted or received. The response of the bandpass filter is displayed in Figure 2.23. In order to predict the response variations due to practical

Figure 2.22 20.3–23.4 kHz channel BPF with tapped RM-4 coils and mica capacitors.

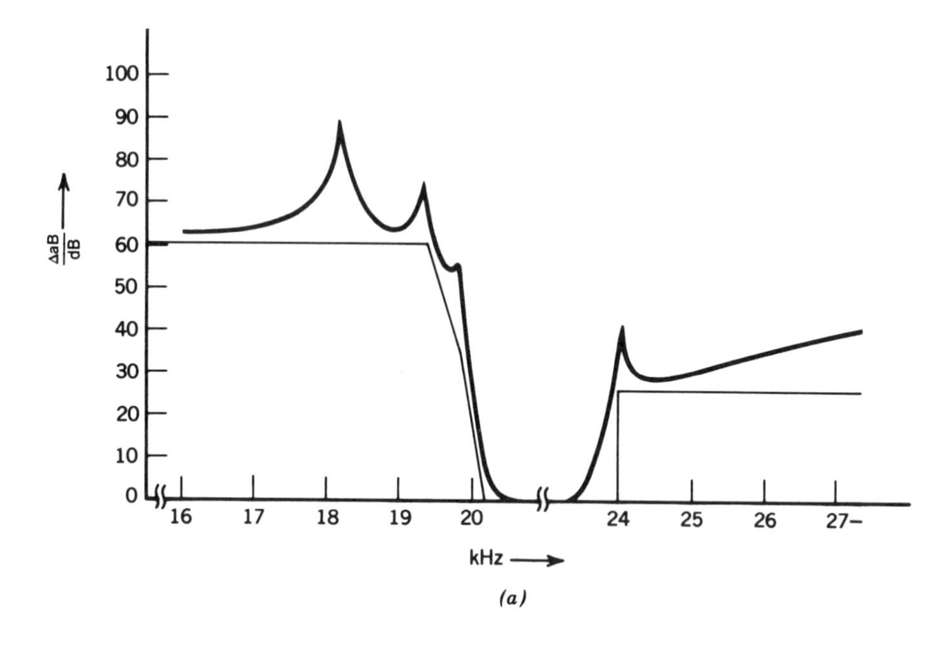

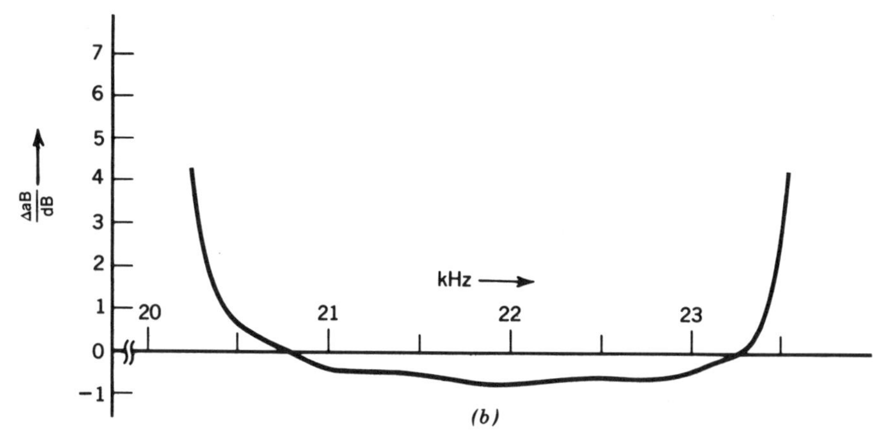

Figure 2.23 Loss responses of a twelfth-order channel BPF with tapped coils: (a) stop-band response; (b) pass-band response.

element tolerances, a Monte Carlo analysis was performed. The result is shown in Figure 2.24. The boundaries of this figure indicate the limits within which 98% of all passband response curves will fall even when the temperature rises 23°C, and for a time span of 130,000 h.

The described example demonstrates that by a combination of suitable design and realization methods, and the use of miniaturized components,

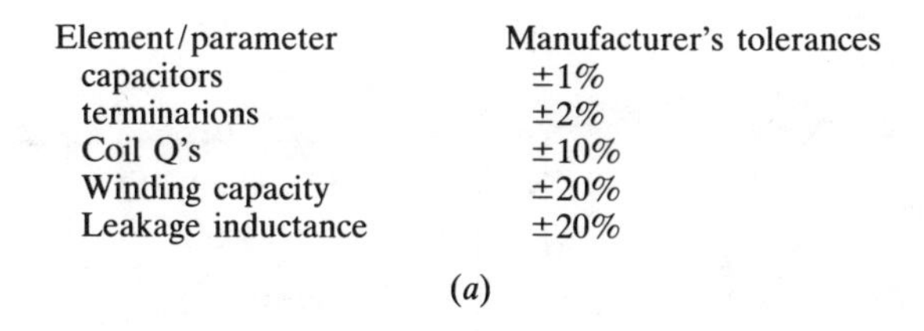

Element/parameter	Manufacturer's tolerances
capacitors	±1%
terminations	±2%
Coil Q's	±10%
Winding capacity	±20%
Leakage inductance	±20%

(a)

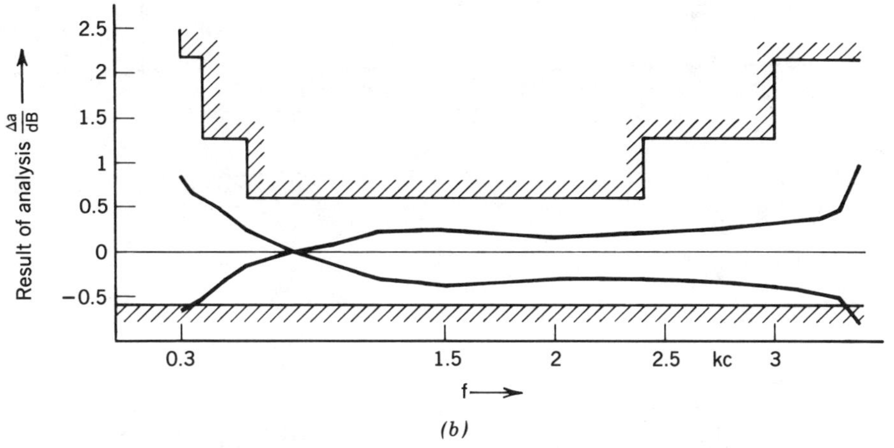

(b)

Figure 2.24 Monte Carlo analysis of channel BPF and LPF.

the conventional LC technology still offers a viable and often competitive alternative to other technologies.

APPENDIX A
COMPLEX POLE REALIZATION IN LC NETWORKS WITHOUT MUTUAL INDUCTANCES

The removal of complex attenuation poles in LC networks was first demonstrated by Darlington and Piloty, almost simultaneously, by means of the basic sections shown in Figure 2.25(a) whose element values are real and satisfy specific coupling conditions [7, 26]. These conditions allow a practical realization using coupled coils as shown in Figure 2.26. The same sections were derived by using other methods in later publications [10, 13, 37]. The removal process, performed in a dual manner with the driving point admittances instead of impedances, leads to the section in Figure 2.25(b). It can be shown that both sections of the figure are equivalent two-ports and are canonical sections for pole quadruplet removals.

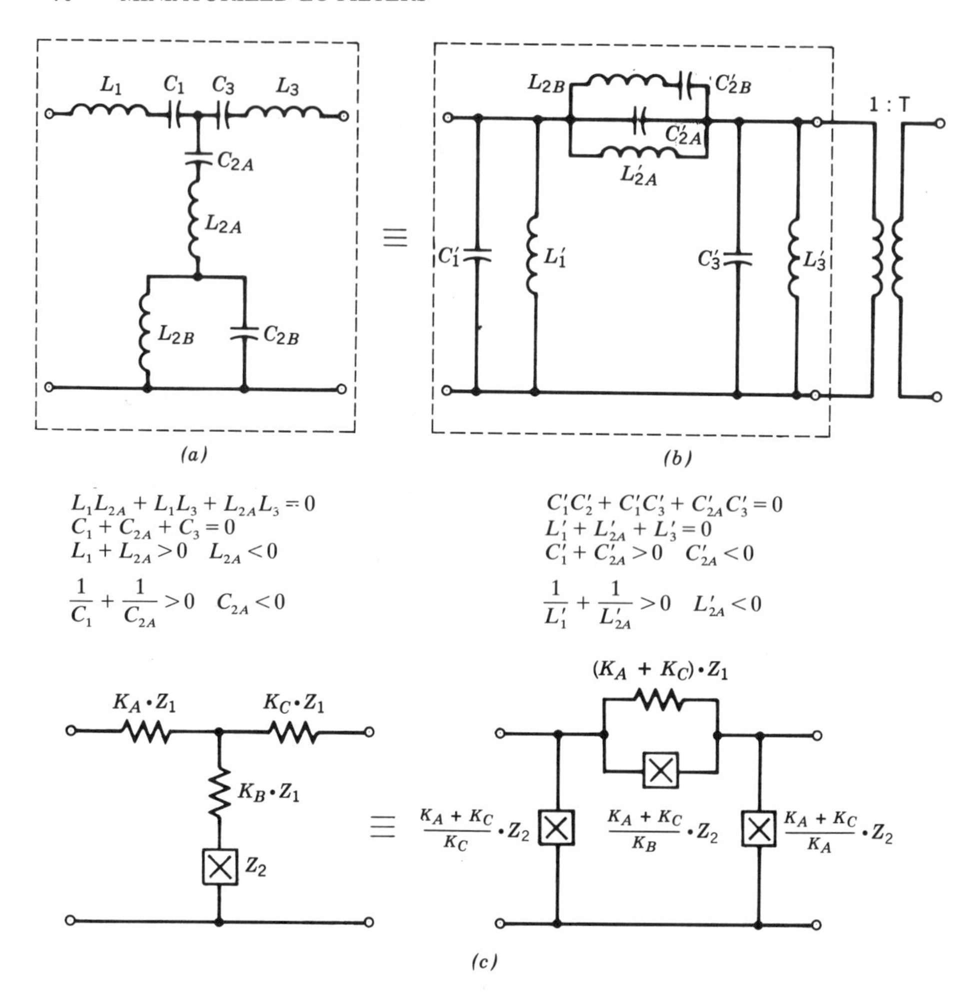

Figure 2.25 Canonical removal sections for pole quadruplets.

When section elements satisfy the coupling conditions

Section (a)	Section (b)	
$L_1L_2 + L_1L_3 + L_{2A}L_3 = 0$	$C'_1C'_2 + C'_1C'_3 + C_{2A}C'_3 = 0$	(A.1)
$C_1 + C_{2A} + C_3 = 0$	$L'_1 + L_{2A} + L'_3 = 0$	(A.2)

and also the conditions

$$L_1C_1 = L_{2A}C_{2A} = L_3C_3 \quad L'_1C'_1 = L'_{2A}C'_{2A} = L'_3C'_3 \tag{A.3}$$

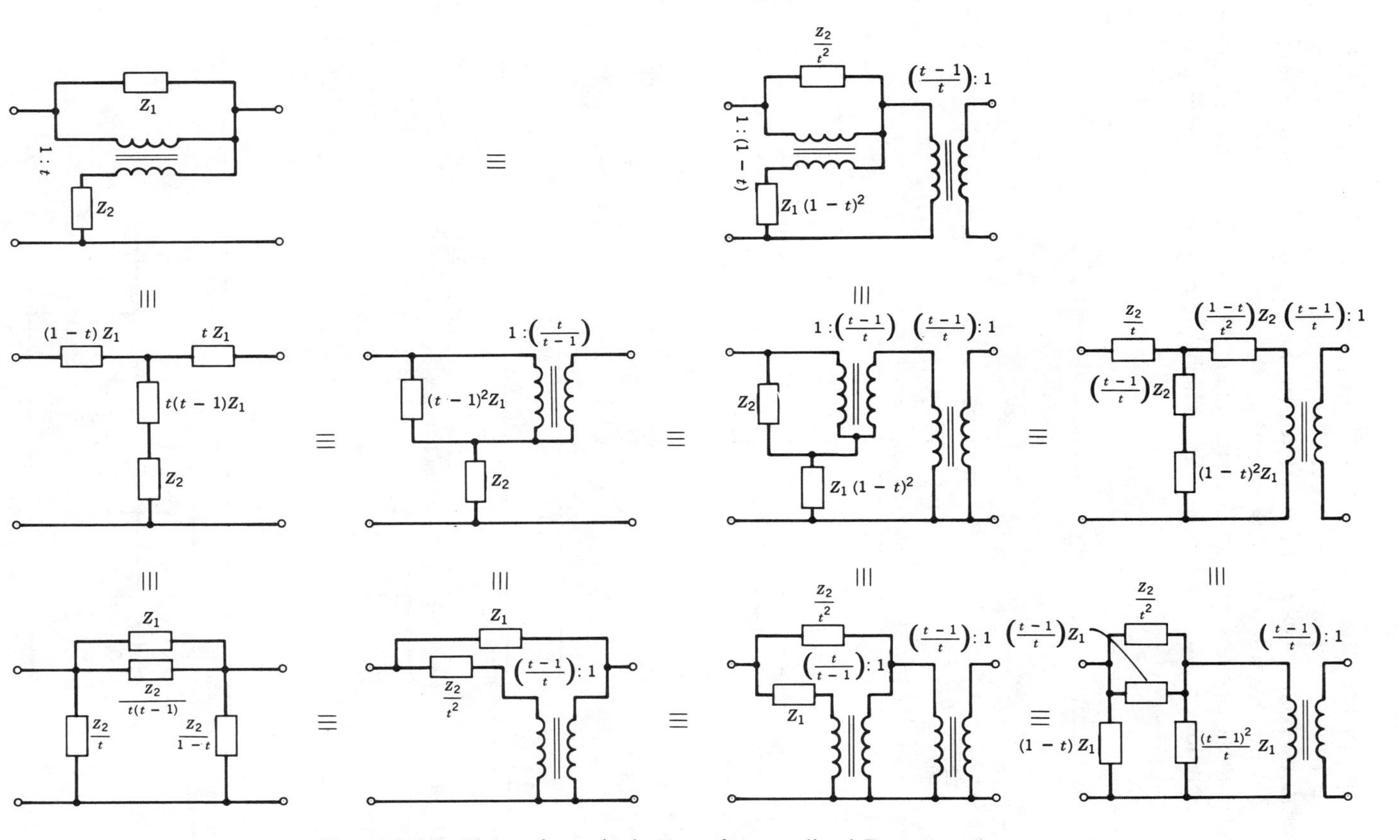

Figure 2.26 Network equivalences of generalized Brune sections.

both sections are directly equivalent, that is, $T = 1.0$ in Figure 2.25(b). Both sections belong to a family of special two-port networks that can be called *generalized Brune sections* of structures shown in Figure 2.25(c), where Z_1 and Z_2 are arbitrary reactances rather than simple Ls and Cs as in conventional Brune sections. The multipliers K_A, K_B, and K_C must satisfy the conditions shown in Figure 2.26, which summarizes the equivalences

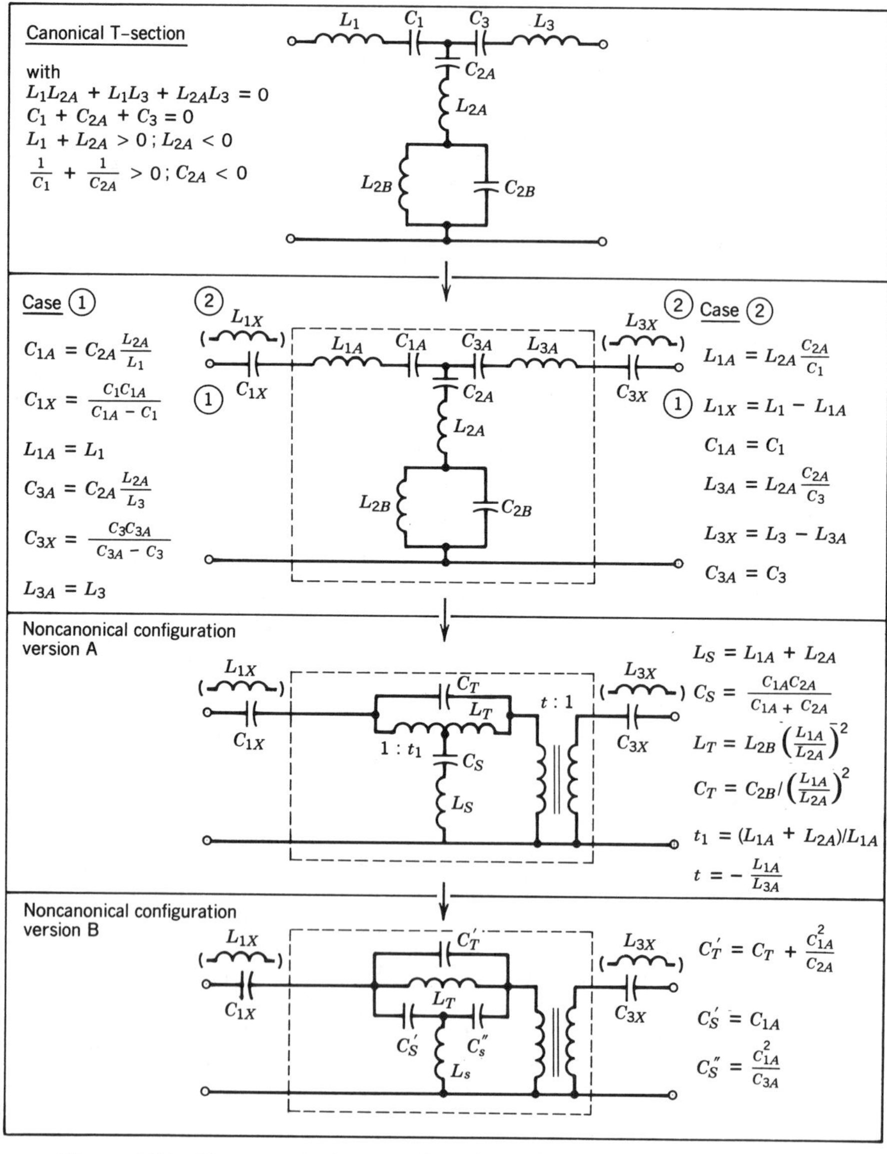

Figure 2.27 Noncanonical removal sections derived from Figure 2.25(a).

between a complete set of Brune sections derived by means of Norton transformations.

In order to apply these equivalence transformations to the removal of sections (a) and (b) in Figure 2.25, first it must be assured that Eq. (A.3) is satisfied. Then in T-section (a), L_{1x} or C_{2x} is removed on the left, and L_{3x} or C_{3x} on the right. The same element type must be removed at either side;

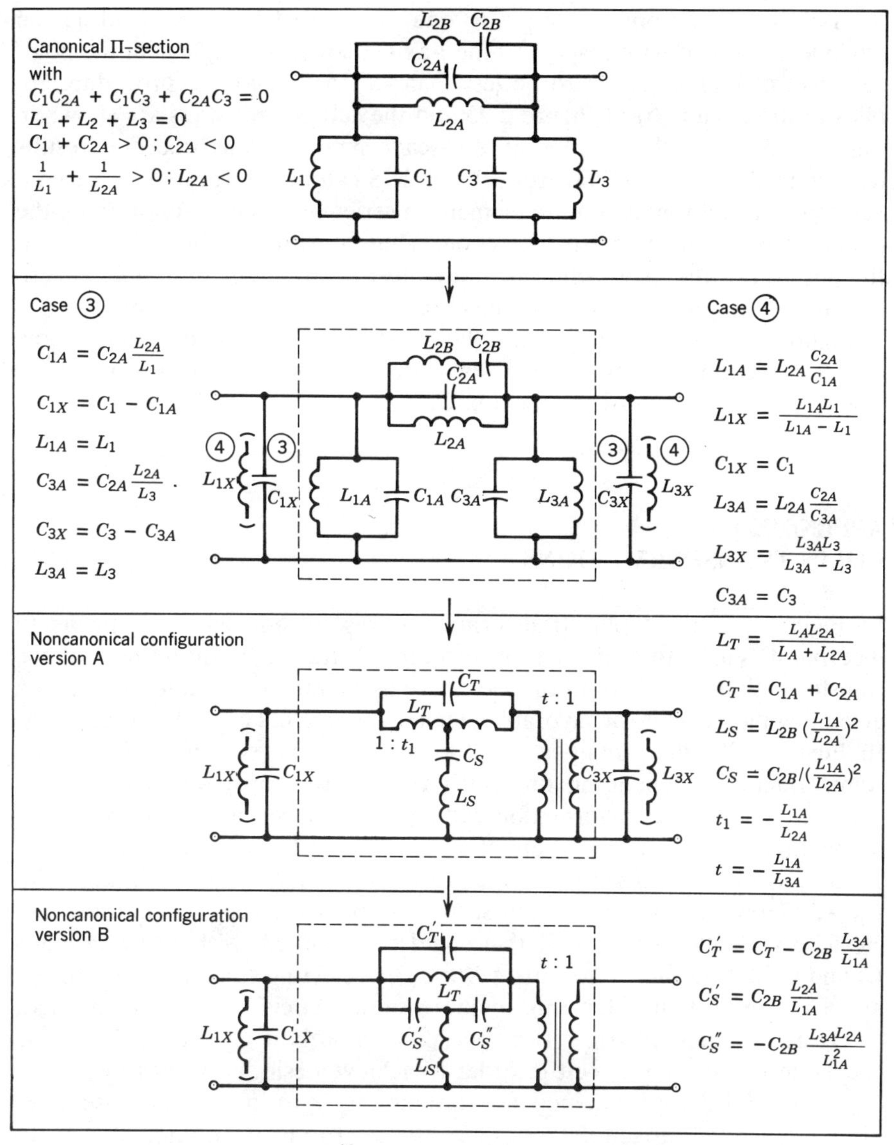

Figure 2.28 Noncanonical removal sections derived from Figure 2.25(b).

they will have opposite signs. Since the removed elements do not enter into the subsequent equivalence transformations, a negative element will appear either preceding or following the removal section. Obviously, the remaining part of the circuit must be capable of absorbing such negative elements. The details of the subsequent quad transformation are shown in Figure 2.27. Guillemin [10] has shown that the condition $C'_T > 0$ is related to location of the attenuation pole quadruplet in the complex plane. According to the cited reference, for C'_T to be positive it is necessary, but not sufficient, that the radius vector from the origin to one member of the pole quad has an angle less than 45° with respect to the j-axis; that is, the angle ϕ in the pole/zero plot of Figure 2.13 must be less than 45°. An analogous procedure applies to the circuit (b) of Figure 2.25 and the subsequent steps are shown in Figure 2.28. From the standpoint of cascade synthesis, the preceding results leading to the circuits of Figures 2.27 or 2.28 can be interpreted as the conventional partial removal of an element in series or in shunt followed by the complete removal of the quad section. Thus, in order to be applicable, at the position of the quad removal, there must be an attenuation pole at zero or infinity still present if noncanonical removal sections are feasible.

Figure 2.16 is an example for the substitution of a noncanonical pole removal section without tapped coils for a canonical section. A second example may be found in Appendix C.

APPENDIX B
COLIN TRANSFORMATIONS

In Figure 2.7, the Colin transformation was introduced as a means to incorporate capacitors into Brune sections. Actually, the usefulness of the transformation goes beyond this objective by making it possible to modify a given low-pass circuit for favorable elements, inductors as well as capacitors. In Figure 2.29, the circuit at the top center is the starting point of the transformation. Depending on which of the two capacitors C_1 or C_3 is positive, the Colin transformation will either produce the circuit on the left or on the right. In cases where the center circuit is a canonical removal section, $K = 0$ and the transformed circuits become Brune sections with $C_P = 0$. However, if the center section is noncanonical, or if there are capacitors parallel to C_3 or C_1 that could be absorbed, within certain limits C_3 and C_1 can be chosen to satisfy $K > 0$. By selecting suitable values for C_1 and C_3, some of the elements of the resulting circuits can be controlled.

Figure 2.29 demonstrates two special cases on how C_1 and/or C_3 can be chosen in the starting circuit in order to achieve desirable values for C_P, or C_P, and C_S. In low-frequency applications this is of interest because quite often the needed capacitors are of large values (up to 100 nF and above). In such cases it is useful to restrict the capacitors to standard values such as

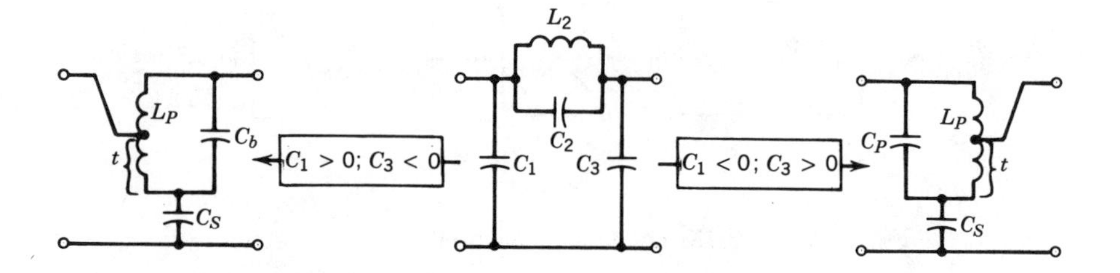

$$C_S = C_1 + C_3$$

$$t = -\frac{C_3}{C_1} \qquad K = \frac{C_1C_2 + C_1C_3 + C_2C_3}{C_3} \geq 0 \qquad t = -\frac{C_1}{C_3}$$

$$t_1 = \frac{C_1}{C_S};\ L_P = L_2 t_1^2;\ C_P = \frac{K}{t_1^2} \qquad = C_2 + \frac{C_1C_3}{C_S} \geq 0 \qquad t_1 = \frac{C_3}{C_S};\ L_P = L_2 t_1^2;\ C_P = \frac{K}{t_1^2}$$

Special cases ($C_1 > 0$):

(a) Specified: C_1, C_2, and C_P; Find: C_S, C_3

$$\underbrace{K = C_P\left(\frac{C_1}{C_S}\right)^2 = C_2 - \frac{C_1^2}{C_S} + C_1}$$

$$C_S^2 - \frac{C_1^2}{(C_1 + C_2)}C_S - \frac{C_PC_1^2}{(C_1 + C_2)} = 0;\ \text{Determine } C_S;\ \text{Then: } C_3 = C_S - C_1$$

(b) Specified: C_2, C_S, and C_P; Find: C_1, C_3

$$\underbrace{K = C_P\left(\frac{C_1}{C_S}\right)^2 = C_2 - \frac{C_1^2}{C_S} + C_1}$$

$$C_1^2 - \frac{C_S^2}{(C_P + C_S)}C_1 - \frac{C_S^2}{(C_P + C_S)}C_2 = 0;\ \text{Determine } C_1;\ \text{then } C_3 = C_S - C_1$$

Figure 2.29 Colin transformations.

33 nF, 47 nF. As the control of element values is limited, the user may have to resort to trial and error methods or implement the resulting circuit by means of optimization programs for standard values. For an example, refer to Figure 2.35 in Appendix C.

APPENDIX C
THE DESIGN OF A DELAY-CORRECTING LOW-PASS FILTER

C.1. DESIGN OBJECTIVE

The following example demonstrates the design of a delay-correcting low-pass filter for an FDM channel unit. Figure 2.30 shows the arrangement of

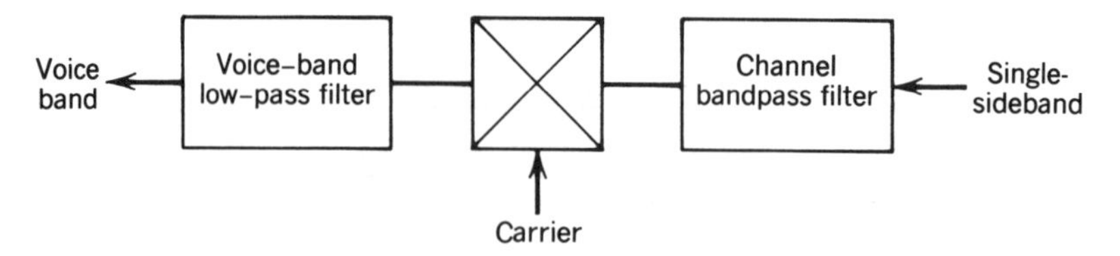

Figure 2.30 FDM channel translation (receive direction).

the LPF, isolating modulator and bandpass filter. The bandpass filter whose delay is to be corrected is similar but not identical to the one shown in Figure 2.6. Also, it is assumed that the FDM system employs outband signaling, in which case the incoming signal passing through the bandpass filter will include a dialing frequency of 3.825 kHz above the carrier fre-

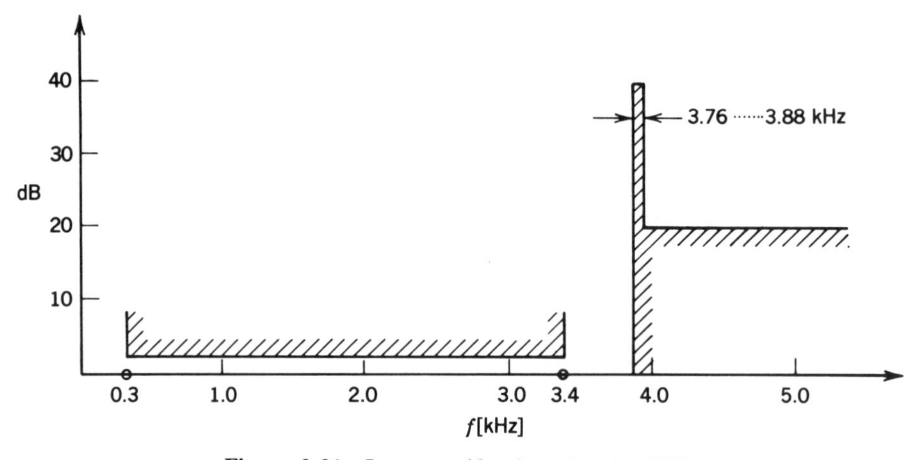

Figure 2.31 Loss specifications for the LPF.

Table C.1

VB Frequency (kHz)	Delay (μs)	VB Frequency	Delay (μs)
1.0	499	1.7	366
1.05	483	1.8	358
1.1	468	1.9	353
1.15	454	2.0	348
1.2	441	2.1	343
1.25	429	2.2	341
1.3	418	2.3	339
1.4	399	2.4	341
1.5	384	2.5	345
1.6	374	2.6	351

quency. This frequency must be blocked out by the low-pass to prevent interference with the voice signal. For this reason, the LPF must satisfy a loss specification as shown in Figure 2.31. In addition to this attenuation specification, the LPF must also have a delay response such that the combined delay distortion of the LPF and BPF is less than $\pm 50\,\mu s$ with respect to a constant delay, in the frequency range 1000–2,600 Hz. The channel response then becomes suitable for data transmission over a cable pair. The measured delay response of the particular FDM BPF over the voice-band (VB) frequencies is listed numerically in Table C.1.

C.2. DESIGN CONSIDERATIONS

For the design of the LPF, it is practical to introduce two sets of attenuation poles [4].

(a) Attenuation poles on the $j\omega$-axis to satisfy the stop-band requirements in the vicinity of the signal frequency. To find suitable locations, one of several computer programs [6, 33] can be employed, or the poles can be placed heuristically. Typical frequency allocations are 3.778 kHz and 3.897 kHz.

(b) To satisfy the delay requirements, it is necessary to optimize the location of the natural modes because they alone are responsible for the group delay. Indirectly, they are optimized by the heuristic placement of one or several quads of complex attenuation poles. For this task, the computer program QUAIL is well suited [6]. It will optimize the composite delay of two filters in a least-square sense by optimizing one of these networks; simultaneously, an equal ripple is always maintained in the pass-band. In the example, the initial location of the quad was chosen to be $(\pm 1.36 \pm j2.34)$ kHz.

C.3. COMPUTER-AIDED DESIGN OF THE LOW-PASS

C.3.1. Input data for QUAIL

```
QUAIL EXAMPLE: DELAY-CORRECTING LOW-PASS FILTER
*
*OBJECTIVE: TO DESIGN A VB LOW-PASS INTENDED TO
*CORRECT FOR THE DELAY DISTORTION OF A SINGLE-
*             SINGLE-SIDEBAND BANDPASS
*
```

```
LEVEL 2
SPECIFIC.  BAND = 0 TO 3400 HZ LOSS = 0.01 DB AT 3400 HZ
POLE  0  3.778KHZ
POLE  0  3.897KHZ
POLE  1.36KHZ 2.34KHZ STEP 100  100
POLE  INFINITY
FREQ UNIT KHZ 1.0   1.05   1.1  1.15   1.2   &
              1.25  1.3    1.4  1.5    1.6   &
              1.7   1.8    1.9  2.0    2.1   &
              2.2   2.3    2.4  2.5    2.6
RESP UNIT USEC 499 483 468 454 441  &
               429 418 399 384 374  &
               366 358 353 348 343  &
               341 339 341 345 351
PROCESS
APPROX MAXEVAL 30
PLOT LOSS 3.4KHZ TO 4.4KHZ STEP 20 HZ
PLOT DELAY 300 HZ TO 3.4KHZ STEP 100 HZ
END
```

C.3.2. Results from QUAIL

The tabulated and plotted delay is shown in Figure 2.32. Optimized locations of poles and zeros, are shown in Table C.2.

Table C.2

	Real part (kHz)	Imaginary part (kHz)
Attenuation poles	0.0	±3.7780
	0.0	±3.8970
	±1.12452	±1.99963
Reflection zeros	0.0	0.0
	0.0	±1.40470
	0.0	±2.33544
	0.0	±3.05659
	0.0	±3.36827
Natural modes	−3.04477	0.0
	−1.03743	±1.77294
	−0.98931	±2.24836
	−0.55919	±3.50586
	−0.10703	±3.54274

FINAL RESULTS - UNITS KHZ AND USEC

FREQUENCY	SPEC DELAY	COMP DELAY	TOTAL DELAY	SPEC TOLER
1.000000	499.000	261.623	760.623	1.000
1.05000	483.000	268.478	751.478	1.000
1.10000	468.000	275.687	743.687	1.000
1.15000	454.000	283.224	737.224	1.000
1.20000	441.000	291.056	732.056	1.000
1.25000	429.000	299.138	728.138	1.000
1.30000	418.000	307.417	725.417	1.000
1.40000	399.000	324.302	723.302	1.000
1.50000	384.000	341.088	725.088	1.000
1.60000	374.000	357.041	731.041	1.000
1.70000	366.000	371.387	737.387	1.000
1.80000	358.000	383.409	741.409	1.000
1.90000	353.000	392.533	745.533	1.000
2.00000	348.000	398.389	746.389	1.000
2.10000	343.000	400.859	743.859	1.000
2.20000	341.000	400.129	741.129	1.000
2.30000	339.000	396.739	735.739	1.000
2.40000	341.000	391.621	732.621	1.000
2.50000	345.000	386.089	731.089	1.000
2.60000	351.000	381.778	732.778	1.000

```
EVALUATE DELAY - UNITS KHZ AND USEC
FREQUENCY RANGE  0.300000      TO    3.40000      STEP  0.100000      LINEAR

   FREQ          RESP  100.00                  500.0                   900.0
                                 300.0                   700.0              1100.

                         - - - - - - - - - - - - - - - - - - - - - - - - - -
  0.30000      204.58    .    *    .         .         .         .         .
  0.40000      208.59    .    *    .         .         .         .         .
  0.50000      213.86    .     *   .         .         .         .         .
  0.60000      220.47    .     *   .         .         .         .         .
  0.70000      228.49    .     *   .         .         .         .         .
  0.80000      237.99    .      *  .         .         .         .         .
  0.90000      249.04    .      *  .         .         .         .         .
   1.0000      261.62    .       * .         .         .         .         .
   1.1000      275.69    .        *.         .         .         .         .
   1.2000      291.06    .         *         .         .         .         .
   1.3000      307.42    .         *         .         .         .         .
   1.4000      324.30    .         .*        .         .         .         .
   1.5000      341.09    .         . *       .         .         .         .
   1.6000      357.04    .         .  *      .         .         .         .
   1.7000      371.39    .         .   *     .         .         .         .
   1.8000      383.41    .         .   *     .         .         .         .
   1.9000      392.53    .         .    *    .         .         .         .
   2.0000      398.39    .         .    *    .         .         .         .
   2.1000      400.86    .         .    *    .         .         .         .
   2.2000      400.13    .         .    *    .         .         .         .
   2.3000      396.74    .         .    *    .         .         .         .
   2.4000      391.62    .         .    *    .         .         .         .
   2.5000      386.09    .         .   *     .         .         .         .
   2.6000      381.78    .         .   *     .         .         .         .
   2.7000      380.58    .         .   *     .         .         .         .
   2.8000      384.66    .         .   *     .         .         .         .
   2.9000      396.56    .         .    *    .         .         .         .
   3.0000      419.63    .         .     *   .         .         .         .
   3.1000      459.21    .         .       * .         .         .         .
   3.2000      526.55    .         .         .*        .         .         .
   3.3000      654.13    .         .         .       * .         .         .
   3.4000      958.84    .         .         .         .         .  *      .
                         - - - - - - - - - - - - - - - - - - - - - - - - - -
```

Figure 2.32 LPF delay response from QUAIL.

C.3.3. Synthesis of the LPF

The resulting reflection zeroes and attenuation poles may serve as input data for suitable synthesis programs, for instance the NC-State program SYNTHESE [4]. The input data for this program is shown in Figure 2.33 and two of many possible low-pass circuits in Figure 2.34. Of these two basic circuits, circuit (a) is a canonical representation for which we may expect the minimum number of circuit elements. However, a close inspection of the element values of the resulting Brune sections indicate that the inductances have unrealistic values if we postulate small physical sizes. Circuit (b) is a noncanonical realization. It has a negative shunt capacitor preceding the first attenuation pole and, consequently, postulates a Brune section. To overcome these contradictory requirements we may want to resort to Colin sections which offer some control over the inductance values.

```
columns 1         2         3         4         5         6         7
1234567890123456789012345678901234567890123456789012345678901234567890123456789

   1 BATCH
 COMPLETE SYNTHESIS                                                       K(S)-DES.
 DELAY-CORRECTING LOWPASS
 DEGREE=  9    REFL.ZEROES         1 AT ZERO          4 PAIRS
               ATTEN.POLES         0 AT ZERO          2 PAIRS             1 QUAD
 REF.FRQ. 3.4          KHZ  LOSS = 0.01       DB   AT       3.4     KHZ
 EVAL      DELAY
 FROM        0.3       KHZ
 TO          3.3       KHZ
 SCALE     LINEAR
 INCREM.    0.1        KHZ
 PLOT MARGINS
 LEFT       0.0        MSEC
 RIGHT      0.8        MSEC
 SUBDIV.    0.2        MSEC
 REFL.ZEROES
 R1         0.0                    1.4047   KHZ
 R2         0.0                    2.3354   KHZ
 R3         0.0                    3.0569   KHZ
 R4         0.0                    3.3682   KHZ
 ATTEN.POLES
 P1         0.0                    3.778    KHZ
 P2         0.0                    3.897    KHZ
 PQ         1.12452                1.9996   KHZ
 REALIZATION DATA
 GENERATOR      600.0           OHM
   2 REALIZATIONS
D -1   -2    4    3
D  1    2    4    3
/*

columns 1         2         3         4         5         6         7
1234567890123456789012345678901234567890123456789012345678901234567890123456789
```

Figure 2.33 Input data for SYNTHESE.

```
            REALIZATION SEQUENCE (   1)

                 REMOVAL SEQUENCE

   D -1  -2   4   3

            DENORMALIZED ELEMENT VALUES

*       *
*       *
*       R          R( 1) =       600.00000000 OHMS
*       *
*       L          L( 2) =        86.10855784 MH
*       *
*       *          C( 3) =         1.63489311 NF
**C***L**          L( 3) =      1085.49367808 MH
*       *          F( 3) =         3.77799964 KHZ
*       *
*       L          L( 4) =       -79.77988800 MH
*       *
*       L          L( 5) =       -31.33503532 MH
*       *
*       *          C( 6) =        18.27129603 NF
**C***L**          L( 6) =        91.28732025 MH
*       *          F( 6) =         3.89699911 KHZ
*       *
*       L          L( 7) =        47.71280039 MH
*       *
****C****          C( 8) =       111.03439187 NF
*       *
****C****          C( 9) =        74.10617409 NF
*       *
****L****          L(10) =        45.29012276 MH
*       *         CP(11) =       -45.26380172 NF
*     *****       LP(11) =       -93.14130928 MH
*     * * L       F1(11) =         2.45116997 KHZ
*     C L *       CS(12) =        50.56829720 NF
*     * * C       LS(12) =       108.65051333 MH
*     *****       F2(12) =         2.14716321 KHZ
*       *
****C****          C(13) =       116.29858762 NF
*       *
****L****          L(14) =        47.85118652 MH
*       *
****R****          R(15) =       669.76949963 OHMS
```

(a) Canonical

Figure 2.34 Resulting realizations.

```
              REALIZATION SEQUENCE (   2)

                     REMOVAL SEQUENCE
     D  1   2   4   3
              DENORMALIZED ELEMENT VALUES

*       *
*       *
*       R         R( 1) =        600.00000000 OHMS
*       *
****C****         C( 2) =        -20.60963714 NF
*       *
*     *****       C( 3) =        280.41692439 NF
*     C   L       L( 3) =          6.32866984 MH
*     *****       F( 3) =          3.77799964 KHZ
*       *
****C****         C( 4) =         75.47369165 NF
*       *
*     *****       C( 5) =        101.84159102 NF
*     C   L       L( 5) =         16.37776507 MH
*     *****       F( 5) =          3.89699911 KHZ
*       *
****C****         C( 6) =         76.07652650 NF
*       *
****C****         C( 7) =         74.10617409 NF
*       *
****L****         L( 8) =         45.29012276 MH
*       *        CP( 9) =        -45.26380172 NF
*     *****      LP( 9) =        -93.14130928 MH
*     * * L      F1( 9) =          2.45116997 KHZ
*     C L *      CS(10) =         50.56829720 NF
*     * * C      LS(10) =        108.65051333 MH
*     *****      F2(10) =          2.14716321 KHZ
*       *
****C****         C(11) =        116.29858762 NF
*       *
****L****         L(12) =         47.85118652 MH
*       *
****R****         R(13) =        669.76930930 OHMS
```

(b) Noncanonical

Figure 2.34 (*Continued*)

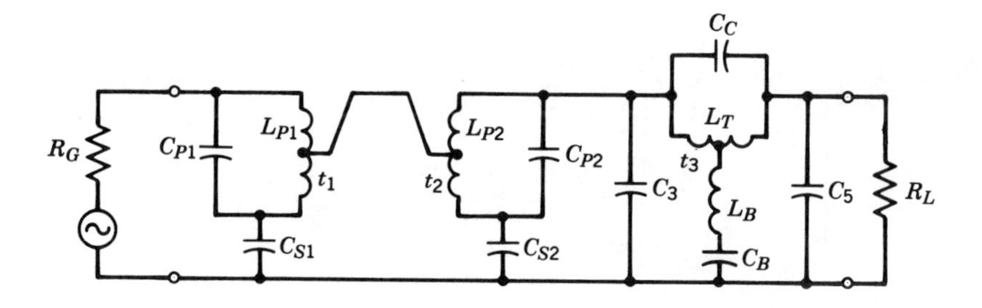

$R_G = 600\ \Omega$					
$L_{P1} = 120.5$ mH	$C_{P1} = 10$ nF	$C_{S1} = 6.13$ nF	$f_{S1} = 3.778$ kHz	$t_1 = 0.7707$	
$L_{P2} = 147.1$ mH	$C_{P2} = 500$ pF	$C_{S2} = 16.3$ nF	$f_{S2} = 3.897$ kHz	$t_2 = 0.6662$	
$C_3 = 89.6$ nF					
$L_T = 88.2$ mH	$C_C = 47.8$ nF	$L_B = 25.7$ mH	$C_B = 214$ nF	$f_B = 2.147$ kHz	$t_3 = 0.5137$
$C_S = 31.5$ nF					
$R_L = 600\ \Omega$					

Figure 2.35 LPF after transformations.

C.3.4. Network Transformations

The objective of network transformations in this particular case is twofold.

(a) Produce a noncanonical removal section. The method described in Figure 2.28 is suitable.

(b) Reduce the inductance values of the removal sections for finite attenuation poles to modify the circuit such that miniaturized components can be used. The Colin transformations of Figure 2.29 are suitable. In both transformations, an attempt was made, by trial and error, to eliminate also some of the capacitors in shunt branches. A designer has a variety of options to carry out such transformations. The resulting low-pass configuration is shown in Figure 2.35.

C.4. OVERALL RESPONSES OF THE CHANNEL UNIT

Figure 2.36 shows the measured overall response of the K-65 Channel Unit manufactured at ITT-Telecommunications in Raleigh, North Carolina. The top part is the delay response demonstrating the effect of the delay-correcting low-pass. The bottom part is a plot of the loss response within the margins of the CCITT tolerance mask. The flatness was accomplished by optimizing the components of the circuit of Figure 2.12. The overall response, delay and loss was accomplished with the addition of only one coil when compared to conventional channel units in LC technology. The delay response makes the unit suited for the transmission of data.

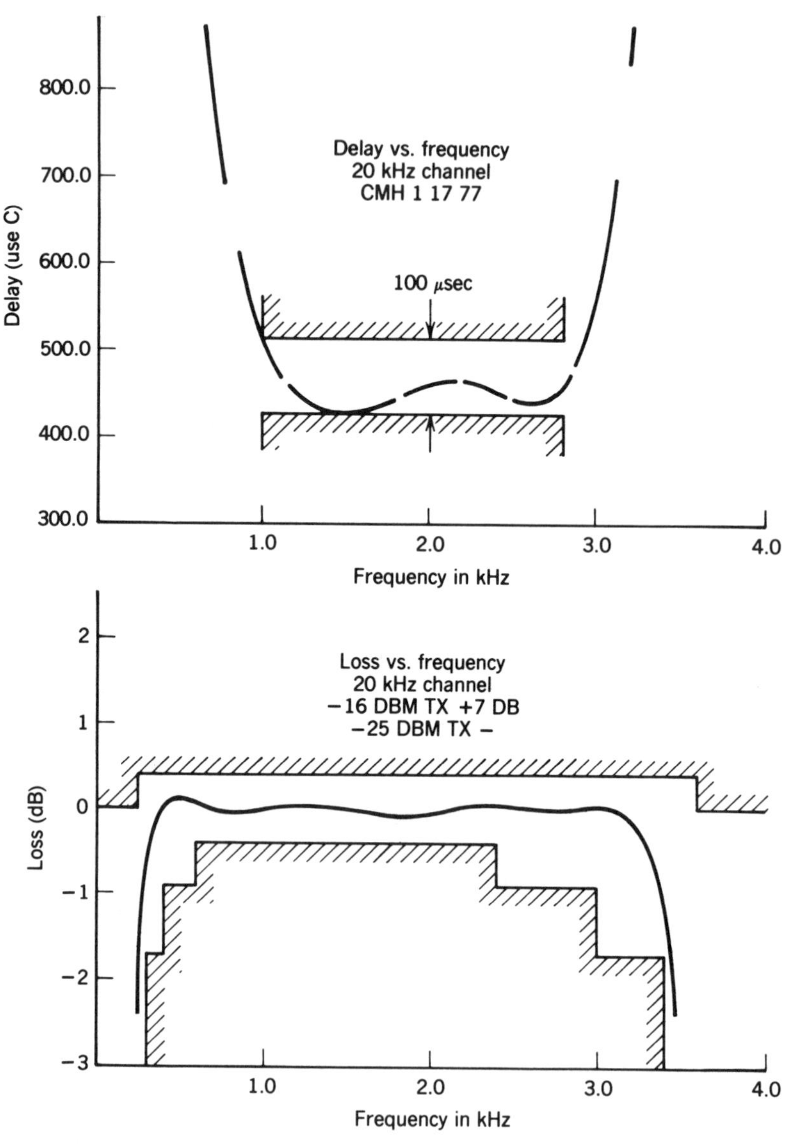

Figure 2.36 Looped response (transmit and receive) of the ITT-K65 channel unit.

REFERENCES

1 K. Antreich et al. (1975) Computer-aided tuning of electrical circuits. *NTZ* **28**, 200–206.

2 K. Antreich et al. (1978) Zur Vereinheitlichung der Toleranz und Empfindlichkeitsanalyse electrischer Netzwerke. *Arch. Elekt. Übertr.* **32**, 369–373.

3 G. W. Bleisch (1972) The A6 Channel Bank. *I.E.E.E. Trans. Commun.* **20**, 196–201.

4 E. Christian (1983) *LC-Filter-Design, Testing and Manufacturing*. Wiley, New York.

5 J. E. Colin (1962) De l'introduction de cristeaux piezoelectriques dans les filtres pass-bas et pass-haut en echelle. *Cables Transm.* **16**, 85.

6 S. Conning (1986) QUAIL (Fortran computer program for optimizing the passband group delay of filters by adjustment of pole quads and real pole pairs); TEMPLATE (Fortran computer program for the adjustment of attenuation poles according to the Laurent–Rumpelt method); PROMETHEUS (Fortran computer program for the synthesis of lossless ladder networks); GREBE (Fortran computer program for the design of all-pass group delay equalizers); EAGLE (Fortran computer program for the design of loss equalizers). All listed computer programs are available commercially for several computer types, including PCs, from: Scientific Computing Services, 26 Hereford Street, Glebe, NSW 2037, Australia.

7 S. Darlington (1939) Synthesis of reactance 4-poles. *J. Math. Phys.* **18**, 257–353.

8 J. Van Dedem (1975) 8 TR 401, a new generation of channel and group translating equipment. *Philips Telecom. Rev.* **33**, 150–158.

9 R. A. Friedenson (1975) RC-active filters for the D3 channel bank. *Bell Systems Tech. J.* **54**, 507–530.

10 E. A. Guillemin (1957) *Synthesis of Passive Networks*. Wiley, New York, pp. 207–210, 247–269, 369–382.

11 H. Gutsche (1973) Approximation of transfer functions for filters with equalized group delay characteristics. *Siemens Forsch. Entwicklungsber.* **2**, 288–292.

12 W. Haas (1973) Channel equipment technology using pregroup modulation. *Electron. Commun.* **48**, 11–15.

13 M. Hibino et al. (1968) Design of Chebyshev filters with flat group delay characteristics. *I.E.E.E Trans. Circuit Theory* **15**, 316–325.

14 Y. Katsuta et al. (1978) Thick film RC active filters for PCM applications. *Proc. I.S.H.M. 78 Intern. Symp. on Microelectronics*; Minneapolis MN.

15 D. Klein et al. (1979) Miniaturized filters for channel translator units by means of LC-technology. *Proc. I.E.E.E.*, **67**, 6–16.

16 H. Kopp (1972) A mechanical filter channel bank. *I.E.E.E. Trans. Comm.* **20**, 64–67.

17 U. Koenig (1974) Ferrite fuer induktive Bauelemente. *Electro-Anzeiger*. **8**.

18 C. F. Kurth (1974) Channel bank filtering in frequency division multiplex communication systems. *Circuits and Systems Magazine, I.E.E.E.* **7**, 5–13.

19 C. F. Kurth and M. L. Liou (1975) Computation of group delay from attenuation characteristics via Hilbert transformation and spline functions and its application to filter design. *I.E.E.E. Trans. Circuits and Systems* **22**, 729–734.

20 C. F. Kurth (1976) Generation of SSB Signals, *I.E.E.E. Trans. Circuits and Systems* **23**, 1–13.

21 C. F. Kurth (1977) Filters. In *Electronics Designer's Handbook*, L. J. Giacoletto (Ed.), Chapter 6. McGraw-Hill.

22 G. Mueller (1979) Interactive circuit optimization and computer aided tuning. Doct. Thesis, Techn. University of Munich, W. Germany.

23 E. L. Norton (1928) U.S. Patent 1681554.

24 H. J. Orchard and G. Temes (1968) Filter design using transformed variables. *I.E.E.E. Trans. Circuit Theory* **15**, 385–408.

25 H. J. Orchard et al. (1978) Image parameter approximations of linear phase filters. *I.E.E.E. Trans. Circuit Theory* **25**, 325–333.

26 H. Piloty (1939) Wellenfilter, insbesondere symmetrische und antimetrische, mit vorgeschriebennem Betriebsverhalten. *T.F.T.* **29**, 10, 363–375.

27 H. Piloty (1940) Kanonische Kettenschaltungen fuer Reaktanzvierpole mit vorgeschriebenem Betriebsverhalten. *T.F.T.* **29**, 9–11, 249–58, 279–90, 320–25.

28 J. F. Pinel (1971) Computer-aided network tuning. *I.E.E.E. Trans. Circuit Theory* **18**, 192–194.

29 R. Saal and E. Ulbrich (1958) On the design of Filters by Synthesis. *I.R.E. Trans. Circuit Theory* **5**, 385–408.

30 T. E. Shea, *Transmission Networks and Wave Filters*. D. van Nostrand Co., New York, 174–284.

31 D. Sheahan (1975) Polylithic Crystal Filters. *Proc. 29th Ann. Symp. Freq. Contr.*, 120–123.

32 T. H. Simmonds (1979) The evolution of the discrete single-side-band filter in the Bell System. *Proc. I.E.E.E.* **67**, 109–115.

33 G. Szentirmai (1977) FILSYN—a general purpose filter synthesis program. *Proc. I.E.E.E.* **65**, 1445–1458.

34 G. C. Temes and S. K. Mitra (Editors) (1973) *Modern Filter Theory and Design*. Wiley, New York.

35 M. Watanabe (1979) Large scale production of a pole electromechanical channel filter. *Proc. I.E.E.E. Int. Symp. on Circuits and Systems*, Tokyo, pp. 1076–1079.

36 H. Watanabe (1960) On the circuit with a minimum number of coils. *I.R.E. Trans. Circuit Theory* **7**, 77–78.

37 D. C. Youla (1961) A new theory of cascade synthesis. *I.R.E. Trans. Circuit Theory* **8**, 244–260.

38 T. Yano et al. (1974) New torsional mode electromechanical channel filters. *Proc. Eur. Conf. on Circuit Theory and Design*, London, pp. 121–126.

39 K. Yakuwa et al. (1974) Development of new channel bandpass filters. *Proc. I.E.E.E. Int. Symp. on Circuits and Systems*, San Francisco, pp. 100–105.

BIBLIOGRAPHY

D. Sheahan (1974) Channel bank filtering at GTE-Lenkurt. *Proc. I.E.E.E. Int. Symp. on Circ. and Syst.*, 115–120.

3 Miniaturized Active RC Filters

ADEL S. SEDRA
University of Toronto, Toronto, Ontario, Canada

In this chapter we shall study methods for the realization of filter transfer functions using circuits composed of operational amplifiers (op amps), resistors and capacitors. Op amp-RC filters are in general easy to design and fabricate. Depending on production volume, they are fabricated using: (a) discrete components on printed-circuit boards (small volumes); (b) thick-film technology (medium volumes); or (c) thin-film technology (large volumes). A number of active RC filter products are currently available from semiconductor manufacturers, such as National Semiconductor; from analog module manufacturers, such as Burr–Brown; and from specialized companies, such as Frequency Devices (see §3.6). More recently, special design techniques have been proposed to permit the fabrication of active RC filters in monolithic form (§3.5). However, these latter methods are still in the research stage.

3.1. ACTIVE RC FILTER DESIGN

3.1.1. The Problem

Two-port RC networks have all their natural modes located on the negative real axis of the s plane [3]; a property that severely limits their use alone to realize selective filter responses. On the other hand, combining RC networks together with amplifiers in feedback structures enables the placement of the poles of the closed-loop circuit at any desired s-plane location. In these circuits, feedback acts to move the natural modes from the negative real axis to complex conjugate locations determined by the filter transfer function to be realized. The active RC filter design problem is that of finding suitable feedback structures that utilize op amps together with RC networks to realize given filter transfer functions.

3.1.2. Overview of Active RC Filter Design Methods

There are basically two approaches to the design of active RC filters. The first is based on the simulation of an LC ladder network that realizes the given transfer function. The second involves the realization of the given transfer function as the cascade of second-order filter sections (and one first-order section for odd-order filters).

The designs based on LC ladder networks are motivated by the observation made in 1966 by Orchard [20] that the doubly-terminated LC ladder network, which is designed for maximum power transfer from source to load, exhibits very low sensitivities (to components changes) in the filter pass-band. Orchard's heuristic argument, which has since been supported by analytic results [26], stimulated considerable interest in finding active RC, SC, and digital filter design methods based on an LC ladder prototype network. All these methods begin by desigining a doubly-terminated LC ladder network that realizes the given transfer function. Then, to obtain an active RC circuit, one can follow either one of two very distinct routes. The first route leads to an active circuit that simulates the internal workings of the LC ladder network. Specifically, the integral i–v relationships that characterize the operation of the individual L and C elements in the LC network are realized in the active circuit using integrators. The Kirchhoff current and voltage relationships that govern the operation of the LC network are simulated in the active circuit using signal summers. Thus the resulting active RC circuit consists of op amp integrators and summers, the latter usually implemented using the virtual-ground inputs of the integrators. The integrator output voltages are the analogs of the inductor currents and capacitor voltages of the LC ladder. We study the operational simulation method in §3.2.

In its most basic form, the other approach to LC ladder simulation is based on replacing the network inductors with active RC circuits having inductive input impedances. This component simulation method of active RC filter design is studied in §3.3.

For filters that do not have stringent specifications, the cascade design method can be used. Although cascade filters have higher sensitivities than filters designed using ladder simulation, they are easy to design and their performance is acceptable in many applications. Cascade design is the subject of §3.4.

The design methods of §3.2–§3.4 lead to op amp RC circuits that can be assembled either on a printed circuit board, or using hybrid thin- or thick-film technologies. The hybrid technologies yield miniaturized filters that are adjustable using laser trimming techniques. The unit price, however, is high compared to what could be achieved if a monolithic integrated-circuit realization were available. These circuits are not directly suitable for monolithic integration for two reasons: the requirement of large capacitance values (especially for audio-frequency applications) and the need for accurate RC time constants. A recently proposed method that seeks to adapt these designs to monolithic IC realization using MOS technology is described in §3.5. An alternative approach is the switched capacitor technique presented in Chapter 5.

As already mentioned, a variety of active RC filters are currently available as off-the-shelf components. Representative examples of such products are mentioned in §3.6.

3.2. DESIGN BASED ON SIMULATING THE OPERATION OF LC LADDER NETWORKS

The design of active RC filters based on simulating the operation of LC ladder networks is illustrated using a fifth-order elliptic-function low-pass filter example. Figure 3.1(a) shows an LC ladder realization of this filter.

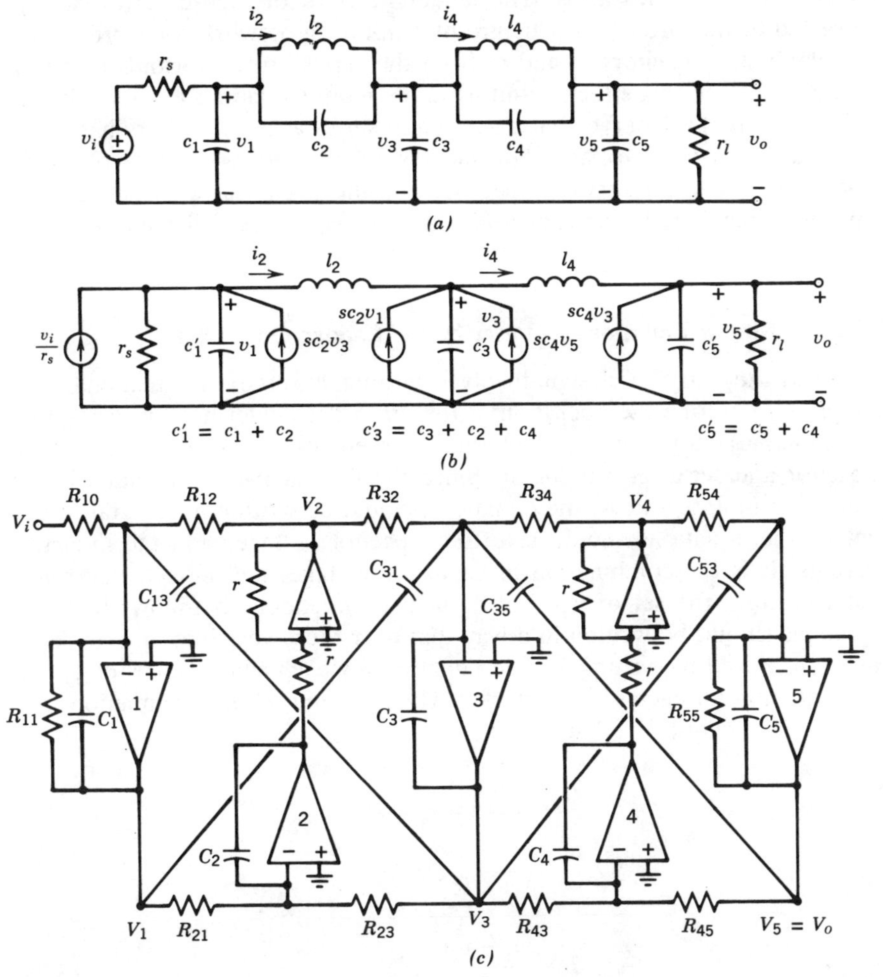

Figure 3.1 Obtaining an active RC circuit that simulates the operation of the fifth-order elliptic low-pass filter in (a).

The component values of the ladder network can be found using either filter design tables [28] or filter design computer programs [25]. A remark on notation is in order here: lowercase symbols will be used to denote component values, voltages, and currents of the LC ladder network; uppercase letters will be used for the elements and voltages of the op amp-RC circuit. Since throughout this chapter we deal mostly with frequency-domain quantities, no confusion should result from adopting this notation.

3.2.1. Preparing the Ladder Network

For reasons that will become clear shortly, the most convenient ladder structure to simulate is that consisting of capacitors in the shunt arms and inductors in the series arms. The ladder network of Figure 3.1(a) can be converted to this preferred structure by employing network transformations that eliminate capacitors c_2 and c_4 from the series arms. Norton's theorem can be applied to this effect, resulting in the modified ladder network shown in Figure 3.1(b). Observe that each of the series-arms capacitors has been replaced with two voltage-controlled current sources and two shunt capacitors. The latter appear in parallel with the already existing shunt-arms capacitors, resulting in the changes in capacitance values indicated in Figure 3.1.

3.2.2. Writing *n* Equations to Describe the Ladder Operation

The next step in the design involves writing a set of n equations that completely describe the operation of the nth order ladder network ($n = 5$ in our example). Specifically, we write one equation for each shunt-arm capacitor and series-arm inductor. Since the final active RC circuit will be realized using integrators, the equations should be written in integral form; that is, the equations should give the capacitor voltages and the inductor currents. It is the combination of capacitor voltages and inductor currents that constitute the set of n variables in the simulation. It follows that the equations should be written in a way that only these variables appear. This can be achieved by writing a node equation for each shunt capacitor and a loop equation for each series inductor. The resulting set of five equations for the ladder network of Figure 3.1(b) is

$$v_1 = \frac{\left(\frac{v_i}{r_s}\right) - i_2 + sc_2v_3}{sc_1' + \frac{1}{r_s}} \tag{3.1}$$

$$i_2 = \frac{v_1 - v_3}{sl_2} \tag{3.2}$$

$$v_3 = \frac{i_2 - i_4 + sc_2v_1 + sc_4v_5}{sc_3'} \tag{3.3}$$

$$i_4 = \frac{v_3 - v_5}{sl_4} \tag{3.4}$$

$$v_5 = \frac{i_4 + sc_4v_3}{sc_5' + \dfrac{1}{r_l}} \tag{3.5}$$

3.2.3. Obtaining the Active RC Circuit

Next we find an active RC circuit whose operation is described by five equations (n for a filter of order n) identical in form to Eqs. (3.1)–(3.5). It follows that the active RC circuit must consist of five integrators with provision for signal summing at their inputs. By utilizing op-amp integrators, signal summing can be readily implemented at their virtual-ground inputs. The interconnections between the five integrators are specified by the numerator terms of Eqs. (3.1)–(3.5), and the voltages at the outputs of the five integrators, V_1–V_5, should correspond (except for possible inversion) to the capacitor voltages and the inductor currents of the LC ladder (v_1, $i_2, \ldots, v_5$). It will become apparent shortly, that the resulting op amp-RC circuit will consist of $(n-1)$ loops, with each loop consisting of two integrators, one integrator simulating the operation of a shunt capacitor and the other integrator simulating the operation of a series inductor. To keep the gain of each loop negative, the two integrators must have opposite signs. Therefore, we shall use inverting integrators to simulate the capacitor operation (i.e., Eqs. (3.1), (3.3), and (3.5)) and noninverting integrators to simulate the inductor operation (i.e., Eqs. (3.2) and (3.4)). The result is the op amp-RC circuit shown in Figure 3.1(c). This circuit is obtained by sketching the five integrators and then interconnecting them so as to realize the numerators of Eqs. (3.1)–(3.5). To see how this is done, consider integrator no. 1 which realizes Eq. (3.1). The output of this integrator, V_1, is the analog of the voltage v_1 in the passive circuit. From the numerator of Eq. (3.1) we see that the input to integrator no. 1 is the sum of three signals: the first is proportional to v_i and is implemented in the active circuit by feeding V_i to the virtual-ground input of the op amp via resistor R_{10}; the second is proportional to i_2 and is implemented in the active circuit by feeding V_2 to the virtual-ground input via resistor R_{12}; and the third is sc_2v_3 which is implemented in the active circuit by feeding V_3 to the virtual ground input of integrator no. 1 via capacitor C_{13}. As another example, consider integrator no. 2. It realizes Eq. (3.2) whose numerator indicates that the integrator input consists of two signals: the first is proportional to v_1 and is implemented by feeding V_1 to the virtual-ground input of integrator no. 2 via resistor R_{21}; and the second is proportional to v_3 and is implemented by feeding V_3 to the virtual-ground input of integrator no. 2 via resistor R_{23}.

In the description above it appears as if we have paid no attention to signal polarities. The fact is, however, that if one uses the structure of alternating integrator polarities, as in Figure 3.1(c), correct signal polarities result, and the designer does not need to keep track of signal signs. To verify this, and to determine the component values of the active circuit, we write the five equations that describe its operation

$$V_1 = -\frac{\dfrac{V_i}{R_{10}} + \dfrac{V_2}{R_{12}} + sC_{13}V_3}{sC_1 + \dfrac{1}{R_{11}}} \tag{3.6}$$

$$V_2 = \frac{\dfrac{V_1}{R_{21}} + \dfrac{V_3}{R_{23}}}{sC_2} \tag{3.7}$$

$$V_3 = -\frac{\dfrac{V_2}{R_{32}} + \dfrac{V_4}{R_{34}} + sC_{31}V_1 + sC_{35}V_5}{sC_3} \tag{3.8}$$

$$V_4 = \frac{\dfrac{V_3}{R_{43}} + \dfrac{V_5}{R_{45}}}{sC_4} \tag{3.9}$$

$$V_5 = -\frac{\dfrac{V_4}{R_{54}} + sC_{53}V_3}{sC_5 + \dfrac{1}{R_{55}}} \tag{3.10}$$

Comparison of each of these equations to the corresponding equation is the set (3.1)–(3.5) reveals the following consistent set of correspondences and the following component values for the active circuit

$$V_i \Leftrightarrow v_i; \quad -V_1 \Leftrightarrow v_1; \quad -V_2 \Leftrightarrow i_2; \quad V_3 \Leftrightarrow v_3; \quad V_4 \Leftrightarrow i_4; \quad -V_5 \Leftrightarrow v_5$$

$$R_{10} = r_s; \quad R_{12} = 1\,\Omega; \quad C_{13} = c_2; \quad C_1 = c_1'$$

$$R_{11} = r_s; \quad R_{21} = 1\,\Omega; \quad R_{23} = 1\,\Omega; \quad C_2 = l_2$$

$$R_{32} = 1\,\Omega; \quad R_{34} = 1\,\Omega; \quad C_{31} = c_2; \quad C_{35} = c_4$$

$$R_{43} = 1\,\Omega; \quad R_{45} = 1\,\Omega; \quad C_4 = l_4; \quad R_{54} = 1\,\Omega$$

$$C_{53} = c_4; \quad C_5 = c_5'; \quad R_{55} = r_l$$

With this set of component values the active circuit realizes a transfer function

$$\frac{V_5}{V_i} = \left(\frac{-v_5}{v_i}\right) = -\frac{v_5}{v_i}$$

which is equal to that of the LC ladder except for a sign inversion. If desired, the gain of the active filter can be increased to any required value by appropriately changing the value of the input feed-in resistance, R_{10}. Specifically, if it is desired to make the gain of the active filter k times that of the LC ladder, R_{10} should be changed to

$$R_{10} = \frac{r_s}{k}$$

Two additional steps are required in order to obtain a practical active filter circuit: scaling the component values to maximize the filter dynamic range and scaling to obtain practical component values.

3.2.4. Dynamic Range Scaling

To understand the dynamic range scaling process consider the circuit in Figure 3.1(c) with the input being a single-frequency sinusoid having a 1-V amplitude. Let the input frequency be swept over the range of interest while observing the signals at the outputs of the five integrators. Denote the peaks attained by these signals $\hat{V}_1, \hat{V}_2, \ldots, \hat{V}_5$. These peaks will of course occur at different frequencies. While the peak value of V_5 is determined by the filter transfer function, and is thus fixed for a given transfer function, this is not the case for the other spectral peaks. The signal handling capability of the filter can be maximized by making all spectral peaks equal. To see how this can be achieved, consider integrator no. x. To change the spectrum peak of its output from $\hat{V}_x$ to $\hat{V}_n$ we need to multiply all its feed-in resistances R_{xy} by $(\hat{V}_x/\hat{V}_n)$ and all its feed-in capitances C_{xy} by $(\hat{V}_n/\hat{V}_x)$. However, in order not to change the filter transfer function all components connected to the output of integrator no. x must be changed as follows: all resistances R_{yx} must be multiplied by $(\hat{V}_n/\hat{V}_x)$ and all capacitances C_{yx} must be multiplied by $(\hat{V}_x/\hat{V}_n)$. It follows that to scale the circuit in Figure 3.1(c) for maximum dynamic range, every resistance R_{xy} should be multiplied by $(\hat{V}_x/\hat{V}_y)$ and every capacitance C_{xy} should be multiplied by $(\hat{V}_y/\hat{V}_x)$. (Recall that R_{xy} denotes the resistance connecting the output of integrator no. y to the input of integrator no. x, C_{xy} denotes the capacitance connecting the output of integrator no. y to the input of integrator no. x; and $\hat{V}_x$ and $\hat{V}_y$ are the spectral peaks of the outputs of integrators no. x and no. y, respectively.) Observe that all integrator capacitors remain unchanged. In addition, the damping resistors R_{11} and R_{55} do not change value. Finally, the input feed-in resistance R_{10} is multiplied by $(\hat{V}_1/\hat{V}_n)$.

3.2.5. Scaling for Practical Component Values

The final step in the design involves scaling the component values to obtain practical sizes*. This scaling is performed on each integrator block at a time, and is based on the fact that the impedances associated with an op amp can all be multiplied by any arbitrary factor without changing the transfer functions from the inputs of this op-amp block to its output. Consider for example integrator no. 1 in the circuit of Figure 3.1(c). If the resistances R_{10}, R_{11}, and R_{12} are multiplied by an arbitrary factor k_1 and the capacitances C_1 and C_{13} are divided by the same factor, the transfer function (V_1/V_i) and (V_1/V_2) remain unchanged. The value of k_1 is chosen to yield practical component values for this block. For example, to change C_1 from its initial value of c_1' to a practical value C_{11_p} we choose

$$k_1 = \frac{c_1'}{C_{11_p}}$$

The process is then repeated for integrator no. 2. The scaling factor k_2 does not, of course, need to be equal to that of integrator no. 1. For integrator no. 2 note that the value of the resistance r utilized in the inverter is arbitrary and can be set to any desired practical value (e.g., $r = 10\,\mathrm{k}\Omega$). Finally, it should be noted that in many instances it is practical to set all n integrator capacitors to the same value. Imposing such a constraint enables the designer to determine the value of $k_1, k_2, \ldots, k_n$.

3.2.6. Sensitivity Performance

As already mentioned, doubly-terminated LC ladders feature low sensitivities to component variation at passband frequencies. This low-sensitivity performance is carried over to the resulting active filter. The active filter, however, suffers from the excess phase shift of each of the integrators. Excess phase shift comes about because of the finite bandwidth of the op amps and gives rise to a change in the filter magnitude response given approximately by [22]

$$\Delta G \approx -4.34(\Phi_I + \Phi_N)\omega\tau(\omega)\ \mathrm{dB} \tag{3.11}$$

where Φ_I and Φ_N are the excess phase shift of the inverting and the noninverting integrators, respectively, and $\tau(\omega)$ is the group delay of the filter. Since the group delay reaches its peak near the edge of the passband, the maximum deviation occurs there.

In the circuit of Figure 3.1(c), the inverting integrator is implemented

* For audio-frequency filters, practical capacitor values are in the 0.1–10 nF range and practical resistor values are in the 1–100 kΩ range.

using a Miller circuit while the noninverting integrator is formed as the cascade of a Miller circuit and an inverter. Assuming op amps with single-pole frequency response it can be shown that [22]

$$\Phi_I \approx -\frac{\omega}{\omega_t} \tag{3.12}$$

and

$$\Phi_N \approx -3\frac{\omega}{\omega_t} \tag{3.13}$$

where ω_t is the unity-gain bandwidth of the op amp. Substituting these values in Eq. (3.11) we obtain

$$\Delta G \approx 17.4\left(\frac{\omega}{\omega_t}\right)\omega\,\tau(\omega)\text{ dB} \tag{3.14}$$

Observe that since Φ_I and Φ_N are of the same sign (negative, signifying phase lag) they add up, resulting in a rather large deviation ΔG. This deviation can be reduced using one of two possible approaches. The first involves introducing compensating phase lead by connecting capacitors across the input resistors of the integrators. This is known as passive compensation. Alternatively, the noninverting integrator can be implemented using the circuit in Figure 3.2 which can be shown to have excess phase lead of

$$\Phi_N = \frac{\omega}{\omega_t} \tag{3.15}$$

This excess phase is equal in magnitude and opposite in sign to the excess phase of the inverting Miller integrator, resulting in a phase compensation. This active compensation method has the advantage of requiring no additional components. More importantly, the compensation is dependent only on the matching and tracking of ω_t of the op amps, a condition that can be met using multiple IC op amps in the same package.

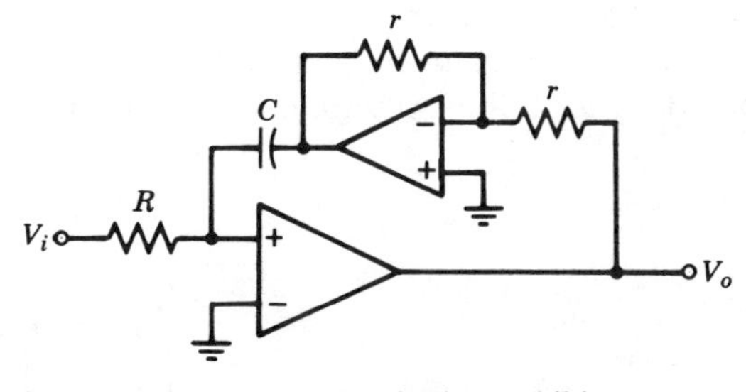

Figure 3.2 A noninverting integrator circuit that exhibits excess phase lead when the finite bandwidth of the op amps is taken into account.

3.2.7. Application to Other Filter Types

The operation simulation design method can also be applied to bandpass filters. It is especially suited to those bandpass filters that are designed using the low-pass to bandpass transformation (see [22] for details). It is not, however, well suited to the design of high-pass filters.

As a final comment, designs based on simulating the operation of LC ladder networks require a relatively large number of op amps. In some cases realizations with fewer op amps can be obtained using the inductor replacement method discussed next.

3.3. DESIGN BASED ON REPLACING THE INDUCTORS OF LC LADDER NETWORKS

A more direct method for the design of active filters based on an LC ladder prototype consists of replacing each inductor in the ladder network by an op amp-RC network having an inductive input impedance (see Figure 3.3).

3.3.1. Inductance Simulation Circuits

A high-performance inductance simulation circuit, due to Antoniou [2] and known as the Antoniou circuit, is shown in Figure 3.4. Assuming ideal op amps, it can be shown that the input impedance of this circuit is given by

$$Z_{in}(s) = sC_4R_1R_3R_5/R_2 \tag{3.16}$$

Thus the circuit realizes an inductance with one grounded terminal (a grounded inductance) of value L

$$L = C_4R_1R_3R_5/R_2 \tag{3.17}$$

and an infinite Q factor. At high frequencies where the finite bandwidth of op amps must be taken into account, analysis [22] shows that the Q factor remains high and can be made infinite by the relative adjustment of R_2 and R_3. It is important to note that this excellent performance is achieved without requiring the op amps to be matched.

The nominal design of the Antoniou circuit is based on Eq. (3.17) together with

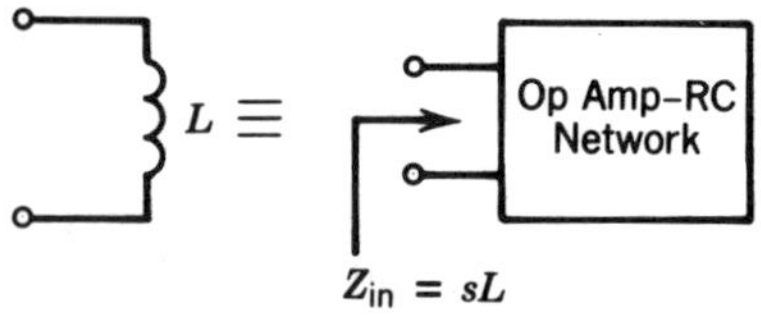

Figure 3.3 Replacing an inductor with a simulated inductance.

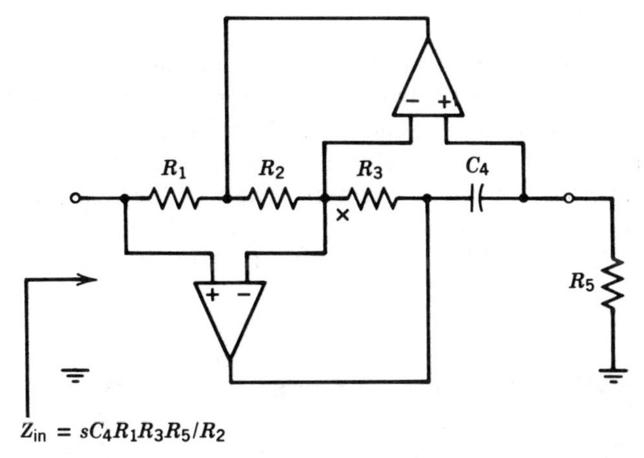

Figure 3.4 The Antoniou inductance simulation circuit.

$$R_2 = R_3 \tag{3.18}$$

$$R_5 = \frac{1}{\omega_c C_4} \tag{3.19}$$

where ω_c is the frequency at which the filter is most sensitive to its component values; usually at or near the edge of the passband where the rate of change of attenuation is maximum and the group delay peaks.

3.3.2. High-pass Filters

The inductance replacement design method is best suited for high-pass filters such as that shown in Figure 3.5. This is a fifth-order elliptic-function filter. Observe that all inductors are grounded and thus can be replaced with the simulated inductances realized with the Antoniou circuit.

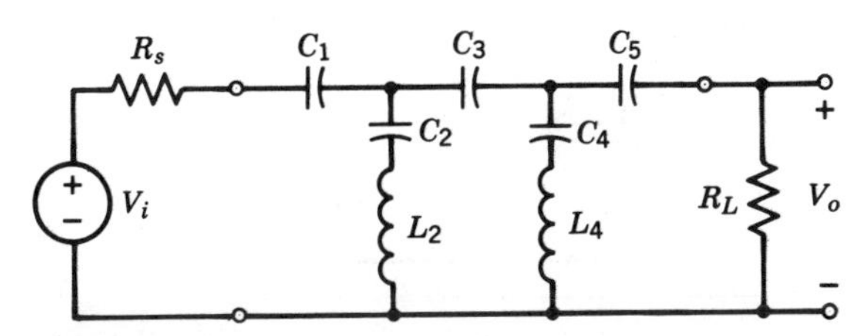

Figure 3.5 A fifth-order elliptic high-pass filter. An active realization of this filter can be obtained by simply replacing L_2 and L_4 with simulated inductances obtained using the Antoniou circuit of Figure 3.4.

3.3.3. Other Filter Types

Filters other than high-pass (and a very special case of bandpass) contain floating inductors in their LC ladder realization. Unfortunately, high quality simulation circuits for floating inductors do not exist. Nevertheless, ingenious techniques have been developed for obtaining component-simulation designs for filters with floating inductors. One such technique, due to Bruton [5], is illustrated in the following.

Consider the LC ladder network shown in Figure 3.6(a). It realizes a

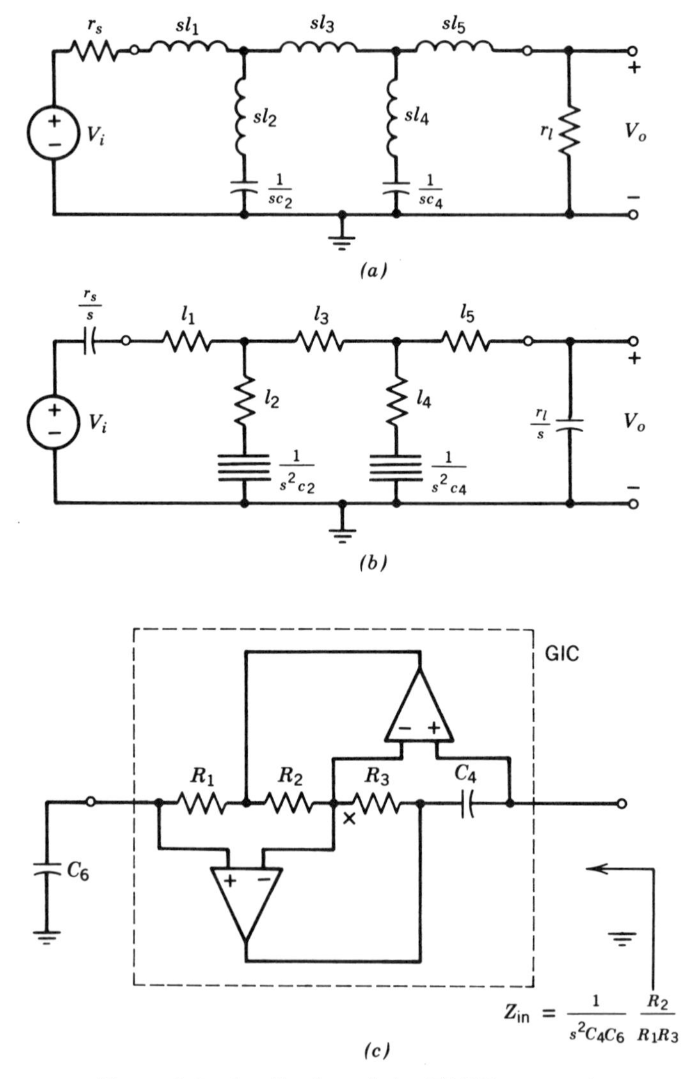

Figure 3.6 Application of the FDNR approach.

fifth-order elliptic-function low-pass filter and contains five floating inductors. All the floating inductors can be eliminated by dividing the impedances of the ladder network by the complex frequency variable s. This transformation, known as the Bruton transformation, preserves the voltage transfer function and yields the network shown in Figure 3.6(b). Observe that the ladder inductors have been transformed into resistors, the termination resistors have been transformed into capacitors, and the ladder capacitors have turned into elements with impedances of the form $(1/s^2C)$. The latter elements are known as frequency-dependent negative resistors or FDNRs. Since the voltage transfer function V_o/V_i remains unchanged, the circuit of Figure 3.6(b) is a perfect realization of the filter provided that we can find suitable realizations for the FDNRs. Such a realization is shown in Figure 3.6(c). Observe that the part of this circuit contained within the box is identical to that used in the inductance simulation circuit of Figure 3.4. This part of the circuit is a versatile circuit building block, known as the generalized impedance converter (GIC). A hybrid IC realization of the GIC is available from National Semiconductor.

Assuming ideal op amps, it can be shown that the input impedance of the circuit in Figure 3.6(c) is given by

$$Z_{in}(s) = \frac{1}{s^2 C_4 C_6} \frac{R_2}{R_1 R_3} \tag{3.20}$$

As in the case of the inductance simulation circuit, the dependence on the high-frequency response of the op amps is minimized by the relative adjustment of R_2 and R_3. Nominally

$$R_2 = R_3 \tag{3.21}$$

and

$$R_1 = \frac{1}{\omega_c C_6} \tag{3.22}$$

Equations (3.20), (3.21), and (3.22) are the three design equations for the FDNR. More practical details on the design of low-pass filters utilizing FDNRs are given by Sedra and Brackett [22]. The application of the GIC to the design of bandpass filters is also given by Sedra and Brackett [22].

3.3.4. Sensitivity Performance

Filters designed using the component simulation approach preserve the low-sensitivity performance of their LC ladder prototypes [7]. The sensitivity formulae of LC ladder networks [26] can be used to predict the performance of the active RC filters. For this purpose it can be shown [22] that at the critical frequency ω_c, inductances realized using the Antoniou circuit will exhibit a deviation in their nominal value L of

$$\frac{\Delta L}{L} \simeq 4\left(\frac{\omega_c}{\omega_t}\right) \tag{3.23}$$

where ω_t is the unity-gain frequency of the op amps. An identical formula applies for the FDNR circuit. Finally, it should be mentioned that both the inductance simulation circuit of Figure 3.4, and the FDNR circuit of Figure 3.6(c) are sensitive to the stray capacitance between the node labelled X and ground. The resulting deviation in the filter attenuation response can be minimized by connecting a resistor between node X and ground (see [22]).

3.4. CASCADE DESIGN AND BIQUAD CIRCUITS

Cascade design offers the easiest method for realizing a filter transfer function using op amp-RC circuits. The transfer function is factored into the product of second-order, and one first-order for odd-order filters, functions. The second- and first-order functions in the product are then realized using op amp-RC networks, and the resulting circuits (which usually have low output impedances because their outputs are taken at op amp output terminals) are connected in cascade.

In a cascade filter each pair of complex conjugate poles is realized in a separate section and thus can be easily adjusted. Cascade filters, however, are more sensitive, in the passband, to changes in the component values than are the ladder simulation filters discussed in the two previous sections [7]. Nevertheless, the simplicity of their design makes cascade filters an excellent choice for the less stringent filtering applications.

First-order functions are easily realizable using a single op amp, a single capacitor, and one or more resistors, as shown in Figure 3.7. These circuits are simple and will not be discussed any further. Rather, we shall concentrate on the realization of second-order or biquadratic transfer functions.

3.4.1. Biquadratic Transfer Functions

In its most general form, a biquadratic transfer function can be expressed as

$$T(s) = \frac{n_2 s^2 + n_1 s + n_0}{s^2 + s\left(\frac{\omega_0}{Q}\right) + \omega_0^2} \tag{3.24}$$

Here ω_0 and Q determine the pair of complex conjugate poles, as indicated in Figure 3.8. The numerator coefficients determine the transmission zeroes and hence the type of second-order filter function. Notable special cases, illustrated in Figure 3.9, are:

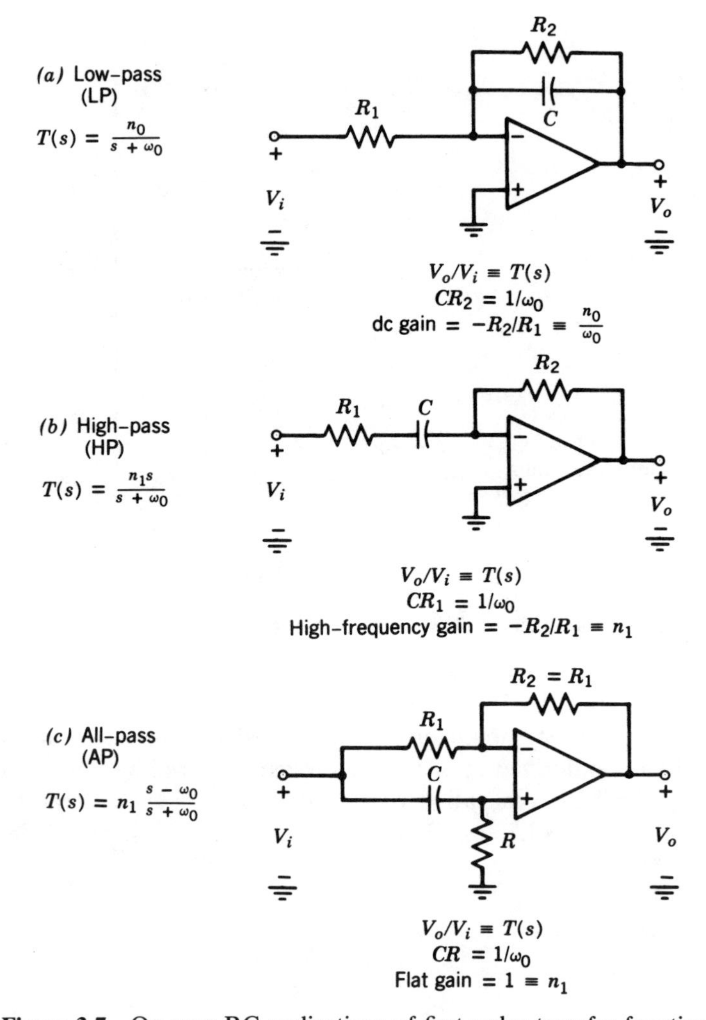

Figure 3.7 Op amp-RC realizations of first-order transfer functions.

(a) low-pass (LP): $n_1 = n_2 = 0$
(b) bandpass (BP): $n_0 = n_2 = 0$
(c) high-pass (HP): $n_0 = n_1 = 0$
(d) notch: $n_1 = 0$; $\frac{n_0}{n_2} = \omega_n^2$, where ω_n is the notch frequency
(e) all-pass (AP): $\frac{n_0}{n_2} = \omega_0^2$, $\frac{n_1}{n_2} = -\frac{\omega_0}{Q}$

In the following we shall present some of the widely used circuits for the realization of biquadratic transfer functions. Such circuits are usually known as *biquads*.

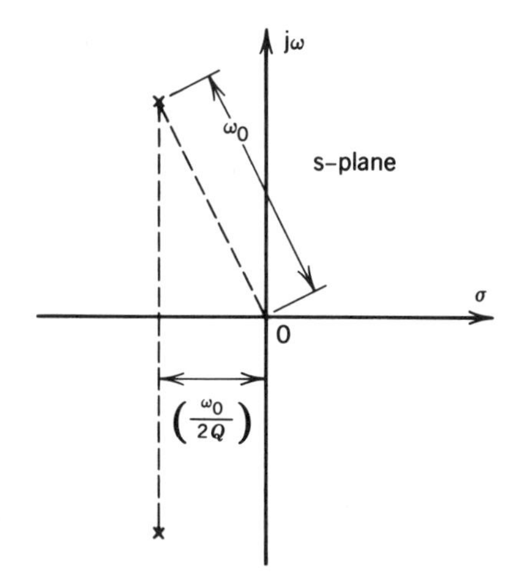

Figure 3.8 Definition of ω_0 and Q of a pair of complex-conjugate poles.

3.4.2. The State-Variable or Two-Integrator-Loop Realization

This is the most popular circuit for realizing a second-order filter transfer function. In its most basic form the circuit consists of two integrators, one inverting and the other noninverting, connected in a single feedback loop. The circuit can be derived in one of two ways; both will be illustrated here.

Consider the RLC realization of a second-order low-pass transfer function, shown in Figure 3.10(a). The operation of this circuit can be described by the two equations

$$i_l = \frac{v_i - v_c}{sl + r} \tag{3.25}$$

$$v_c = \frac{i_l}{sc} \tag{3.26}$$

Following the operation simulation approach described in §3.2, we obtain the active RC circuit shown in Figure 3.10(b). Here integrator no. 1 is a noninverting damped integrator simulating Eq. (3.25) and integrator no. 2 simulates Eq. (3.26). Output voltage V_1 is the analog of i_l and output voltage V_2 is the analog of v_c. Thus (V_2/V_i) is a second-order low-pass function. Also, since V_1 is proportional to (V_2/s), the function (V_1/V_i) is a second-order bandpass. Thus the active circuit simultaneously realizes both the low-pass and the bandpass functions.

The two-integrator-loop biquad is redrawn in Figure 3.10(c) in the form commonly used in the literature. Also given in this figure are the nominal component values employed in typical design. As shown, the circuit realizes the bandpass function

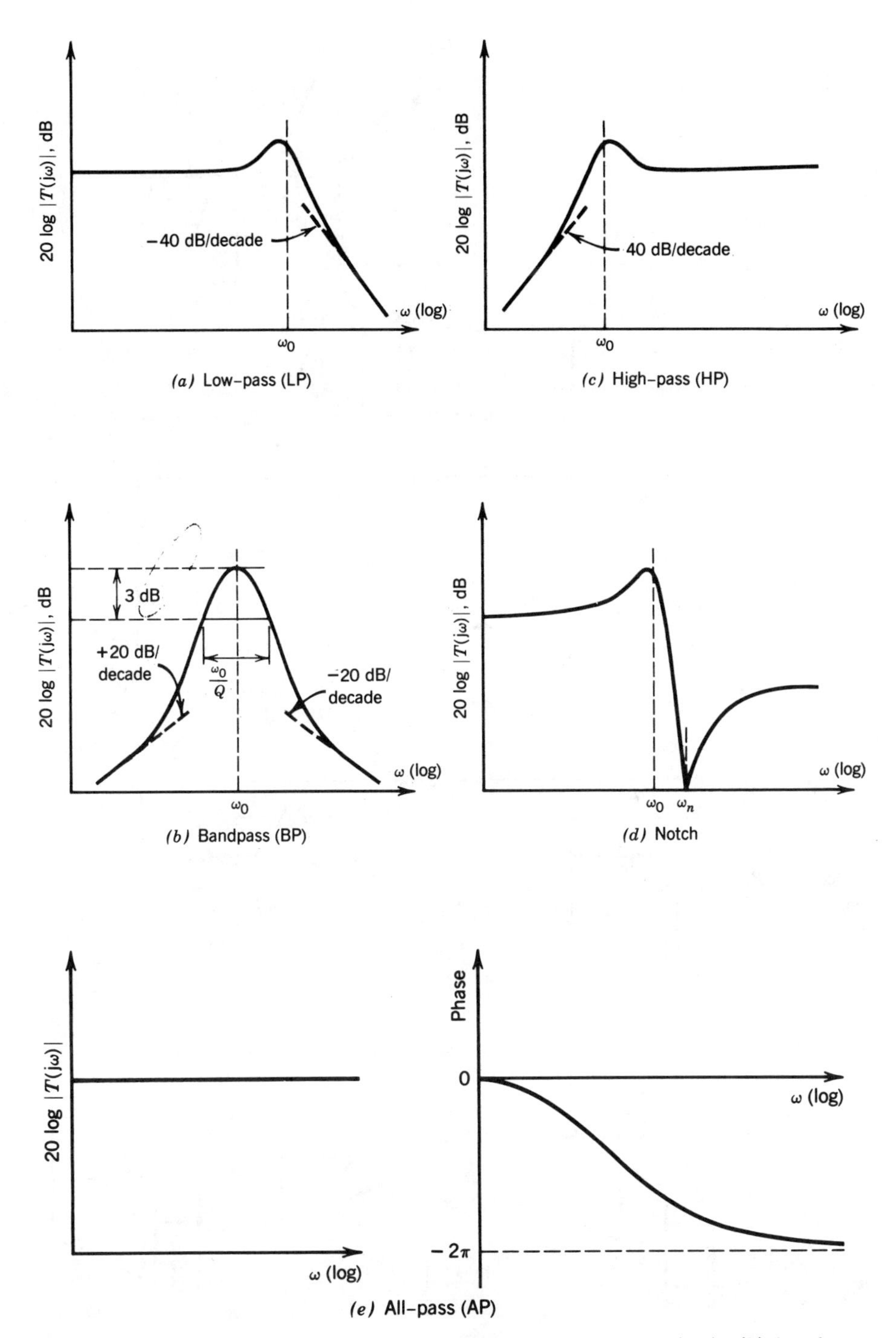

Figure 3.9 Special second-order filter responses. (a) Low-pass (LP); (b) bandpass (BP); (c) high-pass (HP); (d) notch; (e) all-pass (AP).

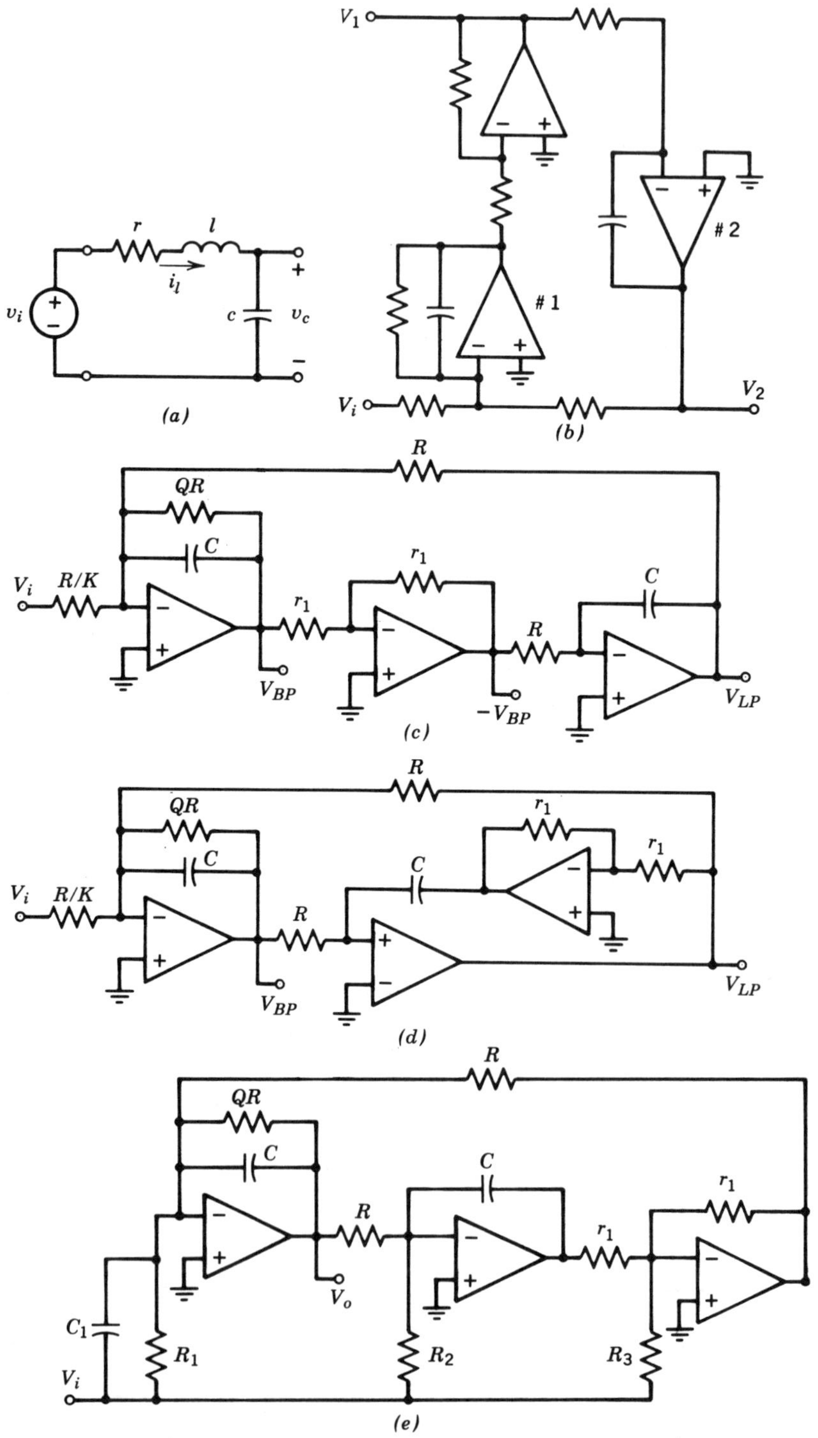

Figure 3.10 Derivation of the two-integrator-loop biquad from an RLC resonator.

$$\frac{V_{BP}(s)}{V_i(s)} = -\frac{sK\omega_0}{s^2 + s(\omega_0/Q) + \omega_0^2} \tag{3.27}$$

where

$$\omega_0 = 1/CR$$

and K is a constant that determines the center-frequency gain; and the low-pass function

$$\frac{V_{LP}(s)}{V_i(s)} = -\frac{K\omega_0^2}{s^2 + s(\omega_0/Q) + \omega_0^2} \tag{3.28}$$

As mentioned in §3.2, the two-integrator loop suffers from the fact that the excess phase shifts of the two integrators add up. Assuming all three amplifiers to be identical with a single-pole frequency response and a unity-gain bandwidth ω_t, the total loop phase shift can be obtained by adding the values in Eqs. (3.12) and (3.13)

$$\Phi_{\text{excess}} \simeq -4\left(\frac{\omega_0}{\omega_t}\right) \tag{3.29}$$

Excess phase causes the Q factor of the realized poles to change from the nominal value Q to the actual value Q_a

$$\frac{1}{Q_a} = \frac{1}{Q} + \Phi_{\text{excess}} \tag{3.30}$$

Thus

$$Q_a = \frac{Q}{1 - 4Q\left(\frac{\omega_0}{\omega_t}\right)} \tag{3.31}$$

which indicates a severe Q-enhancement effect, leading to possible instability. The Q enhancement can be substantially reduced by employing one of the compensation schemes (passive or active compensation) discussed in §3.2. The circuit is shown in one of its possible active-compensated forms in Figure 3.10(d). Here the combination of the inverter and the second Miller integrator has been replaced with the phase lead integrator of Figure 3.2. The compensated circuit of Figure 3.10(d) is known as the Akerberg–Mossberg biquad [1]. Care must be taken in using this circuit because the additional poles of the op-amp frequency response could provide sufficient extra phase shift to cause oscillations (see [16]).

In addition to affecting the Q factor, the finite bandwidth of the op amps causes a fractional shift in ω_0 of [22]

$$\frac{\Delta\omega_0}{\omega_0} \simeq -1.5\left(\frac{\omega_0}{\omega_t}\right)$$

Filter transfer functions other than LP and BP can be realized using the feedforward technique illustrated in Figure 3.10(e). The transfer function of this circuit is given by

$$\frac{V_o}{V_i} = -\frac{s^2\left(\frac{C_1}{C}\right) + s\,\frac{1}{CR}\left(\frac{R}{R_1} - \frac{r_1}{R_3}\right) + \frac{1}{C^2RR_2}}{s^2 + s\,\frac{1}{QCR} + \frac{1}{C^2R^2}} \tag{3.32}$$

which can be used to obtain the nominal component values given in Table 3.1 (in all cases, r_1 is arbitrary and $CR = 1/\omega_0$).

There is a popular alternative form of the two-integrator-loop biquad that simultaneously realizes the high-pass, bandpass and low-pass transfer functions. This biquad, which is commercially available as a hybrid integrated circuit from a number of manufacturers (National, Burr–Brown, etc.), is known as a universal active filter. To derive this circuit consider the second-order high-pass transfer function with high-frequency gain K

$$\frac{V_{HP}}{V_i} = \frac{Ks^2}{s^2 + s\left(\frac{\omega_0}{Q}\right) + \omega_0^2} \tag{3.33}$$

Cross-multiplying, dividing both sides of the resulting equation by s^2, and expressing V_{HP} as a function of V_i, (V_{HP}/s) and (V_{HP}/s^2) results in

$$V_{HP} = KV_i - \frac{1}{Q}\left(\frac{V_{HP}}{s/\omega_0}\right) - \frac{V_{HP}}{(s^2/\omega_0^2)} \tag{3.34}$$

Utilizing two integrators of transfer function $-1/(s/\omega_0)$, Figure 3.11(a) shows a direct implementation of Eq. (3.34). Note the need for a weighted

TABLE 3.1 Design Data for the Biquad Circuit of Figure 3.10(e)

LP	$C_1 = 0,\ R_1 = R_3 = \infty,\ R_2 = R/\text{dc gain}$
Positive BP	$C_1 = 0,\ R_1 = R_2 = \infty,\ R_3 = Qr_1/\text{center-frequency gain}$
Negative BP	$C_1 = 0,\ R_1 = QR/\text{center-frequency gain},\ R_2 = R_3 = \infty$
HP	$C_1 = C \times \text{high-frequency gain},\ R_1 = R_2 = R_3 = \infty$
Notch	$C_1 = C \times \text{high-frequency gain},\ R_1 = R_3 = \infty,\ R_2 = \left(\frac{\omega_0}{\omega_n}\right)^2\left(\frac{R}{\text{HF gain}}\right)$
AP	$C_1 = C \times \text{gain},\ R_1 = \infty,\ R_2 = \frac{R}{\text{gain}},\ R_3 = \frac{Qr_1}{\text{gain}}$

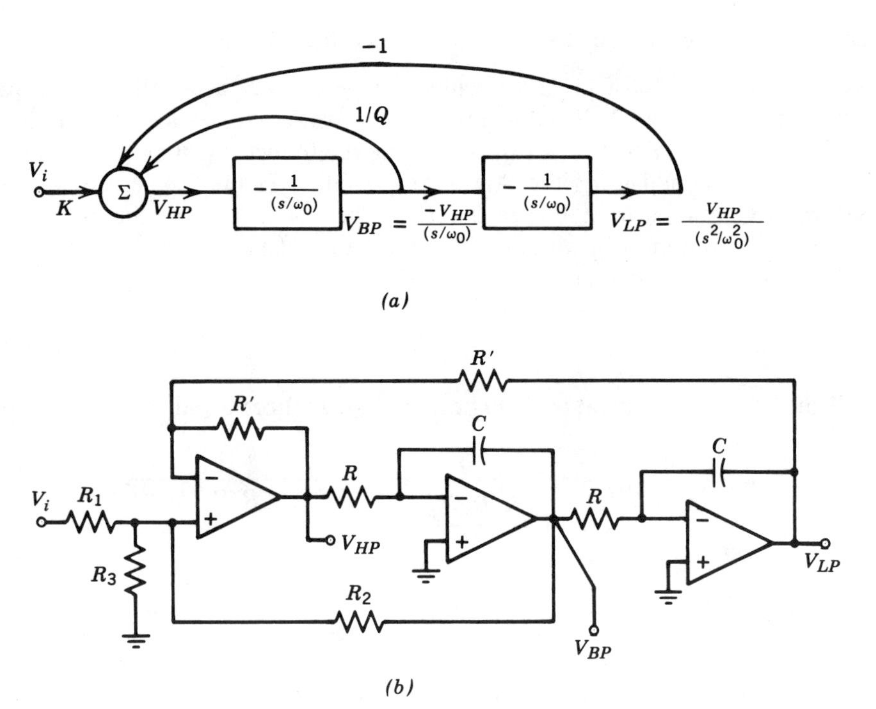

Figure 3.11 Derivation of an alternative form of the two-integrator-loop or state-variable biquad.

summer capable of assigning both positive and negative polarities to its input signals. Such a summer can be implemented by utilizing both input terminals of the op amp, as shown in Figure 3.11(b) which depicts the complete biquad circuit. Its nominal design is based on the following equations

$$\left.\begin{aligned} CR &= 1/\omega_0 \\ R_2 &= KQR_1 \\ R_3 &= \frac{R_1}{\frac{2}{K} - 1 - \frac{1}{KQ}} \end{aligned}\right\} \quad R' \text{ and } R_1\text{: arbitrary} \tag{3.35}$$

This biquad circuit was first proposed by Kerwin, Huelsman and Newcomb [14] and is known as the KHN biquad. It has the advantage of simultaneously realizing the HP, BP, and LP functions. Other filter functions can be obtained by using an additional op amp to provide a weighted sum of the HP, BP, and LP functions. Like the original two-integrator-loop biquad, the KHN circuit suffers from an identical Q-enhancement effect. Thus compensation, either passive or active [6], is called for.

3.4.3. Biquads Based on the Antoniou Generalized Impedance Converter

An extensive family of biquad circuits can be derived using the Antoniou GIC discussed in §3.2. To illustrate the derivation, consider the parallel RLC resonator of Figure 3.12(a). Replacing the inductor L with a simulated inductance obtained using the Antoniou circuit of Figure 3.4 results in the op amp-RC resonator of Figure 3.12(b). The voltage across the resonant circuit, V_r, is buffered by an amplifier with gain K (shown with broken line) to provide a low-impedance filter output

$$V_o = KV_r \tag{3.36}$$

The active RC resonator of Figure 3.12(b) realizes a pair of poles with

$$\omega_0 = \frac{1}{(LC)^{1/2}} = \frac{1}{\left[C\dfrac{R_1R_3R_5}{R_2}C\right]^{1/2}} = \frac{1}{C(R_1R_3R_5/R_2)^{1/2}} \tag{3.37}$$

and a Q factor

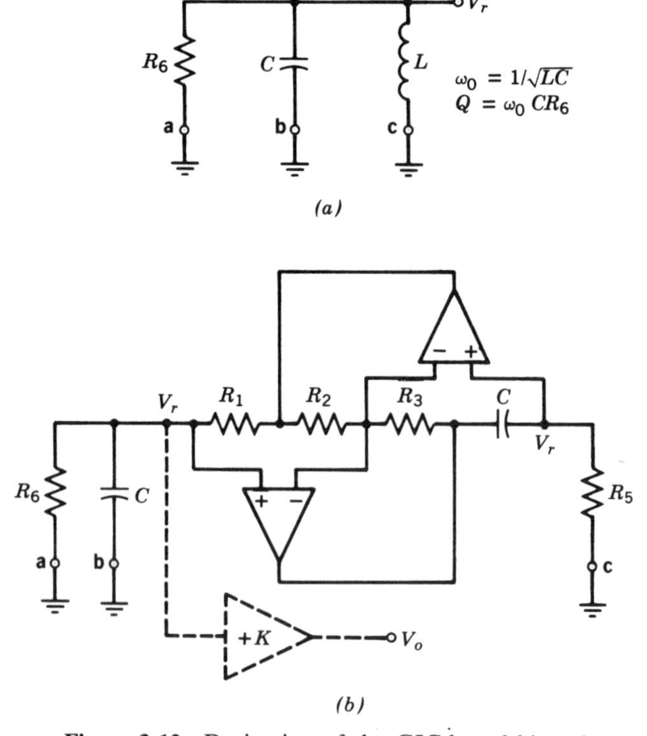

Figure 3.12 Derivation of the GIC-based biquad.

$$Q = \omega_0 C R_6 = \frac{R_6}{(R_1 R_3 R_5 / R_2)^{1/2}} \tag{3.38}$$

The circuit is quite insensitive to the finite bandwidth of the op amps as long as $R_2 = R_3$ (recall that this was the condition to obtain an infinite Q factor for the simulated inductance). The change in the value L as a result of the finite op amp bandwidth (Eq. (3.23)) causes the pole frequency ω_0 to deviate from its nominal value by [22]

$$\frac{\Delta\omega_0}{\omega_0} \simeq -2\left(\frac{\omega_0}{\omega_t}\right)$$

The nominal design of the biquad circuit of Figure 3.12(b) is based on

$$\left.\begin{aligned} &R_1 = R_2 = R_3 = R_5 = R \\ &CR = 1/\omega_0 \\ &R_6 = QR \end{aligned}\right\} \tag{3.39}$$

To realize a bandpass function, node **a** is disconnected from ground and the input signal is fed through R_6. The bandpass function obtained will have a center-frequency gain of K. For a high-pass function, node **b** is disconnected from ground and the input signal is fed through capacitor C. The high-pass function obtained will have a high-frequency gain K. The low-pass function is realized by disconnecting node **c** from ground and feeding the input signal through R_5. The resulting low-pass will have a dc gain K.

A low-pass notch (LPN: $\omega_n \geq \omega_0$) can be obtained using the passive circuit shown in Figure 3.13(a), for which

$$\omega_n = 1/\sqrt{LC_1} \tag{3.40}$$

$$C_1 + C_2 = C \tag{3.41}$$

The corresponding active circuit is shown in Figure 3.13(b). Similarly, a high-pass notch (HPN: $\omega_n \leq \omega_0$) is realized as in Figures 3.13(c) and 3.13(d).

An all-pass circuit can be obtained as follows. First, a bandpass of gain +2 is obtained. Then, the complementary function $[1 - T(s)]$, which is an all-pass, is realized by interchanging the input terminal with ground. The resulting circuit is shown in Figure 3.13(e).

Before leaving this topic we wish to point out that it is possible to obtain two-op amp GIC-based biquads (i.e., circuits in which the buffer amplifier can be dispensed with) by employing an alternative GIC circuit. The resulting circuits, however, are more sensitive to the op amp ω_t. Specifically, their high-frequency performance depends on the matching and tracking of ω_t of dual op amps (see [22] for details).

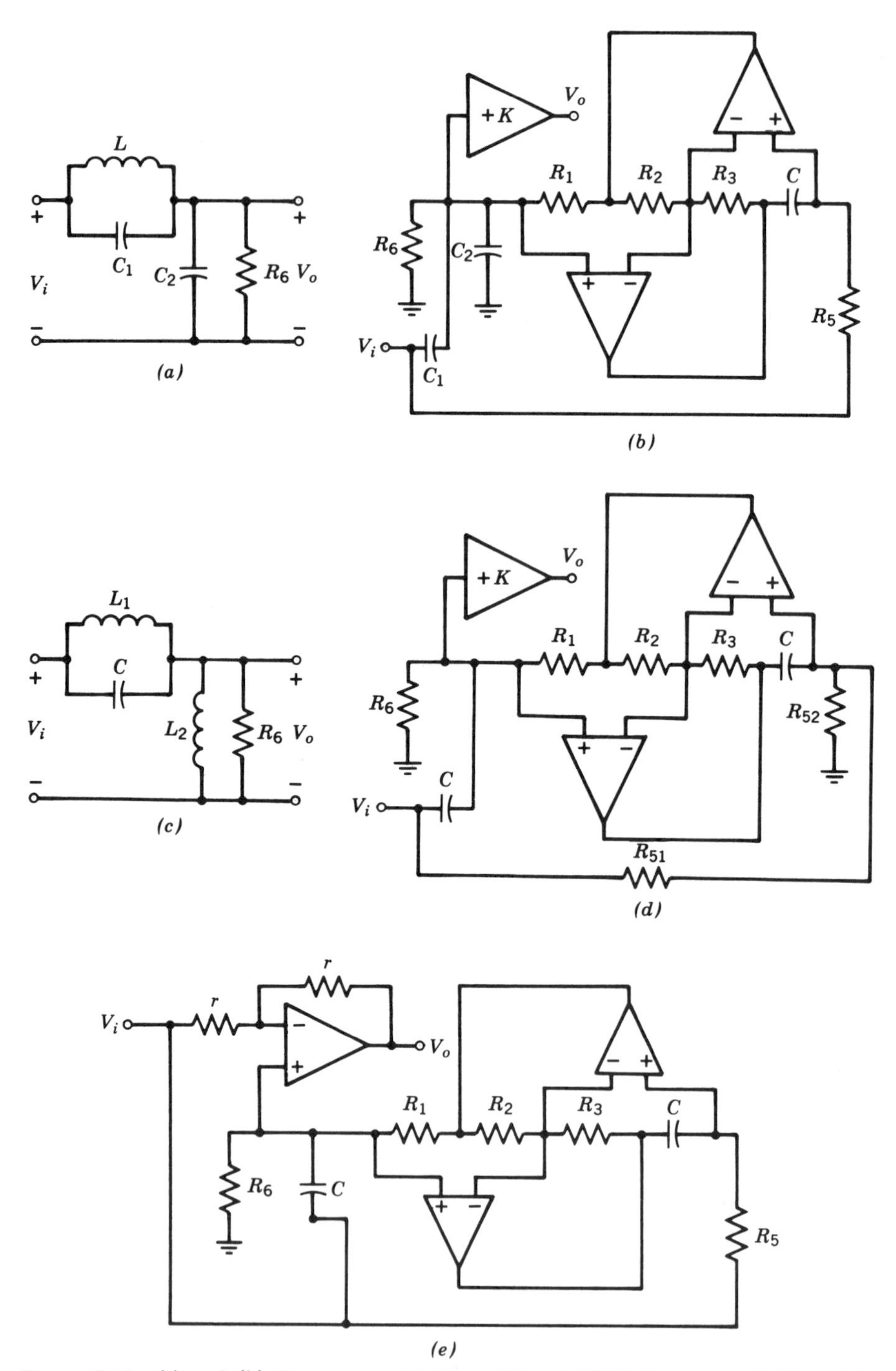

Figure 3.13 (a) and (b) low-pass notch filter; (c) and (d) high-pass notch filter; (e) all-pass filter.

As mentioned earlier GIC circuits are commercially available in hybrid integrated-circuit form. Alternatively, dual or quad op-amp packages can be used in assembling the circuits discussed above in whatever technology desired (discrete, thick-film or thin-film).

3.4.4. Single-Amplifier Biquads

The biquad circuits discussed so far use 3 or 4 op amps. Their performance is excellent in the sense that the sensitivities of ω_0 and Q to changes in the passive component values and in ω_t of the op amps are very low. Nevertheless, if one is considering the cascade design of a high-order filter then the total number of op amps will be high. Not only does this increase cost but the power requirement can become excessively high. The latter can be a major problem if battery operation is contemplated. It follows that there is interest in biquad circuits that use the minimum possible number of op amps, one. It will be shown here, however, that single-amplifier biquads (SABs) are useful only in the less stringent filter applications; those for which the highest pole-Q is no greater than about 30. Otherwise, the circuit becomes quite sensitive.

The simplest SAB circuit is obtained by placing an RC bridged-T network in the negative feedback path of an op amp, as shown in Figure 3.14(a). The resulting feedback loop realizes a pair of complex-conjugate poles that, for an infinite-gain op amp, are characterized by

$$\omega_0 = \frac{1}{(C_1 C_2 R_3 R_4)^{1/2}} \tag{3.42}$$

$$Q = \frac{1}{(C_1 C_2 R_3 R_4)^{1/2}} \Bigg/ \left[\left(\frac{1}{C_1} + \frac{1}{C_2}\right)\frac{1}{R_3}\right] \tag{3.43}$$

The nominal design is

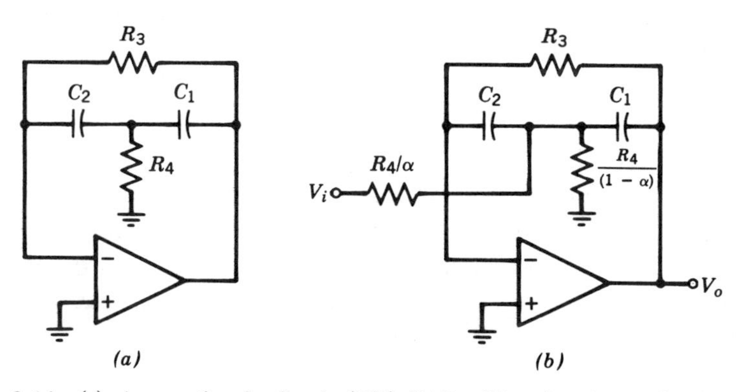

Figure 3.14 (a) A negative-feedback (NF) SAB; (b) a bandpass circuit obtained from the NF SAB in (a).

$$\left.\begin{aligned} &C_1 = C_2 = C \\ &R_3 = R, \quad R_4 = \frac{R}{4Q^2} \\ &CR = \frac{2Q}{\omega_0} \end{aligned}\right\} \tag{3.44}$$

A bandpass transfer function can be obtained from this SAB by feeding the input signal through part or all of R_4. The resulting circuit is shown in Figure 3.14(b) and has the transfer function

$$\frac{V_o}{V_i} = -\frac{s\dfrac{\alpha}{C_1R_4}}{s^2 + s\left(\dfrac{1}{C_1} + \dfrac{1}{C_2}\right)\dfrac{1}{R_3} + \dfrac{1}{C_1C_2R_3R_4}} \tag{3.45}$$

Note that the value of α ($\alpha \leq 1$) determines the center-frequency gain.

The SAB circuit of Figure 3.14(a), known as a negative feedback (NF) SAB, exhibits low sensitivities relative to the values of the passive components. It is, however, very sensitive to the finite gain and bandwidth of the op amp. Specifically, the sensitivity of ω_0 and Q to ω_t are [24]

$$S_{\omega_t}^{\omega_0} \simeq Q\left(\frac{\omega_0}{\omega_t}\right) \tag{3.46}$$

$$S_{\omega_t}^{Q} \simeq -Q\left(\frac{\omega_0}{\omega_t}\right) \tag{3.47}$$

The fact that the sensitivity values depend on the value of Q, limit the utility of this circuit to low-Q applications, for example, $Q \leq 5$. For higher values of Q, $5 < Q \leq 30$, it has been found that the application of some resistive positive feedback to the SAB circuit can be quite beneficial.

Figure 3.15(a) shows the negative-feedback SAB after the addition of resistive positive feedback supplied by the two conductances G_a and G_b. Observe that setting $K = 1$ results in $G_a = 0$ and the elimination of the positive feedback. This reduces the SAB to its previous situation (that in Figure 3.14(a)), and for the component values shown the poles realized have frequency ω_0 and a Q factor Q_0. It can be shown that application of the positive feedback (i.e., $K > 1$) moves the poles on a circle in the s plane of radius ω_0, as shown in Figure 3.15(b). Thus while ω_0 remains constant, Q is enhanced according to the relationship

$$Q = \frac{Q_0}{1 - 2(K-1)Q_0^2} \tag{3.48}$$

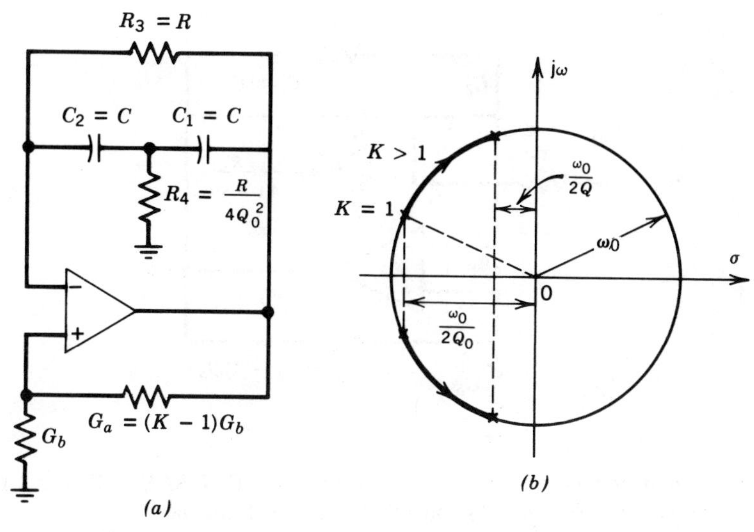

Figure 3.15 (a) An enhanced-Q negative feedback (ENF) SAB; (b) increasing K above unity enhances the Q-factor from Q_0 to Q while keeping ω_0 constant.

For this reason, the SAB circuit of Figure 3.14(a) is known as an enhanced-Q NF realization (an ENF SAB). Equation (3.48) can be used to find the value of K required to obtain a given Q for a given value of the initial Q factor (denoted Q_0) as

$$K = 1 + \frac{1}{2Q^2}\left(1 - \frac{Q_0}{Q}\right) \tag{3.49}$$

The ENF SAB has the following sensitivities

$$S_{\omega_t}^{\omega_0} \simeq Q_0\left(\frac{\omega_0}{\omega_t}\right) \tag{3.50}$$

$$S_{\omega_t}^{Q} \simeq -\,Q_0\left(\frac{\omega_0}{\omega_t}\right) \tag{3.51}$$

Thus, the sensitivities to ω_t are no longer proportional to Q but to the smaller value Q_0 which is a parameter under the designer's control. One then is tempted to use a low value for Q_0. Unfortunately, however, the sensitivities to the passive components are inversely proportional to Q_0. For example

$$S_{R_3}^{Q} = \frac{Q}{Q_0} - \frac{1}{2} \tag{3.52}$$

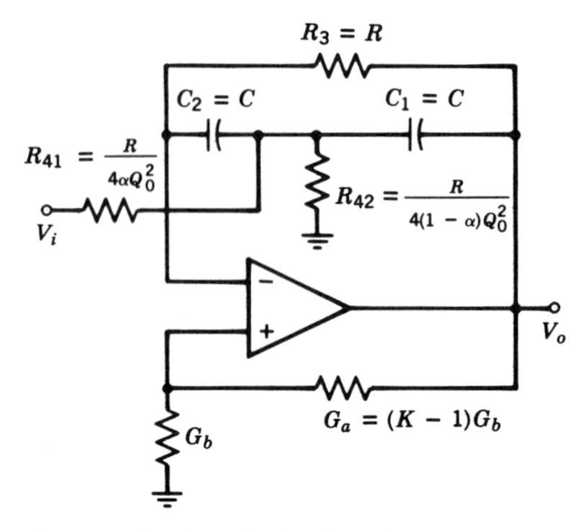

Figure 3.16 A bandpass realization obtained from the ENF SAB of Figure 3.15(a). Here the parameter α $(0 < \alpha \leq 1)$ controls the center-frequency gain.

Thus, there is a trade-off between the active and passive sensitivities. As a rule of thumb, one selects Q_0 in the range of 2–5 (see [9]).

The enhanced-Q NF circuit can be used to obtain a bandpass realization as shown in Figure 3.16. The ENF SAB has been successfully employed in the design of the STAR hybrid IC filter building block [10]. Also, application of the complementary transformation [24] to the ENF SAB of Figure 3.15(a) results in the enhanced positive-feedback (EPF) SAB of Figure 3.17(a). The latter can be directly used to realize a high-pass filter, as shown in Figure 3.17(b). A complete set of enhanced SAB circuits together with the details of their design and performance prediction are given by Sedra et al. [24].

Before leaving the subject of cascade design it should be pointed out that the filter dynamic range can be maximized by the proper pairing of transfer function poles and zeroes to form second-order functions, and by the proper sequencing of the biquad sections to form a cascade [18, 22].

3.4.5. Coupled-Biquad Structures

As mentioned before, active RC filters designed by cascading biquad sections exhibit high sensitivities to component variations. These sensitivities can be considerably reduced by coupling the biquad sections using resistive negative feedback. A variety of such coupled structures have been proposed [12]. Interestingly, but not surprisingly, the realization obtained by simulating the operation of LC ladder networks (§3.2) can be drawn as a coupled biquad structure. Such a structure is known in the literature as the leap-frog realization [13].

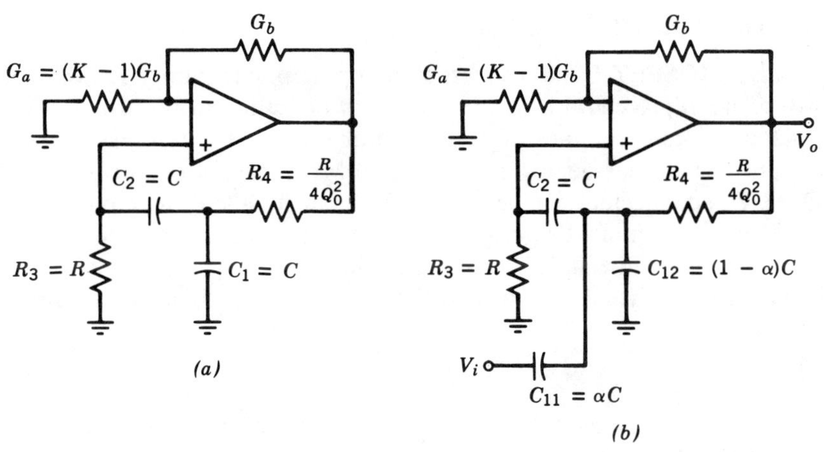

Figure 3.17 (a) An enhanced-Q positive-feedback (EPF) SAB generated from the ENF SAB of Figure 3.15(a) via the complementary transformation. (b) A high-pass circuit obtained from the EPF SAB of (a); the high-frequency gain is determined by α $(0 < \alpha \leq 1)$.

3.5. MONOLITHIC REALIZATIONS

The circuits described in the previous sections are not suitable for fabrication in fully-integrated form. This is because they require relatively large valued resistances and capacitances, especially in audio-frequency applications. Such large components would occupy impractically large chip areas. Furthermore, the precision of the realized frequency response is a direct function of the precision to which RC products are implemented. This entails either the use of very precise resistors and capacitors, which is done in discrete implementations, or the trimming of resistance values, as is done in hybrid implementations where laser trimming techniques are frequently employed. Resistors and capacitors fabricated using monolithic IC technology exhibit such large tolerances that they are practically useless as components of precision filters.

Integrated circuit technology does of course offer the ultimate in miniaturization. The size and cost advantages that IC technology affords have for many years motivated the search for active RC designs that can be monolithically fabricated. Furthermore, as VLSI technology pervades electronic systems, it is natural to seek ways for implementing precision analog filters on the same chip, side-by-side with digital circuits. In this section we shall discuss a method for the design of precision analog filters utilizing MOS transistors, MOS op amps and capacitors. The resulting circuits are known as MOSFET-C filters* [27] and can be fabricated using standard CMOS technology.

* These circuits are usually referred to as continuous-time filters to distinguish them from SC filters which are discrete-time circuits (see Chapter 5).

3.5.1. The MOSFET-C Integrator

Basically, MOSFET-C filters consist of op-amp integrators in which the resistors are replaced with MOS transistors operating in the triode region. To illustrate, consider the Miller integrator circuit of Figure 3.18 in which the input resistor has been replaced with a MOS transistor. The gate of the MOSFET is connected to a dc voltage V_G that, for the device to operate in the triode region, must be greater than the voltages at the source and drain by at least V_t, the threshold voltage of the MOSFET (typically 1–2 V). Since the source is at virtual ground and the drain is at v_I, then we require that

$$V_G > V_t \quad \text{and} \quad V_G > v_I + V_t$$

The substrate of the MOSFET (not shown) is assumed to be connected to the most negative supply voltage (e.g., −5 V in a ±5-V system). The drain current i will be given by [23]

$$i = K[2(V_G - V_t)v_I - v_I^2] \tag{3.53}$$

where

$$K = \frac{1}{2}(\mu_n C_{ox})\left(\frac{W}{L}\right) \tag{3.54}$$

Here μ_n is the electron mobility in the channel, C_{ox} is the gate capacitance per unit area (typically, $\mu_n C_{ox} = 30\mu\ \text{A/V}^2$), L and W are the channel length and width, respectively.

Equation (3.53) clearly indicates that the MOSFET is a nonlinear resistor whose use would result in nonlinear distortion. For the MOSFET to be useful as a resistor replacement some means has to be found for cancelling the nonlinear term in Eq. (3.53). An ingeneous way for accomplishing this linearization is depicted in Figure 3.19. The circuit uses a differential-output op amp that must be fully balanced; that is, it has a terminal connected to ground, and the two outputs are centered around ground. The equivalent

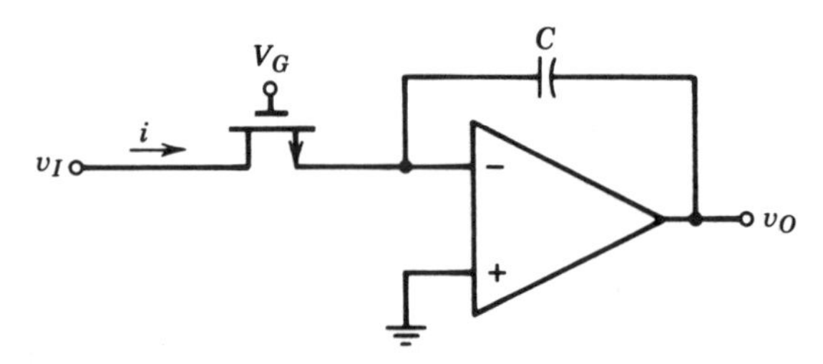

Figure 3.18 A Miller integrator in which the input resistor is replaced with a MOSFET operated in the triode region. The FET nonlinear operation results in a nonlinear integrator.

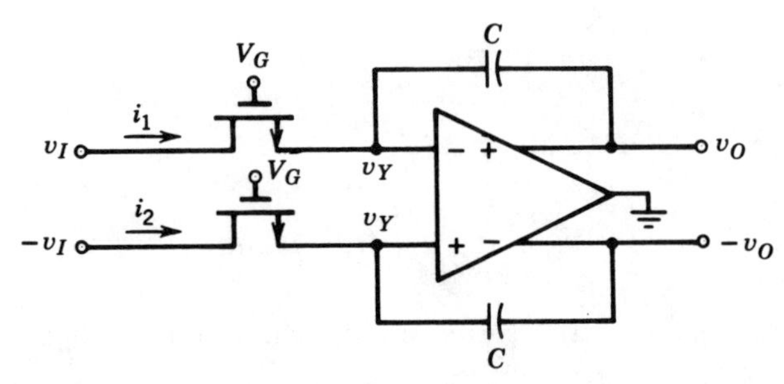

Figure 3.19 A linear integrator circuit is realized by utilizing a differential output op amp.

circuit for such an op amp is shown in Figure 3.20 where A denotes the open-loop gain. The differential-output op amp can be realized using a single-ended op amp and an inverter, as shown in Figure 3.21. There are, however, more direct realizations of balanced differential-output op amps in CMOS technology.

We shall now show that the differential circuit of Figure 3.19 realizes a linear integrator. Assuming an ideal op amp, a virtual short circuit appears between its input terminals. If we denote the common-mode voltage at the op amp input v_Y, then we can write for the two output voltages, v_O and $-v_O$

$$v_O = v_Y - \frac{1}{C}\int i_1\, dt \qquad (3.55)$$

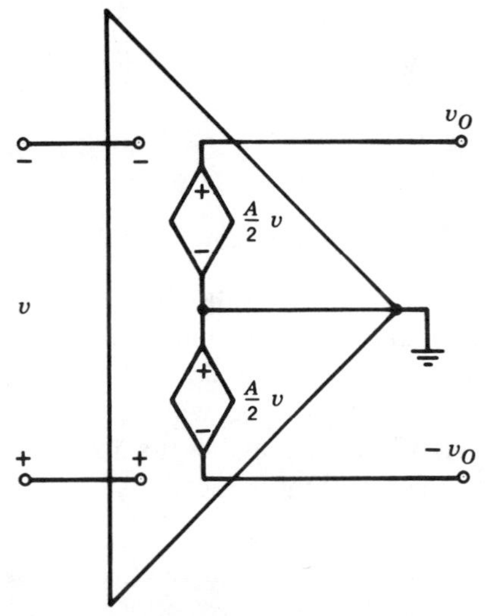

Figure 3.20 Ideal equivalent circuit (except for the finite gain A) of a balanced differential-output op amp.

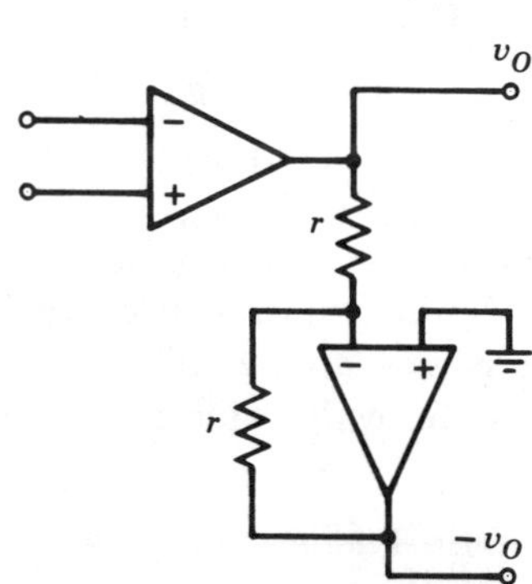

Figure 3.21 A differential output op amp realized using two single-ended-output op amps.

$$-v_O = v_Y - \frac{1}{C}\int i_2\, dt \tag{3.56}$$

Subtracting (3.56) from (3.55) gives

$$v_O = -\frac{1}{2C}\int (i_1 - i_2)\, dt \tag{3.57}$$

Now assuming that the transistors are matched and are operating in the triode region we can write

$$i_1 = K[2(V_G - v_Y - V_t)(v_I - v_Y) - (v_I - v_Y)^2] \tag{3.58}$$

$$i_2 = K[2(V_G - v_Y - V_t)(-v_I - v_Y) - (-v_I - v_Y)^2] \tag{3.59}$$

Subtracting (3.59) from (3.58) gives

$$i_1 - i_2 = 4K(V_G - V_t)v_I \tag{3.60}$$

Substituting (3.60) into (3.57) results in

$$v_O = \frac{-1}{C\left[\dfrac{1}{2K(V_G - V_t)}\right]}\int v_I\, dt \tag{3.61}$$

Thus the circuit implements a linear integrator with an integrator resistance R

$$R = \frac{1}{2K(V_G - V_t)} \tag{3.62}$$

Substituting for K from (3.54) yields

$$R = \frac{(L/W)}{(\mu_n C_{ox})(V_G - V_t)} \tag{3.63}$$

Observe that R is determined by the transistor aspect ratio (L/W) and by the gate voltage V_G. This latter property makes it possible to adjust the resistor values and hence tune the filter by adjusting the value of V_G. Furthermore, this adjustment can be made automatic, as will be explained shortly. It is this ability to tune the IC filter that makes its realization feasible in spite of the wide tolerances in component values.

3.5.2. On-Chip Tuning

Figure 3.22 illustrates a mechanism for performing on-chip automatic tuning of MOSFET-C filters. Two of the integrators are used to form an oscillator whose frequency is compared to that of a stable external clock. The comparison circuit, usually a phase detector, generates a voltage V_G that is

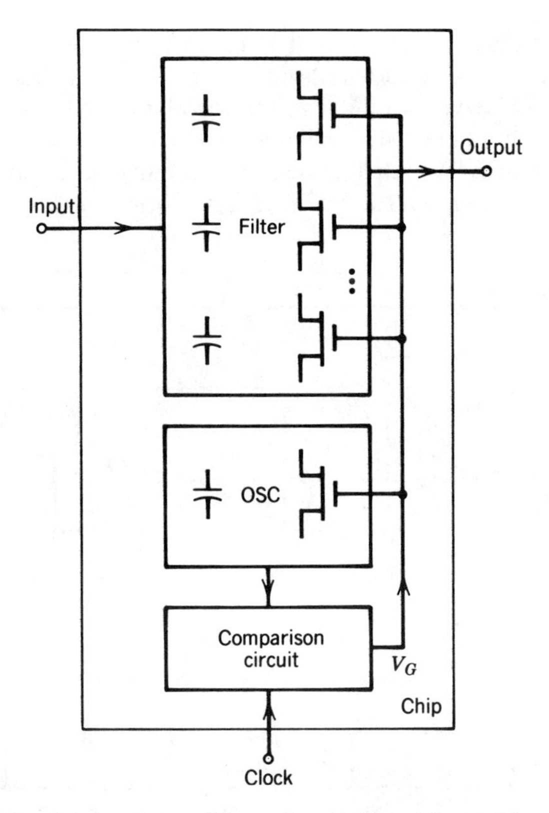

Figure 3.22 Automatic on-chip tuning (Adapted from reference [27]).

applied to the gates of the MOSFETs in the oscillator circuit and thus controls the values of their resistance and in turn the integrator time constants and hence the oscillator frequency. In this way, the time constants change until the oscillator frequency equals the clock frequency. At this point the time constants of the integrators in the oscillator circuit are stabilized at desired predetermined values. Now by applying the same voltage V_G to all other MOSFETs on the chip their resistances become known ratios of the resistances of the MOSFETs in the oscillator circuit; the ratios are determined by the (L/W) values. This statement assumes that all devices on the chip are matched and track. The same applies to the capacitors. It follows that once the tuning loop stabilizes the RC products of the integrators in the oscillator circuit, all the integrators on chip will have stabilized, well-determined time constants. Furthermore, this self-tuning is automatic, operating against changes in temperature, aging, and so on.

3.5.3. Obtaining MOSFET-C Filter Circuits

Any active RC filter circuit that uses integrators as building blocks can be converted to a MOSFET-C circuit. To illustrate, consider the two-integrator-loop biquad studied in §3.4 and redrawn in Figure 3.23(a). Replacing each integrator with a differential-output one results in the the circuit of Figure 3.23(b). Observe that the inverter is no longer needed since both V_{LP} and $-V_{LP}$ are simultaneously available at the output of the second integrator. The MOSFET-C implementation can now be obtained by simply replacing each resistor with a MOSFET, as shown in Figure 3.23(c). An

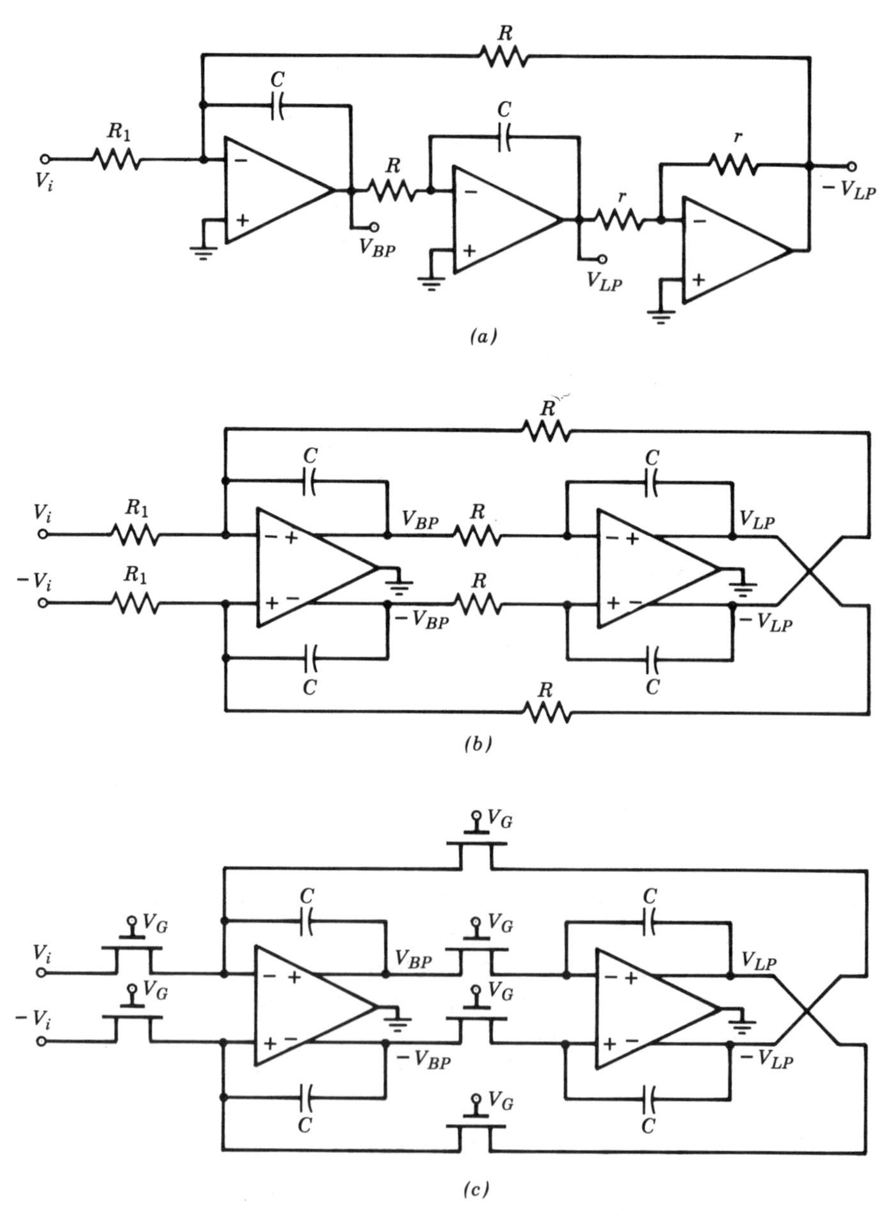

Figure 3.23 Derivation of a two-integrator-loop MOSFET-C biquad.

identical approach can be used to obtain MOSFET-C implementations of the ladder simulation circuits of §3.2.

In conclusion, the MOSFET-C approach could provide a viable alternative to the switched capacitor technique (Chapter 5) for the implementation of fully-integrated precision analog filters.

Another approach for the realization of active RC filters in monolithic form deserves mention. It is based on the use of a transconductance amplifier together with a capacitor to form the basic integrator block. Geiger and Sanchez-Sinencio [11] give an excellent tutorial on the subject. Examples of the application of this approach in industry are given by Masuda and Kitamura [15] and Plett et al. [21].

3.6. COMMERCIALLY AVAILABLE ACTIVE RC FILTERS

Compared to other filter types, the active RC filter is easy to design and construct. For small production volumes, one assembles the active filter using integrated-circuit op amps, metal-film resistors and polystyrene capacitors on a printed-circuit board, and one usually opts for dual or quad op-amp packages. For larger volumes, hybrid thick- or thin-film technology is usually employed [17]. There are situations, however, in which one needs only a few units of a relatively simple filter. For such an application, the system designer may opt for one of the active RC filter packages available as off-the-shelf components from a number of manufacturers. In the following we discuss briefly representative examples of such products.

By far the most widely available active RC filter product is that based on the two-integrator-loop biquad, especially in its KHN form. Known as a *universal active filter*, this package usually contains three op amps, two of which have a capacitor in the negative feedback path. A number of resistors are also included in the hybrid thick-film circuit. The user usually requires four additional resistors to be connected externally to configure the circuit in the desired form. The external resistors determine the intergrator time constants, and hence the pole frequency. They also determine the pole Q. Such a package is capable of realizing the LP, BP and HP filtering functions. Notch and all-pass functions can be realized by summing appropriate outputs using an additional op amp. Some filter packages include a fourth, uncommitted op amp to be used as a summer.

An example of a universal active filter is the UAF11 manufactured by Burr–Brown [4]. Utilizing 1 nF integrator capacitors, it can be used to realize pole frequencies in the range 0.001 Hz to 20 kHz and pole Qs of 0.5–500. The UAF21 can be used for f_0 up to 200 kHz. The device data sheets provide detailed application information, enabling the user to obtain a design with minimum effort. Another example of a similar product is the AF100 manufactured by National Semiconductor [19]. These universal active filters can of course be cascaded to realize transfer functions of order greater than two.

In addition to the universal active filters based on the two-integrator-loop biquad topology, National Semiconductor offers a two-op amp GIC implementation similar to that discussed earlier in this chapter. This is the AF120 (see [19]).

A different and more sophisticated approach to the provision of active-filter products has been adopted by Frequency Devices Inc. [8]. They offer filters of order up to eight realizing a variety of classical filter functions such as Butterworth (maximally-flat magnitude response) and Bessel (maximally-flat delay response). Such a filter can be specified to have any desired pass-band edge frequency within a certain range. For instance, the 790 Series of low-pass filters have 3-dB frequencies in the range 100 Hz to 50 kHz that can be specified by the user to within three digits of accuracy. Such filters do not require any external components and thus are the easiest to use. With this ease of application, however, one by necessity loses flexibility. Some flexibility is restored by making the filter programmable, as in the 848 Series of digitally-programmable low-pass active filters.

3.7. CONCLUDING REMARKS

Except for monolithic realizations, active RC filters represent a mature technology. They achieve a reasonable degree of miniaturization when fabricated in thin- or thick-film technology. Their design theory is well developed and they attain high levels of performance. They are best suited for applications ranging in frequency from low audio to 100 kHz. A great deal of work remains to be done on the fully-integrated realization of active RC filters. In this direction the MOSFET-C approach discussed above is the most viable so far. It offers greater dynamic range than currently achieved with SC filters, as well as the prospect of higher operating frequencies.

REFERENCES

1 D. Akerberg and K. Mossberg (1975) A versatile active-RC building block with inherent compensation for the finite bandwidth of the amplifier. *I.E.E.E. Trans. Circuits Systems* **22**, 407–415.

2 A. Antoniou (1969) Realization of gyrators using operational amplifiers and their use in RC-active network synthesis. *Proc. I.E.E.* **116**, 1838–1850.

3 N. Balabanian (1958) *Network Synthesis*. Prentice-Hall, Englewood Cliffs, NJ.

4 *Product Data Book* (1984) Burr–Brown Corporation, Tucson, AZ.

5 L. T. Bruton (1969) Network transfer functions using the concept of frequency-dependent negative resistance. *I.E.E.E. Trans. Circuit Theory* **16**, 406–408.

6 L. Brown and A. S. Sedra (1977) New multifunction biquadratic filter circuit with inherently stable *Q*-factor. *Electron. Lett.* **13**, 719–721.

7 L. T. Bruton (1980) *RC-Active Circuits*. Prentice-Hall, Englewood Cliffs, NJ.

8 Frequency Devices Catalogue (1984) Frequency Devices, Haverhill, MA.

9 P. E. Fleischer (1976) Sensitivity minimization in a single-amplifier biquad circuit. *I.E.E.E. Trans. Circuits Systems* **23**, 45–55.

10 J. J. Friend, C. A. Harris and D. Hilberman (1975) STAR: an active biquadratic filter section. *I.E.E.E. Trans. Circuits Systems* **22**, 115–121.

11 R. L. Geiger and E. Sanchez-Sinencio (1985) Active filter design using operational transconductance amplifiers: a tutorial. *I.E.E.E. Circuits Devices Mag.* **1**, 20–32.

12 M. S. Ghausi and K. Laker (1981) *Modern Filter Design*. Prentice-Hall, Englewood Cliffs, NJ.

13 F. E. J. Girling and E. F. Good (1970) Active Filters 12—The leap-frog or active ladder synthesis. *Wireless World* **76**, 341–345.

14 W. J. Kerwin, L.P. Huelsman and R. W. Newcomb (1967) State-variable synthesis for insensitive integrated circuit transfer functions. *I.E.E.E. J. Solid-State Circuits* **2**, 87–92.

15 S. Masuda and Y. Kitamura (1986) Design considerations of monolithic continuous-time filters. *Proc. I.E.E.E. Int. Symp. on Circuits and Systems*, San Jose, CA, pp. 1165–1168.

16 K. Martin and A. S. Sedra (1977) On the stability of the phase-lead integrator. *I.E.E.E. Trans. Circuits Systems* **24**, 321–324.

17 G. S. Moschytz (1974) *Linear Integrated Networks: Fundamentals*. Van Nostrand, New York.

18 G. S. Moschytz (1975) *Linear Integrated Networks: Design*. Van Nostrand, New York.

19 *Hybrid Products Databook* (1982) National Semiconductor Corporation, Santa Clara, CA.

20 H. J. Orchard (1966) Inductorless filters. *Electron. Lett.* **2**, 224.

21 C. Plett, M. A. Copeland and R. A. Hadaway (1986) Continuous-time filters using open loop tunable transconductance amplifier. *Proc. I.E.E.E. Int. Symp. on Circuits and Systems*, San Jose, CA, pp. 1172–1176.

22 A. S. Sedra and P. O. Brackett (1978) *Filter Theory and Design: Active and Passive*. Matrix, Portland, OR.

23 A. S. Sedra and K. C. Smith (1987) *Microelectronic Circuits*, 2nd edn. Holt, Rienhart and Winston, New York.

24 A. S. Sedra, M. A. Ghorab and K. Martin (1980) Optimum configurations for single-amplifier biquadratic filters *I.E.E.E. Trans. Circuits Systems* **27**, 1155–1163.

25 W. M. Snelgrove (1980) *FILTOR 2—A Computer-Aided Filter Design Package*. University of Toronto.

26 G. C. Temes and H. J. Orchard (1973) First-order sensitivity and worst-case analysis of doubly-terminated reactance two-ports. *I.E.E.E. Trans. Circuit Theory* **20**, 567–571.

27 Y. Tsividis, M. Banu and J. Khoury (1986) Continuous-time MOSFET-C filters in VLSI. *I.E.E.E. Trans. Circuits Systems* **33**, 125–140.

28 A. J. Zverev (1967) *Handbook of Filter Synthesis*. Wiley, New York.

4 Mechanical Filters

ROBERT A. JOHNSON
Filter Products, Rockwell International

This chapter is about electromechanical bandpass filters, which in the electronic industry are better known simply as mechanical filters. Mechanical filters are electromechanical in nature because they are driven by electrical signals, filter mechanically, and then convert the remaining signals back to their electrical form. The mechanical elements internally propagate acoustic (mechanical) waves, that is, mechanical vibrations, as opposed for instance to electromagnetic waves in microwave cavity filters. Due to the low velocity of mechanical waves, a wavelength in a mechanical system is on the order of 10^5 smaller than a wavelength in an electromagnetic system of the same frequency. This allows small mechanical filters to be built at low frequencies.

Of course filters can be built at low frequencies out of inductors and capacitors as described in Chapter 2, but not with the high Q values of mechanical resonators, which are on the order of 10,000–30,000. In addition, the LC components are often costly and, at very low frequencies, quite large. Along with the high Q of the mechanical resonators, their stability with both temperature and age allows mechanical filters to be built with bandwidths as narrow as 0.1% of their center frequency. In summary, the mechanical filter has found and maintained a place in the electronics industry principally because it is realizable in small sizes, at low frequencies (below 600 kHz), and with narrow bandwidths.

4.1. WHAT IS A MECHANICAL FILTER?

In this section, we will first look at the filter as a system. Then we will study its general characteristics and how these characteristics have led to applications ranging from watch resonators to telephone filters.

4.1.1. Basic Concepts

The mechanical filter is close in concept to the LC ladder filter or maybe even closer to a microwave cavity filter [10]. The mechanical filter is a passive device requiring no external energy except for the driving signal, as opposed, for example, to active RC filters. Its elements are linear within the operating voltage range of the filter, which is from sub-microvolts to a volt or so. Both the filter and its elements are bilateral, in other words, they operate identically in both directions, the filter can be turned around and the input transducer becomes the output transducer, and so on. Like the microwave cavity filter, the filter elements are distributed rather than lumped, as in the LC filter case. Finally, the mechanical filter is a recursive filter in that signals are flowing back and forth between both the internal sections and the source and load resistances as opposed to a non-recursive type, such as a digital or SAW filter where the signal ideally goes from input to output, never to return.

4.1.1.1. Component Parts. Before treating the filter as a whole, let us look at the individual elements that comprise a mechanical filter. Figure 4.1 shows the electrical, electromechanical, and mechanical sections of the filter in block diagram form.

The electrical circuit components of the input and the output of the filter perform a variety of functions. Considering a filter with piezoelectric ceramic transducers as an example, inductors or transformers may be used to tune or partially tune the capacitance of the transducer. This makes it possible to realize wider bandwidth filters. A transformer replacing the tuning coil can be used for impedance matching, terminating a balanced modulator, or as a parallel tapped connection to another filter. A shunt capacitor, either alone or in parallel with the coil, can be used for temperature compensation. The capacitor, coil, transducer-capacitance circuit can also provide additional selectivity to the filter response by acting as an additional resonator.

Electromechanical energy conversion takes place in the composite transducer resonator. The term composite is used because in most cases the transducer is comprised of piezoelectric or magnetostrictive material bonded to metal. A piezoelectric ceramic for example, provides the energy conversion while an iron-nickel metal alloy provides the frequency stability and increased Q of the resonator. The transducer is also a resonator; it is tuned inside or near the passband of the filter and adds to the selectivity of

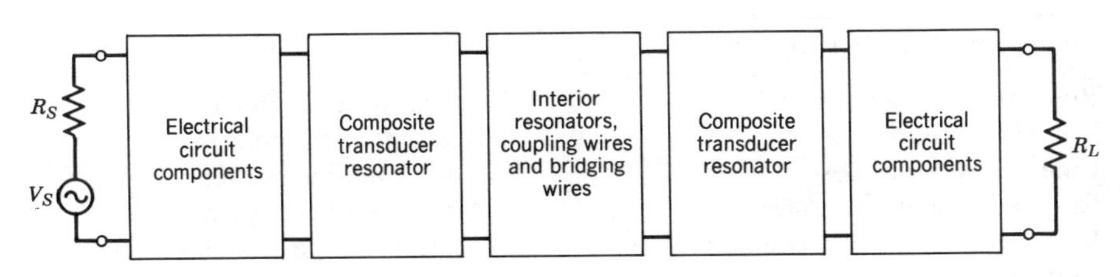

Figure 4.1 Basic mechanical filter elements.

the filter. In an untuned two-resonator filter, the transducer resonators provide all of the selectivity.

Using a loose comparison, the input transducer acts like a motor and the output transducer like a generator. As in the case of a motor, if the mechanical elements are clamped the effect is seen on the electrical side of the circuit. In the motor case, the current increases and the windings may burn up. This means that the mechanical impedance is passed into the electrical circuit and similarly, in the case of a generator, the electrical load is passed into the mechanical circuit. As a mechanical-filter transducer-resonator is a bilateral device, both effects take place; the mechanical circuit affects the input and output impedance of the filter and the electrical circuit source and load resistances affect the mechanical network and provide terminations for the filter.

In summary, the transducer-resonator provides the following:

(1) electromechanical energy conversion;
(2) selectivity from the mechanical resonance circuit and, if tuned, selectivity from the electrical circuit;
(3) a means for the electrical source and load resistance to terminate the mechanical coupled-resonator circuit.

The remaining set of component parts are the interior resonators, coupling wires, and bridging wires. Beginning with the transducer resonators, each resonator is coupled to its adjacent resonator through a small diameter wire or array of parallel wires. The stiffness of the wire and the so-called *equivalent mass* of the resonator at the point of attachment determine the filter bandwidth. The filter center frequency is determined by the frequencies of the resonators. The number of resonators determines the selectivity (shape-factor), and the relative coupling between pairs of resonators determines the amount of passband ripple. If wire coupling between nonadjacent resonators is used, then transmission zeroes can be realized above and below the filter passband. Summarizing, the resonators and coupling wires perform the functions shown in Table 4.1.

TABLE 4.1

Component Parts/Parameters	Functions
Resonator frequencies	Establish the center frequency
Number of resonators	Determines the selectivity
Coupling wire stiffness	Affects the bandwidth
Resonator equivalent mass	Affects the bandwidth
Relative adjacent pair coupling	Determines the passband shape
Bridging coupling	Determines the stopband shape

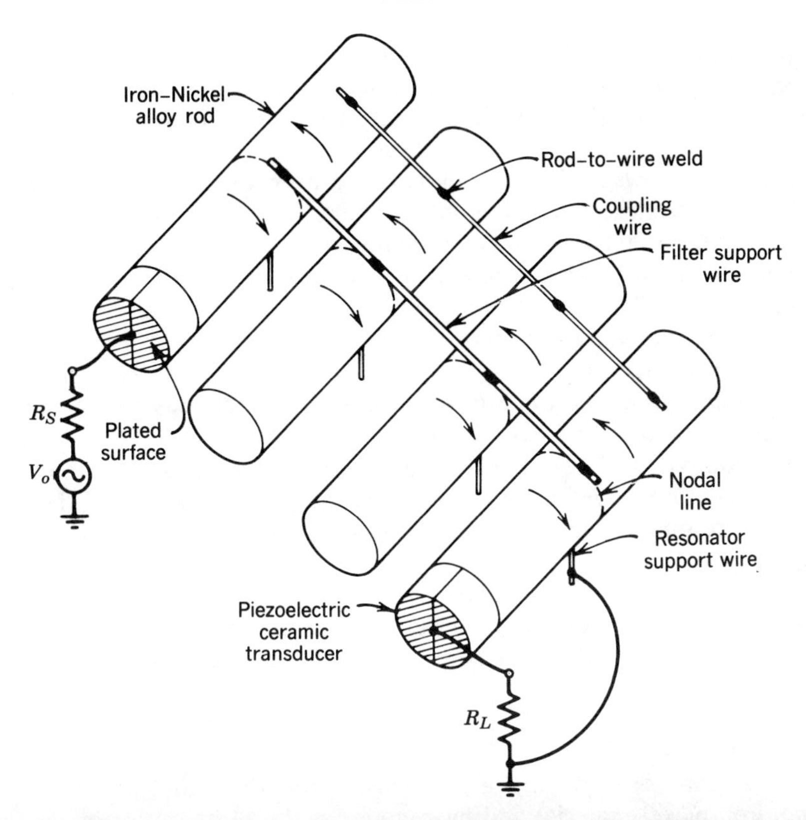

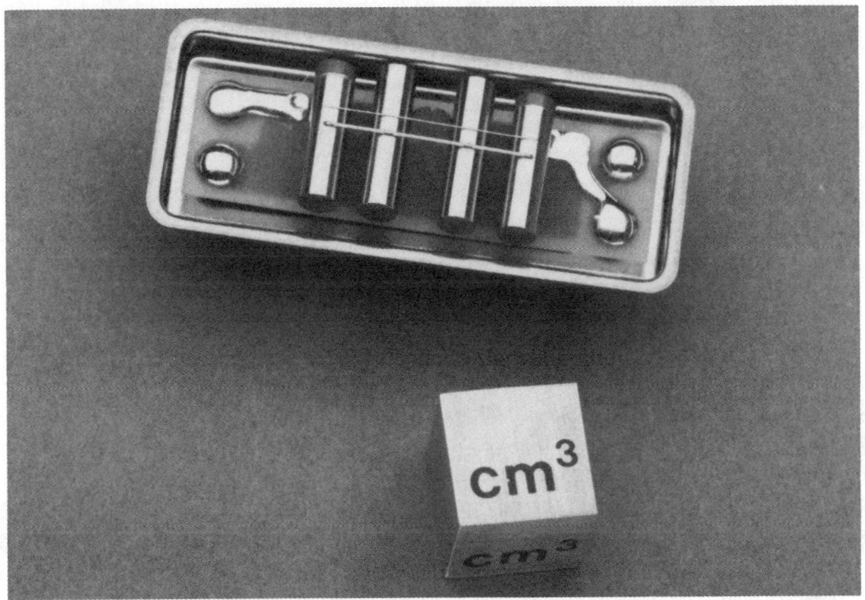

Figure 4.2 Telephone signaling filter.

To this point, we have been somewhat abstract in not having described any specific transducer resonator, interior resonator, or coupling configuration. Let us remedy this with a concrete example that we can use to illustrate general principles of operation.

Figure 4.2 shows both a photograph and a drawing of a four-resonator mechanical filter used in telephone signaling circuits. Typical 3-dB bandwidths are 150–200 Hz at a center frequency of 131.825 kHz. These filters are considered narrowband filters and require no electrical tuning with inductors or transformers. As no tuning coils are needed, the filters can be packaged in a volume of 3 cm^3.

The transducer resonators are composed of a piezoelectric ceramic that is solder- or epoxy-bonded to an iron-nickel alloy rod that has a low temperature coefficient of resonance frequency. The piezoelectric ceramic transducer is composed of two half discs oppositely polarized to cause torsional motion when a voltage is applied between the plated surface and the alloy rod. The resonator vibrates in a half-wavelength mode at its resonance frequency. This means that there is a nodal plane, of no motion shown by the dashed lines, through the approximate center of the resonator. Torsional motion is in opposite directions on either side of the nodal line (plane) as shown by the arrows. As there is no motion at the nodal line, the resonator is supported by a wire at that point. The other end of the support wire is attached to a more massive support or base by means of welding, soldering, or epoxy bonding.

The two interior resonators are also made from the high-Q, temperature-stable, iron-nickel material and have nodes at the exact center of the bar. The four resonators are mechanically coupled by means of a small diameter iron-nickel alloy wire. This coupling wire is attached to each resonator by capacitor discharge resistance welding. The wire length between welds, the diameter of the wire, the wire material, and the position of the wire with respect to the end of each rod are all factors in determining the filter bandwidth and the shape of the passband frequency response.

4.1.2. Principles of Operation

4.1.2.1. One way of understanding how a mechanical filter works is to trace a signal from the generator to the output resistor. From Figure 4.2, the generator voltage V_0 minus the drop across R_s is applied across the piezoelectric ceramic transducer. The resulting electric field causes the ceramic to twist in the direction of the torsional motion of the transducer resonator therefore exciting the entire resonator. Near the lower edge of the filter passband all of the resonators vibrate in-phase. The first resonator drives the second through the wire welded to the two resonators while the second drives the third, and so on, causing the end resonator to vibrate in a torsional mode. The torque applied to the output ceramic transducer causes strains in the material resulting in an electric field due to the piezoelectric effect and a resulting voltage across the output terminals.

As the generator frequency is varied across the passband, the phase relationship between adjacent resonators changes. Near the lower frequency edge of the passband all resonators vibrate in-phase and at the high-frequency edge of the passband each adjacent resonator is out-of-phase. These are the same phase relationships that are present in a system where the resonators are represented by a spring and a mass and the coupling wire by a spring.

4.1.2.2. Equivalent Circuits. A second means of describing the operation of a mechanical filter is to represent the filter by lumped-element mechanical or electrical circuits. Figure 4.3 shows an electromechanical and an all-electrical pair of equivalent circuits corresponding to the filter of Figure 4.2.

In Figure 4.3(a), the electrical part of the electromechanical equivalent circuit is simply the capacitance of the transducer between its plated surface and the metal rod. Next in the equivalent circuit is an electromechanical transformer that both converts electrical and mechanical energy (signals) and acts as an impedance transformer. Internally it acts like an inverter in cascade with a gyrator [13], which means that when we look into the electrical terminals the filter behaves like an electrical circuit, and when we observe from the mechanical side of the circuit, the electrical circuit elements behave like mechanical elements.

In the mechanical portion of the equivalent circuit, the resonators are represented by springs and masses, and the coupling wires between resonators are represented by springs. If we included losses, we would add

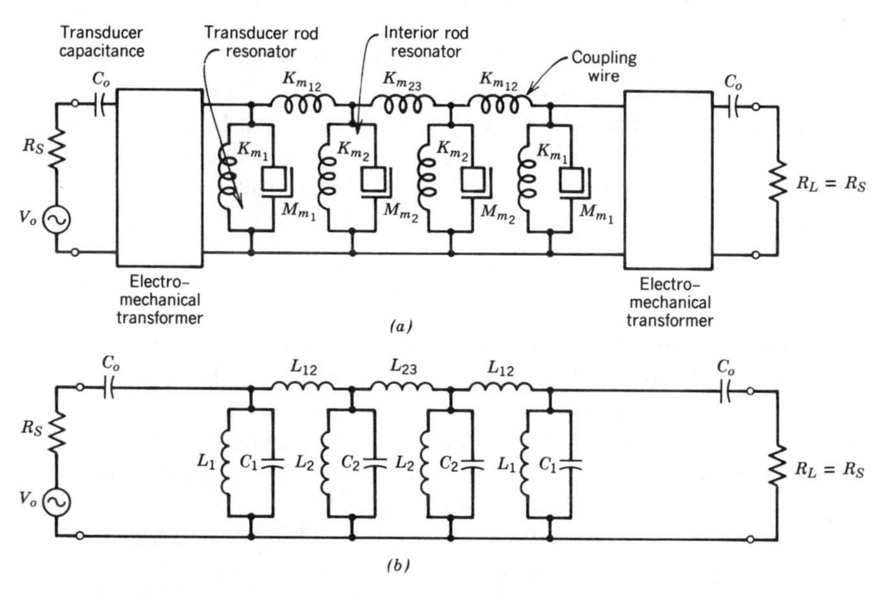

Figure 4.3 (a) The electromechanical equivalent circuit and (b) the electrical equivalent circuit of the signaling filter of Figure 4.2.

dashpot damping elements in parallel with the springs and masses of the resonators. The resultant circuit is a first-order approximation of the actual distributed-element lossy filter.

In Figure 4.3(b), we have an all-electric equivalent circuit of the mechanical filter. This circuit was generated by moving one of the electromechanical transformers through the filter to the other side, where it combines with the other transformer. The back-to-back connection of the electromechanical transformers cancels their effect. As the transformer passes through each mechanical element, it converts it to its electrical analog. For example, when the electromechanical transformer on the left-hand side passes through K_{m1} and M_{m1} of the left resonator, the electrical circuit composed of L_1 and C_1 results. The electromechanical analog resulting from this process is called the mobility analog and is described in more detail in §4.2. In this analogy, compliance (reciprocal of the stiffness K_m) is analogous to inductance, mass M_m is analogous to capacitance, and the network topology of the two circuits are identical.

An equivalent circuit like that of Figure 4.3(b) is valuable for two reasons. First, many filter designers have been trained as electronic engineers so the use of electrical elements instead of mechanical elements makes the design process less difficult. Second, there is a vast body of knowledge on filter design and analysis using electrical elements that can be tapped and used directly. In the present example, the circuit of Figure 4.3(b) is called a narrowband or three-element section bandpass filter, and its design equations can be found in many reference books [3]. This type of filter can be easily analyzed by most computer network-analysis programs. We will make frequent use of electrical analogies in this chapter.

4.1.2.3. Transfer Functions. A third way of describing the operations of a mechanical filter is to analyze the transfer function of the network. In this chapter we will use the attenuation function $H(s)$, which is the ratio of a reference voltage to the voltage across the load resistance expressed in the complex frequency variable $s = \sigma + j\omega$. Analyzing the network of Figure 4.3(b), we can write the attenuation transfer function in the form

$$H(s) = \frac{a_{10}s^{10} + a_9 s^9 + \cdots + a_0}{s^3} \tag{4.1}$$

The roots of the numerator polynomial of Eq. (4.1) are the natural frequencies of the network and are shown as zeroes in Figure 4.4(a). The three attenuation function poles at $s = 0$ and the seven poles at $s = \infty$ are shown as x. The selectivity in the stopband is determined by the number of poles at zero and infinity. The four zeros determine the behavior in the passband.

If parallel coils had been used to tune the static capacitance C_0 at the input and output of the filter, and if the coupling between the resonators had been increased, the pattern of natural frequencies would generally resemble that of Figure 4.4(b).

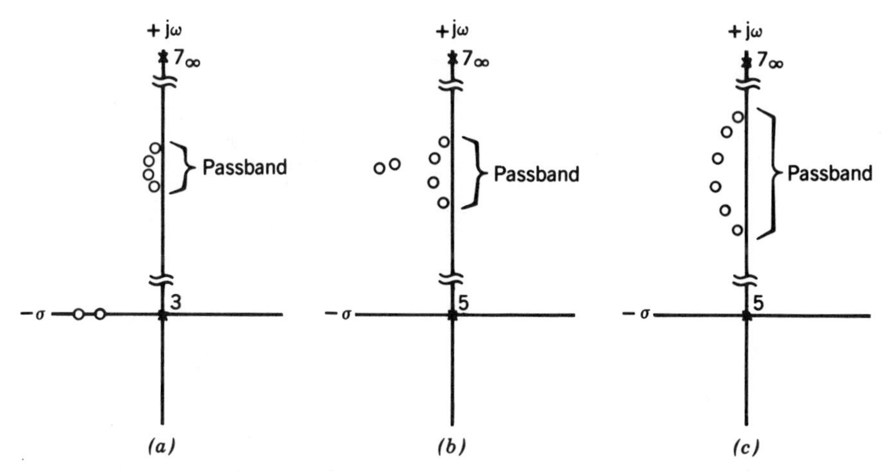

Figure 4.4 Attenuation transfer function zeroes for: (a) the narrowband four-resonator filter of Figure 4.3(b); (b) an intermediate-bandwidth filter; (c) a wide-bandwidth mechanical filter.

The two zeroes that are quite removed from the axis are the result of tuning the static capacitances C_0. These natural resonances, when positioned in this way, have only a marginal effect on the filter selectivity, but (i) allow a wider bandwidth filter to be realized or (ii) make it possible to improve the stability of the composite transducer resonators by reducing the ratio of the more unstable piezoelectric ceramic material to the stable metal alloy. If the coupling between resonators is further increased, or if the ceramic-to-metal ratio is reduced (or if additional capacitance is added to C_0), the wide-bandwidth filter zero pattern of Figure 4.4(c) is obtained. The electrical tuned circuits now add two resonators worth of selectivity to the filters frequency response, but the stability of the response may be marginal because of the relative instability of the tuning inductance and capacitance.

Although maximum stability and performance can be achieved using the intermediate-bandwidth design, the additional ferrite cup-core inductors on the input and output increase the filter size and cost. Assuming that minimum size is a requirement, the filter designer is challenged to design highly frequency-stable composite transducer resonators with high electromechanical coupling. Generally, high coupling and good stability are not compatible; therefore, the challenge. The subject of the electromechanical coupling will be discussed in detail in §4.2.1.

4.1.3. Applications of Mechanical Filters

The constraints on mechanical filter performance determine the applications in which mechanical filters can be used. The major performance constraints are those of realizable center frequency and bandwidth, which are primarily governed by resonator size and electromechanical coupling.

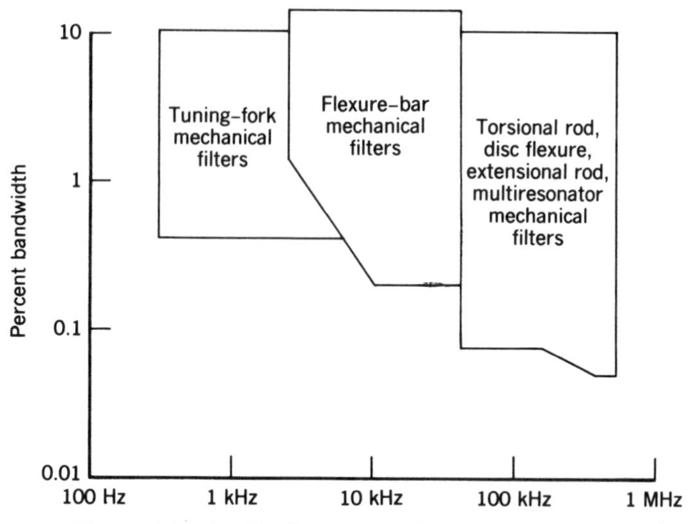

Figure 4.5 Application range of mechanical filters.

4.1.3.1. Bandwidth versus Center Frequency. Figure 4.5 shows the performance limits of various types of mechanical filters. The lower limits of the center frequency are based on a maximum resonator length of 50 mm, in the case of torsional and extensional resonator filters, or a combination of the above length (or diameter in the disc case) and a practical thickness in the flexural resonator and tuning-fork cases. Examples of various resonator modes of vibration are shown in Figures 4.6 and 4.8. The high frequency limits are based on practical fabrication and assembly tolerances.

The minimum bandwidth limits are determined by a resonator stability of 20% of the filter bandwidth over a temperature range of 0°C to +60°C or a maximum insertion loss per resonator of 1.5 dB. The upper bandwidth limits are determined by electromechanical coupling limitations of the composite transducer resonator or unwanted (spurious) resonances in or near the filter passband. Let us look at the types of equipment that use mechanical filters.

4.1.3.2. Types of Equipment. Applications of mechanical filters can generally be divided into two categories. One is for systems that have low data rates and therefore only need narrow bandwidth filters, the second category is for voice and multiple channel data.

The narrowband applications generally involve a single frequency tone that is set at the specified center frequency of the filter. Although the narrow-bandwidth filters have greater frequency shifts relative to the bandwidth, this is usually not an important consideration because of the single frequency nature of the tone and the detection schemes used. In the case of two-tone frequency-shift keying (FSK) systems [8], tolerances on center frequency and bandwidth are slightly more stringent. An additional characteristic of these applications is that they generally require only a few resonators, often only a single resonator and usually not more than four. Most often the filters are coil-less and are quite small. Some specific applications are for out-of-band signaling and pilot-tone filtering in telephone systems, front-end filtering in Omega and very low-frequency (VLF) navigation systems [4], as well as use in FSK modems [8]. In addition, there are many applications in the instrumentation and control field where simply the presence of a tone at a specified frequency needs to be detected. Often the tones are generated by an oscillator that uses a single-pole tuning-fork filter in its feedback loop as the frequency determining element as practiced in wrist watch circuits.

The best known application of mechanical filters is in voice bandwidth communication systems such as high-frequency (HF) radios or FDM telephone systems [7]. These systems require single-sideband (SSB) filtering that often involves data transmission as well as voice transmission. The data requirements often make it necessary to tightly control both the amplitude and delay responses of the filters. In addition, there has been a strong emphasis on size reduction of both telephone equipment and radios. This has challenged mechanical filter designers to realize order-of-magnitude size reductions of their filters. Besides the SSB applications in radios, mechanical filters are used for continuous wave (CW) and amplitude modulation (AM) radio reception. Most of the mechanical filters used in high-performance receivers have 6–12 resonators. An exception are two-resonator mechanical filters used in very inexpensive AM radios. Small, high-performance telephone and radio filters are described in Sections §4.3 and §4.4.1.

4.2. TRANSDUCERS, RESONATORS, AND COUPLING WIRES

The principal components in a mechanical filter are the transducers, resonators, and coupling wires. Rather than attempt to discuss all types of these components, we will emphasize only those most often used. A broader study of this subject is given by Johnson [7].

4.2.1. Transducers

Two types of transducers are used in mechanical filters: magnetostrictive and piezoelectric. As the piezoelectric transducer filters are built in greater volume than the magnetostrictive types, we will concentrate on the piezoelectric and specifically on piezoelectric ceramic composite transducer resonators. Figure 4.6 shows five popular composite transducers.

The Langevin extensional-mode transducer of Figure 4.6(a) is simply a piezoelectric ceramic disc sandwiched between two iron-nickel alloy rods. A sinusoidal voltage applied across the terminals causes expansion and contraction in the disc thickness which, in turn, causes the entire composite

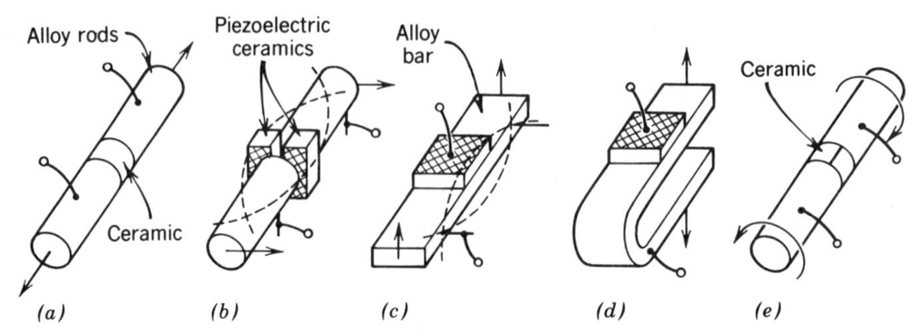

Figure 4.6 Composite transducer resonators: (a) Langevin extensional-mode; (b) modified Langevin flexural-mode; (c) bar-flexure; (d) (tuning) fork; (e) torsional.

structure to vibrate along its length. The arrows show the direction of motion of the ends of the resonator during one half of the vibration cycle. The modified Langevin transducer of Figure 4.6(b) operates on a similar principle except that, instead of a single ceramic disc, two oppositely polarized blocks are bonded to the metal rods. The electric field through the ceramics causes one to expand when the other contracts resulting in a flexure motion of the composite structure. The dashed lines show the relative amplitude of the resonator's vibration at its lowest resonance frequency.

The operation of the third transducer resonator, Figure 4.6(c) is based on the fact that the piezoelectric ceramic changes its length as well as its thickness when a voltage is applied across its major surfaces, in this case, between the top plated surface and the metal bar. As the ceramic length expands and contracts, the composite structure flexes. By moving the ceramic of the bar-flexure resonator nearer to the end of the metal bar and then bending the metal into a U-shape, we obtain the (tuning) fork resonator of Figure 4.6(d).

The transducer resonator of Figure 4.6(e) is simply the torsional-mode piezoelectric ceramic of Figure 4.2 sandwiched between two alloy rods, as opposed to being bonded to the end of a rod. By placing the ceramic in the center of the resonator, the electromechanical coupling is at its maximum value.

At this point you may be asking the question of why one transducer is used instead of the others. The characteristics of the transducer that are important in answering that question include: size, shape, ability to wire couple to other resonators, frequency, mechanical Q, and electromechanical coupling. In other words, there are physical and electromechanical as well as manufacturing parameters that determine the choice. Let us first look at the electromechanical characteristics and then, in §4.2.2, we will look closely at the physical or dimensional characteristics.

4.2.1.1. Equivalent Circuits. The key to an analytical understanding of a mechanical filter is understanding the relationship between four transducer parameters: (i) mechanical resonance frequency; (ii) mechanical Q; (iii) static capacitance; and (iv) electromechanical coupling coefficient.

Let us start by making electrical measurements across the two terminals of any of the five transducer resonators of Figure 4.6. If we ignore the effect of mechanical losses in the resonator and measure the electrical driving-point (input) impedance of the resonator, we obtain the reactance curve of Figure 4.7(a). Two electrical circuits have this particular reactance; one of those is shown in Figure 4.7(b). We will use this circuit because it is compatible with the topology of the mechanical network.

The capacitance C_0 in Figure 4.7(b) is called the static capacitance and is the capacitance measured across the transducer electrical terminals at a frequency far below the mechanical resonance. The second parameter is the open circuit mechanical resonance frequency f_p, that is, the frequency of the pole of the reactance curve. It is called the mechanical resonance frequency because, under the conditions of the electrical terminals being open-circuited, the natural mechanical resonance of the transducer resonator is at f_p. In other words, if one physically strikes the resonator, as in the case of a tuning fork with its electrical leads not connected, the resonator will vibrate at f_p as well as at the other natural frequencies (which we will ignore). From the electrical equivalent circuit we also see that $f_p = 1/[2\pi(L_1C_1)^{1/2}]$. But

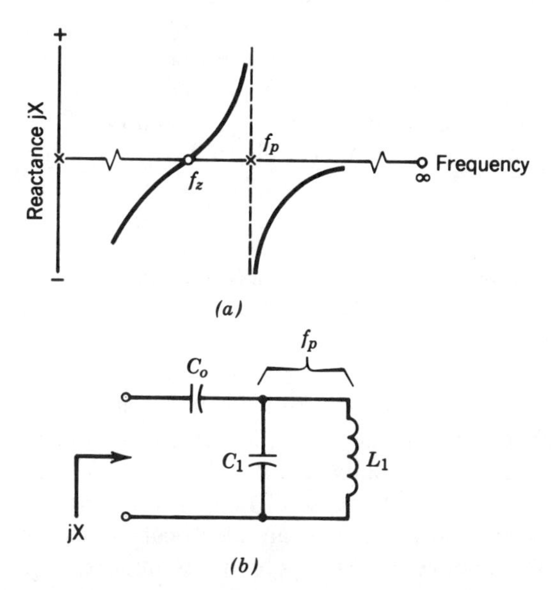

Figure 4.7 (a) Reactance of the driving-point impedance across the electrical terminals of the transducer resonators of Figure 4.6 and (b) the lossless electrical equivalent.

what if the terminals are shorted? In this case the natural resonance frequency of the freely vibrating resonator decreases to a value f_z; f_z is both the zero of the electrical driving-point reactance and is the frequency of the circuit C_0 in parallel with C_1 and L_1 obtained by shorting the electrical terminals of Figure 4.7(b). With regard to frequency, our simple three element electrical equivalent circuit is compatible with both electrical and mechanical measurements.

The frequency difference between f_z and f_p, is called the pole-zero spacing of the transducer resonator and is related to the electromechanical coupling coefficient k_{em} and the circuit elements of Figure 4.7(b) by

$$k_{em}^2 \simeq \frac{C_0}{C_1} \simeq \frac{2(f_p - f_z)}{f_z} \tag{4.2}$$

The parameter k_{em} can be viewed the same as electrical or mechanical circuit coupling coefficients. As the coupling is increased, the realizable bandwidth of the filter is increased. As an example, the maximum bandwidth of a narrowband (no coils) mechanical filter is

$$B_{\max} = q_1(f_p - f_z) \tag{4.3}$$

where q_1 is a parameter related to the passband ripple and the number of resonators [3]. Finally, if we add a resistor in parallel with L_1 and C_1 of our equivalent circuit, we can take into account the mechanical $Q(Q_m)$ of the transducer resonator.

Referring back to Figure 4.3(b), we see that we have now established a means for finding the element values C_0, L_1, and C_1 of our electrical equivalent circuit by measuring C_0, f_p, and f_z (and Q_m if we want to include losses). But how are the elements L_1, C_1, and K_{m1}, M_{m1} in Figure 4.3(a) and (b) related?

4.2.1.2. The Mobility Analogy. The mobility analogy relating electrical and mechanical circuit parameters is based on the following: (i) the differential equations describing the electrical and mechanical systems have the same form, (ii) the circuit topologies are the same, and (iii) electrical and mechanical variables across and through the systems are proportionally related to each other. As an example, voltage is measured across (or between) two points as is velocity. Table 4.2 provides a brief summary of the mobility analogy [7]. As in the case of lumped elements, it is easy to show that distributed elements such as an electrical transmission line and a torsional rod are analogous. Although this is important in concept, most often the mechanical filter designer approximates the distributed element with an equivalent lumped element, as in Figure 4.3.

TABLE 4.2. Electromechanical Mobility Analogy

	Mechanical	Electrical
Variables		
Across	Velocity (V_m) Angular velocity ($\dot{\theta}_m$)	Voltage (V)
Through	Force (F_m) Torque (τ_m)	Current (I)
Lumped network elements	Damping (D_m) Stiffness (K_m) Mass (M_m) Mass moment of inertia (J_m)	Resistance^{-1} ($1/R$) Inductance^{-1} ($1/L$) Capacitance (C) Capacitance (C)
Topology	Loop Node	Loop Node

4.2.2. Resonators

The most commonly used resonators are torsional rods, flexural bars, circular discs vibrating in flexure, and forks. In each case, these are distributed element resonators that, because of the narrow bandwidth nature of mechanical filters, can be represented by lumped elements (springs and masses).

4.2.2.1. Lumped-Element Equivalent Circuits. Figure 4.8 shows a torsional resonator and its spring-mass equivalent circuit. The torsional resonance frequency of the resonator in its lowest natural mode is a function of its length l, the shear modulus of the metal rod G_m and the density of the rod ρ_m. The rod diameter d has virtually no effect on the resonance frequency. Finding the resonance is, therefore, quite simple both in concept and in practice. But what are M_{eq}, K_{eq}, and how are they measured or calculated?

The concept of equivalent mass M_{eq} or equivalent stiffness K_{eq} is very important in the design of a mechanical filter. By the term *equivalent mass* we mean the mass of a spring-mass resonator that has equivalent characteristics to a distributed-mass resonator at a specific point on the resonator, in a specific direction and near its resonance frequency. For example, if we weld a coupling wire at point A on the resonator of Figure 4.8(a) at a distance x from the end of the rod and observe the characteristics of the resonator at point A in the direction of the wire, we can, in theory, measure the equivalent mass at that point in that direction. It is as though the resonator is in a black box and as we drive it, we do not know whether we are driving a spring-mass resonator or a distributed resonator or something

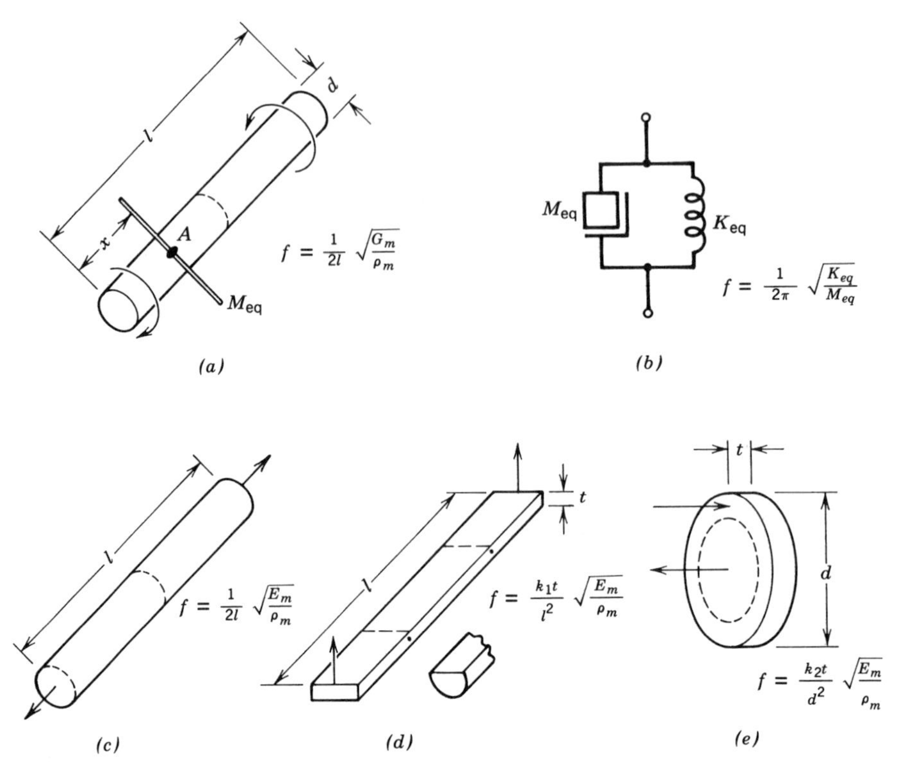

Figure 4.8 Mechanical filter interior resonators: (a) torsional resonator; (b) the lumped-element resonator equivalent circuit; (c) extensional-mode resonator; (d) flexure-mode resonators; (e) a disc flexure-mode resonator.

even more complex. As the point A moves toward the nodal point (dashed line), the equivalent mass increases and becomes infinite at the node; for a finite driving force the velocity is zero at the node.

Equations for equivalent mass can be derived by either (i) equating the slope of the force over velocity (F_m/V_m) versus frequency curves of the distributed resonator and the spring-mass resonator at their resonance frequency or (ii) by equating the kinetic energy in each system [7]. In either method, the lumped equivalent mass can be found in terms of the distributed resonator dimensions and density, and the point of measurement. For the torsional resonator of Figure 4.8 the equivalent mass is

$$M_{eq} = \frac{\rho_m \pi l d^2/4}{4\cos^2(\pi x/l)} = \frac{\text{static mass}}{4\cos^2(\pi x/l)} \tag{4.4}$$

The equivalent stiffness K_{eq} of the resonator can be found from the equation of Figure 4.8(b).

Knowing the equivalent mass of each resonator, including the transducer resonators, allows us at this point to specify everything but the coupling wire stiffnesses K_{12} and K_{23} in Figure 4.3(a). Based on the mobility analogy, the electrical equivalent circuit values have the same ratios as the mechanical values, therefore L_2 and C_2 can be easily calculated.

Three other popular resonators are shown in Figure 4.8. The equivalent mass of each can be found by dividing the total kinetic energy in the resonator by one-half of the velocity squared as was done in the torsional case.

4.2.3. Coupling and Bridging Wires

Having described, measured, and calculated transducer and resonator characteristics, we must do the same for coupling and bridging wires. First we will look at coupling-wire equivalent circuits and then at resonator-wire coupled systems.

4.2.3.1. Equivalent Circuits. The most common coupling-wire modes of vibration are extensional and torsional and can be analyzed using basic transmission line theory. Flexural coupling is considerably more complex and is beyond the scope of this chapter, but is described in great detail by Konno et al. [9].

From transmission line theory, as applied to a lossless slender wire vibrating in an extension mode, we can write

$$\begin{bmatrix} V_{m1} \\ F_{m1} \end{bmatrix} = \begin{bmatrix} \cos \alpha l & jZ_{mo} \sin \alpha l \\ \dfrac{j \sin \alpha l}{Z_{mo}} & \cos \alpha l \end{bmatrix} \begin{bmatrix} V_{m2} \\ F_{m2} \end{bmatrix} \tag{4.5}$$

where l is the wire length, Z_{mo} is the characteristic mobility $1/A(\rho_m E_m)^{1/2}$, α is the propagation constant ω / ν_e, ν_e is the propagation velocity $(E_m/\rho_m)^{1/2}$ of an extensional wave, A is the cross-sectional area of the wire, ρ_m is density and E_m is the elastic modulus of the wire, and V_{m1}, V_{m2}, F_{m1}, and F_{m2} are the velocities and forces at the ends of the coupling wire. The equivalent circuit corresponding to the transmission matrix of Eq. (4.5) is shown in Figure 4.9(a). In this figure, Y_m is the mechanical impedance F_m/V_m. Similar torsional line circuits are described by Johnson [7]. The equivalent circuits and equations described above can also be used to model bridging wires, that is, wires that are used to couple non-adjacent resonators in order to produce transmission zeroes in a filter's frequency response.

For ease of discussion, we will assume that the coupling wires of Figures 4.2 and 4.3 are less than one-eighth wavelength ($\lambda = \nu_e/f$) at the filter center frequency. This means that we can use the equivalent circuits of Figure 4.9(b) and (c) in the equivalent circuits of Figure 4.3 by combining the parallel arm masses and capacitance of the wire with those of the resonator,

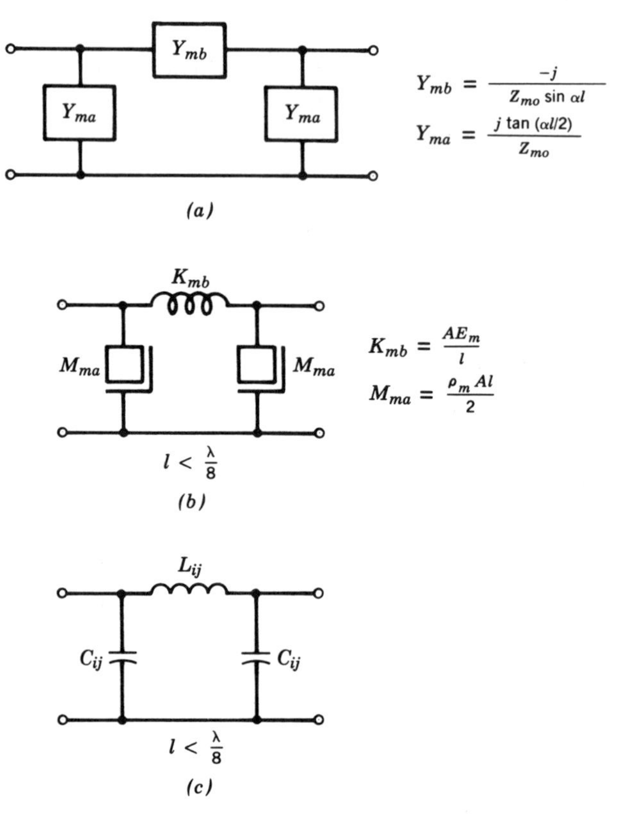

Figure 4.9 Coupling wire equivalent circuits for the extensional mode of vibration: (a) general two-port pi network; (b) short line lumped element circuit; (c) short line mobility analogy.

thus completing the description of the filter; that is, we now have enough equations and equivalent circuits to analyze or conversely design an existing filter comprised of transducer resonators, interior resonators, and coupling wires. The design process involves realizing the actual parts from derived equivalent LC circuits such as shown in Figure 4.3(b). In other words, we design the LC filter and then substitute elements. Let us next look at adjacent resonator coupling.

4.2.3.2. Coupled Systems. The same equations that apply to coupled circuits in electrical filters also apply to coupled mechanical resonators. From ITT tables [3] and the use of the mobility analogy, we can write for the coupling wire stiffness

$$K_{mij} = k_{ij} \frac{B}{f_0} (K_{mi} K_{mj})^{1/2} \tag{4.6}$$

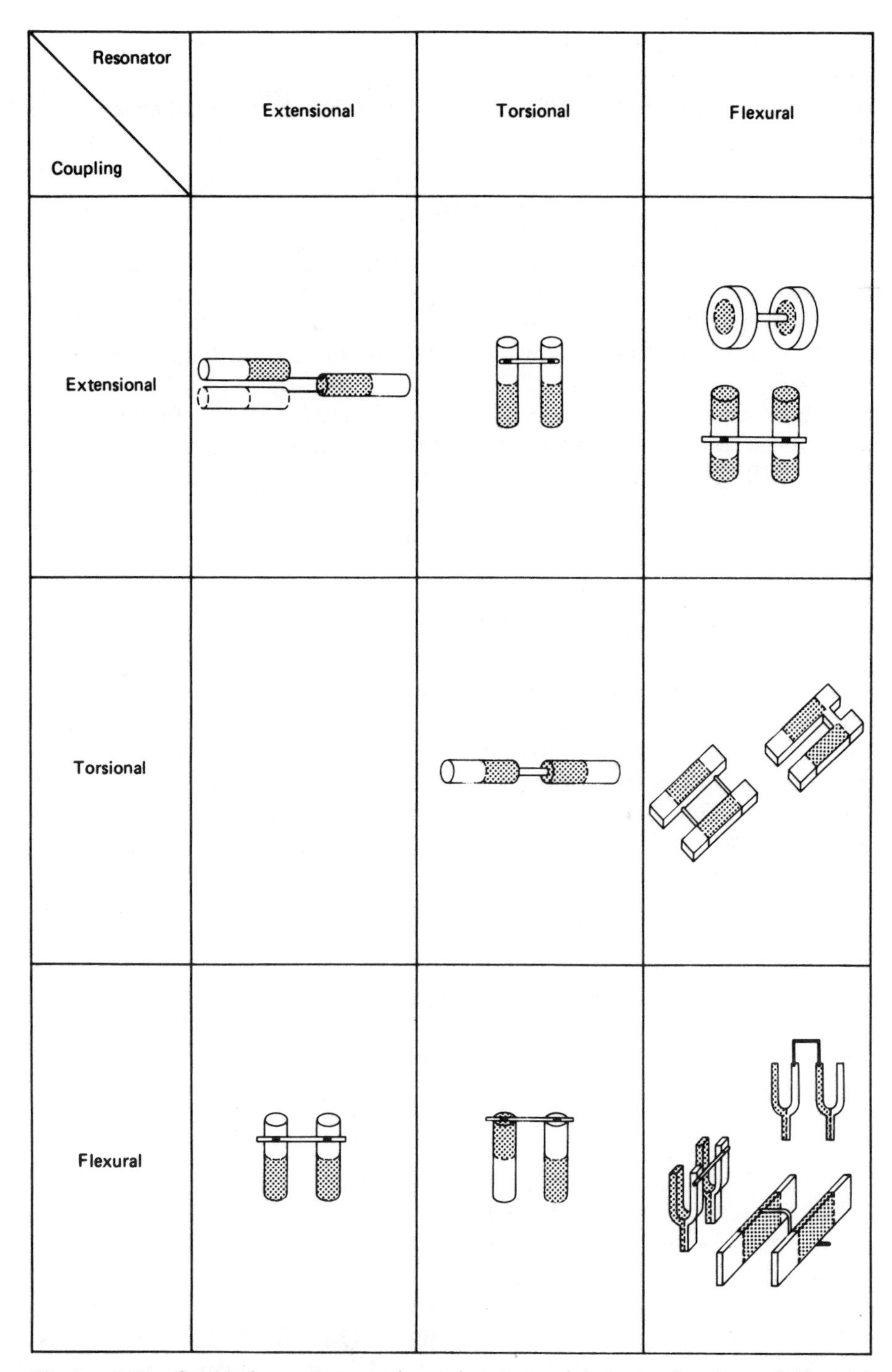

Figure 4.10 Coupled resonator-pairs using extensional, torsional, and flexural resonators and coupling wires. Displacements in the shaded areas are in-phase at the lower natural resonance of the pair.

where k_{ij} is the normalized coupling coefficient, B is the filter 3 dB bandwidth, f_0 is the filter center frequency, and K_{mi} is the equivalent stiffness K_{eq} of resonator i, where $K_{mi} = (2\pi f_i)^2 M_{mi}$. From the above, an alternate form of Eq. (4.6) is

$$k_{ij}\frac{B}{f_0} \propto \frac{K_{mij}}{(M_{mi}M_{mj})^{1/2}} \tag{4.7}$$

In other words, the coupling between resonators i and j is proportional to the equivalent stiffness of the coupling wire (or bridging wire) divided by the square root of the product of the equivalent masses of the coupled resonators. There is no equation more basic to mechanical filter design than Eq. (4.7).

Use of the concepts, equivalent circuits, and equations that we have derived can be applied to a variety of filter structures such as those shown in Figure 4.10.

4.2.3.3. Bridging Wires. The idea of bridging over resonators is to produce transmission zeroes (attenuation poles) in the filter stopband (e.g., as in the filter response of Fig. 4.13). The zeroes are the result of the signal through the bridging wire cancelling the signal through the main coupling wire(s). By bridging across two resonators, for example, from resonator one to resonator four, transmission zeroes can be realized in the upper and lower stopbands. By bridging across a single resonator, a zero can be generated below or above the filter passband. Various bridging methods for realizing transmission zeroes are described by Johnson [5, 7] and Sheahan and Johnson [11]. Essentially these methods involve realizing phase inversion in the bridging or main signal paths to cause cancellation of the signals at the node where they are terminated.

4.2.4. Size Conditions

It is possible to design and manufacture a single-resonator fork mechanical filter for wrist watches that has a volume of 0.003 cm^3 [1]. Two-resonator filters for AM radios are packaged in 0.21 cm^3. Telephone channel filters, high-performance radio filters, and low-frequency flexure-mode filters have package volumes as low as 6 cm^3. What are the constraints on minimizing the size of a mechanical filter?

In the most simple terms, the constraints on mechanical filter size reduction are those of strength, performance, and the limitations dictated by the equations relating frequency and resonator size. Let us look at this size constraint.

4.2.4.1. Reducing Resonator Length. Figure 4.11 shows the useful frequency ranges of various resonators presently used in manufactured mechanical filters. The equations relating frequency and physical dimensions and elastic

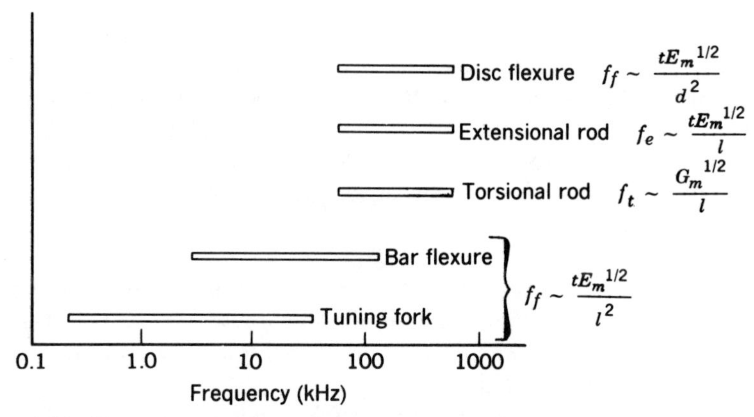

Figure 4.11 Frequency ranges of resonators used in manufactured mechanical filters.

constants are also shown. We have little control over the density or the constants relating to the order of the mode of vibration (such as k_1 of the equation shown in Figure 4.8(d)), therefore we have not included these in the equations. Let us examine first the extensional and torsional modes.

For a given filter center frequency and resonator material, the torsional and extensional-mode resonator lengths are fixed, but since the elastic modulus E_m is roughly 2.75 times as large as the shear modulus G_m, torsional-mode resonators will be 0.6 times as long as extensional resonators at the same frequency. This smaller length has been a major factor in the popularity of torsional filters for FDM telephone systems.

By machining the resonator as a nonuniform line, for instance with a necked-down region in the center and larger diameter sections on the ends, a mass-loaded or dumbbell type of resonator is created. This reduces the length of the resonator for a given frequency, but at an additional fabrication cost.

Whereas the uniform resonator length is fixed by the frequency and material in the case of the extensional and torsional-mode resonators, the flexural resonator does not have that same limitation. Note, in Figure 4.11, that the disc, bar, and fork flexure mode frequencies are also proportional to the thickness. By reducing the thickness and therefore the length or diameter as well, the resonator can be reduced to any size, the size only being subject to strength or process limitations. For example, a flexural resonator bar tuned at the signaling frequency 3.825 kHz has the dimensions: $l = 31.75$ mm, $t = 0.786$ mm. To reduce the bar length to 1 inch (25.4 mm), the thickness becomes 0.503 mm, which becomes too thin a bar to process without bending the bar beyond its elastic limit. This is an even greater problem with the end composite transducer resonators, because a small amount of bending in the fabrication process will break the piezoelectric transducer/metal-bar solder bond.

If in the above example it were essential that the length be reduced, it would be possible to simply fold the resonator into a U-shape, creating a fork resonator. However, doing this creates potential problems such as difficulty in mounting, lower electromechanical coupling (if needed), and more expensive resonator fabrication. When these problems can be overcome, practical resonators with center frequencies as low as 200 Hz can be manufactured.

4.2.4.2. The Effect of Resonator Size on Performance. Besides physical strength, there are two other major limitations on resonator size: (i) signal level sensitivity and (ii) fabrication and assembly tolerances. Let us first look at signal or drive level sensitivity.

Figure 4.12 shows how the resonance frequency of various resonators is affected by signal level, that is, by the energy per unit volume within a resonator [7]. Each of the curves corresponds to a specific material and a specific mode of vibration. Composite transducer resonator curves fall between the ceramic and alloy curves. Frequency shift with signal level is very important in omega navigation receiver applications where small filter center frequency shifts with signal level result in substantial shifts in the phase of the output signal.

Not only does the frequency of the resonator shift with signal level and energy density but the resonator mechanical Q_m also shifts (Yakuwa and Okuda in [11]). This shift affects not only the filter's insertion loss, but also the filter's intermodulation distortion. This is especially important in the

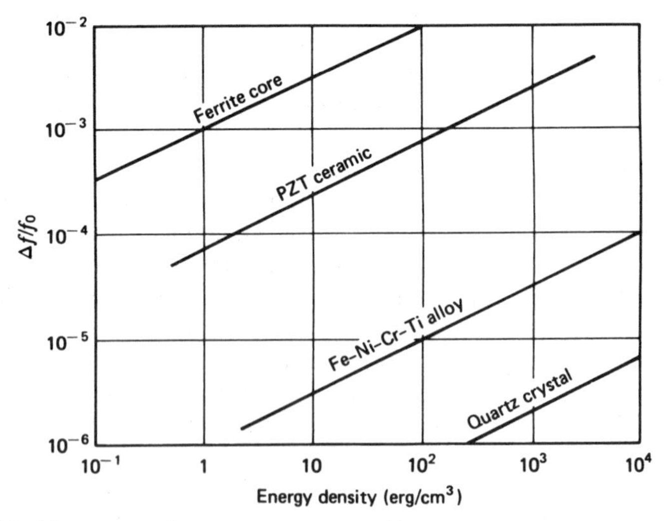

Figure 4.12 Frequency change versus energy density for various resonator materials. (Robert A. Johnson (1978) *Proc. I.E.E.E. Int. Symp. on Circuits and Systems*, New York. © 1978 I.E.E.E.)

design of SSB filters for high performance FDM telephone and HF radio applications.

The second major limitation on resonator size relates to fabrication and assembly tolerances that can be held in a factory environment. In a first-order sense, absolute variations in length and diameter are independent of the length and diameter. For example, this means that the variation of the diameter of a 2.5 mm diameter torsional fabricated by centerless grinding will be about the same as that of a 1.25 mm resonator, but the effect on the change in volume of the part will be four times greater in the small diameter example. Therefore, from Eqs. (4.4) and (4.7), the variation in coupling due to resonator mass variations is greater in the small diameter case.

With regard to assembly tolerance, the absolute variations of position are also independent of resonator size, in critical processes such as coupling-wire to resonator welding. How this affects the coupling between resonators can be shown by an example. Let us look at doubling the center frequency of a torsional filter by halving the length of its rod resonators. From Eq. (4.4) the equivalent mass is halved, as long as x/l remains constant. Also the effect on M_{eq}, of a fixed variation of wire position Δx on the short resonator, is greater than the effect of the variation in the longer resonator case. Because the static mass is halved, the equivalent mass can be held constant by moving the wire position x closer to the nodal point. However, this increases the sensitivity of M_{eq} to weld position variations because of the $\cos^2$ nature of Eq. (4.4). The resultant changes in M_{eq} affect the coupling between resonators according to Eq. (4.7). If we simply allow M_{eq} to be reduced in value, then we must reduce the coupling wire stiffness K_{mij} to keep the coupling between resonators constant, but this means reducing the strength of the filter structure.

4.2.4.3. Methods of Filter Size Reduction. Having minimized the resonator volume while staying within practical limits, we then need to look at filter configurations that result in a minimum filter size.

One method of size reduction is simply folding the filter as the flexure bar resonator can be folded into a fork [7]. For instance, the first and last resonators are placed adjacent to one another, the second and next-to-last, and so on. In this type of filter built in two planes, problems of coupling around the bend and means of resonator support require very creative solutions. A second design is the zig-zag type in a single plane as shown in Figure 4.10 (extensional-extensional). A third type which is folded, but is in a single plane, is described in §4.4.2. Keeping the resonators in a single plane eliminates coupling around the bend and resonator support problems.

Another method of size reduction is to eliminate tuning coils. As was discussed earlier, this is only possible if Eq. (4.3) is satisfied, that is, if the pole-zero spacing of the transducer is large enough to realize a given filter bandwidth.

Last, but not least important, the use of bridging wires steepens the amplitude versus frequency response of the filter between the pass-band and stop-band. This makes it possible to both eliminate resonators and improve the filter group delay. Examples of these filters are shown in §4.3 and §4.4.

4.3. VOICE-BANDWIDTH TELEPHONE CHANNEL FILTER DESIGN

A telephone channel-filter design will be used to illustrate some of the ideas developed so far. Whenever possible we will concentrate on concepts rather than numbers.

4.3.1. Statement of the Design Problem

The design problem is to meet the specification shown in Figure 4.13 with a minimum volume filter and a low manufacturing cost. In addition, attenuation poles both above and below the filter passband are required in order to minimize both the absolute and differential delay responses [7]. The carrier frequency of 128 kHz is the most commonly desired mechanical channel filter frequency and the bandwidth is fixed by international (CCITT) standards. The passband ripple requirement chosen for this example is termed 1/10 CCITT, which in the center region of the passband is ±0.22 dB as defined by the CCITT standards. Although meeting the 1/10 CCITT specification is not a trivial problem, it does allow flexibility in the design over having to meet the more stringent 1/20 CCITT specification (±0.11 dB ripple) that is required in some telephone systems.

Finally, the specification of Figure 4.13 must be met over a temperature range of 0–65°C. Although this range is slightly wider than most telephone

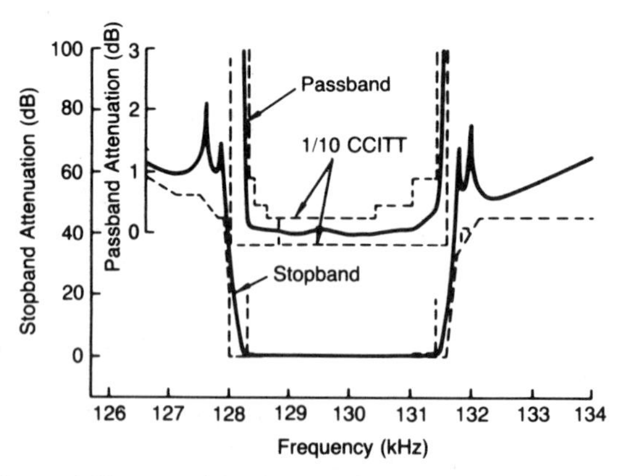

Figure 4.13 Channel filter specifications and frequency response (courtesy of Rockwell International, USA).

filter specifications, it is compatible with most resonator and transducer materials and the passband and stopband requirements.

4.3.2. Design Strategy

Developing a new mechanical filter is not too surprisingly, a long iterative process of trying and rejecting many ideas until finally a device that can be built with realizable processing equipment and materials is developed. Rather than try to describe this lengthy process, let us simply look at the mechanical filter elements (transducers and resonators), one at a time, and make choices based on hindsight.

4.3.2.1. Transducers. Our choice of transducers is primarily based on the need to eliminate tuning coils, in order to reduce the size of the filter. Having made that choice, the bar flexure and fork transducer resonators of Figure 4.6(c) and (d) are eliminated because they cannot meet the pole-zero requirements of Eq. (4.3), that is, $(f_p - f_z) = B_{\max}/q_1$. Remaining are the high coupling extensional-mode Langevin, modified flexure-mode Langevin, and the torsional mode transducer resonator of Figure 4.6(e).

Two disadvantages of the extensional-mode Langevin are: (i) it is roughly 1.7 times as long as the torsional-mode transducer, and (ii) it is difficult to couple extension and torsion without placing the resonators at right angles to one another thus increasing the volume of the filter. Section 4.4.2 discusses a clever method of surmounting the above disadvantages. The major advantage of the Langevin transducer resonator is its simplicity with regard to both the piezoelectric ceramic portion of the transducer resonator and its assembly, resulting in both constant transducer characteristics (f, Q_m, k_{em}, and C_0) and low cost.

The torsional transducer of Figure 4.6(e) uses a complex ceramic disc and, if used with other torsional resonators, suffers from difficulty in realizing transmission zeroes. This is because of the difficulty in realizing phase inversion in the bridging or the main signal path in order to obtain signal cancellation. This latter problem has been solved by the use of angled bridging wires as discussed by Sawamoto et al. [11].

The modified-Langevin flexure-mode transducer of Figure 4.6(b) is chosen because it has the advantage of small size and a more complex vibration pattern that allows simple phase inversion bridging to realize the transmission zeroes. Its disadvantage is that it requires a complex assembly process.

4.3.2.2. Resonators. Because of the importance of filter size, we will eliminate the extensional and disc flexure-mode resonators shown in Figure 4.8(c) and (e). This leaves the torsional and flexural-mode resonators. Yano et al. [11] argue that, because of higher Q and better stability of loss factor $(2\pi f/Q_m)$ with changing signal amplitude, the torsional resonator is better at 128 kHz. In spite of this argument, excellent 128 kHz filters are built with flexure-mode resonators, but at the expense of having to use a pair of dumbbell torsional-mode resonators to realize phase-inversion bridging [2].

Possibly a stronger argument for the torsional resonator is that it is easier to fabricate and only requires a single support wire. In addition, when used with the modified Langevin transducer, out-of-phase (phase inversion) bridging to obtain transmission zeroes is easy to realize.

At frequencies in the range of 25–50 kHz and above, flexure-mode resonators are usually fabricated from rod stock; a flat is surface ground on one side as shown in Figure 4.8(d). The flat is necessary to eliminate degeneracy, which simply means that a round cross-section rod will vibrate in two planes at approximately the same frequency. The flat separates the frequencies of the two modes.

Besides the torsional resonator needing only a single support wire, the support wire is attached to a point of no motion. Conversely, the flexural node is one of flexure but not a node of torsion, and therefore the support system affects the resonance frequency and Q. Considering the need for low-cost construction and highest Q, the torsional resonator is chosen for this example.

In spite of the above arguments, often the decision to use one type of resonator or transducer over another is based on other criteria such as previous experience, available factory tooling, availability and cost of transducer and resonator materials, patent status, and specifications that may be important to one user but not to another.

4.3.3. Filter Design

In order to meet the specifications of Figure 4.13 it is necessary to use the equivalent of nine resonant circuits and four finite-frequency transmission zeroes in the electrical and mechanical filter realizations. This can be determined by the use of approximation programs or by analysis of circuits having different numbers of resonant circuits and transmission zero locations. Because tuning coils will not be used, in order to reduce the size and cost of the mechanical filter, a network such as Figure 4.14(a) is used as a starting point for the design. The element values of the ladder network are found by using the computer program FILSYN [12].

The electrical network can be transformed into various topologies realizable as mechanical filters. One example is a parallel ladder design described by Yano et al. [11]. This is a physically complex filter, but dissymmetrical (about the center frequency) transmission zeroes are realizable. Another possibility is a multiple bridging wire design (see §4.4.2) where a wire that bridges two resonators is inside a wire that bridges four resonators. This design works well with the choice of Langevin transducer resonators and the resulting folded topology. Instead of these options, the more conventional ladder network with two double-resonator bridging sections shown in Figure

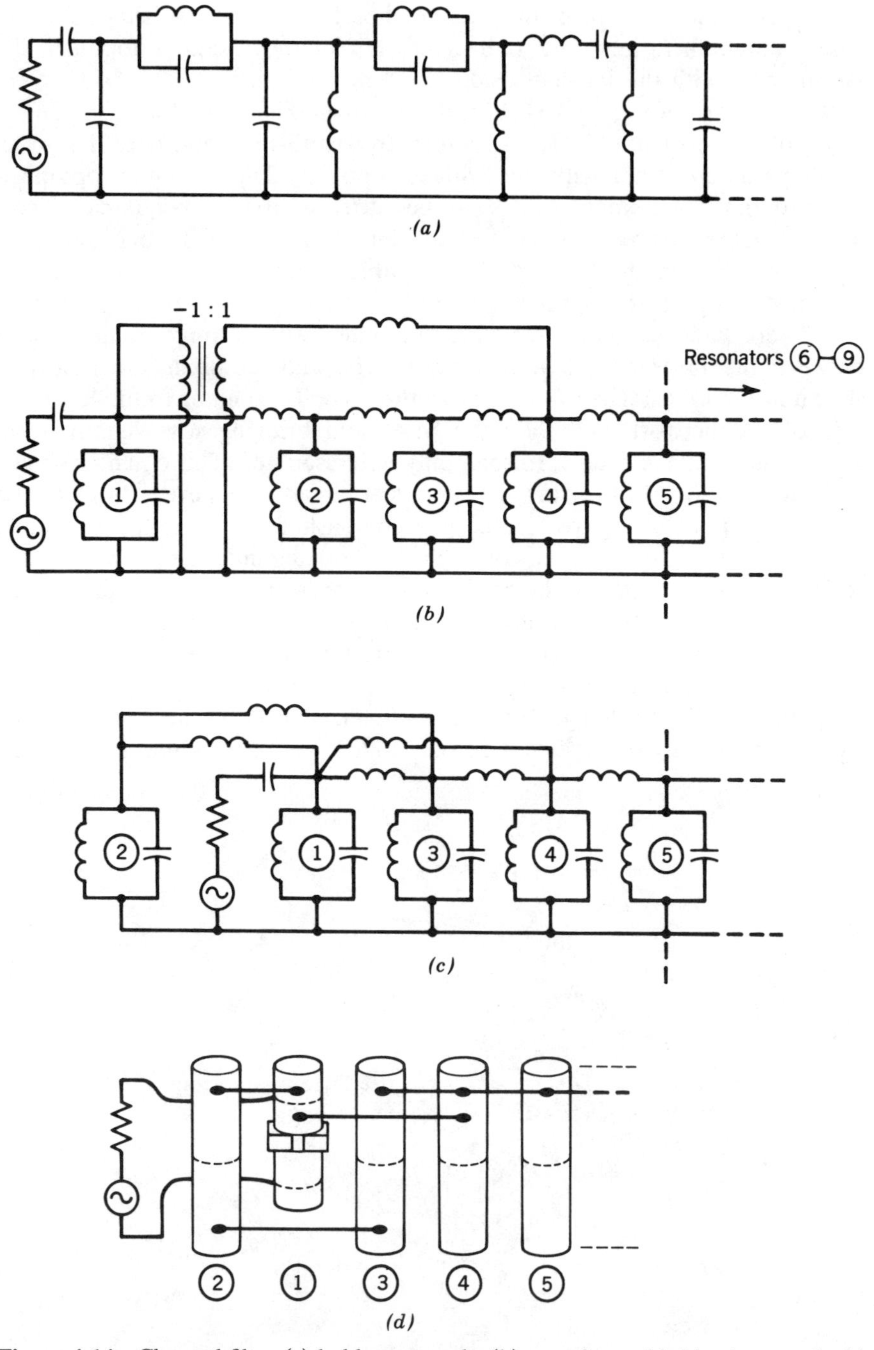

Figure 4.14 Channel filter (a) ladder network, (b) transformed bridged network, (c) shortened bridging element realization and (d) mechanical realization with a modified Langevin transducer resonator and torsional interior resonators.

4.14(b) is chosen for our example. The choice is based on its simplicity and ease of manufacturing. The right-hand half of the circuit is topologically symmetrical with the left-hand side.

The network of Figure 4.14(b) is derived from Figure 4.14(a) by making a series of transformations [7]. An alternative to this method is to design an image-parameter filter with the bridged topology. Then using an optimization program the element values can be varied until a least-squares or least nth fit to the desired passband shape and transmission zero positions is found. Another method is to design a simple monotonic stopband filter and then add bridging and optimize.

Rather than physically realizing the filter with a long bridging wire between the first and the fourth resonators (and the sixth and ninth) as shown in Figure 4.14(b), we can move the second resonator to the left-hand side of the network and then the maximum bridging wire length is no greater than what is needed to span only one resonator. This circuit is shown in Figure 4.14(c). Converting the electrical circuit of Figure 4.14(c) to the mechanical filter of Figure 4.14(d) is accomplished replacing the LC tuned circuits by resonators and the coupling and bridging inductors by wires. The series capacitor, to the right of the source resistance, and the first tuned circuit are replaced by the transducer resonator. Note in Figure 4.14(d) that the coupling and bridging wires are on opposite sides of the nodal line of the transducer resonator. This results in out-of-phase signals through the main coupling and the bridging paths causing cancellation of the signals in the stop-band, that is, transmission zeroes. Figure 4.15 shows the actual filter.

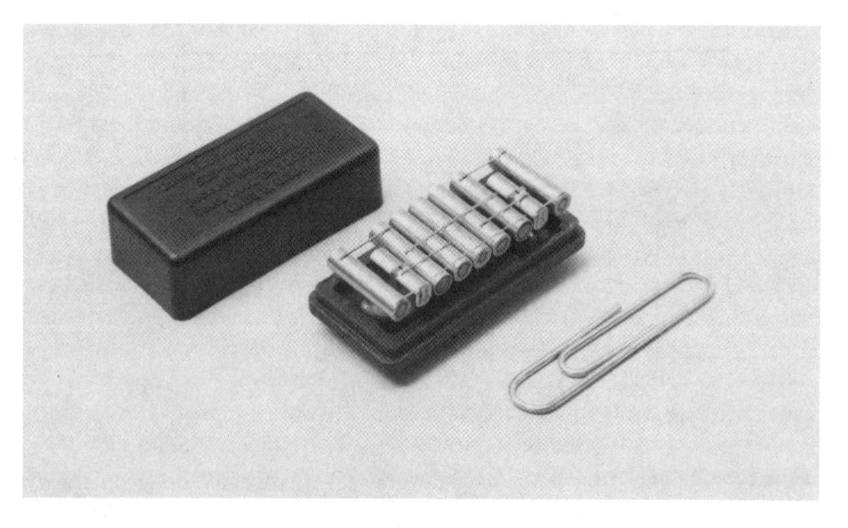

Figure 4.15 Telephone channel filter (courtesy of Rockwell International, USA).

The major task in the physical design of the filter is calculating coupling and bridging wire diameters, lengths, and resonator weld points (such as the position x in Figure 4.8(a)) in order to satisfy the required coupling k_{ij} between resonators. Besides satisfying the required couplings, the filter must be spurious-response free within and near the filter pass-band. Analytical solutions to this problem exist, but are only as good as our ability to measure coupling wire parameters such as diameter and the elastic modulus E_m.

Besides calculating wire diameters, it is necessary to determine metal alloy heat treatment temperatures to minimize the frequency shift of the resonators over temperature. In addition, the transducer must be designed so that the effective Q of the terminating circuit remains constant with varying temperature.

4.3.4. Manufacturing

Manufacturing a telephone channel filter requires a great deal of precision in both the fabrication of the parts and the weld assembly process. Tolerances on the resonator diameters are about ± 3 μm and, on length, about ± 50 μm. The 50-μm tolerance on an 11-mm torsional resonator rod means that the frequency variation of the resonator due to its length (remembering that $f \sim 1/l$) will be about ± 600 Hz at 130 kHz. Including variations in the heat treatment and in G_m due to inhomogeneities of the rod, the frequency variation may be closer to ± 1 kHz. The ± 1 kHz is within the tuning range of a YAG-Nd laser described by Yakuwa et al. [14]. Although the method used of laser drilling holes into the ends of the torsional bars to change frequency is not unique, the idea of using robots to hold and position the resonators when measuring frequency and drilling holes is unique. A computer controlled robot/laser tuner is shown in Figure 4.16. The central idea is to have a machine that can tune any kind of resonators or already assembled filters such as telephone signaling filters without having to modify expensive tooling but simply reprogram the robots. In addition, two robots are used; while one is measuring the frequency of one resonator the other is presenting the second resonator to the laser. On average, two sets of laser shots are needed to tune a 128 kHz resonator to ± 4 Hz of the desired frequency.

The most critical process step is the coupling-wire to the resonator welding. As filters are reduced in size and the performance requirements are increased, the welding becomes more critical. As the position x on the resonator of Figure 4.8(a) varies, the equivalent mass varies according to Eq. (4.4). If the wire is near the end of the resonator, the sensitivity of M_{eq} to wire position is much less than if the wire is near the node where the slope of the $\cos^2$ expression is greater. Optimum welding is accomplished using fixed tooling to hold the resonators and wires. The tooling is mounted on an X–Y table which is positionally controlled by a numerical control unit. The weld unit is a capacitive discharge resistance welder [7].

Figure 4.16 Computer-controlled robot/laser resonator tuning equipment (courtesy of Rockwell International, USA).

4.4. OTHER MINIATURIZED FILTERS

We have concentrated on the torsional telephone signaling filter (Figure 4.2) and the torsional telephone channel filter (Figure 4.15). In this section we will describe two radio filters and two very different signaling and channel filters.

4.4.1. Mechanical Filter for Radios

The first high production mechanical filters were built in 1952. These were the disc-wire type with magnetostrictive alloy-wire transducers. Since that time, the wire transducers have been replaced by both magnetostrictive ferrite [7] and piezoelectric ceramic transducers. Essentially, the compact structure of *stacked* discs coupled by small diameter wires has not changed. Figure 4.17 shows a modern ferrite-transducer version of the disc-wire filter and its frequency response. One reason this filter has maintained its popularity for many years is its high selectivity in a small package size of 13 cm^3.

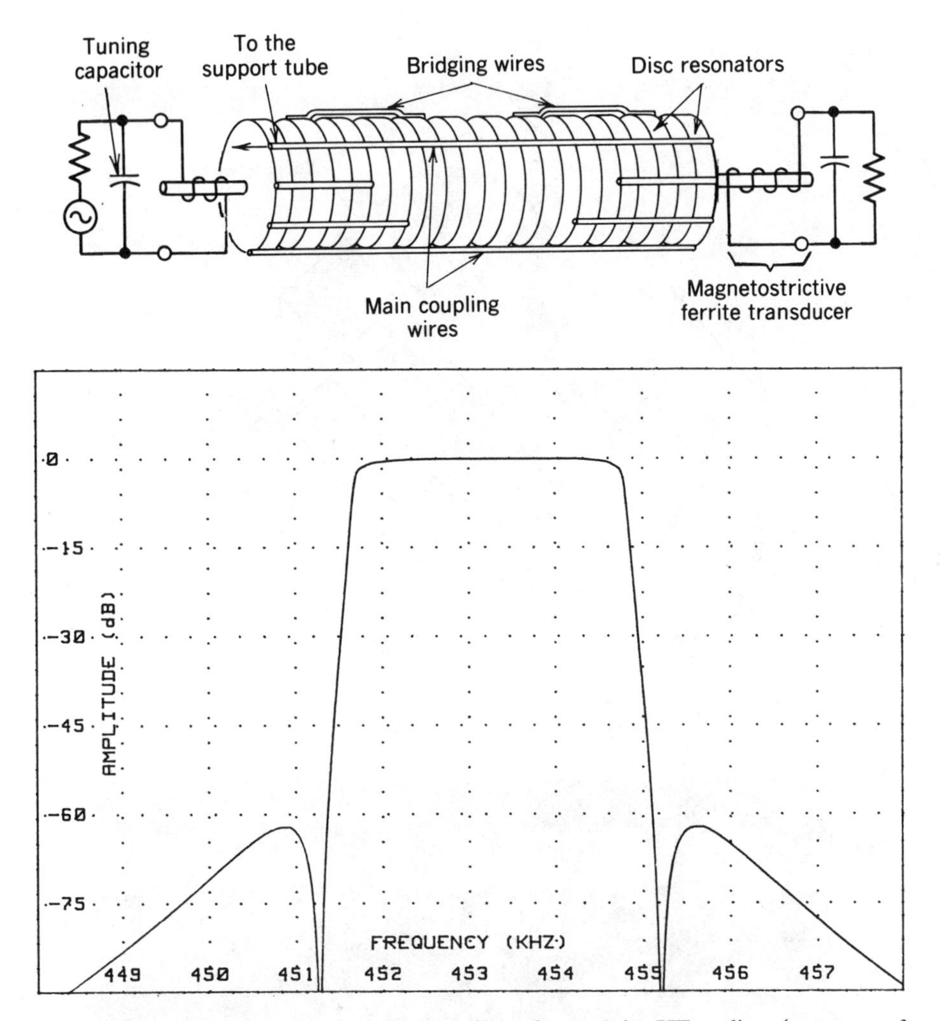

Figure 4.17 SSB disc-wire mechanical filter for use in HF radios (courtesy of Rockwell International, USA).

The disc resonators, which vibrate in a two-circle flexural mode, are driven by full-wavelength extensional-mode magnetostrictive transducers. This filter is an intermediate-bandwidth type having natural frequencies of the type shown in Figure 4.4(b). An NPO ceramic or mica capacitor is used to tune the transducer coil inductance. At 455 kHz, the bridging wires are greater that one-half wavelength long causing phase inversion and the realization of the attenuation poles shown in the stopband of the frequency response curve. These filters commonly maintain a 2 dB passband ripple specification and frequency shift of about 50 Hz over a temperature range of −40°C to +85°C.

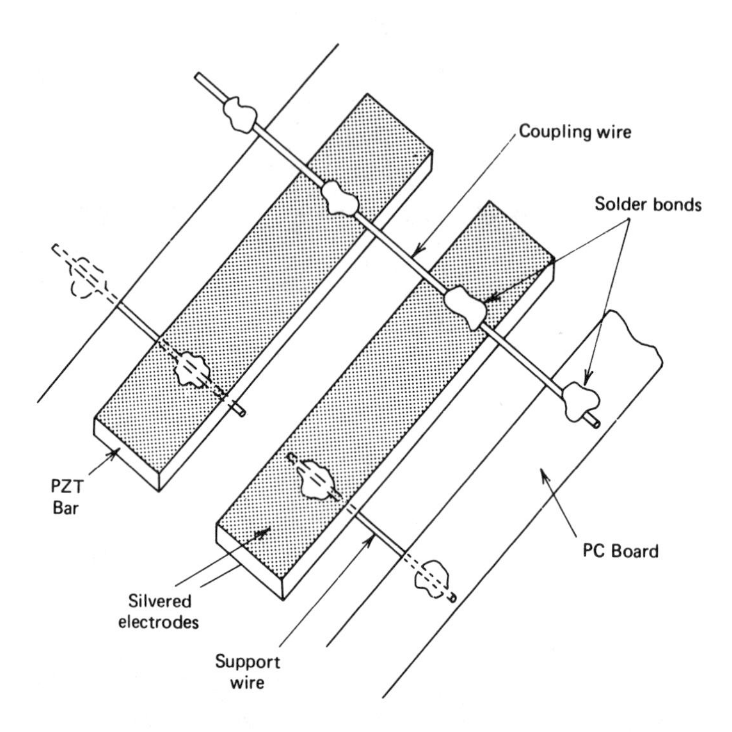

Figure 4.18 Mechanical filter for small-size, low-cost applications (courtesy of Toko, Japan).

Let us next look at a two-resonator, very small, very inexpensive 455 kHz AM mechanical filter designed for car and portable radios and CB transceivers. Figure 4.18 shows the construction of the filter. The piezoelectric ceramic bar resonators, which are silver plated on both top and bottom, vibrate in a half-wavelength mode and are supported by the two support wires and the extensions of the coupling wire. The filter bandwidth is determined by the position of the flexure-mode coupling wire with respect to the resonator nodal lines. The support wires act as the input and output leads, and the coupling wire acts as a common ground lead. The resulting filter has a volume of $0.21\ \mathrm{cm}^3$ and frequency versus temperature characteristics that are an order of magnitude more stable than that of transformer coupled LC filters.

4.4.2. Telephone Signaling and Channel Filters

Some telephone systems are designed with the signaling filter frequency just above the channel filter passband, that is, at 131.825 kHz. In others, such as the one we will describe in this section, the signaling filter is at 3825 Hz. Figure 4.19 shows a two-resonator signaling filter that makes use of the flexural-mode composite transducer resonators shown in Figure 4.6(c). Coupling between resonators is accomplished by means of wires welded to the flexural nodal lines on the top and bottom surfaces of the resonators. The coupling is therefore predominantly torsional with some flexural components due to the fact that the actual nodal point of flexure is through the center of the support bar. The support wires are welded to the bar at the

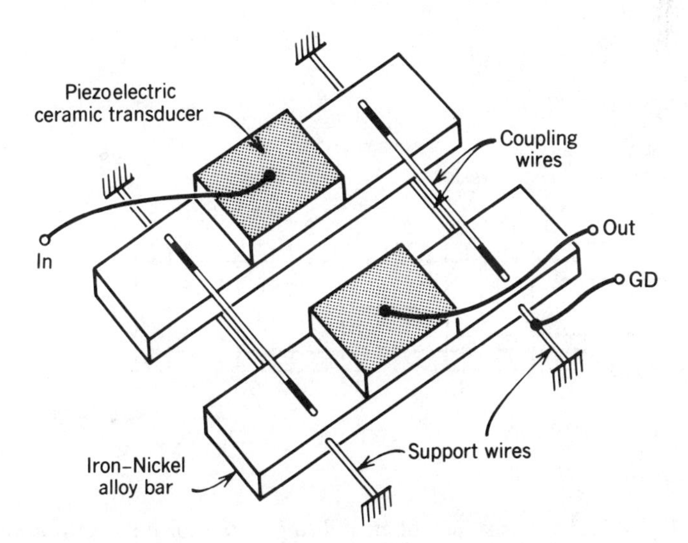

Figure 4.19 Low-frequency, flexure-mode mechanical filter (courtesy of Rockwell International, USA).

actual flexural node. This reduces the effect of the supports on the resonance frequency and Q_m of the resonators. The 3825 Hz signaling filter has a 50 Hz bandwidth and is designed with 32 mm length bar resonators. Any substantial reduction in bar length results in the bar thickness being too small to resist bending beyond the elastic limit of the material during the construction process. Flexural-mode filters have also been designed for train control and navigation receivers. They operate at frequencies between

(*a*)

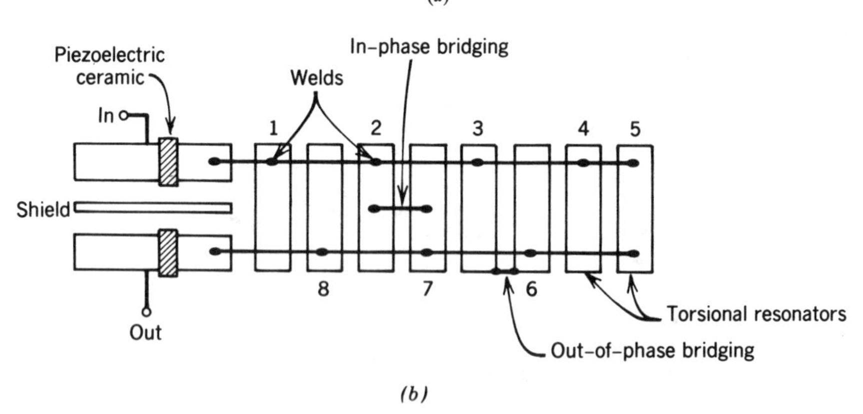

(*b*)

Figure 4.20 Telephone channel filter utilizing extensional-mode transducer resonators and torsional interior resonators in a folded configuration (courtesy of SEL, F.R.G.).

5 kHz and 25 kHz and have bar resonator lengths of less than 25 mm. The amplitude versus frequency response characteristics of these filters range from rounded passband gaussian shapes to equal-ripple response shapes depending on the user's requirements.

The final design we will look at is the telephone channel filter shown in Figure 4.20. This filter was chosen because of its simple, but clever construction.

The transducer resonators used are the high k_{em} Langevin extensional mode type shown in Figure 4.6(a). The high coupling eliminates the need for tuning coils, but the great length of the transducers could have created a filter size problem. The solution to this problem was to plane the transducer resonators side-by-side. To be able to do this, the scheme of Figure 4. 20(b) was devised. Note that the coupling wires are welded to alternate torsional rod resonators, the signal flow being from the input transducer to resonator 1, to resonator 2, and so on, back to resonator 8 and the output transducer. This also accomplished a single plane realization and low profile filter packaging.

An additional benefit of this design topology is that the bridging wires are coupled to adjacent resonators, namely resonators 3 and 6 and resonators 2 and 7. The out-of-phase bridging required across two resonators (4 and 5) is accomplished by welding a wire between adjacent points on the end surfaces of the resonators 3 and 6. To produce two additional attenuation poles, low-coupling in-phase bridging is required between resonators 2 and 7. This is achieved by welding a wire between the two upper surfaces of the adjacent resonators at a point slightly away from the nodal point.

4.4.3. Further Size Reductions

Regarding further size reductions, the only basic limitation is one of the non-linearity of the resonator and transducer materials as a function of the signal level. Use of the filter at lower power levels and the development of new materials with more linear properties will ease this constraint.

REFERENCES

1 S. Fujishima, H. Nonaka, T. Nakamura and H. Nishiyama (1977) Tuning fork resonators for electronic wrist watches using ZnO sputtered film. *1st Meeting on Applications of Ferro-Electric Materials*, Japan.

2 A. Günther, H. Albsmeier and K. Traub (1979) Mechanical channel filters meeting CCITT specification. *Proc. I.E.E.E.* **67**, 102–108.

3 ITT Staff (1974) *Reference Data for Radio Engineers*. Howard Sams, Indianapolis, IN.

4 R. Johnson (1977) Mechanical filters take on selective jobs. *Electronics* **Oct 13**, 81–85.

5 R. Johnson (1975) The design of mechanical filters with bridged resonators. *Proc. I.E.E.E. Int. Symp. on Circuits and Systems*, Boston, pp. 313–316.

6 R. Johnson (1978) Miniaturized mechanical filters. *Proc. I.E.E.E. Int. Symp. on Circuits and Systems*, New York, pp. 330–335.

7 R. Johnson (1983) *Mechanical Filters in Electronics*. Wiley, New York.

8 T. Kawana and H. Kawahata (1979) Preshift mechanical filter for voice frequency telegraph transmission system. *Proc. Ultrason. Symp.*, New Orleans, pp. 119–122.

9 M. Konno and H. Nakamura (1965) Equivalent electrical network for the transversely vibrating uniform bar. *J. Acoust. Soc. Am.* **38**, 614–622.

10 G. Matthaei, L. Young and E. M. T. Jones (1964) *Microwave Filters, Impedance Matching Networks, and Coupling Structures*. McGraw-Hill, New York.

11 D. Sheahan and R. Johnson (1977) *Modern Crystal and Mechanical Filters*. I.E.E.E. Press, New York.

12 G. Szentirmai (1978) Interactive filter design by computer. *Circuits Systems* **12**, 1–13.

13 G. Temes and S. Mitra (1973) *Modern Filter Theory and Design*, Chapter 5. Wiley, New York.

14 K. Yakuwa, T. Kojima, S. Okuda, K. Shirai and Y. Kasai (1979) A 128 kHz mechanical channel filter with finite frequency attenuation poles. *Proc. I.E.E.E.* **67**, 115–119.

5 Switched-Capacitor Filters: Circuits and Applications

ROUBIK GREGORIAN
Sierra Semiconductor, San Jose, California
GABOR C. TEMES
University of California, Los Angeles, California

In this chapter, the basic operation, design and analysis of SC filters will be briefly discussed. The application of these circuits in commercially available integrated devices will also be described, and their potentials as well as limitations analyzed. Finally, a survey of commercially available SC filters will be carried out. The contents of the chapter rely heavily on our recent book [12] and survey paper [11].

5.1. WHY ARE SWITCHED-CAPACITOR FILTERS USED?

Analog filters have traditionally been among the most complex and most accurate components of communication, control, and radar systems. They play a key role in separating signals from noise and from other signals, in restoring distorted waveforms, in providing precise timing, and so on. Usually, the simplest (and, in terms of low sensitivity and noise, often the best) realization of such circuits is in the form of a passive RLC filter, as discussed in Chapter 2. However, in some applications, such as telephone transmission systems, passive filters may occupy a large space, of the order of 10 inch^3 or more per filter, and, for accurate response, these circuits often require a lengthy tuning process.

Alternative filter realizations that allow more compact hardware forms, include mechanical filters (Chapter 4), crystal filters (Chapter 7) and, in suitable frequency ranges, SAW filters (Chapter 8). Active filters (cf. Chapter 3) can also provide, especially at low frequencies, a flexible and physically small realization of analog filtering functions. Such circuits require no inductors, only resistors, capacitors, and operational amplifiers (op amps). Hence, a hybrid arrangement, in which chip capacitors and integrated op amps are soldered on a substrate containing thin- or thick-film resistors, can be used.

However, the possibility of using fully integrated analog filters, preferably sharing the same MOS chip with the digital circuitry needed by the system, remained attractive if elusive. The difficulty was the realization of sufficiently large RC time constants on the small available chip area, with adequate accuracy, linearity, and stability. The sheet resistivity of diffused or polysilicon lines on a MOS chip is of the order of $50\,\Omega$/square. To form a resistor with a value of (say) $1\,M\Omega$ from such a line requires a chip area of around $10^6\,\mu m^2$, which is extravagant. Yet such resistors may be required to form time constants of the order of 0.1 ms (a fairly typical requirement) even when combined with a very large on-chip capacitor. Furthermore, the accuracy of these time constants is relatively poor, since resistors and capacitors have different structures so that their values do not track. To achieve high accuracy, either a sophisticated and expensive trimming process is required, or the resistors must be realized by MOS transistors in conjunction with an on-chip control circuit that adjusts the channel resistances to their desired values [24].

An alternative solution, which gives rise to the circuits discussed in this chapter, is to realize each resistor as a combination of a capacitor and several (two to four) periodically operated switches. A circuit that can accomplish this task is shown in Figure 5.1. Consider the branch of Figure 5.1(a), and assume that switches S_1 and S_4 open and close simultaneously with each other, but in opposite phase with switches S_2 and S_3 (Figure

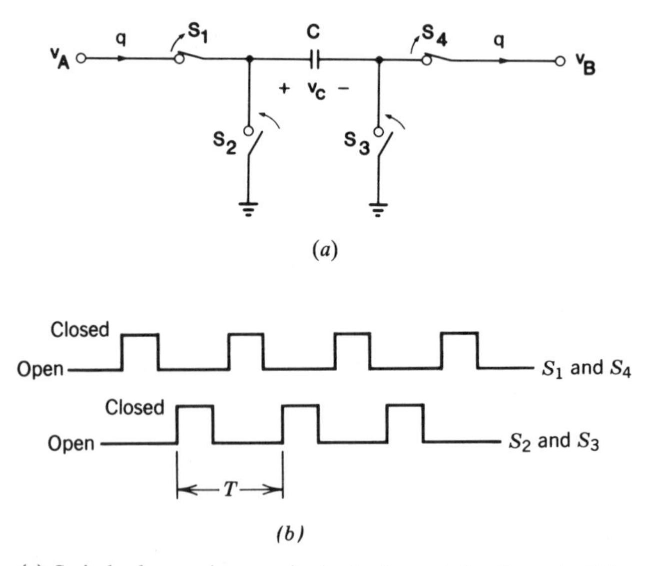

Figure 5.1 (a) Switched-capacitor equivalent of a resistive branch; (b) switch timing diagram.

5.1(b)). Assume also that the nodal voltages $v_A(t)$ and $v_B(t)$ vary slowly with time, so that they can be considered to have constant values during a switching period T (Figure 5.1 (b)). Then the operation is as follows: when S_2 and S_3 are closed, C is discharged; when afterwards S_1 and S_4 close, C charges to a voltage $v_A - v_B$. This causes a charge $q = C(v_A - v_B)$ to flow from the input to the output node of the branch. Since this action is repeated once in every T seconds, the **average** current through the branch is

$$i_{av} = (v_A - v_B)/(T/C) \tag{5.1}$$

This relation is similar to Ohm's law for a resistor of value $R = T/C$. In fact, if the branch current is used to charge a second capacitor C' and the voltage across C' is sampled every T seconds, then these samples will remain unchanged when the branch is replaced by a resistor R. Note, however, that the current waveform for the circuit of Figure 5.1(a) consists of sharp impulses occurring whenever S_1 and S_4 close, as opposed to a continuous current flow as is the case for a real resistor.

It is easy to show that Eq. (5.1) also holds for the circuits of Figure 5.2, which are therefore also equivalent to a resistor of value $R = T/C$, provided that S_1 and S_2 operate alternatively.

An interesting variation in the operation of the circuit of Figure 5.1(a) can be obtained if switches S_1 and S_3 operate together, and S_2 and S_4 in the opposite phase. Then, when S_1 and S_3 are closed, C charges up to $v_C = v_A$; when S_2 and S_4 are closed, C recharges to $v_C = -v_B$. Thus, the charge transferred to the output node is $C(-v_A - v_B)$, and the average current is now $(C/T)(-v_A - v_B)$. Thus, this circuit is equivalent to the cascade of an inverter that generates $-v_A$ and of a resistor with resistance $R = T/C$ (Figure 5.3).

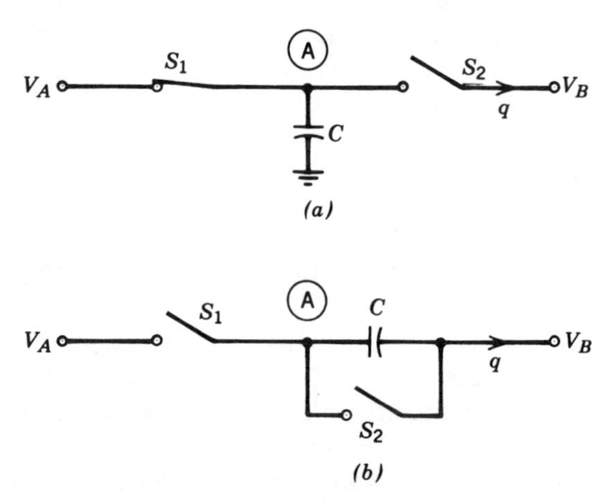

Figure 5.2 Simulated resistors: (a) shunt circuit; (b) series circuit.

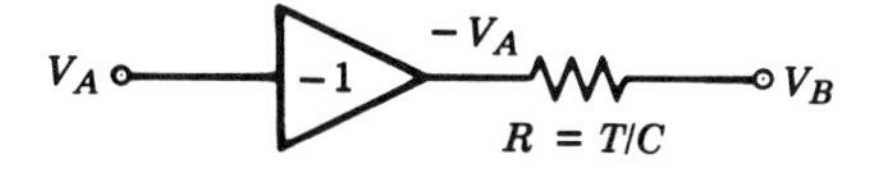

Figure 5.3 Equivalent circuit for the stage of Figure 5.1(a) with modified switch timing.

In most cases, the output node of the circuits of Figures 5.1 and 5.2 is a virtual ground, so that $v_B(t) \equiv 0$. Then, the average current of the branch with the modified switching scheme just described is $-(C/T)v_A$. Hence, in such cases, the branch can be regarded as the equivalent of a **negative resistor**.

The RC time constant of a SC resistor combined with an unswitched capacitor C' is clearly $(T/C)C' = (C'/C)/f_c$, where f_c is the **switching frequency** (more often called the **clock frequency**) of the circuit. Since the clock frequency is usually regulated by a highly accurate quartz crystal, its accuracy and stability can be very high. The other factor regulating the RC time constant is C'/C, the **ratio** of two on-chip capacitances. With appropriate layout techniques, the **matching** accuracy of such capacitors can be within 0.1%, even though the absolute values may be a 100 times less accurate for both. Thus, the accuracy problem is effectively solved by the use of the SC resistor branch.

For typical resistor values (say, in the 1–10 MΩ range), the chip area requirement is also reduced. For a convenient clock frequency (around 100 kHz) the required value for the switched capacitor is $C = 1/(f_c R) = 1 \sim 10$ pF. Since the realization of 1 pF requires a chip area of only around 2000 μm^2, while the area needed by the switches is negligible, the area of the resistor is reduced by a factor of nearly 500 as compared to a direct realization of a 1 MΩ resistor that may need 10^6 μm^2 chip area.

As will be shown in the next two sections, the realization of a filter pole requires one or two SC branches, one op amp and one unswitched feedback capacitor. The area required by all these components is typically 2×10^5 μm^2, while the overall chip may be as large as 5×10^7 μm^2. Thus, if necessary, several hundred filter poles can be realized on a single chip. Since SC circuits are by nature sampled-data systems, many of their components (SC branches, op amps, other memoryless elements) can be multiplexed, that is, time shared, thus extending their capabilities even further.

5.2. COMPONENTS FOR SWITCHED-CAPACITOR CIRCUITS

In this section, the basic building blocks of integrated SC circuits will be described. It will be assumed that the fabrication technology is MOS, although recently the gallium arsenide (GaAs) process has also been used to realize high-speed SC filters. The three most common components of MOS SC filters are **capacitors**, **switches**, and **operational amplifiers**. The most important properties of these are briefly discussed.

5.2.1. MOS Capacitors

Various structures for realizing capacitors in MOS technology are shown in Figure 5.4. The first one (Figure 5.4(a)) uses a heavily doped diffusion region as one of the capacitor plates, and either a metal (usually aluminum) or heavily doped polysilicon layer as the other plate. The dielectric is thermally grown SiO_2. This structure can provide a reasonably well-controlled capacitance, since its bottom plate has a crystalline structure and is smooth, and hence the random variations in the oxide (SiO_2) thickness are relatively small and there are few structural defects. However, if the voltage $V_2 - V_1$ is of such polarity that it depletes the diffusion forming the bottom plate ($V_2 - V_1 < 0$, for an n^+ diffusion), a depletion layer is generated at the surface of the diffusion region that alters the capacitance slightly. The resulting capacitance nonlinearity can be reduced to negligible values by using heavier doping ($N_D > 5 \times 10^{20}\ \text{cm}^{-3}$) and larger oxide thickness.

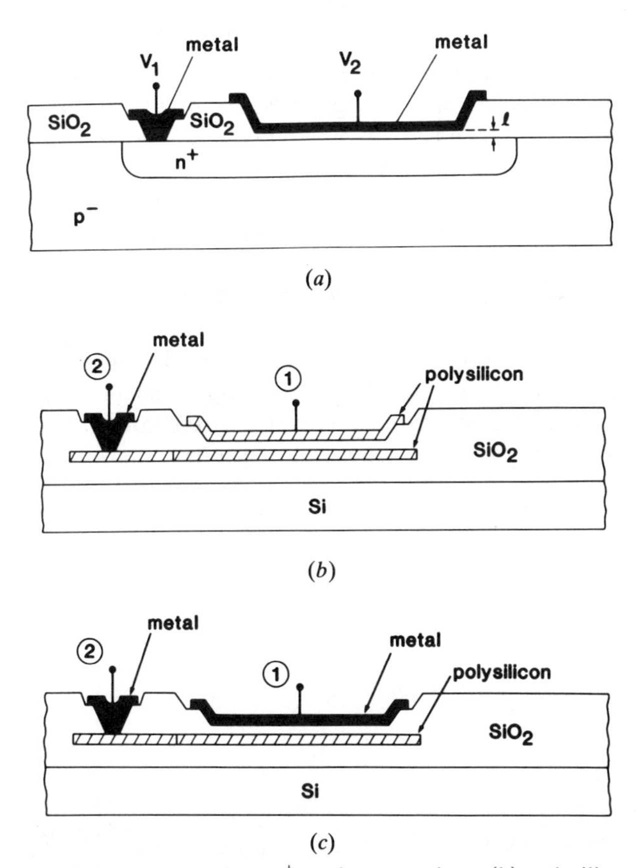

Figure 5.4 (a) Thin-oxide metal-to-n^+-region capacitor; (b) polysilicon-oxide-polysilicon capacitor structure; (c) metal-oxide-polysilicon capacitor structure.

Since the oxide layer used in this structure is the gate (thin) oxide, the capacitance/area ratio can be large. There is a fairly large (10–80 Ω) distributed series resistance associated with the capacitor, due mostly to the resistivity of its bottom plate. In addition, the stray capacitance between the diffusion layer and the substrate is large (up to 10% of the main capacitance) and very nonlinear. It can couple substrate noise into the signal path, unless this is prevented by proper circuit design.

The second type of structure, shown in Figures 5.4(b), and (c), uses metal and/or polysilicon layers for both plates, with a thick oxide layer in between acting as the dielectric. These capacitors are highly linear, and have low (<10 Ω) series resistance. However, due to the thicker oxide used, they require a larger area, and the capacitance value has larger random variations due to the granularity of both plate-to-dielectric surfaces. A large bottom-plate-to-substrate stray capacitances also exists for this structure. However, the noise coupling introduced by this stray capacitance can be now prevented by introducing a grounded diffusion layer under the bottom capacitance plate into the substrate. This added layer acts as a shield, and reduces the noise injection.

The capacitance of a practical MOS capacitor is subject to **random errors**, introduced by surface granularity, jagged edges, quantization errors, and so on. The effects of these can usually only be reduced by increasing the size of the device. Usually, for a capacitance of 0.5 pF or more, the random errors are negligible [22].

In addition, the capacitance value is affected by **systematic errors**. These include undercutting due to poorly controlled lateral etching, and fringing field effects. These effects are associated with the edges of the device; hence, they do not influence the ratio of two adjacent on-chip capacitors, as long as their perimeter/area ratios are the same. This can be achieved by building the capacitors as the parallel connections of smaller equal-sized **unit capacitors**. By arranging these in a **common-centroid** configuration [19], slow variations of oxide thickness or lateral dimensions can also be compensated for (Figure 5.5).

Taking proper precautions, the ratio of two capacitors C_1 and C_2 with a nominal range $C_1/C_2 = 0.1$–10 can be made as accurate as 0.1%. This is satisfactory for most applications.

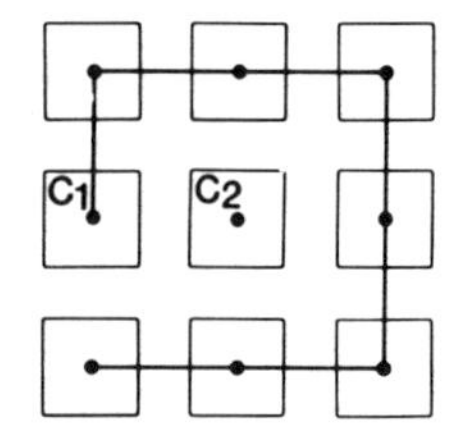

Figure 5.5 Typical common-centroid layout technique for matched capacitors.

5.2.2. MOS Switches

The simplest realization of a switch in MOS technology is a single MOS transistor (Figure 5.6). Its drain and source are connected to the two nodes to be periodically shorted, and the switch control (clock) signal is connected to the gate. When the clock signal is high, a channel is formed with an on-resistance $R_{on} = 1/[2k(|V_{GS}| - |V_T|)]$. Here, $k = (\mu C_{ox}/2)(W/L)$ where μ is the carrier mobility in the channel, C_{ox} the gate-to-channel capacitance per unit area, and W and L are the width and length, respectively, of the channel. V_{GS} is the gate-to-source voltage, while V_T is the threshold voltage. Unless the device is in a well so that its source can be connected to its substrate, there is usually a signal-dependent reverse bias voltage V_{BS} between the source and the substrate. This voltage tends to reduce the channel charge, and hence increase R_{on}. This so-called body effect is usually modeled by considering $|V_T|$ to be an increasing function of $|V_{BS}|$.

It follows from the above considerations that R_{on} depends on the voltage level of the nodes to which the switch is connected. If this level v_s is so high that $|V_{GS}| \sim |V_T|$, then the switch may not have a low enough R_{on} value to equalize its terminal voltages within the available time period. This situation can be remedied by using a CMOS transmission gate (Figure 5.6(c)) instead of a single device as the switch. Here, $\bar{\phi}$ is the complement of ϕ; that is, $\bar{\phi}$ is high when ϕ is low, and vice versa.

A major problem often associated with switches in SC circuits is the effect called **clock feedthrough**. When the clock signal on the gate of a MOS transistor goes high, a channel is established. For example, in an NMOS device this requires negative charges (electrons) to enter the device from the outside circuit connected to the switch. In addition, the gate-to-source and gate-to-drain overlap capacitances must also be charged; this needs addition-

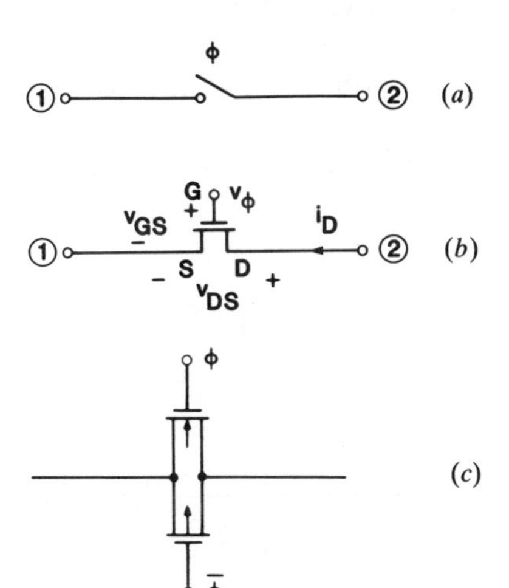

Figure 5.6 MOS switches: (a) symbol; (b) single-MOSFET realization; (c) CMOS transmission gate realization.

al electrons. For a minimum-size (say, $3 \times 3\ \mu m^2$) switch, the total charge entering may be around $q_{sw} = -5\ \text{fC} = -5 \times 10^{-15}\text{C}$. When the switch cuts off, the charging process is reversed, and the device loses its stored charges. Assuming for example that all charges are drawn from and returned to a storage capacitance $C = 0.5$ pF, the resulting voltage steps occurring across C will have an amplitude $|q_{sw}|/C = 10$ mV. This clock-feedthrough noise usually introduces a dc offset error, as well as noise components around f_c, $2f_c, \ldots,$ where f_c is the clock frequency.

It can readily be shown [23] that the charge q_{sw} stored in the channel of the switch transistor and its on-resistance R_{on} are related by the equation

$$q_{sw} R_{on} = \frac{L^2}{\mu} \tag{5.2}$$

where L is the channel length, and μ the carrier mobility in the channel. Hence, it is advantageous to select L as small as the design rules permit. However, having done so, a lower R_{on} can only be obtained at the expense of a larger q_{sw}, and hence more clock feedthrough noise. Since a high clock rate f_c generally requires a low R_{on}, achieving low clock feedthrough noise is a particularly difficult problem for high-frequency SC filters.

For CMOS transmission gates (Figure 5.6(c)), there is an inherent first-order cancellation of the clock feedthrough noise, because charges of opposite polarity enter and leave the two complementary devices at the same time. To improve this cancellation, the q_{sw} values should be the same (for equal C_{ox} values, this requires equal channel areas), and the clock signals ϕ and $\bar{\phi}$ should be accurate complements of each other.

For a single-channel switch, a similar compensation can be achieved by using a **dummy switch** (Figure 5.7). Assuming as a rough estimate that the charge q_{sw} of Q_1 is equally divided between the source and drain terminals during charging and discharging, in order to cancel the clock feedthrough charge at node 2 the dummy switch Q_2 should contribute a charge $-q_{sw}/2$. This can be achieved, for example, if the length of Q_2 is chosen to equal that of Q_1, but its width is only half that of Q_1. Computer simulations may be performed for a given circuit to obtain optimum dimensions for Q_2, taking into account the outside circuitry to which the switch Q_1 is connected. This circuit, as well as the clock waveforms, affect the distribution of the charges stored in Q_1 when it cuts off, and hence the optimum design of Q_2.

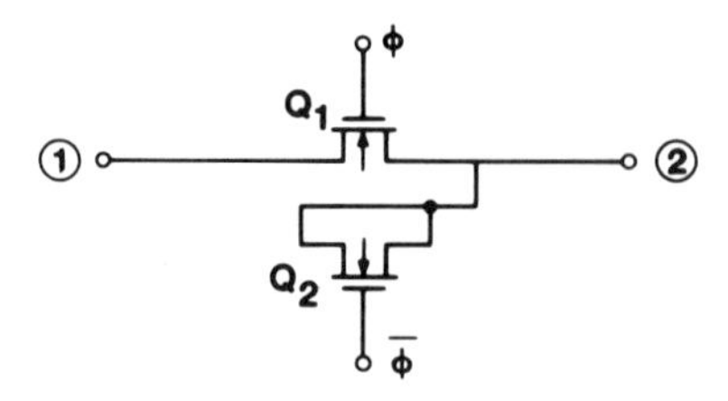

Figure 5.7 Clock feedthrough compensation scheme.

5.2.3. MOS Operational Amplifiers

An ideal operational amplifier is a voltage-controlled voltage source with an output/input relation

$$v_{\text{out}} = A(v_a - v_b) \tag{5.3}$$

(Figure 5.8). Here, v_a is the voltage of the noninverting input terminal and v_b the voltage of the inverting one. The **voltage gain** A is ideally frequency-independent, and infinitely large. The performance of a practical op amp fabricated in state-of-the-art CMOS technology (as of 1987) can be characterized by the following parameters.

(i) *DC Gain*. This is the value of the voltage gain A for a constant input. Typical values are 10^3–2×10^4.

(ii) *Linear Output Voltage Range*. This depends on the dc voltages used to bias the op amp. For, say, ± 5 V supply voltages, the linear range may be $-4\text{ V} < v_{\text{out}} < 4\text{ V}$.

(iii) *Offset Voltage*. As eq. (5.3) shows, for an ideal op amp, if $v_a = v_b$ then $v_{\text{out}} = 0$. This is not true for a practical amplifier, and $v_{\text{out}} = 0$ is achieved for an input voltage $v_{\text{in,off}} = v_a - v_b \neq 0$. Usually, the magnitude of $v_{\text{in,off}}$ (called the **input-referred offset voltage**) is in the 1–15 mV range.

(iv) *Common-Mode Rejection Ratio*. Ideally, the output voltage v_{out} depends only on the **differential** input voltage $v_a - v_b$. In fact, v_{out} is also affected by the average (or **common-mode**) voltage $(v_a + v_b)/2$. The ratio of the op amp gains to the differential and common-mode voltages, usually expressed in decibels, is the common-mode rejection ratio (CMRR). Typically, CMRR = 60–80 dB can be expected. The range of the common-mode input

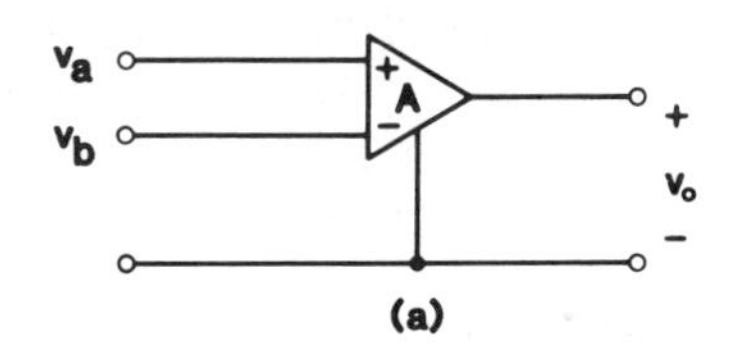

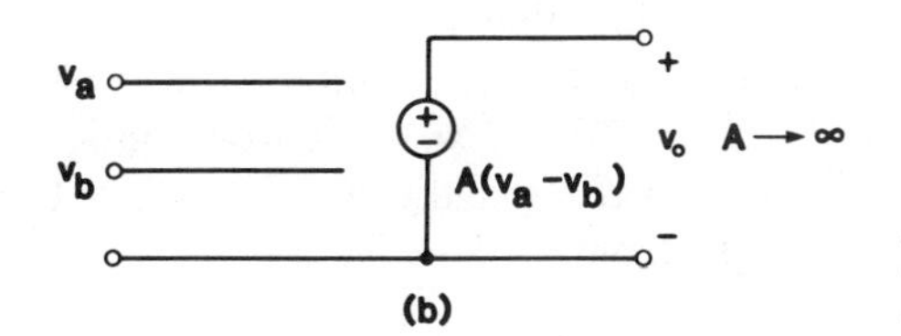

Figure 5.8 (a) Symbol for ideal op amp; (b) equivalent circuit.

voltage in which the CMRR remains high is called the **common-mode range** (CMR). It is typically ±2–3 V.

(v) *Power-Supply Rejection Ratio*. A noise voltage superimposed on the positive or negative supply voltage will inevitably appear in v_{out}. The ratio of the gains to the differential input signal voltage and to the power-supply noise is called **power-supply rejection ratio** (PSRR). Typical values of the PSRR are in the 60–90 dB range at low frequencies.

(vi) *Unity-Gain Bandwidth*. The frequency f_0 at which the gain (which usually decreases in the frequency range of interest with a slope of −6 dB/octave) reaches the value 1, or 0 dB, is called the **unity-gain bandwidth** of the op amp. For simple op amps, f_0 may be in the 1–5 MHz range. For sophisticated folded-cascade or current-mirror type circuits, f_0 can range up to 100 MHz.

(vii) *Slew Rate*. For a large input step, v_{out} cannot follow $v_a - v_b$ immediately. Its maximum rate of change dv_{out}/dt is called the **slew rate** (SR) of the op amp. It is related to the nonlinear performance of the devices, and is **not** determined by f_0. For simple op amps, SR = 1–10 V/μs can be achieved; and by using added slew-enhancement circuitry, the value of the SR can be increased by one or even two orders of magnitude.

(viii) *Noise and Dynamic Range*. The noise generated in the transistors of the op amp (especially in the input devices) appears in an amplified form at v_{out}. The output noise is especially high at low frequencies, where the $1/f$ (flicker) noise dominates. For a wide-band CMOS op amp, the input-referred noise is of the order of 10–100 μV. The ratio of the maximum undistorted output voltage (determined by the linear voltage range) and the minimum discernible output signal voltage (determined by the total output noise) is the **dynamic range** of the op amp. Under open-loop conditions, the dynamic range of a CMOS op amp is only about 30–50 dB. However, negative feedback can reduce the output noise without affecting the maximum output signal, thus enhancing the dynamic range. This makes it possible to achieve dynamic range values up to 100 dB for feedback systems containing MOS op amps (such as SC filters).

(ix) *Settling Time*. This is the time required for v_{out} to approximate its final value to within a specified error (usually 0.1% or 1%) when the input voltage changes. It is assumed that this change is by a small amount so that the amplifier remains in its linear range of operation. The settling time is related to f_0, to the output impedance and the load of the op amp, to the magnitude of the input step, and so on. Typically, it ranges from 0.05 to 5 μs.

(x) *DC Power Dissipation*. This is the total average power provided by the dc power supplies. Generally, the faster the op amp, the more dc power it requires. Internal amplifiers with capacitive loads need only a relatively small dc current (0.1–1 mA) and hence little (0.5–5 mW) dc power. Amplifiers that have to drive resistive and/or very large capacitive loads, that is, they function as output buffers, may need much more dc power.

(xi) *Chip Area*. The area occupied on the chip by the op amp. For a simple interstage amplifier, this may be $4 \times 10^4\ \mu m^2$; for a highly sophisticated fast circuit, $10^5\ \mu m^2$ or more.

5.3. SWITCHED-CAPACITOR INTEGRATORS AND BIQUADS

In this section, the simplest SC stages, integrators and biquads, will be discussed. These stages are easy to design, and they can be used to construct more complicated filter circuits.

5.3.1. Switched-Capacitor Integrators

Using the simulated resistors of Figure 5.1, an active RC integrator (shown in Figure 5.9(a)) can be transformed into the SC integrator of Figure 5.9(b). The timing of the switches is illustrated in Figure 5.9(c), and the waveform of v_{out} in Figure 5.9(d). The input-output relation for the circuit of Figure 5.9(a) is

$$v_{out}(t) = -\frac{1}{R_1 C_2} \int_{-\infty}^{t} v_{in}(\tau)\, d\tau \tag{5.4}$$

in the time-domain. This leads to the transfer function

$$H_a(j\omega) = \frac{V_{out}(j\omega)}{V_{in}(j\omega)} = \frac{-1/(R_1 C_2)}{j\omega} \tag{5.5}$$

in the frequency-domain. For the SC integrator of Figure 5.9(b) (with the clock phases chosen as shown **without parentheses**), a difference equation can be constructed in the time-domain. The basis of the equation is the charge conservation law (CCL) applied to node © in the form

$$\left.\begin{matrix}\text{charge entering via } C_1\\ \text{per clock period } T\end{matrix}\right\} = \left\{\begin{matrix}\text{charge leaving via } C_2\\ \text{per clock period } T\end{matrix}\right.$$

or, in mathematical form

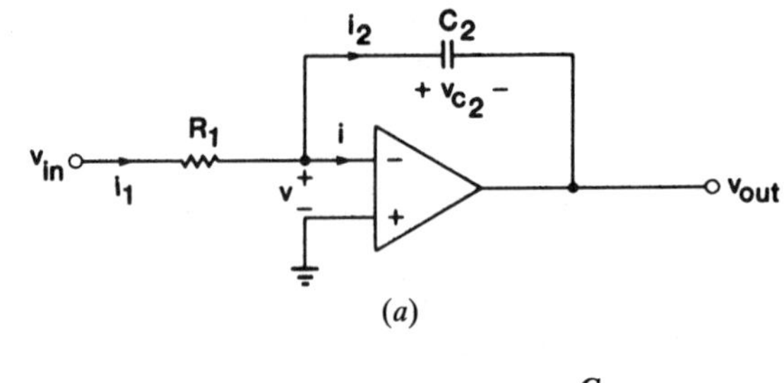

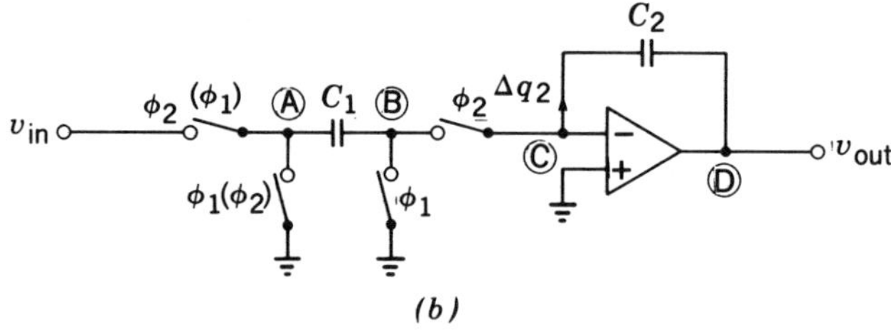

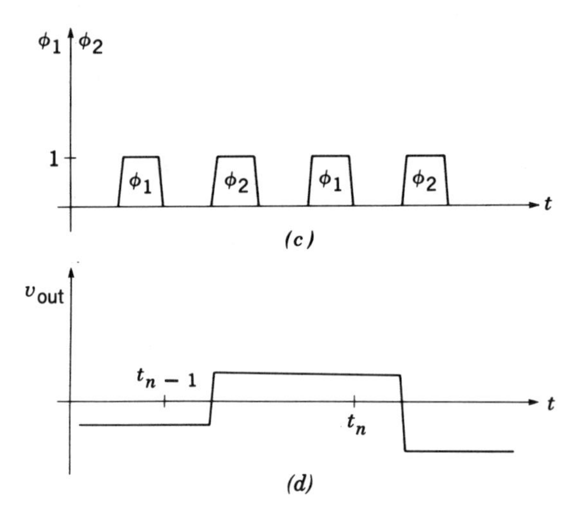

Figure 5.9 (a) active RC integrator; (b) stray-insensitive switched-capacitor integrator circuit; (c) clock signals; (d) output voltage waveform.

$$
\begin{aligned}
C_1 v_{\text{in}}(nT) &= C_2\{[0 - v_{\text{out}}(nT)] - [0 - v_{\text{out}}(nT - T)]\} \\
&= C_2\{v_{\text{out}}(nT - T) - v_{\text{out}}(nT)\}
\end{aligned} \tag{5.6}
$$

where $nT = t_n$ and $(nT - T) = t_{n-1}$ are illustrated in Figure 5.9(d). Using z-transformation, Eq. (5.6) gives the z-domain transfer function

$$
H(z) = \frac{V_{\text{out}}(z)}{V_{\text{in}}(z)} = \frac{-C_1/C_2}{1 - z^{-1}} \tag{5.7}
$$

The frequency response is obtained by setting $z^{-1} = \exp(-j\omega T)$. For the usual case when the signal frequency $f = \omega/2\pi$ is much lower than $f_c = 1/T$, $\omega T = 2\pi f/f_c \ll 1$ holds and hence the approximation $z^{-1} \cong 1 - j\omega T - \omega^2 T^2/2$ may be used. Substituting these values into $H(z)$, Eq. (5.7) gives

$$H(\omega) = \frac{-C_1/C_2}{1 - e^{-j\omega T}} \cong \frac{-C_1/(C_2 T)}{j\omega + \omega^2 T/2} \tag{5.8}$$

A comparison with Eq. (5.5) shows that if $R_1 = T/C_1$ and $\omega T/2 \ll 1$, then the two frequency responses will be close to each other. The error term $\omega^2 T/2$ can be represented by a resistor with the frequency-dependent value $2/(\omega^2 T C_2)$ connected in parallel with C_2 in Figure 5.9(a). It causes an error $\tan^{-1}(\omega T/2) \cong \omega T/2$ in the phase shift (nominally equal to 90°) of the integrator.

If the clock phases shown **in parentheses** are used in the SC integrator of Figure 5.9(a), then a derivation similar to that giving Eq. (5.7) yields the transfer function given below

$$H(z) = (C_1/C_2)\,\frac{z^{-1}}{1 - z^{-1}} \tag{5.9}$$

This leads to the frequency response

$$H(\omega) = \frac{C_1/C_2}{e^{j\omega T} - 1} \cong \frac{C_1/(C_2 T)}{j\omega - \omega^2 T/2} \tag{5.10}$$

The error term is now $-\omega^2 T/2$, and it can be represented by a resistor with a negative frequency-dependent resistance $-2/(\omega^2 T C_2)$ connected across C_2. It causes a phase error $-\omega T/2$, equal in magnitude to that occurring in Eq. (5.8) but of opposite sign. Thus, a cascade connection of two integrators of the form shown in Figure 5.9(b), with different phasings for the two stages behaves (to a very good approximation) as a cascade of two ideal active RC integrators. This fact is utilized in both cascade and ladder SC filter design, as will be shown later in this chapter.

One major advantage of the integrators shown in Figure 5.9(b) is that they are insensitive to the stray capacitances that unavoidably load every node in an MOS integrated circuit. For example, at node Ⓐ of the circuit of Figure 5.9(b), the stray capacitance includes the drain- (or source-) to-substrate capacitances of the MOS transistors realizing the two switches connected to the node; also, the capacitance between the left-side plate of C_1 and the substrate; and the capacitance between the leads connecting the switches to C_1 (and to each other) and the substrate. The total stray capacitance may be of the order of 0.05–0.1 pF. This capacitance is periodically charged to v_{in}, and then discharged to ground. Neither action affects the signal charge that flows to C_1 and C_2, and thus the value of v_{out}.

Similarly, the stray capacitance loading node Ⓑ is alternatively connected to true ground and to virtual ground. This involves no net charge-flow to the outside circuit, and hence it does not alter v_{out} either. The stray capacitance at node Ⓒ is connected between ideally equipotential points and hence is always uncharged. Finally, the stray capacitances at the input and output nodes merely increase the settling times of the op amps that charge them, but do not affect the signal charges in C_1 and C_2. Thus, the operation of these integrators is **stray insensitive**, as long as the op amp gain is sufficiently high.

Note that SC integrators can also be constructed from the circuits of Figure 5.2. However, for these circuits the stray capacitance loading node Ⓐ does influence the signal charge q leaving the branch. Hence, the resulting integrators are stray sensitive. This restricts their practical usefulness to less critical applications, where the inaccuracy and nonlinearity caused by the stray capacitances are acceptable, and the saving in the number of switches and lines due to the simpler circuitry is important.

It is also possible to design integrators that are insensitive to op amp offset and finite gain [13].

5.3.2. Switched-Capacitor Biquads

By cascading the two stray-insensitive integrators shown in Figure 5.9(b), and adding a few feedback and feedforward branches, it is possible to construct SC circuits that realize the biquadratic transfer function

$$H(z) = -\frac{a_2 z^2 + a_1 z + a_0}{b_2 z^2 + b_1 z + b_0} \tag{5.11}$$

This transfer function is usually obtained from an analog transfer function $H_a(s_a)$ using the bilinear s_a–z transformation

$$s_a = \frac{2}{T}\frac{z-1}{z+1} \tag{5.12}$$

[12]. For stability, the poles of $H(z)$ must be inside the unit circle (described by $|z| = 1$). Hence, assuming complex poles, the condition $|b_0/b_2| < 1$ must hold.

A circuit with a transfer function of the form given in Eq. (5.11) is shown in Figure 5.10(a); the clock and signal waveforms are illustrated in Figure 5.10(b). It is assumed that v_{in} is a sampled and held signal, which changes only when ϕ_2 rises. Writing the charge conservation relations for the inverting input nodes of the two op amps, and using z-transformation to solve them, we obtain the transfer function

$$H(z) = \frac{V_{out}(z)}{V_{in}(z)} = -\frac{(C_1' + C_1'')z^2 + (C_1 C_3 - C_1' - 2C_1'')z + C_1''}{(1 + C_4)z^2 + (C_2 C_3 - C_4 - 2)z + 1} \tag{5.13}$$

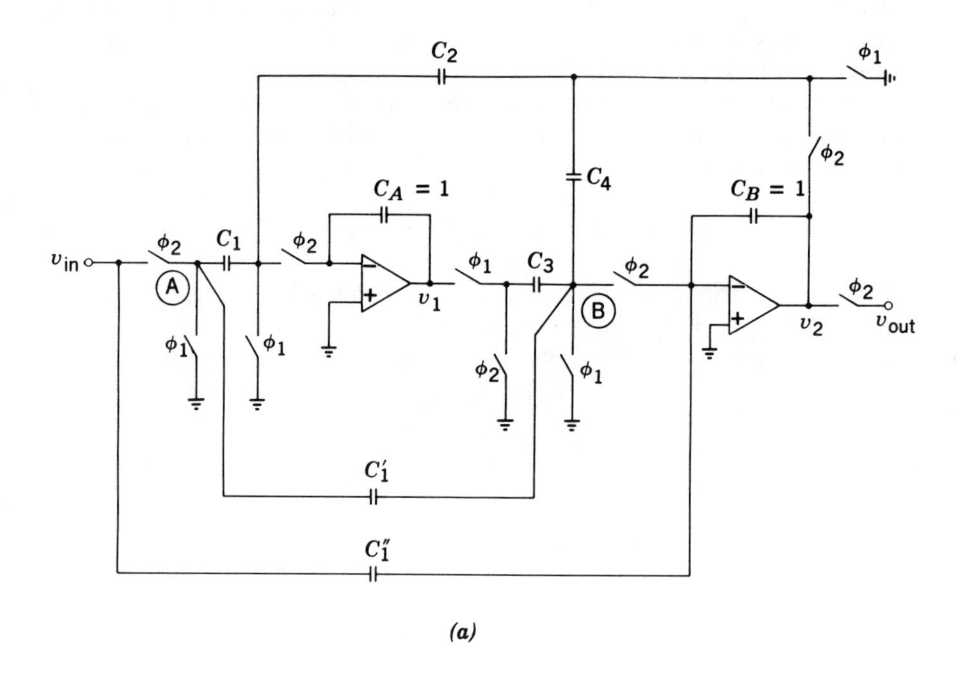

(a)

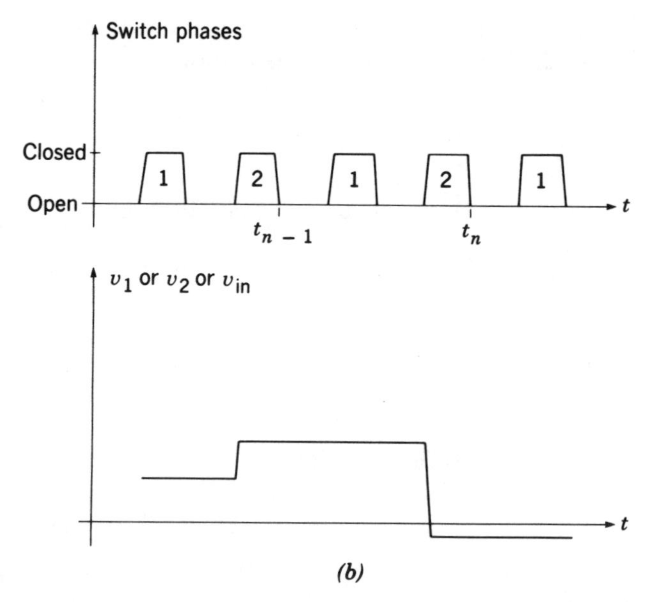

(b)

Figure 5.10 (a) Low Q switched-capacitor biquad; (b) clock and signal waveforms.

Clearly, here $|b_0/b_2| = 1/(1 + C_4) < 1$, so the circuit is stable for any capacitance values as long as $C_4 > 0$. (Note that all capacitance values are normalized to $C_A = C_B$!)

To obtain an estimate for the element values, we can use Eq. (5.12) to express the denominator polynomial of $H(z)$ in terms of the analog frequency variable s_a. The result is

$$D(s_a) = (4 + 2C_4 - C_2C_3)y^2 + C_4y + C_2C_3 \tag{5.14}$$

Here, the notation $y \triangleq s_aT/2$ was used. Let the poles s_{ap}, defined by the relation $D(s_{ap}) = 0$, be characterized by the *pole-Q*

$$Q \triangleq \frac{|s_{ap}|}{2|\text{Re}\, s_{ap}|} \tag{5.15}$$

and the *pole frequency*

$$\omega_0 = |s_{ap}| \tag{5.16}$$

It follows from Eq. (5.14) that

$$(\omega_0T)^2 \cong \frac{4C_2C_3}{4 + 2C_4 - C_2C_3} \tag{5.17}$$

Since for all practical transfer functions, the poles satisfy $(\omega_0T)^2 \ll 1$, we have $C_2C_3 \ll 1 + C_4/2$, thus to a good approximation

$$(\omega_0T)^2 \cong \frac{C_2C_3}{1 + C_4/2} \tag{5.18}$$

From Eqs. (5.14) and (5.18), the pole-Q can be obtained

$$Q = \omega_0T\,\frac{4 + 2C_4 - C_2C_3}{4C_4} \cong \frac{[C_2C_3(1 + C_4/2)]^{1/2}}{C_4} \tag{5.19}$$

Q increases with decreasing C_4; for $C_4 = 0$, Q is infinite. For usual values of Q, $C_4/2 \ll 1$ so that the approximations

$$(\omega_0T)^2 \cong C_2C_3 \ll 1 \tag{5.20}$$

$$Q \cong \frac{\omega_0T}{C_4} \tag{5.21}$$

can be used. The dc gain of the stage is

$$H(1) = H(z)|_{z=1} = \frac{C_1C_3}{C_2C_3} = \frac{C_1}{C_2} \tag{5.22}$$

Since (as Eq. (5.20) shows) C_2 and C_3 are normally among the smallest capacitors in the stage, and their product is specified, it is advantageous to choose $C_2 = C_3 \cong \omega_0 T$. Then $C_1 = \omega_0 T H(1)$ and $C_4 \cong \omega_0 T/Q$. Clearly, if $Q > 1$ then C_4 will be even smaller than C_2 and C_3, and the capacitance spread will be increased. Hence, the circuit of Figure 5.10 is more suitable for low Q ($Q < 1$) applications than for high Q ones. Since $C_A = C_B = 1$, the capacitance spread is $C_{max}/C_{min} \cong C_A/C_2 \cong 1/(\omega_0 T)$.

The exact element values can be found by equating the coefficients of $H(z)$ in Eq. (5.13) with those in Eq. (5.11) where $b_0 = 1$ is assumed. The results are

$$\left.\begin{aligned} C_1'' &= a_0 \\ C_1' &= a_2 - a_0 \\ C_4 &= b_2 - 1 \\ C_2 C_3 &= 1 + b_1 + b_2 \\ C_1 &= (a_0 + a_1 + a_2)/C_3 \end{aligned}\right\} \tag{5.23}$$

As explained above, normally

$$C_2 = C_3 = (1 + b_1 + b_2)^{1/2} \tag{5.24}$$

is an optimal choice. Using Eqs. (5.23) and (5.24), all capacitances values can be obtained one-by-one from the a_i, b_j values.

Usually, not all of the input capacitors (C_1, C_1', C_1'') are needed. If, for example, the transmission zeroes of the biquad are on the unit circle, then $C_1' = 0$ and hence it can be omitted. If both zeroes are at $z = 1$ (i.e., at dc) then C_1 is also unnecessary. Transmission zeroes outside the unit circle can be obtained by changing the phases of the clock signals for the switched input capacitor C_1' [12].

As was already pointed out, the circuit of Figure 5.10(a) is better suited for low pole-Q ($Q < 1$) applications. For high Q values, the biquad shown in Figure 5.11(a) can be used. An analysis simular to that performed for the circuit of Figure 5.10(a) shows that for this biquad the approximating relations

$$(\omega_0 T)^2 \cong C_2 C_3 \ll 1 \tag{5.25}$$

$$Q \cong \frac{1}{C_4} \tag{5.26}$$

$$H(z)|_{z=1} = \frac{C_1}{C_2} \tag{5.27}$$

hold. Choosing again $C_2 = C_3 \cong \omega_0 T$, $C_1 \cong \omega_0 T H(1)$ and $C_4 \cong 1/Q$ result.

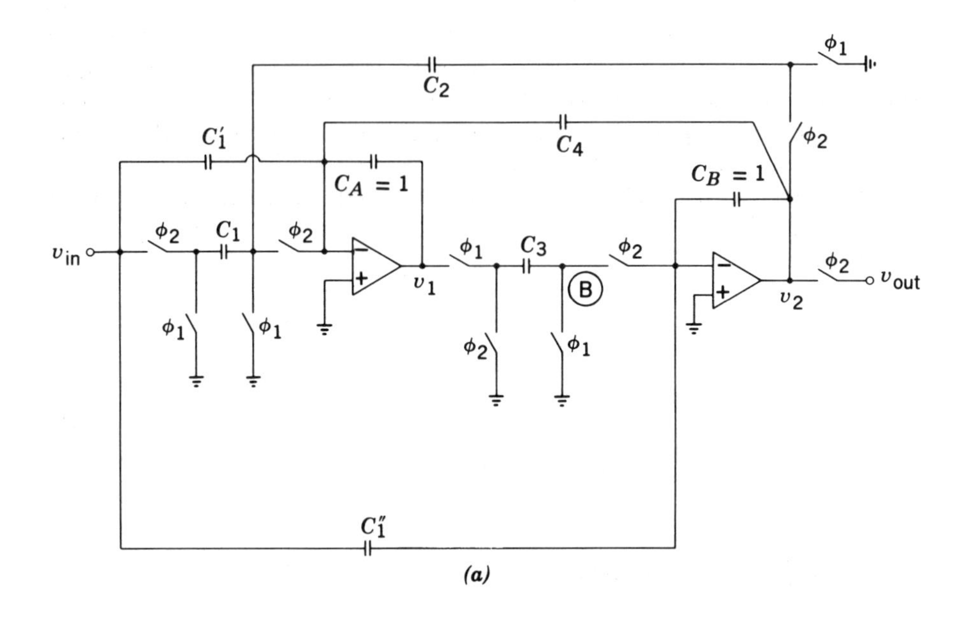

(a)

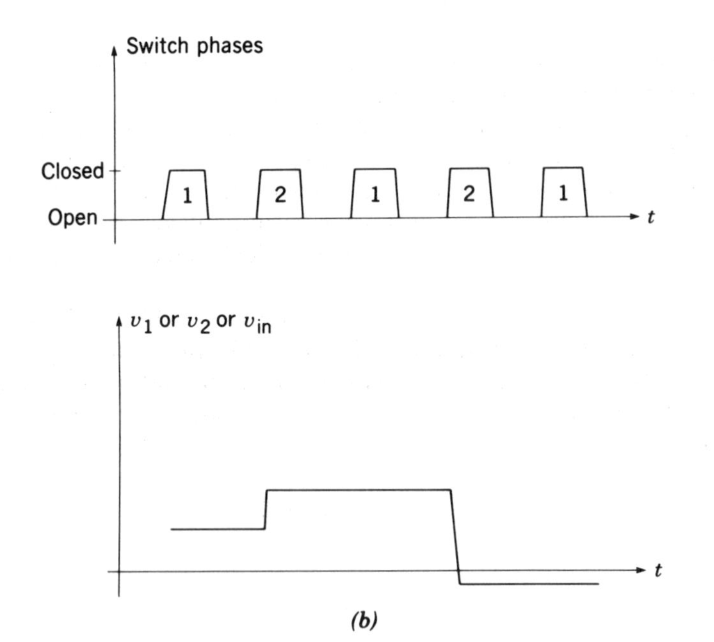

(b)

Figure 5.11 (a) High Q switched-capacitor biquad; (b) clock and signal waveforms.

For reasonable Q values ($Q < 100$), $C_4 > C_2 = C_3$ and the capacitance spread is again around $1/(\omega_0 T)$.

The transfer function of the biquad of Figure 5.11(a) is

$$H(z) = \frac{V_{out}(z)}{V_{in}(z)} = -\frac{C_1''z^2 + (C_1C_3 + C_1'C_3 - 2C_1'')z + (C_1'' - C_1'C_3)}{z^2 + (C_2C_3 + C_3C_4 - 2)z + (1 - C_3C_4)} \quad (5.28)$$

Comparison with Eq. (5.11), using $b_2 = 1$, gives the design equations

$$\left.\begin{aligned} C_1'' &= a_2 \\ C_1' &= \frac{a_2 - a_0}{C_3} \\ C_1 &= \frac{a_0 + a_1 + a_2}{C_3} \\ C_4 &= \frac{1 - b_0}{C_3} \\ C_2C_3 &= 1 + b_0 + b_1 \end{aligned}\right\} \quad (5.29)$$

Choosing $C_2 = C_3 = (1 + b_0 + b_1)^{1/2}$, all capacitance values can be obtained. As before, some of the input capacitors may be omitted: for unit-circle transmission zeroes, $C_1' = 0$; for zeroes at $z = 1$, $C_1 = C_1' = 0$, and so on. Changing the clocking of C_1', zeroes outside the unit circle can again be generated [12].

The circuit of Figure 5.11(a) has some attractive properties. For the case when $\omega_0 T \ll 1$ and $Q \gg 0$, the poles are near the $z = 1$ point, and hence the denominator polynomial is approximately equal to $(z - 1)^2 = z^2 - 2z + 1$. Thus, by Eq. (5.28), $C_3C_4 \ll 1$ and $C_2C_3 \ll 1$. However, this does not mean that the individual element values C_2, C_3, and C_4 need to be very small. Thus, for example, denominator $z^2 - 1.98z + 0.99$ can be realized using $C_2 = C_3 = C_4 = 0.1$, which is easily achieved. Also, the sensitivities $\partial b_1/\partial C_i$ and $\partial b_2/\partial C_j$ are low, thanks to the product form in which all C_i enter the denominator of $H(z)$. In both biquads, the pole-Q is determined by a single element, C_4. This makes the realization of variable-Q circuits simple.

Another simple SC stage is that corresponding to the bilinear transfer function

$$H(z) = \frac{a_1 z + a_0}{b_1 z + b_0} \quad (5.30)$$

Some possible realizations of this function are shown in Figure 5.12. For the circuit of Figure 5.12(a), if the clock phases *outside the parentheses* are used at the input, then the transfer function is

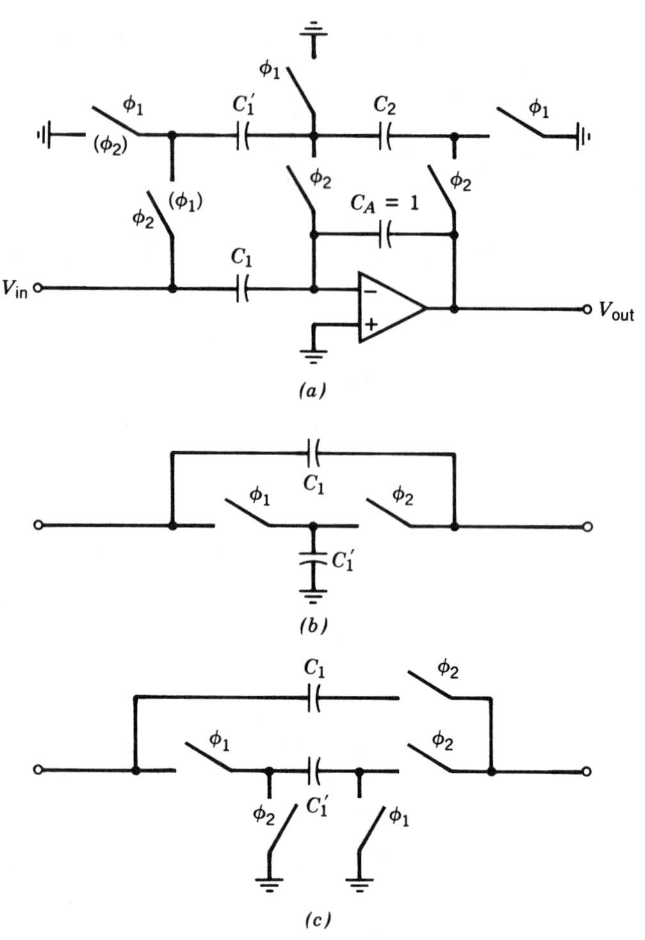

Figure 5.12 Switched-capacitor linear section: (a) complete circuit; (b) and (c) alternative input branches.

$$H(z) = \frac{(C_1 + C_1')z - C_1}{(1 + C_2)z - 1} \tag{5.31}$$

Hence, this circuit can realize a positive real zero and pole, both inside the unit circle. The element values can be obtained simply by equating the corresponding coefficients in Eqs. (5.30) and (5.31). It is assumed that $v_{in}(t)$ changes only when phase 2 goes high.

If the clock phases shown *within parentheses* are used, then the transfer function becomes

$$H(z) = -\frac{C_1 z - (C_1 + C_1')}{(1 + C_2)z - 1} \tag{5.32}$$

The pole realized by this function is as before; however, the zero has now a positive value larger than 1.

A negative real zero can be obtained by using the branch shown in Figure 5.12(b) as an input branch. The numerator polynomial then becomes $C_1 z + (C_1' - C_1)$. For $C_1' > C_1$, the zero will be negative. By using a similar branch (but with interchanged clock phases) in the feedback path, a negative pole can also be realized. Note that the branch of Figure 5.12(b) is stray sensitive: stray capacitance affects the value of C_1'. However, for usual values, the resulting shift in the value of the negative zero (or pole) has a minor effect in the low-frequency region of the response. In fact, it normally introduces only a small additional flat gain which is in most cases unimportant. If necessary, the stray-insensitive branch of Figure 5.12(c) can be used as an input branch. Assuming now that $v_{in}(t)$ changes only when phase 1 rises, the numerator polynomial realized here is $-[(C_1' - C_1)z + C_1]$. For $C_1' > C_1$, this polynomial contributes a negative real zero, as well as an additional phase inversion.

If a high-order transfer function $H(z)$ is to be realized, it can be factored into biquadratic and (if the order of $H(z)$ is odd) bilinear factors. Each biquadratic factor can then be realized by a biquad, and the bilinear factor (if there is one) by one of the first-order stages of Figure 5.12. Since all these stages have low output impedances, they provide buffered output voltages. Thus, they can be cascaded to provide the desired high-order transfer function. This cascade realization gives an SC filter circuit that is relatively easy to design, to lay out for fabrication, to test, and to troubleshoot if an error occurs. This is because of its modular structure which allows section-by-section design and testing. On the other hand, the sensitivity of the cascade circuit to element-value variations is fairly high. Typically, in the filter pass-band, a 1% change in a capacitance value results in a gain change around 0.2 dB. Thus, for a biquad containing eight capacitors, each with about a 0.3% uncorrelated error, an overall gain error of $\sqrt{8} \times 0.3 \times 0.2\,\text{dB} \cong 0.17\,\text{dB}$ may be expected at a pass-band frequency. In many applications, this is tolerable. If lower sensitivities are required, then the SC *ladder filters* to be discussed in the next section, may be used.

5.4. SWITCHED-CAPACITOR LADDER FILTERS

It has long been recognized [20] that a doubly-terminated reactance filter that provides matching in its pass-band has very low pass-band gain sensitivities to reactance value variations. This is because any such change decreases the gain that is maximum (or near maximum) for nominal element values. The gain sensitivities to variations of the terminating resistances are not low, but they are frequency independent, and hence such variations result in a constant gain change that is usually tolerable.

Based on this observation, a design technique can be developed for low-sensitivity SC filters. The design is based on starting with a doubly terminated LC filter model, and then constructing an SC filter that has the same transfer function, and whose element values influence the voltage gain in the same way as those of the LC filter. In fact, of course, neither of these conditions can be exactly satisfied, since the reactance filter is a continuous-time circuit, while the SC filter is a sampled-data one. However, an approximate solution can be found that still provides low pass-band sensitivities, typically around 0.05 dB/1% capacitance change, for the SC circuit. The process will be illustrated by a simple example.

Consider the third-order LC ladder filter shown in Figure 5.13(a). Between the Laplace-transformed state variables V_1, I_2, and V_3, the following equations can be constructed by applying the KCL to nodes ① and ③, and the KVL to the C_1–L_2–C_3 loop

$$\left.\begin{aligned} -V_1 &= \frac{-1}{s(C_1 + C_2)}\left[\frac{V_{\text{in}} - V_1}{R_S} + sC_2V_3 - I_2\right] \\ -I_2 &= \frac{-1}{sL_2}[V_1 - V_3] \\ V_3 &= \frac{-1}{s(C_2 + C_3)}\left[-sC_2V_1 - I_2 + \frac{V_3}{R_L}\right] \end{aligned}\right\} \quad (5.33)$$

These equations have been organized in such a way that each state variable is expressed as the negative integral of a weighted sum of the state variables and V_{in}. A block diagram representation of Eqs. (5.33) is shown in Figure 5.13(b); an active RC realization in Figure 5.13(c). In the last circuit, all impedances in the central stage have been multiplied by an (arbitrary) resistance to obtain correct dimensions for all parameters.

Finally, all resistors can be replaced by the equivalent SC branches shown in Figure 5.1(a). The resulting SC ladder filter is illustrated in Figure 5.13(d). The capacitance values can be obtained by the equations

$$\left.\begin{aligned} C_S &\cong T/R_S \\ C_A &= C_1 + C_2 \\ C &\cong T \\ C_B &= L_2 \\ C_C &= C_2 + C_3 \\ C_L &\cong T/R_L \end{aligned}\right\} \quad (5.34)$$

The replacement of the active RC integrators in Figure 5.13(c) by active SC integrators introduces phase errors, as explained in §5.3.1. However, since the two integrator types (inverting and noninverting) alternate in the

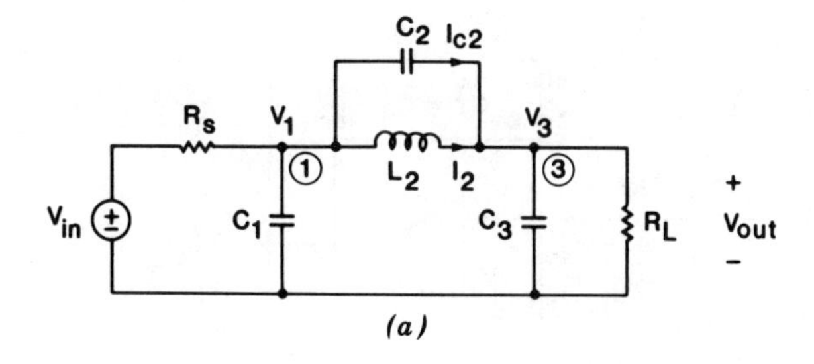

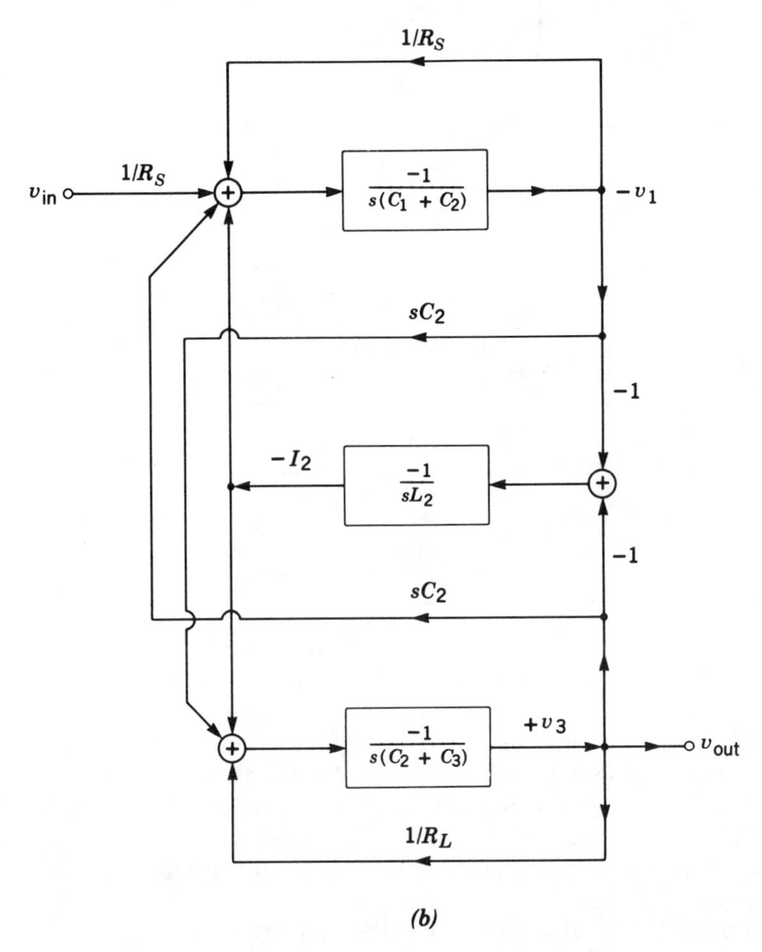

Figure 5.13 Low-pass switched-capacitor filter with finite transmission zeroes: (a) LCR prototype circuit; (b) block diagram; (c) active RC realization; (d) SC circuit.

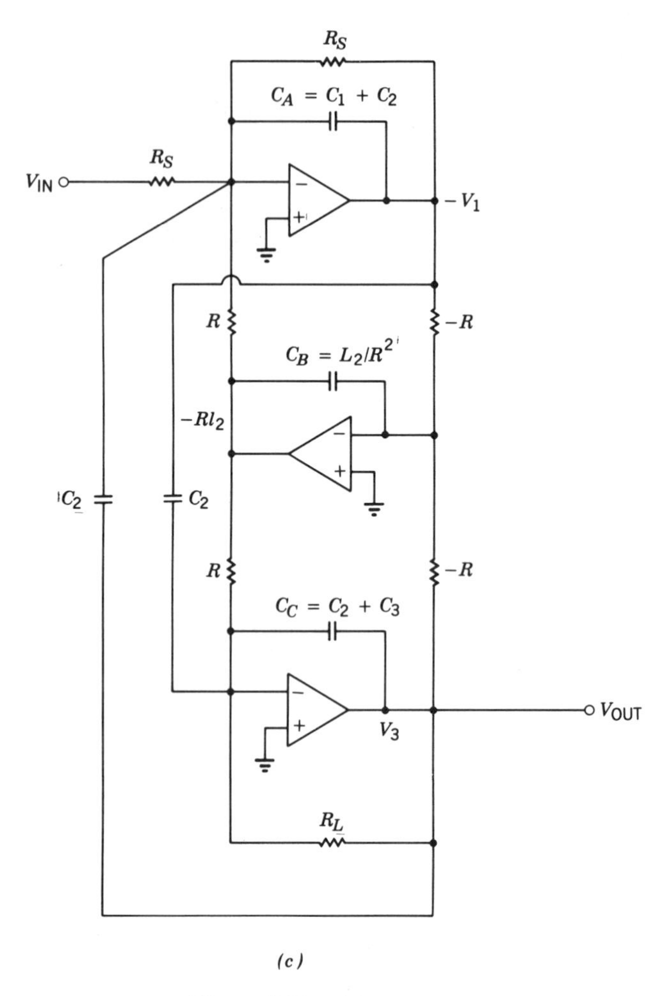

Figure 5.13. (*Continued*)

circuit, and since the phase errors $\pm\omega T/2$ for these integrators are of equal magnitude but opposite signs, these errors cancel to a first-order approximation. Thus, the overall gain response will be close to that of the LC filter model, for frequencies that satisfy $f \ll f_c$. The error introduced in the pass-band gain is usually of the order of 0.1–0.5 dB if the upper pass-band edge frequency is less than $f_c/20$. If such errors are not acceptable, then either computer-aided optimization can be used to refine the capacitance values, or an exact design technique applied. Such a technique is now briefly discussed.

It is possible to design an SC filter from an LC model filter such that the transfer function $H(z)$ of the former equals the transfer function $H_a(s_a)$ of the latter if the bilinear mapping (5.12) is used. This also means that the

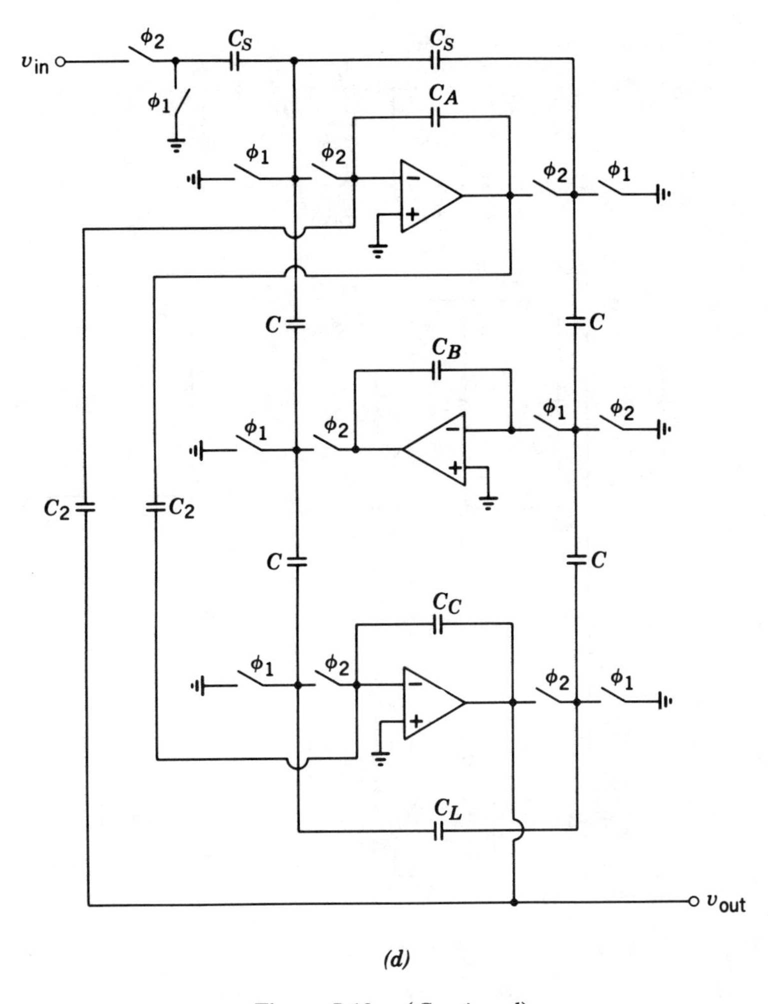

Figure 5.13. (*Continued*)

frequency response of the SC filter $H(e^{j\omega T})$ is the same as the response of the LC filter, except that the frequency scale is warped according to the relation

$$\frac{\omega T}{2} = \tan^{-1} \frac{\omega_a T}{2} \tag{5.35}$$

The derivation of the exact design equations for this case is straightforward but lengthy [12] and will be omitted. The resulting circuits and design relations are shown in Figures 5.14, 5.15 and 5.16, for the input section, for an internal section, and for the output section, respectively. Additional details are given by Datar and Sedra [5] and Gregorian and Temes [12].

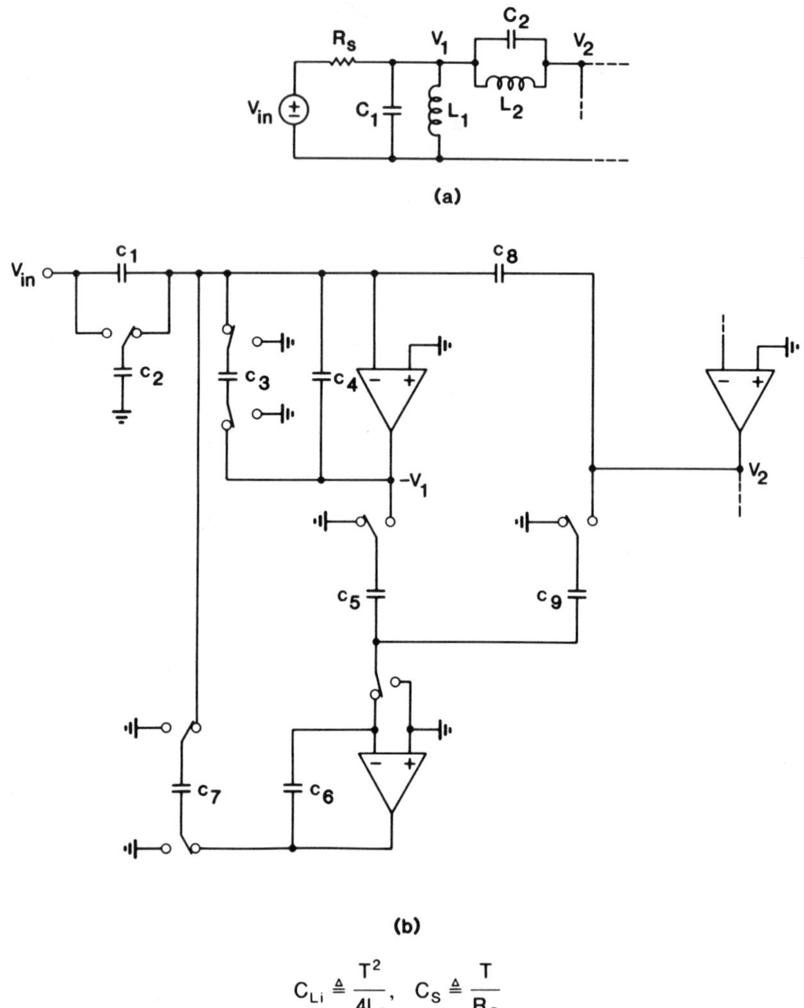

$$C_{Li} \triangleq \frac{T^2}{4L_i}, \quad C_S \triangleq \frac{T}{R_S}$$

$$c_1 = \frac{C_S}{2}, \; c_2 = c_3 = C_9$$

$$c_4 = C_1 + C_2 + C_{L1} + C_{L2} - C_S/2$$

$$c_5 = 4\frac{c_6}{c_7}(C_{L1} + C_{L2})$$

$$c_8 = C_2 + C_{L2}$$

$$c_9 = 4\frac{c_6}{c_7}C_{L2}$$

c_6, c_7 arbitrary

(c)

Figure 5.14 Bilinear ladder, input section: (a) LCR prototype; (b) SC realization; (c) design equations.

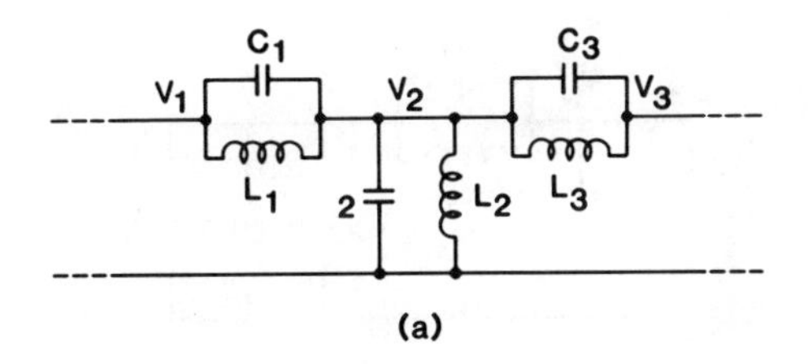

(a)

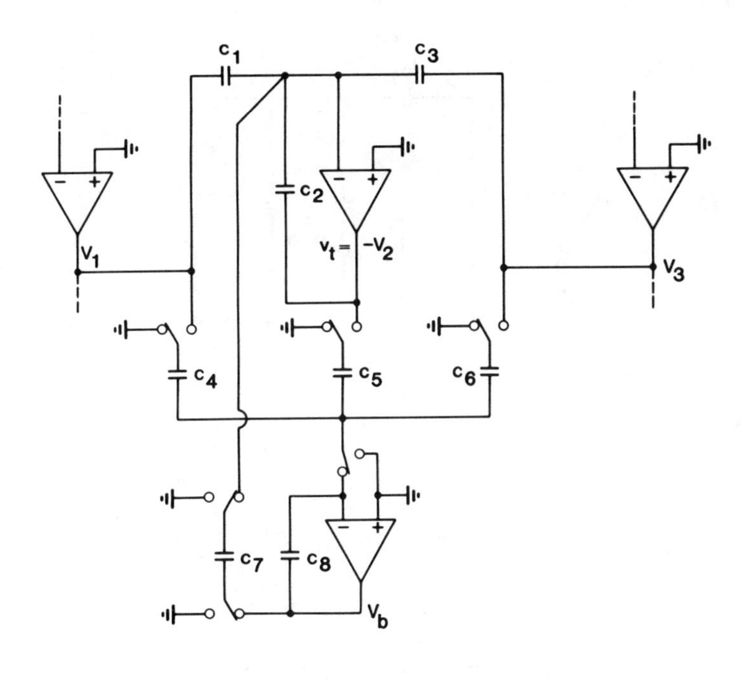

(b)

$$C_{Li} = T^2/(4L_i)$$
$$c_1 = C_1 + C_{L1}$$
$$c_2 = C_1 + C_2 + C_3 + C_{L1} + C_{L2} + C_{L3}$$
$$c_3 = C_3 + C_{L3}$$
$$c_4 = 4\frac{c_8}{c_7}C_{L1}$$
$$c_5 = 4\frac{c_8}{c_7}(C_{L1} + C_{L2} + C_{L3})$$
$$c_6 = 4\frac{c_8}{c_7}C_{L3}$$

c_7, c_8 arbitrary

(c)

Figure 5.15 Bilinear ladder, internal section: (a) LC prototype; (b) SC realization; (c) design equations.

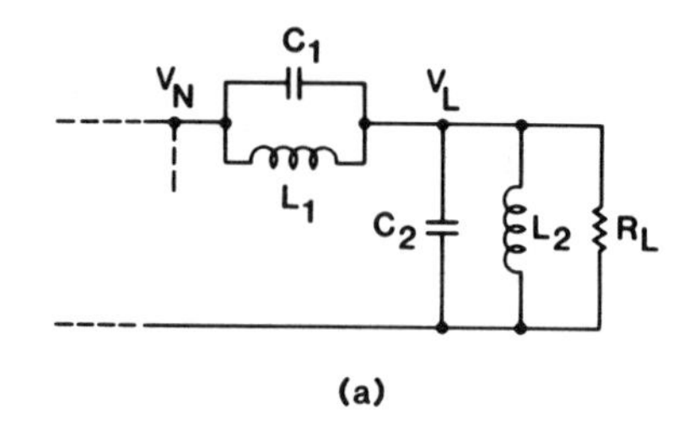

(a)

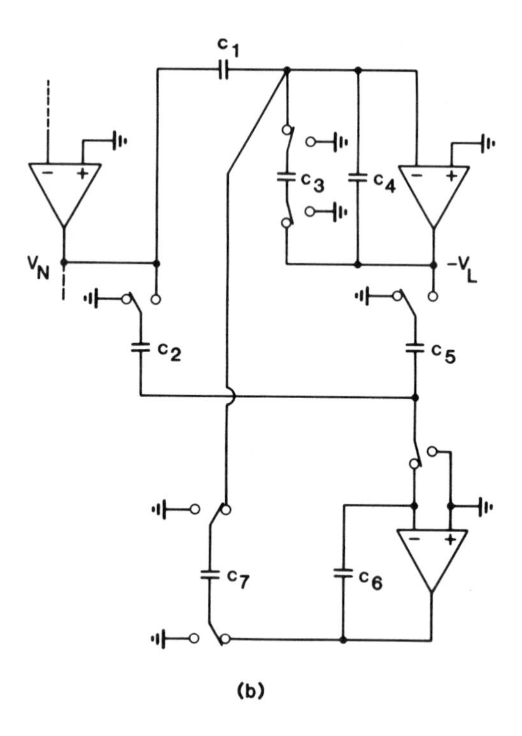

(b)

$$C_{L1} \triangleq \frac{T^2}{4L_i}, \quad C_L \triangleq \frac{T}{R_L}$$

$$c_1 = C_1 + C_{L1}$$

$$c_2 = 4\frac{c_6}{c_7}C_{L1}$$

$$c_3 = C_L$$

$$c_4 = C_1 + C_2 + C_{L1} + C_{L2} - C_L/2$$

$$c_5 = 4\frac{c_6}{c_7}(C_{L1} + C_{L2})$$

c_6, c_7 are arbitrary

(c)

Figure 5.16 Bilinear ladder, output section: (a) LCR prototype; (b) SC realization; (c) design equations.

5.5. THE SCALING OF SWITCHED-CAPACITOR FILTERS

The dynamic range of an SC filter can usually be significantly improved by a scaling process based on the following theorem. Let all impedances connected to the output node of the *i*th op amp in a stray-insensitive filter be multiplied by a constant k. Thus, in Figure 5.17, let all capacitances in branches F_4, F_5, and F_6 be divided by k. Then, the current in F_4 remains unchanged; V_i becomes kV_i; and the currents in the output branches F_5 and F_6 remain unchanged. Thus, all currents flowing into other stages stay the same and hence so do all other op amp output voltages. Using this process, we can therefore scale all output voltages V_i individually.

If the peak value $V_{i,\max}$ of $V_i(\omega)$, occurring at some frequency ω_i, is now larger than the maximum value $V_{\text{out},\max}$ of the output voltage V_{out} of the filter, then the linear range of the filter can be enlarged by reducing $V_{i,\max}$. Thus, the dynamic range can be improved by reducing all $V_{i,\max}$ to (or below) $V_{\text{out},\max}$. Let now $V_{i,\max} < V_{\text{out},\max}$. Then the gain $\beta_{i,\text{in}} \stackrel{\Delta}{=} V_i/V_{\text{in}}$ can be increased without reducing the linear range of the SC filter. Since the product of $\beta_{i,\text{in}}$ and $\beta_{i,\text{out}} \stackrel{\Delta}{=} V_{\text{out}}/V_i$ is a constant, that is, $\beta_{i,\text{in}}\beta_{i,\text{out}} = V_{\text{out}}/V_{\text{in}}$, if $\beta_{i,\text{in}}$ increases then $\beta_{i,\text{out}}$ is reduced. But $\beta_{i,\text{out}}$ is the voltage gain of the noise generated in the *i*th op amp OA_i. Hence, the output noise is reduced, and the dynamic range is improved, by this change. In conclusion, for optimum dynamic range, the op amp voltages should be scaled such that all $V_{i,\max}$ equal $V_{\text{out},\max}$. This is achieved, as discussed above, if the capacitances in all branches connected (or switched) to the output terminal of OA_i are multiplied by $V_{i,\max}/V_{\text{out},\max}$, for all op amps.

It can be shown [21] that the described scaling process also minimizes the sensitivities of the pass-band gain response to finite op amp gain effects. The logarithmic sensitivities to capacitance variations, on the other hand, remain unaffected by the scaling.

A second scaling step, which leaves the dynamic range and all sensitivities unchanged but which can reduce the total capacitance needed by the circuit, should also be carried out. This step consists of multiplying all capacitances connected to the **input** terminal of the op amp OA_i (in branches F_1, F_2, F_3,

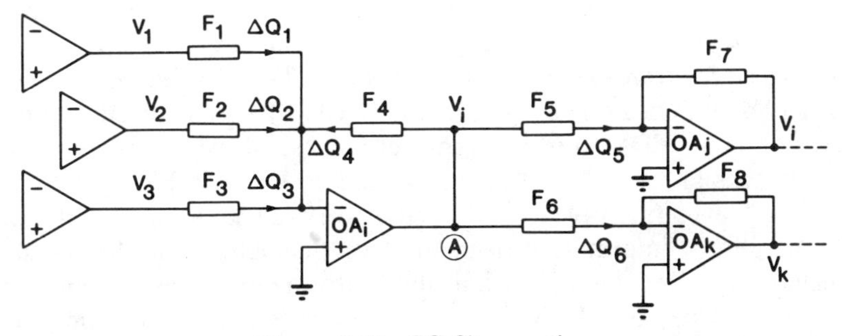

Figure 5.17 SC filter section.

and F_4 in Figure 5.17) by a constant l. This causes all currents in these branches to be multiplied by l; however, V_i remains unchanged, since the current in F_4 was multiplied while the branch impedance was divided by l. Hence, assuming ideal amplifiers, all op amp voltages (including V_i) remain unchanged by this step. This degree of freedom is usually utilized to make the value $C_{i,\min}$ of the smallest capacitor connected to the input node equal to $C_{\min}$, the minimum capacitance value allowed by the technology used. (Typically, $C_{\min}$ is in the range 0.1–1 pF.) The resulting saving in the total capacitance can be significant, especially for narrow-band bandpass filters.

5.6. COMMERCIALLY AVAILABLE SWITCHED-CAPACITOR FILTERS

In most commercial signal processing applications, SC filters form a part of a complex system that performs a number of functions and is integrated on a single silicon chip. In a few cases, stand-alone programmable SC filters have also been fabricated and marketed to serve as flexible building blocks for various system applications. A few typical examples, taken mostly from the authors' first-hand experience, are described in this section.

5.6.1. Switched-Capacitor Filters in Digital Voice Communication Systems

In spite of being basically analog signal processors, SC circuits have played important roles in digital voice communication systems. In fact, even in such a digital system, many signals are processed in analog form. They are converted into digital format by one of several modulation schemes, such as pulse-width modulation (PWM), pulse-amplitude modulation (PAM), delta modulation (DM) and pulse-code modulation (PCM). Of these, PCM is the most widely used scheme. The PCM signal is obtained by sampling, quantizing and coding an analog signal. The resulting digital binary bit stream can be applied either directly, or after further modulation steps, to the line. In order to retain all information contained in the analog signal, the sampling must be performed at a rate that equals at least twice the bandwidth of the signal.

Many practical PCM systems use time-division multiplexing (TDM). In such a system, samples are taken alternately from several voice channels, and interlaced for transmission on a single line. The International Telegraph and Telephone Consultative Committee (CCITT) has in recent years approved two main TDM transmission hierarchies, on which many national PCM networks are now based. The first one is shown in Figure 5.18. This system interleaves (multiplexes) 32 digital works at its input. Of these, 30 contain amplitude information from 30 voice channels, while the remaining two include signaling and synchronization information. This system is used mainly in Europe, South America, Australia, and Africa. The second hierarchy (Figure 5.19) interleaves 24 speech signals. It is used mainly in

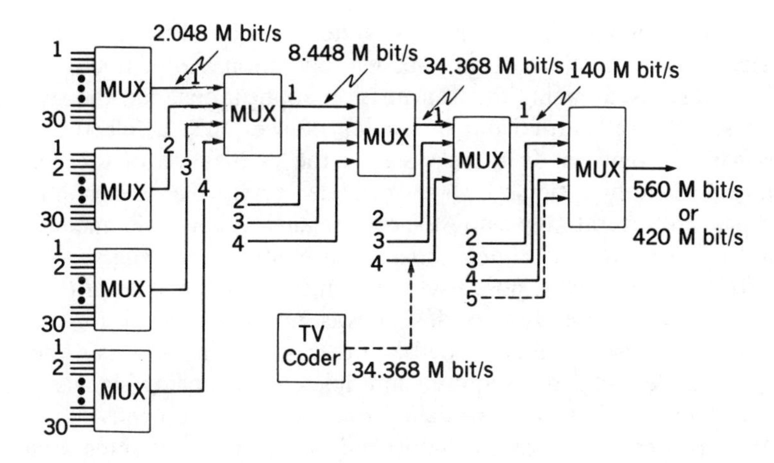

Figure 5.18 The PCM hierarchy used in Europe, Africa, Australia, and South America.

North America and Japan. (For simplicity, these diagrams ignore signaling and synchronization.)

The two main functions performed in a TDM system are digital voice transmission and digital switching. Transmission involves sending voice data from one location to another; switching involves establishing the connection between two digital channels.

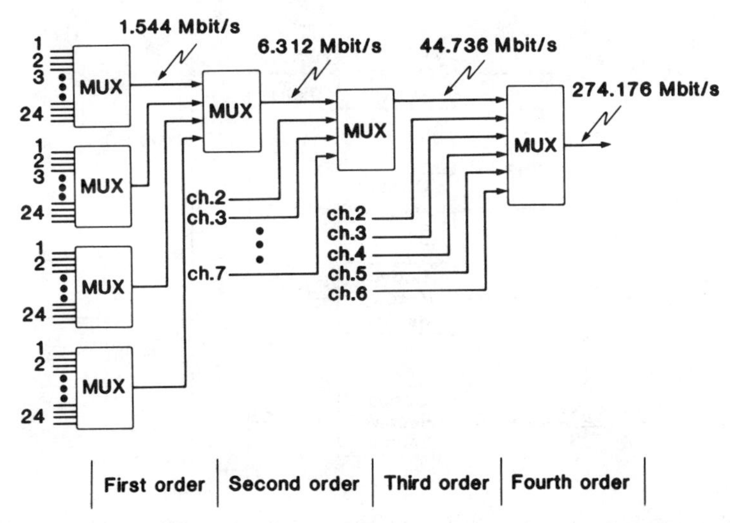

Figure 5.19 The PCM hierarchy used in the United States, Canada, and Japan.

A basic functional unit in a PCM system is the *codec* (**co**der-**dec**oder). It carries out the functions of pulse-code modulation and demodulation. Early codecs were used mainly for transmission applications, and the switching of voice signals was carried out by analog devices such as electromechanical crossbar mechanisms. In these systems, the digitization of the voice signal was performed by a single high-speed codec circuit that was time-multiplexed between several (typically, 24 or 30) analog voice channels. The high speed required for the codec and the problems of keeping the line-to-line crosstalk low, however, posed such difficulties that (in spite of the saving in hardware costs) this system design was not economical for subsequent generations. Also, a single codec failure resulted in the loss of service for many lines. Finally, the required multiplexing of analog signals was more difficult and less flexible than digital ones would have been.

With the reduced cost and improved performance of integrated circuits over the years, it subsequently became practical to allocate a codec to each voice channel. The signal of each subscriber line was then digitized first, and switching and multiplexing both became digital functions, performed by integrated digital logic and memory circuits. This scheme resulted in a considerable reduction of crosstalk and noise, and in savings in cost and size.

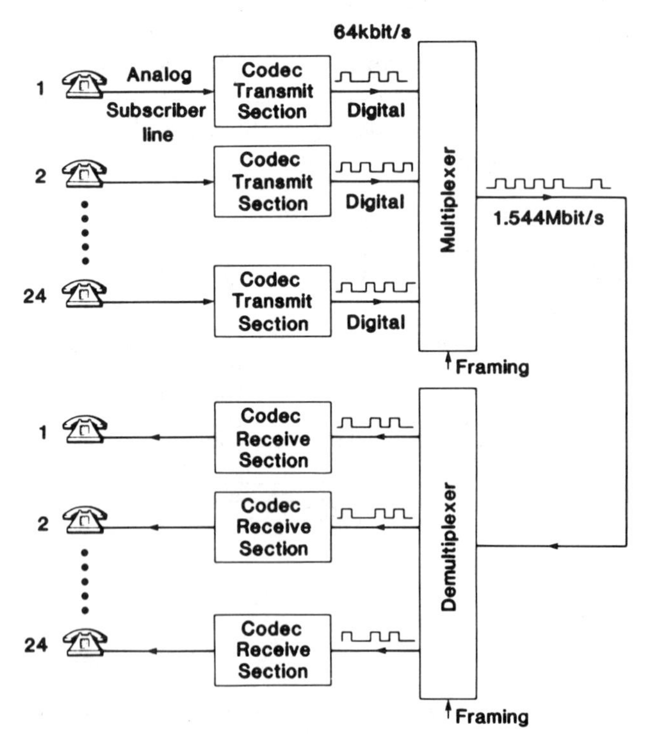

Figure 5.20 A 24-channel time-division multiplexed PCM telephone system.

The pass-band of a standard voice channel is between 300 Hz and 3,400 Hz; CCITT recommends a sampling rate of 8,000 Hz. Thus, at intervals of 125 μs, a sample must be taken and converted into an 8-bit digital word. This word is then transmitted serially (i.e., one bit at a time) to the multiplexer. The multiplexer interleaves 24 (or 32) words within a 125-μs **frame**, and transmits the resulting bit stream to the line.

In addition to the voice data, each frame must also include some extra bits to identify the beginning of the frame. For example, for the 24-channel AT&T D4 PCM system, each frame contains $24 \times 8 = 192$ voice bits, plus one framing bit. Thus, for an 8,000 Hz sampling rate, the transmitted bit rate is $193 \times 8{,}000 = 1.544 \times 10^6$ bits/s. Figure 5.20 illustrates such a system. Figure 5.21 shows the signal path through a typical PCM system. In the transmit direction, the analog voice signal is applied to the codec through the subscriber line-interface circuit (SLIC). The SLIC performs also the two-to-four wire conversion, battery feed, line supervision, ringing access and overvoltage protection functions. The signal then passes through the transmit filter, which removes its high frequency contents above 4 kHz, and low frequencies below 300 Hz. The former operation is necessary to allow the subsequent sampling at 8 kHz; the latter helps to suppress the 60 Hz (or 50 Hz) line-frequency noise. The resulting bandlimited signal is then sampled and encoded at an 8 kHz rate into 8-bit nonlinear PCM data (Figure 5.22). These data are multiplexed and transmitted to the receiver. There, the incoming data are demultiplexed and decoded. The resulting analog signal first enters a sample-and-hold (S/H) stage, and then the receiver filter. The latter is a low-pass filter that smoothes the staircase output signal of the S/H stage, and also compensates for the sin x/x amplitude distortion introduced by the S/H operation.

In the last few years, several approaches were used to realize fully integrated low-cost per-channel codecs. In the earliest one [4], the on-chip filters were realized with CCDs. With the development of SC filters, they were used to carry out the filtering tasks. In one type of SC-based implementation [2, 7, 8, 15], two chips were used for each codec. One contained the transmit and receive filters, while the other the encoder and decoder circuits. In another approach [14], the complete transmit path was integrated on one chip, and the receive path on another. This architecture reduced both crosstalk and noise in asynchronous operation.

As an example, Figure 5.23 illustrates the block diagram of the Intel 2912 PCM channel filter [7]. The signal to be transmitted is first applied to an input op amp that (with two external resistors) provides amplification. The output then passes through a second-order active RC low-pass filter, which bandlimits it to allow subsequent sampling. The following SC filters remove line-frequency noise, as well as all frequencies above 4 kHz. The resulting signal is then smoothed by a second active RC low-pass filter.

The receiver filter contains a low-pass filter that is essentially identical to the transmit low-pass filter. However, in addition, a first-order equalizer

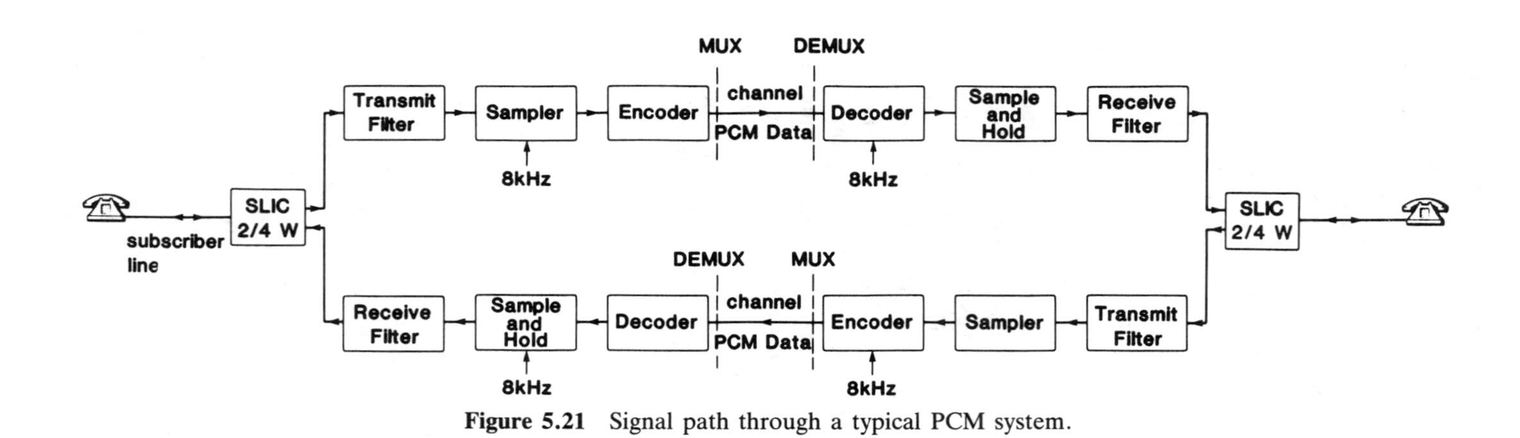

Figure 5.21 Signal path through a typical PCM system.

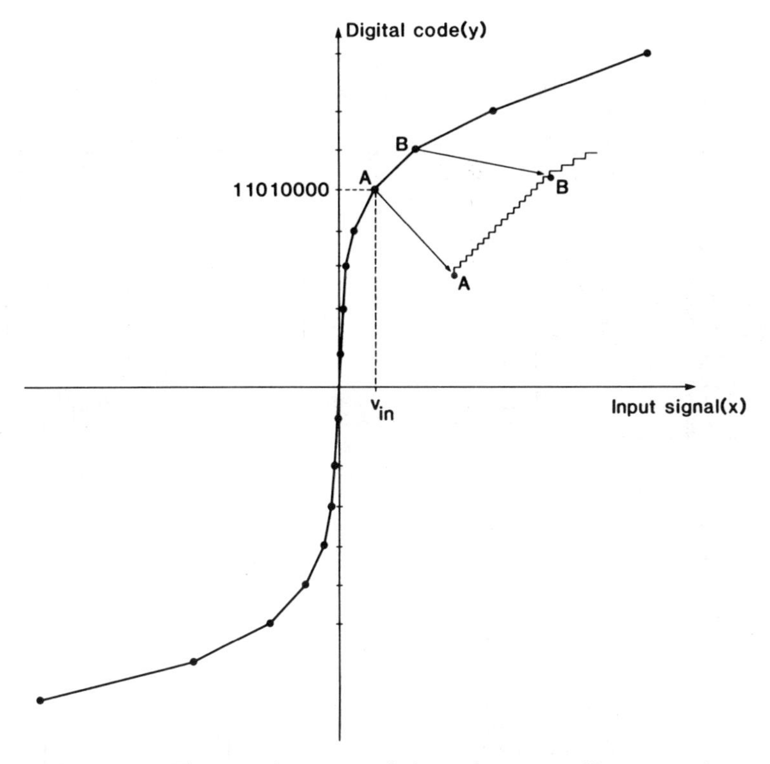

Figure 5.22 The transfer curve of the μ-law nonuniform encoder.

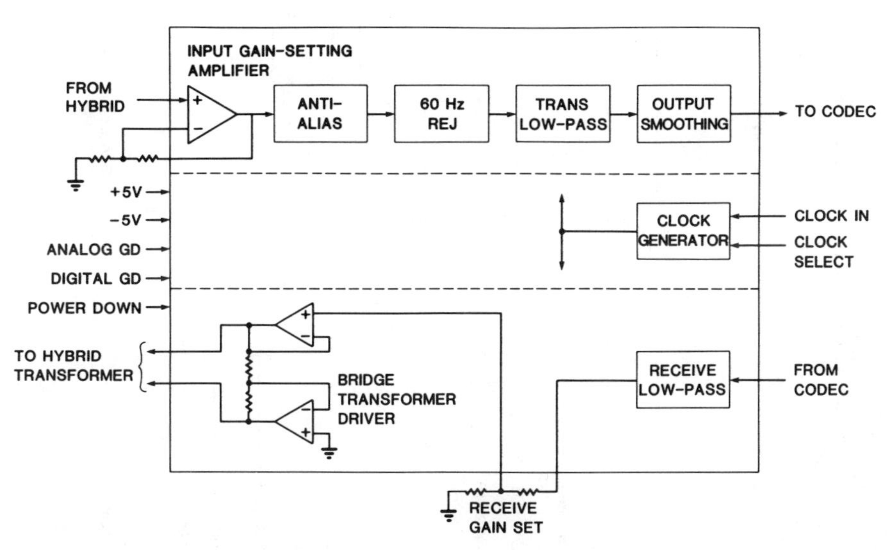

Figure 5.23 The block diagram of the Intel 2912 channel filter (from Gray et al. (1979) *I.E.E.E. J. Solid-State Circuits* **14**, 981–990 © 1979 I.E.E.E.).

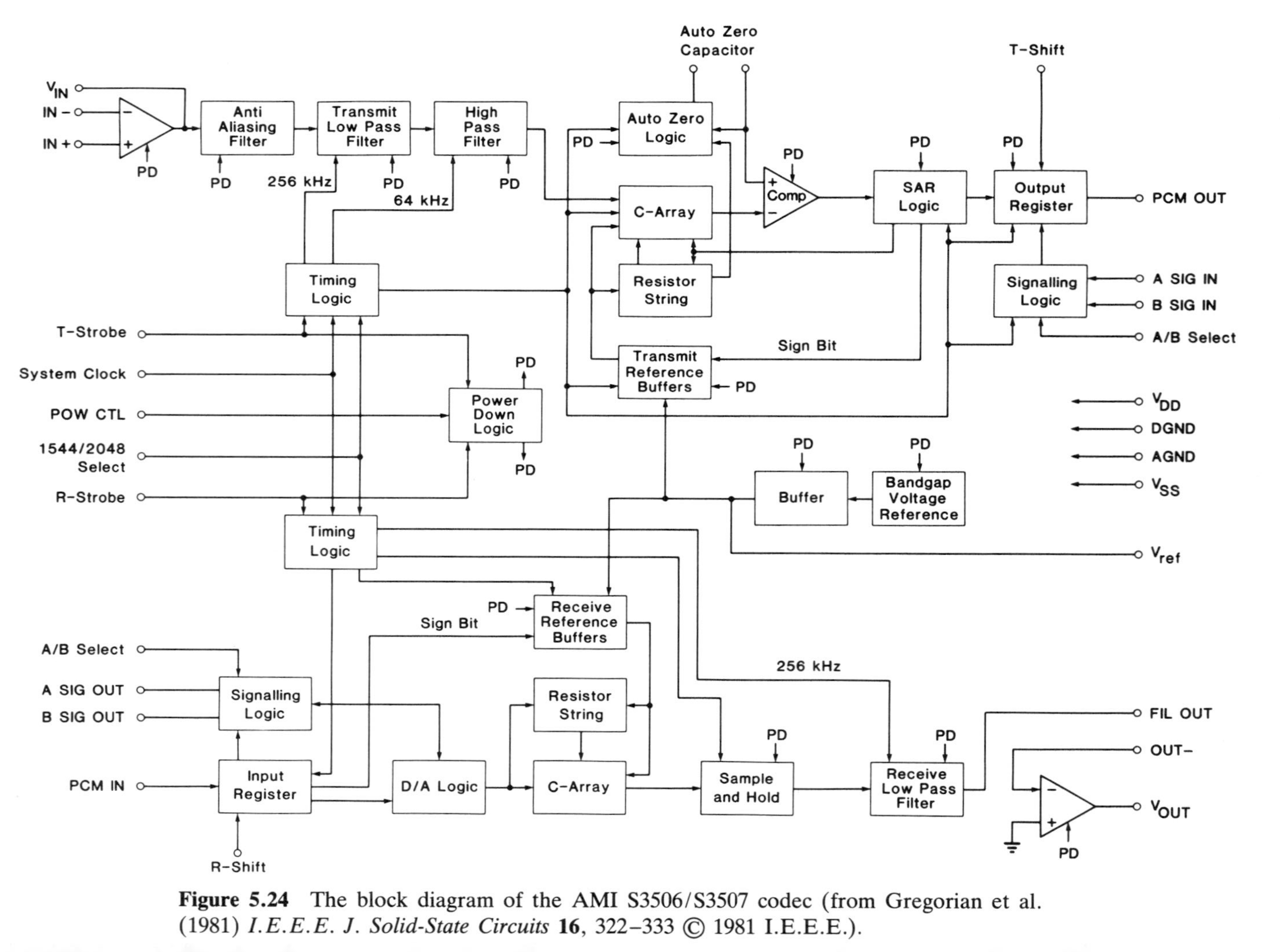

Figure 5.24 The block diagram of the AMI S3506/S3507 codec (from Gregorian et al. (1981) *I.E.E.E. J. Solid-State Circuits* **16**, 322–333 © 1981 I.E.E.E.).

stage is cascaded at its input. The equalizer provides a 2 dB gain peak that compensates for the sin x/x amplitude response of the decoder. The receiver filter is followed by a transformer driver that consists of two high-current amplifiers connected in a bridge configuration.

The channel filter was fabricated in NMOS technology. It contains a total polysilicon/polysilicon capacitance of 400 pF and a total polycrystalline resistance of 600 kΩ. It also includes 20 op amps. The die size of the chip is $150 \times 166\,\text{mil}^2$. The performance (for ±5 V supply voltages and at 25°C) is summarized in [7].

In order to reduce the cost, as well as the space and dc power requirements, second-generation single-clip LSI codecs have been developed. These codecs integrated all functions needed for the voice-signal to PCM conversion, including the filters, encoders/decoders and voltage references on a single chip [10, 16, 18, 26]. As a result, there exist several commercially available single-chip MOS codecs, such as the AMI S3506/S3507, the Hitachi HD 44231A/44233A, the National Semiconductor TP 3052, TP 3053, TP 3054, and TP 3057, and the Motorola MC 14400, MC 14401, and MC 14402.

As an example of the architecture and circuitry of these codecs the detailed structure of the AMI S3506/S3507 is discussed [10]. The block diagram of the codec is shown in Figure 5.24. In the transmit direction, the voice signal V_{in} is amplified using an op amp. The gain can be set by external resistors connected to the chip. The amplifier is followed by the active RC anti-aliasing low-pass filter, which limits the signal band to 508 kHz; by a decimating fifth-order SC low-pass ladder filter that samples its input signal at a 512 kHz rate, and provides a 256 kHz S/H output signal (Figure 5.25); and by a third-order SC high-pass filter (Figure 5.26). The low-pass filter restricts the bandwidth to below 3.4 kHz. The high-pass filter passes only signal frequencies above 300 Hz; it is clocked at 64 kHz.

The output signal of the high-pass filter is resampled and held at a rate of 8 kHz, and then sent to the encoder. The encoder is a nonlinear (μ 255) A/D converter that utilizes a binary-weighted capacitor array and a resistor chain to implement a piecewise-linear characteristic for the decision-level generator. The conversion operation and hence the process is performed in nine steps. In the first step, the sign of the signal is determined; in the next eight steps, the seven bits giving the appropriate segment and step on the piecewise-linear characteristic are determined and then clocked out serially. The clock frequency can vary between 64 and 2,048 kHz.

In the receive direction, the decoder accepts the PCM words and performs the D/A conversion. The simplified circuit diagram of the μ-law decoder used is shown in Figure 5.27. It functions as follows. The sign bit of the incoming digital word is used to set the sign of the reference voltage V_{REF}. Then, the other bits are used to turn on an appropriate combination of switches in the capacitor array, so as to develop the required analog voltage at the top plates of the array capacitors. Also, during this period,

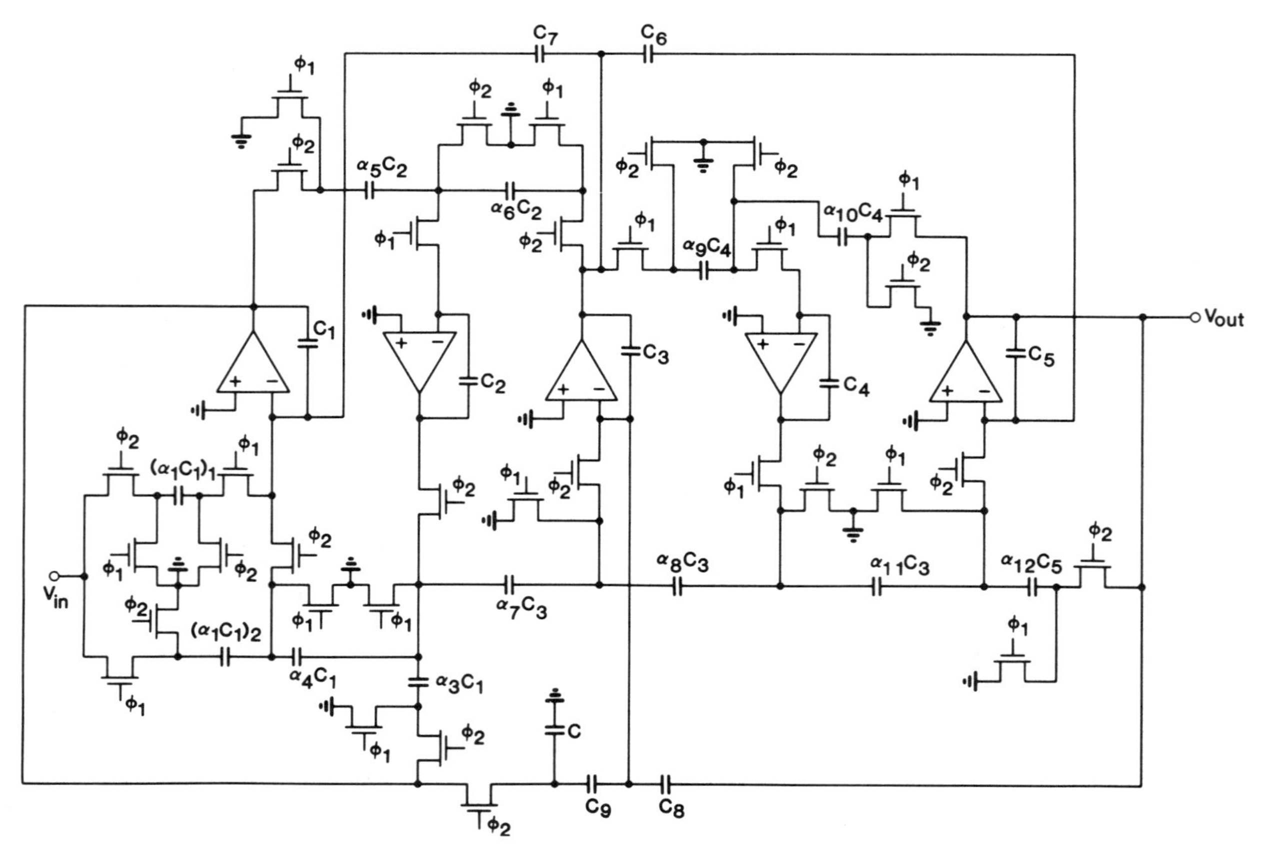

Figure 5.25 The transmit low-pass filter of the AMI S3506/S3507 codec (from Gregorian et al. (1981) *I.E.E.E. J. Solid-State Circuits* **16**, 322–333 © 1981 I.E.E.E.).

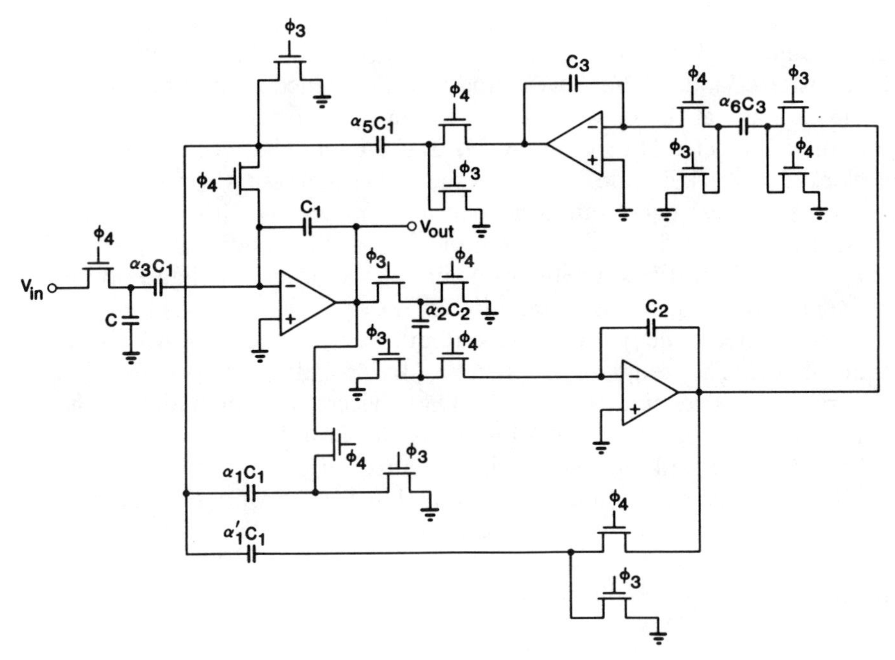

Figure 5.26 The transmit high-pass filter of the AMI S3506/S3507 codec (from Gregorian et al. (1981) *I.E.E.E. J. Solid-State Circuits* **16**, 322–333 © 1981 I.E.E.E.).

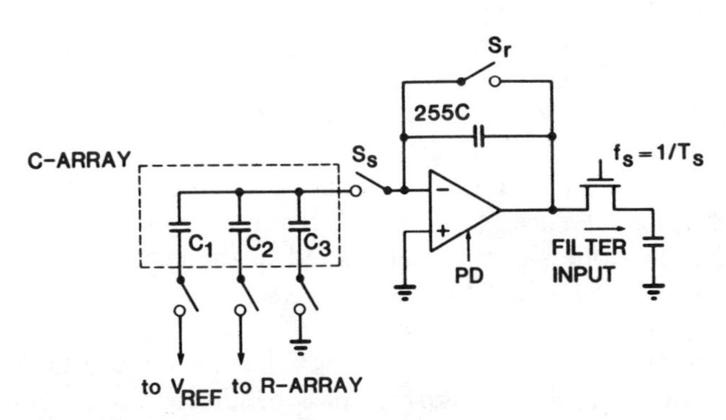

Figure 5.27 The simplified schematic of the μ-law decoder used in the AMI S3506/S3507 codec (from Gregorian et al. (1981) *I.E.E.E. J. Solid-State Circuits* **16**, 322–333 © 1981 I.E.E.E.).

the feedback capacitor of the op amp is discharged by closing S_r. Next, S_r is opened and S_s is closed, causing the charge stored in the capacitor array to be transferred into the feedback capacitor, and hence generating the proper analog voltage at the op amp output terminal. This voltage is then entered into the SC receive filter via an 8 kHz S/H circuit. The receive filter, which is clocked at 256 kHz rate, smoothes the 8 kHz staircase waveform, and also performs the loss equalization needed to eliminate the $\sin x/x$ S/H distortion. The receive filter is also a fifth-order ladder filter, as the transmit low-pass filter, and has a similar circuit configuration but slightly different element values to provide the required $x/\sin x$ response. The output voltage can be obtained directly, or via the output op amp, which can drive a hybrid transformer. The required temperature-stable reference voltage V_{REF} is generated by a circuit that uses MOS devices in combination with the parasitic *npn* bipolar transistors formed in the CMOS process by the *n*-type substrate, the *p*-well and the n^+ drain/source diffusion.

The described codec chip was fabricated in 5-μm CMOS technology. The die size was $212 \times 217\ \text{mil}^2$. The capacitors were realized by polysilicon/polysilicon devices to achieve low voltage sensitivity; the resistors were realized by doped polysilicon lines.

5.6.2. Switched-Capacitor Filters in Tone Receivers

Signaling with tones (i.e., pure sine waves) is a common practice in communication systems. The generation of tones is relatively simple; however, their detection and decoding is quite complex. Typical signaling applications include paging, multifrequency signaling between switching systems, and the dual-tone multifrequency (DTMF) signaling used between push-button subscriber telephones and the central office. DTMF signaling is a high-volume application, and hence a very attractive one for integrated filters. The general principles and some specific examples of DTMF chips are discussed next.

The DTMF system uses eight voice-band frequencies for dialing. These are selected to avoid harmonically related interference from speech signals. They are divided into two groups, one containing the four low-frequency tones, the other the four high-frequency tones. When a push-button is depressed on the subscriber set, two tones are generated simultaneously, one for each group. Hence, $4 \times 4 = 16$ different combinations are possible. The resulting dual-tone signal is transmitted to the central office, where it must be correctly interpreted even in the presence of overlapping dial-tone and speech signals.

There exist several possible approaches for detecting the DTMF signal. In an early system [1] a low-pass/high-pass band-separating filter pair was used to separate the low and high tones. Each tone was then hard-limited to reduce the effects of level variations. Each of the resulting square waves was fed to a bank of four bandpass filters tuned to the four possible tone

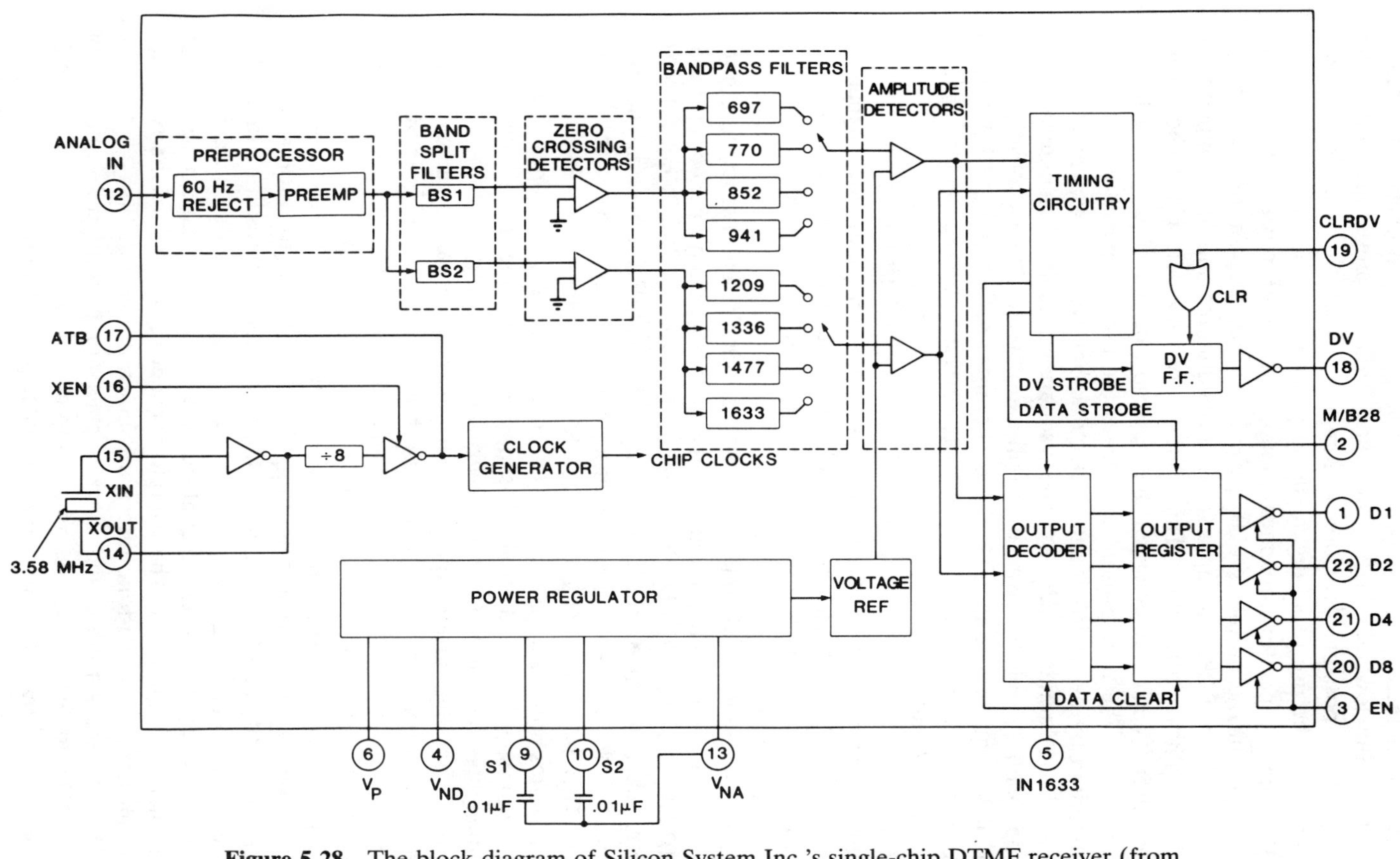

Figure 5.28 The block diagram of Silicon System Inc.'s single-chip DTMF receiver (from White et al. (1979) *I.E.E.E. J. Solid-State Circuits* **14**, 991–997 © 1979 I.E.E.E.).

frequencies. The output of each filter was monitored by an envelope detector. A detected digit was regarded as valid only if exactly one output from both the low and high groups had sufficient amplitude and proper timing. Other possible detection techniques include the counting of zero-crossings, or the use of all-digital signal processing.

A switched-capacitor implementation of the DTMF receiver was used by Silicon System Inc. in the SSI 201 [25]. The block diagram of the system is shown in Figure 5.28. The input signal is bandlimited and then applied to an SC band-splitting filter pair. This contains two six-pole Chebyshev band-reject filters, each implemented as a cascade of SC biquads. Each filter rejects one of the tone groups and passes the other. The filter outputs are hard-limited by two zero-crossing detectors that produce two square waves at the received tone frequencies. These square waves are then entered into a bandpass filter bank followed by amplitude detectors that choose the low- and high-frequency tone. Based on the information obtained from the detectors, the four bits of the digital word corresponding to the tone pair are found by the output decoder.

The complete receiver realizes 36 filter poles, and uses 37 op amps. It was integrated on a single $211 \times 223\,\text{mil}^2$ chip. The timing and clock signals are derived from a 3.58 MHz crystal oscillator and count-down circuits.

Other tone receivers use SC band-separating filters to perform the front-end functions, and then employ digital zero-crossing counters to identify the transmitted tones [3, 6, 9]. Several SC band-separating filters followed by digital detectors are available commercially, including the SC 11202-204 and SC 11270 by Sierra Semiconductor, and the Mitel MI 8870. The block diagram of the Sierra SC 11204 DTMF receiver is shown in Figure 5.29. The dual-tone *analog in* input signal is applied to the dial tone filter. This is a fifth-order high-pass SC filter, clocked at 55.93 kHz. It incorporates notches at 350 Hz and 450 Hz to reject any dial tones present in the input signal. In addition, it suppresses any line frequency and dc components and provides the lower skirt of the low-group path gain response. The high-group/low-group filter pair carries out the actual separation of the tones in the two groups. Both are sixth-order elliptic SC filters, with high-pass and low-pass characteristics, respectively. Both are clocked at a 55.93 kHz rate. The output of these filters are resampled at a four times higher (223.6 kHz) clock rate, and then filtered by two second-order low-pass smoothing filters. The smoothing filters contain auto-zeroing feedback to eliminate any dc offset voltage in their outputs. The resulting tones then enter two zero-crossing detectors. The hard-limited square waves from the detectors are applied to two digital frequency detectors that determine their fundamental frequencies. The overall frequency responses of the two channels (including the dial-tone filter) are shown in Figure 5.30.

The necessary timing and clock signals are generated by an on-chip 3.579 MHz oscillator that uses an off-chip crystal, and by on-chip frequency dividers. The complete chip realizes 28 filter poles, and uses 28 op amps for

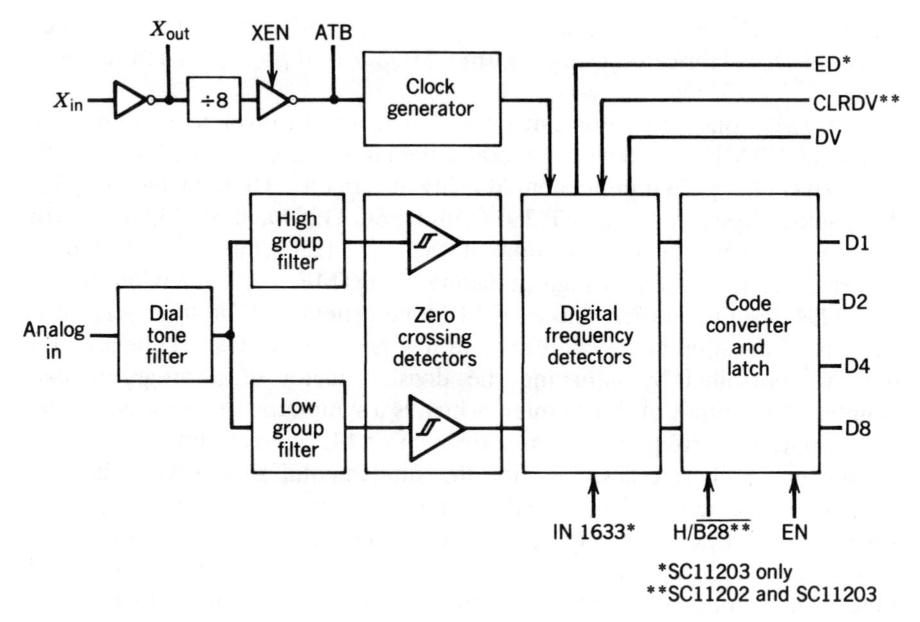

Figure 5.29 Block diagram of the SC 11202, SC 11203, and SC 11204 DTMF receivers.

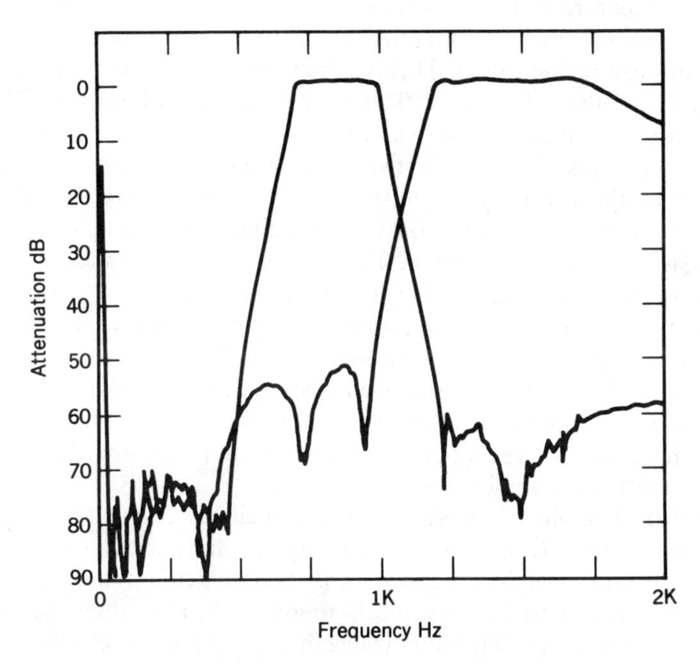

Figure 5.30 Frequency response of the band-split filter in the SC 11204 DTMF receiver.

this purpose. It also contains two comparators and two digital frequency detectors. It is fabricated using a 3 μm CMOS n-well process. The total chip area is $132 \times 132\ \text{mil}^2$.

A DTMF tone generator can be added to the receiver to implement a complete DTMF transceiver. Several commercial single-chip MOS DTMF transceivers have been implemented using SC circuits. These include the SSI 289 (Silicon Systems Inc.), MT 280 (Mitel), SC 11289 and SC 11280 (Sierra Semiconductor). The organization of the SC 11280 chip is illustrated in Figure 5.31. It contains a high-performance DTMF receiver similar to the SC 11204 describe earlier, and a DTMF tone generator. The tone generator consists of programmable counters and low-pass filters. The frequency of a tone is controlled by adjusting the division factor of a programmable counter. The output of the counter, which is a symmetric square wave of the appropriate tone frequency, passes through an SC low-pass filter. The filter attenuates the harmonics, but not the fundamental sine wave. The frequency of the clock signal used by the SC filter is 32 times the tone frequency, and hence can be generated by the same counter as the tone.

Two similar circuits are employed to produce the low-band and high-band tones. These tones are then combined, using another SC filter clocked at a high rate. The amplitudes of the high-group tones are higher by about 2 dB than those in the low group, to compensate for the larger loss at higher frequencies on long line loops.

A special feature of the transceiver is the "Call Progress" mode. In this mode the various tones that identify the progress of a telephone call across the network can be identified. This is achieved by passing these tones (which normally fall in the 300 Hz to 650 Hz range) through a bandpass filter. The filter consists of a cascade combination of a low-band low-pass filter and a dial-tone high-pass filter, with the clock frequency scaled to shift the pass-band to the call-progress tone frequency band. Input signals whose frequencies are within the filter pass-band are hard-limited by a comparator, and appear at an output terminal. This output voltage can then be analyzed by a microprocessor to determine the status of the call.

In addition to multifrequency signaling, single-frequency (SF) signaling is also used extensively in telephone systems for line supervision. The most commonly used SF signaling tone frequency is 2,600 Hz. Whenever the line between two switching sites is idle (i.e., not used for voice transmissions) a 2,600 Hz tone with a large amplitude is connected to the line. When the line is in use, the tone is removed.

The block diagram of an SF signaling circuit used to implement such a supervisory system is shown in Figure 5.32. In normal operation, the four-wire transmission line between the two switching sites A and B is continuously monitored. If the line is found to be idle, then the circuit at each site applies a 2,600 Hz to its transmit (TRX) port by closing switch C. The idleness of the channel can now be established by the energy detect circuit, on the basis of the output amplitudes of the two preceding filters.

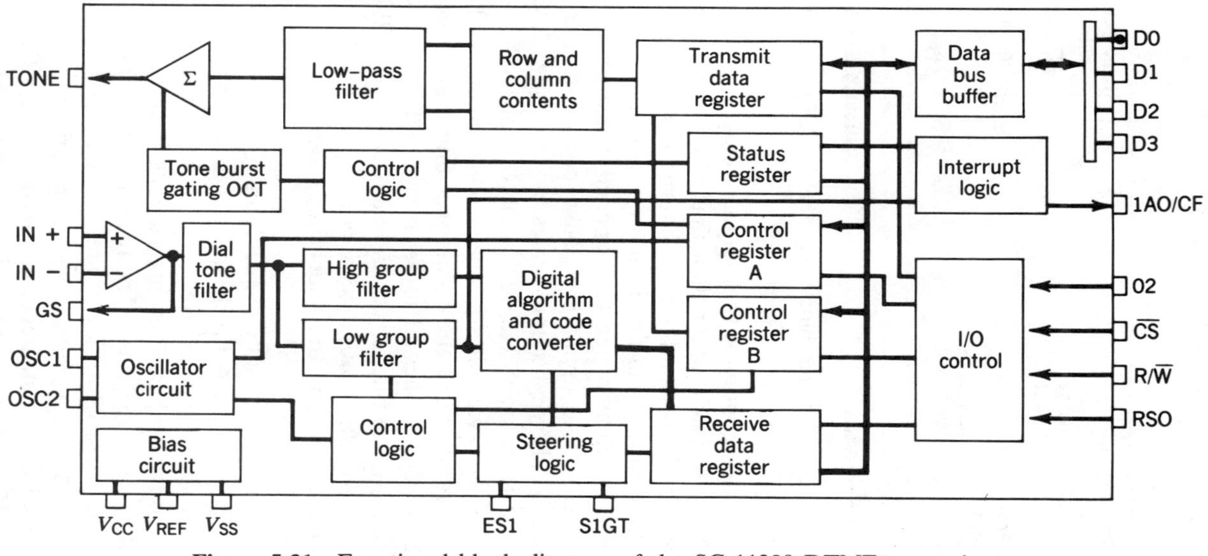

Figure 5.31 Functional block diagram of the SC 11280 DTMF transceiver.

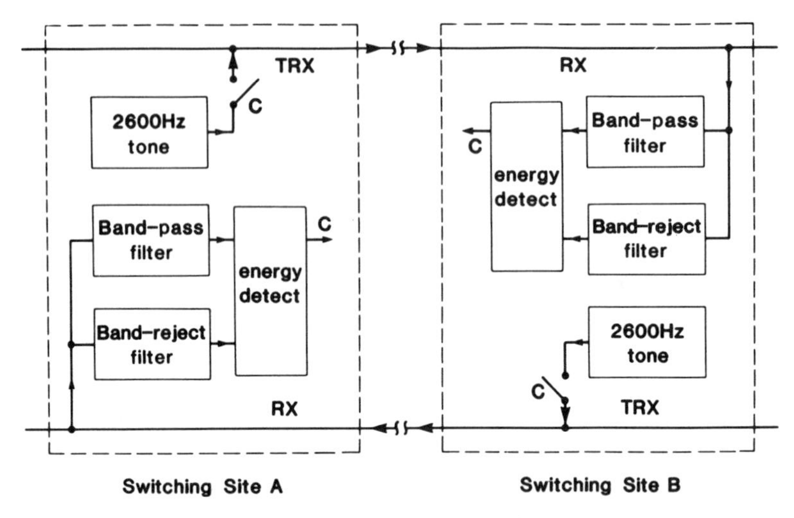

Figure 5.32 The functional block diagram of a SF signaling system.

Specifically, under this condition the output voltage of the bandpass filter (tuned to 2,600 Hz) is much larger than that of the bandreject filter which suppresses 2,600 Hz. This condition is interpreted by the energy detect circuit as an indication that the line is unused. If now a call is initiated, for example from site A, then switch C is opened at site A and the 2,600 Hz tone is removed from the upper line. Now both filter outputs at site B become very small. This condition indicates to the energy detect circuit that the lines are about to be used. Hence, switch C is opened at site B as well, and now both lines are available for voice transmission. While the lines are being used, the outputs from the bandreject filters are much larger than those from the bandpass ones. This is interpreted by the energy detect circuits as an indication of line usage, and the C switches will be kept open. The purpose of this elaborate detection system is to avoid premature activation or deactivation of the supervisory system that may otherwise occur.

An example of a commercially available integrated SC filter chip for SF signaling is the AMI S3526. A block diagram is shown in Figure 5.33. The bandpass filter is a sixth-order SC Butterworth filter realized as a cascade of three biquads. The bandreject filter also consists of three cascaded biquads, but it has an elliptic loss response. The measured responses of the filter are illustrated in Figure 5.34. Both filters are clocked at $f_c = 128$ kHz. All time base and clock signals are generated by an on-chip crystal oscillator (which requires an external crystal element) and frequency dividers. The chip also contains a 2,600 Hz tone generator that is basically a second-order SC bandpass filter tuned to 2,600 Hz and driven by a 2,600 Hz square wave generated by the frequency dividers. The bandpass filter rejects the third- and higher-order harmonics of the square wave, and thus its output is the

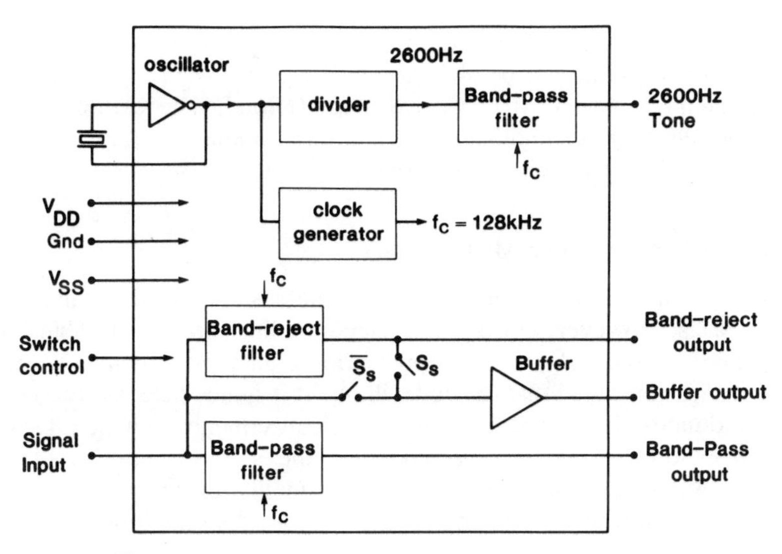

Figure 5.33 The AMI S3526 SF signaling filter unit.

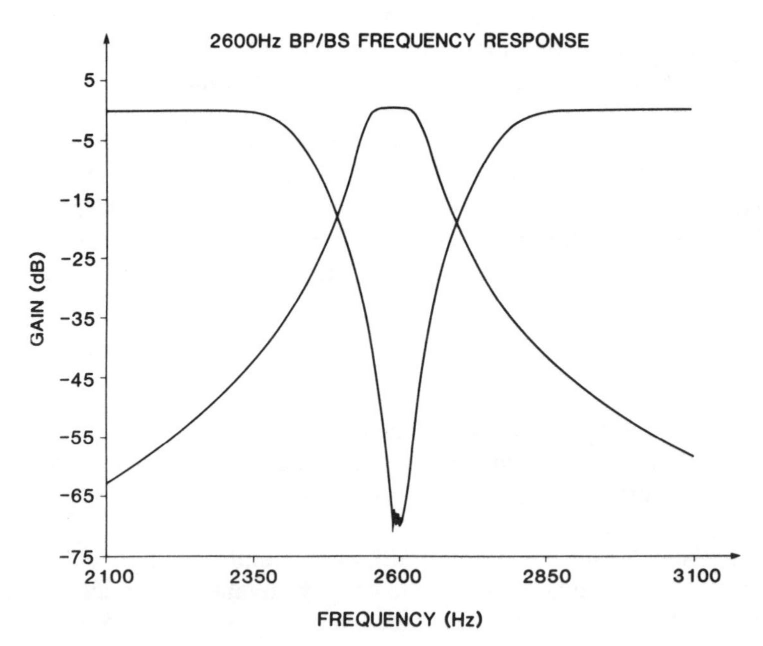

Figure 5.34 Typical frequency response of the 2600-Hz bandpass/bandreject filters in the AMI S3526 signaling filter unit.

desired sine-wave signal. In addition, an on-chip buffer is also provided, whose input terminal can be switched either to the signal input or to the bandreject filter output. It can drive a 600 Ω load. The complete chip is fabricated in CMOS technology. It contains 16 op amps, and realizes a total of 12 filter poles. The chip area is $163 \times 131\ \text{mil}^2$.

5.6.3. Switched-Capacitor Modem Filters

Telephone transmission facilities are now increasingly used for data transmission tasks. However, voice-grade telephone lines are not suitable for the transmission of the pulse trains representing binary bit streams. To permit data transmission over voice-grade facilities, it is hence necessary to convert (i.e., modulate) the digital data into a voice-frequency signal. Of course, this conversion has to be reversed (i.e., the signal demodulated) after transmission. The circuit performing the **mo**dulation-**dem**odulation operations is called a **modem**. The most commonly used modulation schemes are **frequency-shift keying** (FSK) and **phase-shift keying** (PSK). In an FSK modem, the two binary states of the data are represented by two different carrier frequencies. In a two-phase PSK modem, by contrast, a single carrier frequency is used, and its two possible phase angles represent the two binary states.

A large part of the circuitry (and hence of the cost) of a modem is needed to carry out filtering. In early modems, these filters utilized discrete active RC circuitry that required a lengthy (and costly) trimming process to achieve the specified accuracy. Recently, however, SC filters have been employed to implement fully-integrated untrimmed modems.

Several SC filters are now commercially available for 1,200 bits/s modem applications. These include the Reticon R5632, the AMI S3522 and S35212, and the Sierra SC 11005. As an example, the latter will be briefly described. The block diagram of the SC 11005 is shown in Figure 5.35. The filter can be used in a two-wire full-duplex system, where the modem must transmit and receive data simultaneously. It must also meet the appropriate specifications (Bell System 212A and CCITT V.22) for the modem. The chip includes a high-band (2,400 Hz) and a low-band (1,200 Hz) filter, and amplitude as well as group-delay equalizers for both filters. To meet the CCITT V.22 requirements, a notch filter that can be programmed for either a 550 Hz or an 1,800 Hz notch frequency is provided. Two unassigned op amps, that can be used as part of anti-aliasing filters or gain control, are also included. The chip also supports analog loop-back (ALB) and call-progress monitoring (CPM) operations.

The low-band path contains a tenth-order bandpass filter with a center frequency of 1,200 Hz. This is followed by a tenth-order all-pass delay equalizer that compensates for the pass-band delay variations of the filter, as well as those of a half-length compromise line. In the originate mode, the low-band filter is used in the transmit direction; in the answer mode, it is

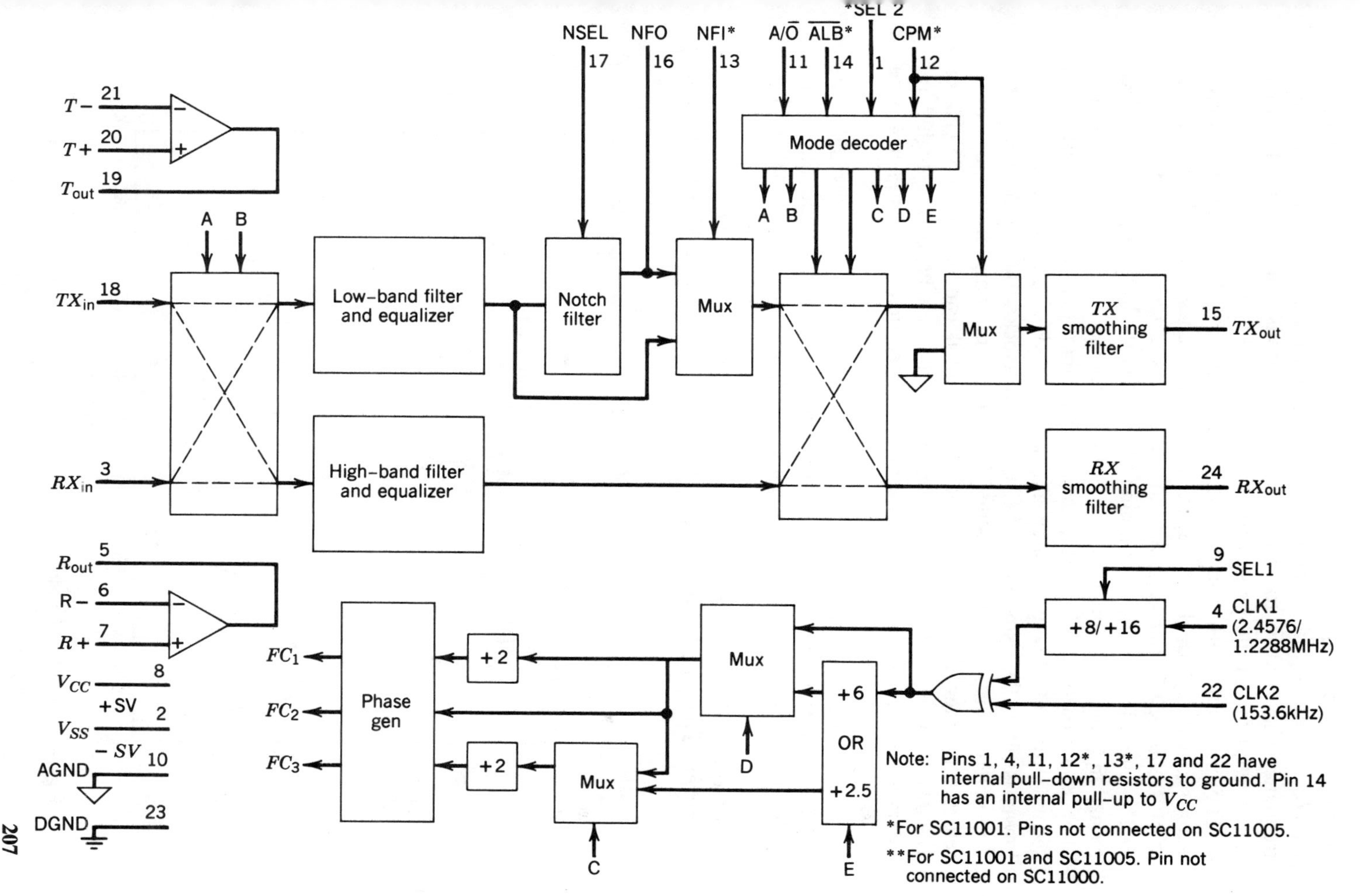

Figure 5.35 The functional block diagram of the SC 11005 modem filter.

used in the receive direction. In the CPM mode, the clock frequency of the filter/equalizer is scaled down by a factor of 2.5, so that the center frequency shifts down to 480 Hz, and the various tones (dial tone, busy signal, etc.) fall in the pass-band.

The high-band path also consists of a tenth-order bandpass filter and a tenth-order delay equalizer, but the center frequency is now 2,400 Hz. The clock frequency is 76.8 kHz in both the high- and low-band paths.

The chip also contains smoothing filters in both the transmit and receive paths, and the V.22 notch filter mentioned above. The transmit smoothing filter is a first-order SC filter; the receive smoothing filter consists of a second-order SC low-pass filter and a cascaded second-order active RC low-pass filter. The smoothing SC filters are clocked at a 153.6 kHz rate. The notch filter is a programmable SC biquad. The center frequency can be selected through the notch-select (NSEL) pin; it can be either 550 Hz or 1,800 Hz. Typical amplitude responses for the low-band and high-band filters are shown in Figure 5.36 and the equalized delay responses in Figure 5.37.

Another example of 1,200 bits/s PSK modem filter chips is the AMI S3522. This chip contains a high-band filter and a low-band filter. The high-band filter is a tenth-order bandpass SC filter centered at 2,400 Hz. It is followed by a fourth-order SC delay equalizer. The low-band filter contains a 14th-order SC bandpass filter centered at 1,200 Hz, and a sixth-order delay equalizer. The pass-bands of both bandpass filters are 800 Hz wide. Both

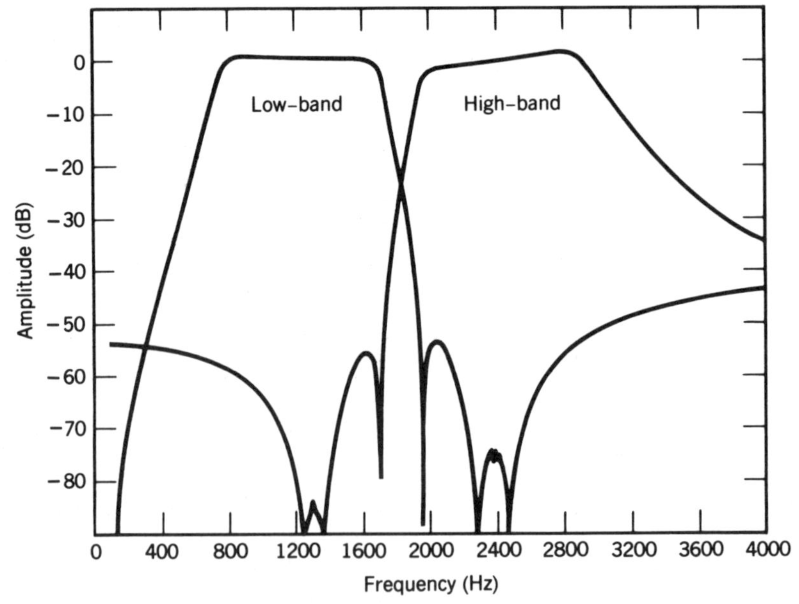

Figure 5.36 High-band and low-band filter responses for the SC 11005.

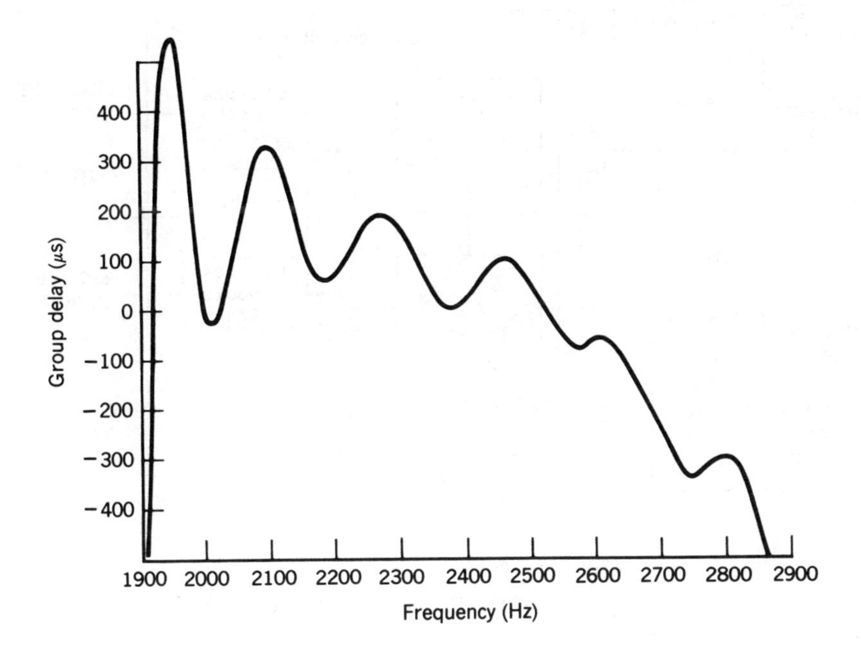

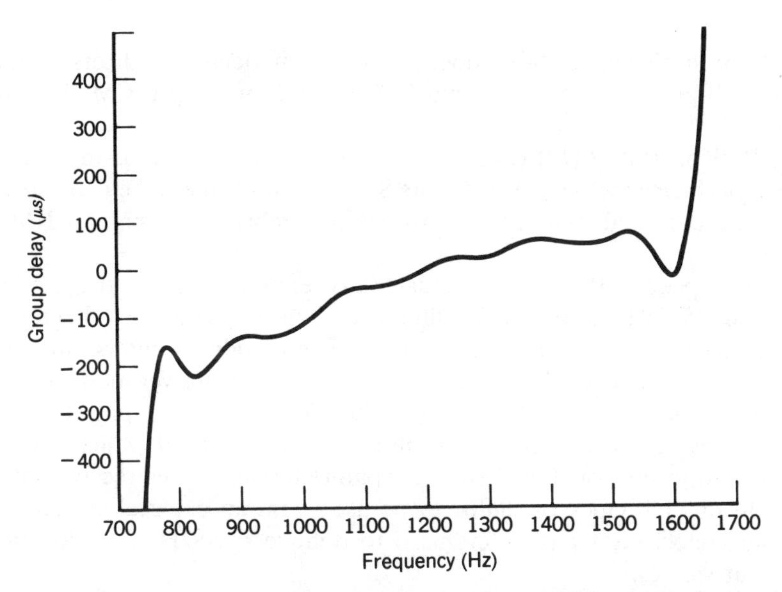

Figure 5.37 High-band and low-band group-delay responses.

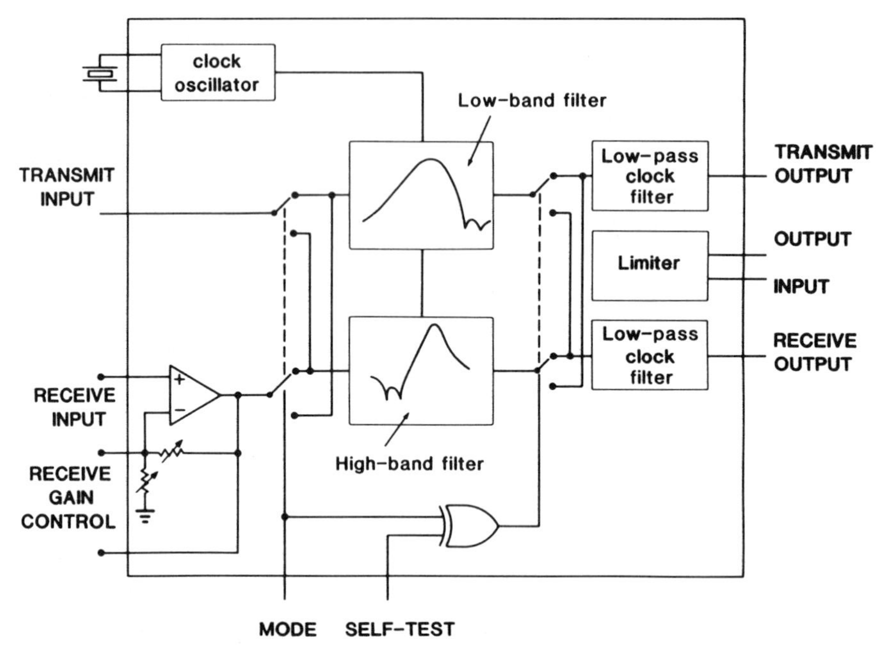

Figure 5.38 The block diagram of the Reticon R5630 and R5631 300-baud modem filter sets.

filters were designed as SC ladder filters; both delay equalizers contain cascaded all-pass biquads. The chip is fabricated in a 5 μm CMOS *p*-well process.

The Reticon R5632 [17] is somewhat similar to the SC 11005 in architecture and performance. However, all its SC filters as well as all equalizers are realized by cascaded biquads, and the chip is fabricated using an NMOS process.

For low-speed (300 bits/s) originate/answer applications, the Reticon R5630 and R5631 band-separation filters can be used. The block diagram of these devices is shown in Figure 5.38. The bandpass filters are both tenth-order SC circuits. The choice of the originate or answer mode is made by a TTL-compatible signal entered at the "Mode control" pin. The chip also contains a receive gain control stage, the gain of which can be externally adjusted from 0 to 30 dB; a separate limiter for the receive output signal; an on-chip clock oscillator to be used with an external crystal; and smoothing (clock) low-pass filters used to remove the S/H distortion from the output signals.

5.6.4. Single-Chip Modems

The first single-chip modems to become commercially available were 300 bits/s devices, satisfying the Bell System 103 and CCITT V.21 specifications. They included the TMS 99532 (Texas Instruments), the 74 HC 942-3 (National Semiconductor), and the SC 11002-3 (Sierra Semiconductor). All use SC techniques to perform the FSK modulation and demodulation tasks, and all are full-duplex modems. The block diagram of the SC 11002-3 is shown in Figure 5.39. Included on-chip are the high-band and low-band filters, an FSK modulator and demodulator, a line driver for driving directly a 600 Ω telephone line, and another op amp that can be connected as a two-wire to four-wire converter.

In the transmit path, the digital input data arriving at the TXD pin are FSK modulated. To achieve the separation of the transmitted and received signals, the originating modem transmits in the low band, while the answering one transmits in the high band. The modulated carrier is then applied to the transmit filter, which bandlimits and smooths it. The center frequency of this filter is either 2,125 Hz or 1,170 Hz, depending on whether the modem is used in the answering or originating mode, respectively. The transmit filter is an eighth-order SC bandpass filter, realized by four cascaded biquads. The output of the filter appears at the output buffered by the line driver.

In the receive direction, the modulated analog signal is received and buffered by a two-wire to four-wire converter. It then enters the receive filter, which is a bandpass filter similar to the transmit filter. However, if the

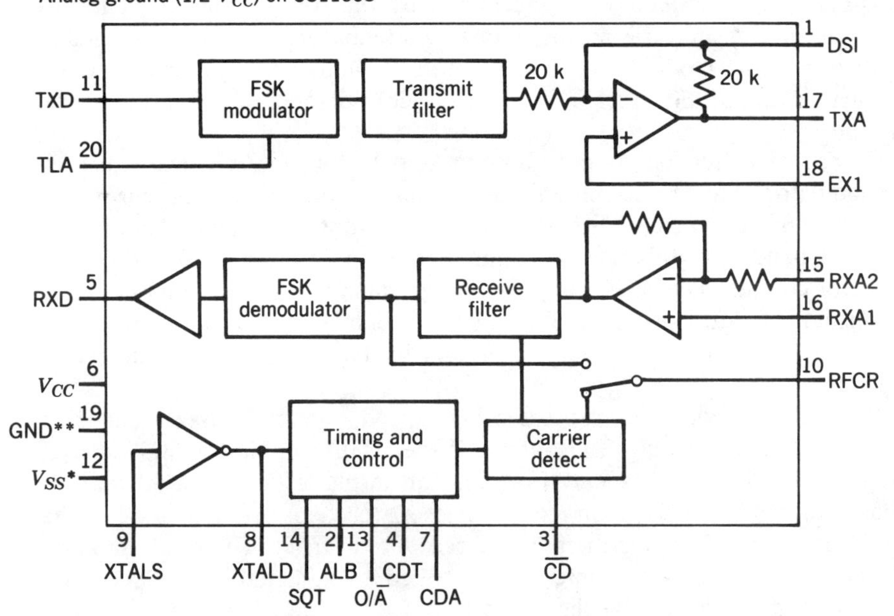

Figure 5.39 The functional block diagram of the SC 11002 and 11003 modems.

transmit filter is centered at 2,125 Hz, then the receive filter is at 1,170 Hz, and vice versa. The output signal of the receive filter is hard-limited, and then applied to the FSK demodulator, which recovers the digital data. The chip is fabricated in 3 μm *n*-well CMOS process. Its area is $150 \times 150\,\text{mil}^2$.

The next generation of single-chip modems consisted of full-duplex, 300/1,200 bits/s systems satisfying the Bell System 212A and the CCITT V.22 requirements. These chips contain all circuits necessary for FSK and PSK modulation and demodulation, high- and low-band filtering, and so on.

Among the commercially available single-chip 300/1,200 bits/s modems are the Sierra Semiconductor SC 11004/SC 11014, the Silicon Systems SSI K212, and the Fairchild μA212AT. As an example, the organization of the SC 11004/14 is briefly discussed. Its block diagram is shown in Figure 5.40. The transmit path contains a 1,200 bits/s PSK modulator and a 300 bits/s FSK modulator. The PSK modulator consists of a data buffer, a synchronizer (i.e., asynchronous-to-synchronous converter), a data scrambler, and a four-phase PSK modulator satisfying the V.22 and 212A specifications. The FSK modulator satisfies the V.21 and 103 recommendations. In the high-speed (1,200 bits/s) mode, the PSK modulator is used; in the low-speed (300 bits/s) mode, the FSK modulator is connected.

The transmit and receive filters are similar to those contained in the SC 11005 described earlier. The transmit side of the chip also contains several tone generators, including a DTMF generator, an 1,800/550 Hz guard tone generator and a 2,100/1,300 Hz tone generator. These generators are all realized by digital frequency synthesizers, followed by SC filters whose pass-band edge frequency is determined by the tone frequency. The last stage of the transmitter is a programmable attenuator that can be changed in 3-dB steps. The modulators, tone generators, filters and attenuator require altogether about 50 filter poles and a considerable amount of digital circuitry.

The receiver section contains an energy detector, an AGC stage, a PSK demodulator, an FSK demodulator, a descrambler and a synchronous/asynchronous converter. The received signal is routed through the appropriate (low-band or high-band) filter, and applied to both the AGC and energy detector circuits. The AGC stage is a programmable-gain SC amplifier. The gain is adjusted to a preset level by a circuit containing two comparators and some digital logic. The energy detector is based on a peak detection algorithm.

The PSK demodulator processes high-speed signals. It uses a coherent demodulation technique. The output of the AGC amplifier is applied to a dual-phase splitter that separates it into an in-phase (I) and a 90° out-of-phase quadrature (Q) component. These are then demodulated to the base-band, by multiplying them by the recovered carrier. The resulting I and Q signals are filtered, summed and applied to a digital phase-lock loop (DPLL) that produces the digital data in a dibit (i.e., two-bit) code. The carrier is recovered by another DPLL that uses the difference of the I and Q signals.

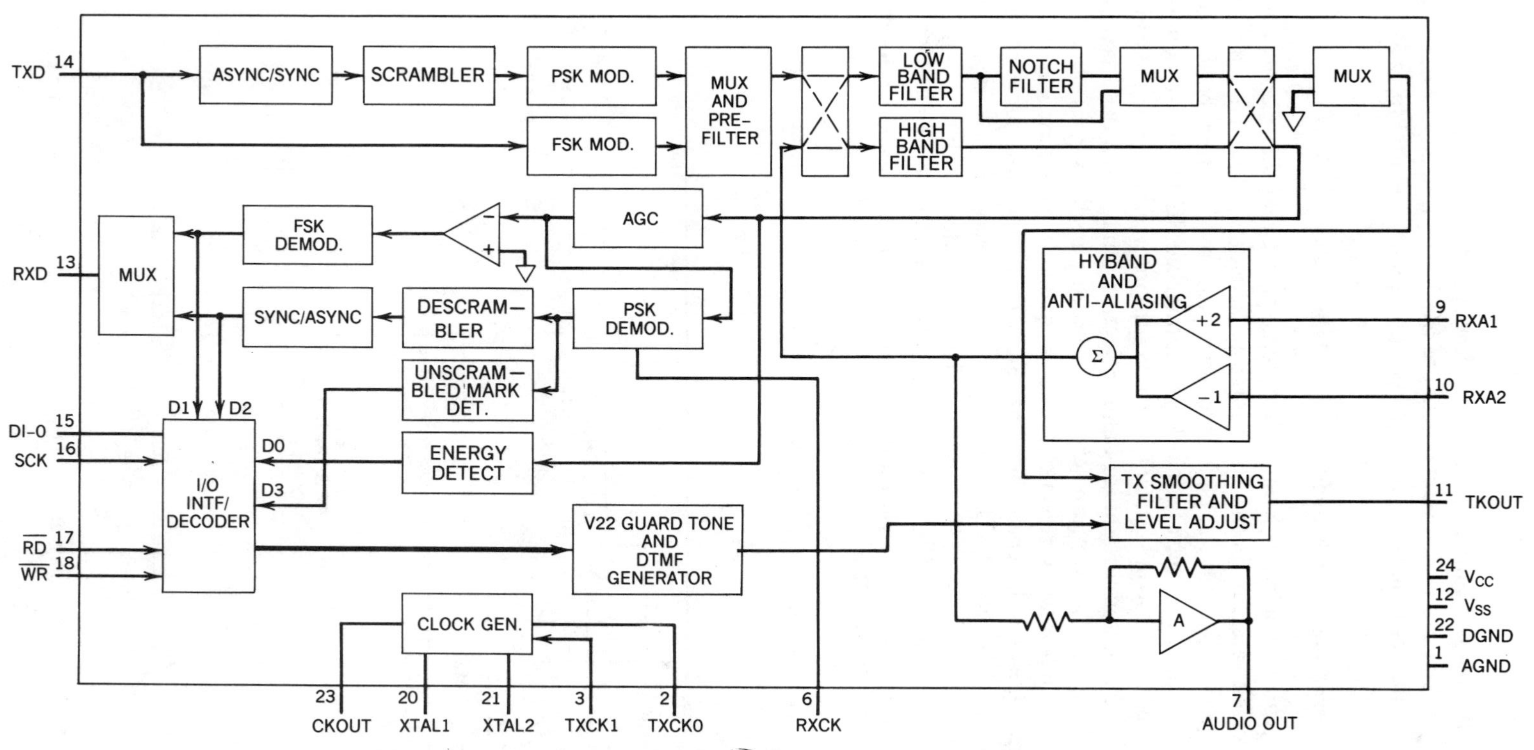

Figure 5.40 The functional block diagram of the SC 11004/14 modems.

The FSK demodulator is used only in low-speed operation. Its operation is similar to that of the demodulator of the SC 11003, discussed earlier. The two demodulators require 12 filter poles, and assorted digital logic.

5.6.5. Programmable Switched-Capacitor Filters

SC filters can be made programmable by changing either their capacitances, or their clock rates, or both. The simplest technique is to change only the clock frequency, f_c. The complete frequency response will then be scaled by the same factor as f_c. Several clock-programmable SC filters are commercially available. Reticon's R5609 is a seventh-order elliptic low-pass filter, with a 75-dB minimum out-of-band rejection and a 0.5-dB maximum pass-band ripple. The Reticon R5613 is a linear-phase low-pass filter, with over 60 dB out-of-band rejection. The frequency characteristics of both filters are adjustable by changing their clock frequency. Both are fabricated in a double-polysilicon NMOS process. The Reticon R5611 is a clock-programmable fifth-order Chebyshev high-pass filter, with a 30-dB per octave roll-off and a 0.6-dB maximum pass-band ripple. Reticon's R5612 is a fourth-order notch filter with a tunable notch frequency. At the notch, at least 50 dB rejection is guaranteed.

The AMI S3528 is a programmable SC low-pass filter with a seventh-order quasi-elliptic response. Its block diagram is shown in Figure 5.41. The

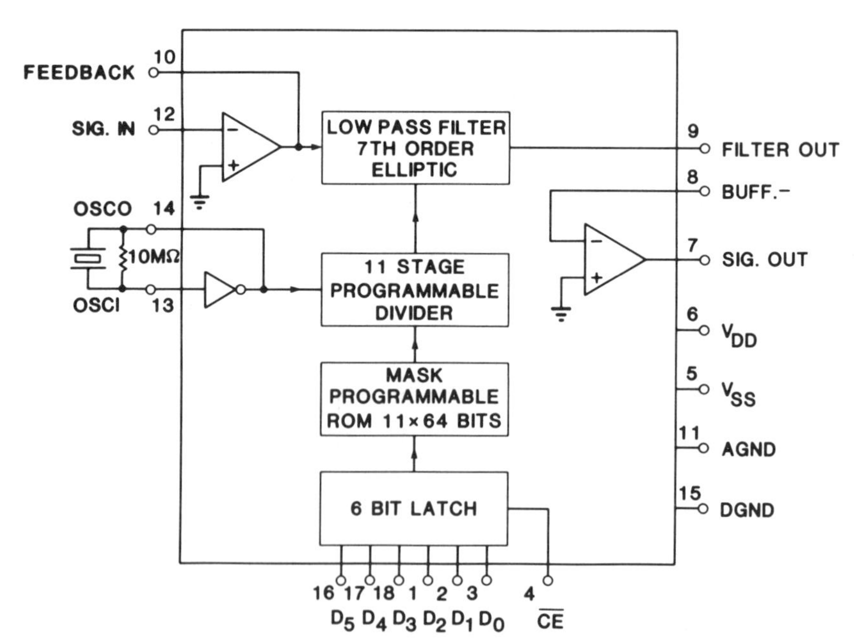

Figure 5.41 The block diagram of the AMI S3528 programmable low-pass filter.

analog signal is entered through an op amp which, with some external resistors and capacitors, can amplify or low-pass filter the signal. The output of the op amp passes through the SC filter. The pass-band edge of the filter is at $f_c/40$, where f_c is the programmable clock frequency. The stop-band frequency is around $f_c/30$; the pass-band ripple is around 0.1 dB and the out-of-band rejection over 51 dB. The circuit was designed as a bilinear SC ladder network (cf. §5.4). Its typical frequency response is shown in Figure 5.42. If necessary, the output signal of the filter can be connected to an on-chip buffer amplifier that is capable of driving a 600 Ω load.

The clock frequency f_c of the filter is obtained by driving the frequency of a crystal oscillator with an off-chip crystal element. The setting of f_c is achieved by programming the divider via a six-bit input digital word. The input word is translated by an on-chip ROM into a 11-bit divider ratio. Thus, f_c can assume any one of $2^6 = 64$ possible values. For proper operation, f_c must be below 500 kHz. The overall dimension of the chip is 130×103 mil^2. It is fabricated in a 5 μm CMOS technology.

A recent entry to the programmable SC filter market is National Semiconductor's MF8. It contains two SC bandpass biquads and an unassigned op-amp. The two biquads are identical. They can be used, for example, as two tracking filters, or in cascade as a fourth-order bandpass filter. The block diagram of the MF8 is shown in Figure 5.43. The band center frequency of the filters can be controlled by an external clock, and may be set anywhere between 0.1 Hz and 20 kHz. The ratio of the clock rate to the center frequency can be selected either as 100:1 or as 50:1. By using

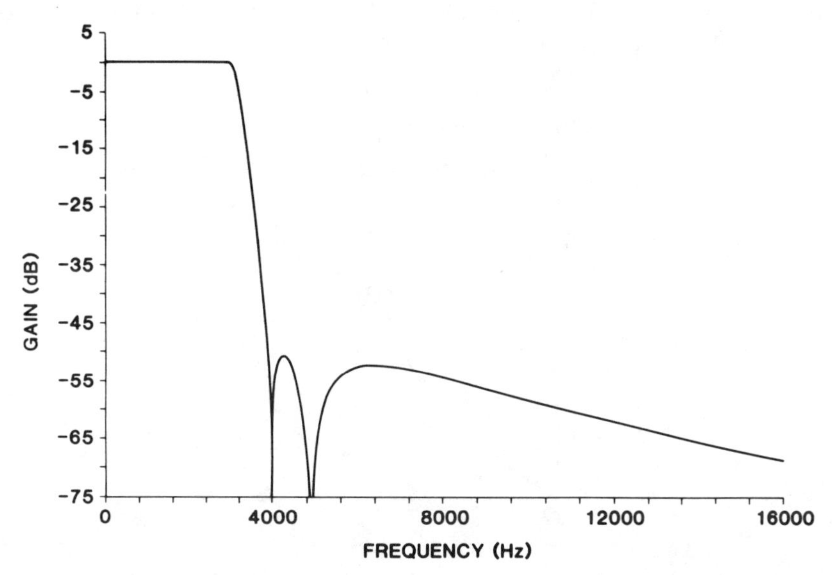

Figure 5.42 Measured frequency response of the AMI S3528 programmable low-pass filter.

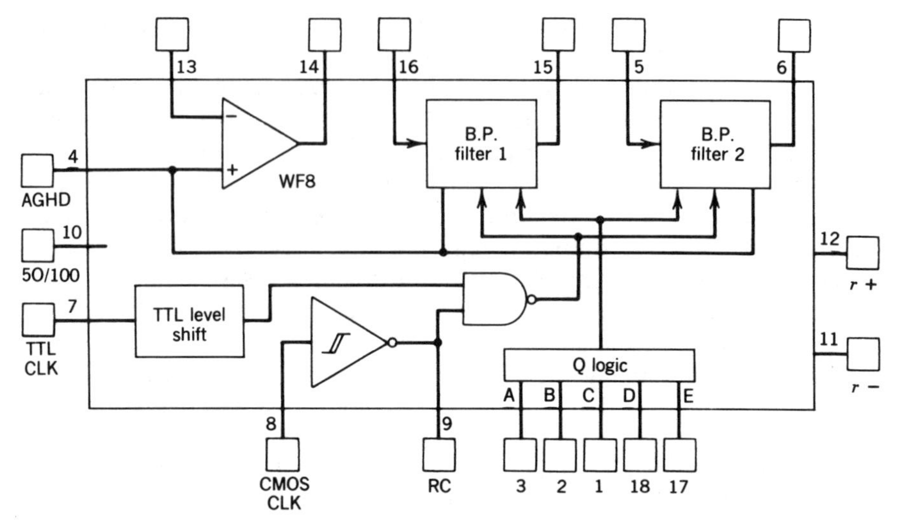

Figure 5.43 Simplified block diagram of the MF8 programmable SC filter. (Reprinted with permission of National Semiconductor Corp.)

the five Q-logic pins, the pole-Qs of the biquads can also be programmed to any of 31 available values, ranging from 0.5 to 90. Figure 5.44 illustrates the gain responses for four typical Qs. To achieve higher-order responses, the biquads of one or more MF8 chips may be combined either directly or by using the on-chip uncommitted op amp and external resistors. As an example, Figure 5.45 illustrates the block diagram of an eighth-order bandpass filter obtained by using two MF8s.

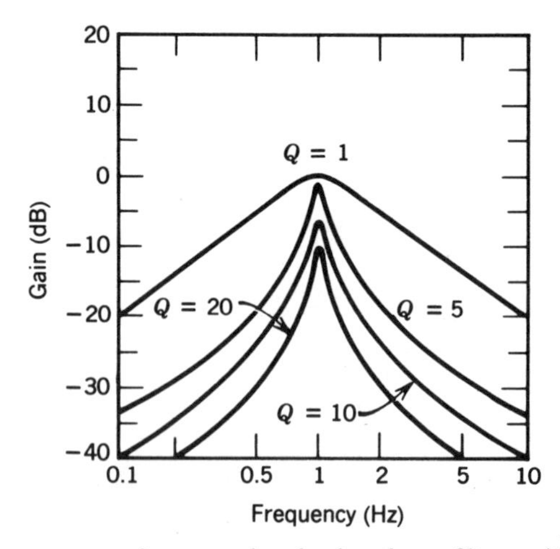

Figure 5.44 Gain responses for second-order bandpass filters with various values of Q. (Reprinted with permission of National Semiconductor Corp.)

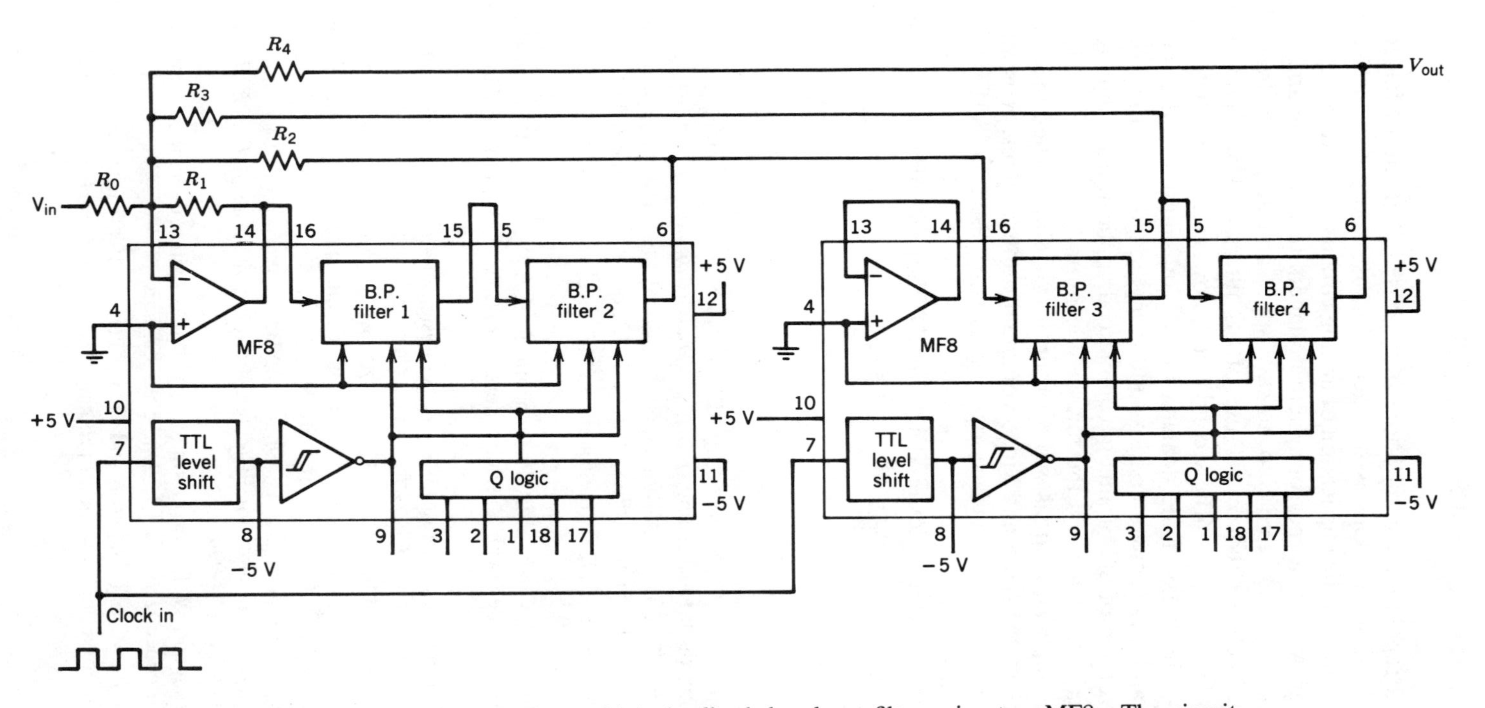

Figure 5.45 Eighth-order multiple-feedback bandpass filter using two MF8s. The circuit shown accepts a TTL-level clock signal and has a clock-to-center frequency ratio of 100:1. (Reprinted with permission of National Semiconductor Corp.)

5.7. FUTURE APPLICATIONS FOR SC FILTERS

In trying to extrapolate the applications of SC-based filters in the future, two trends become apparent. One is the increasing use and potential of digital filters, which restricts the application of analog circuits in many systems. The other is the growing tendency of system designers to put many analog and digital functions on the same chip, to save space and dc power drain. The first trend is likely to reduce the use of SC filters as stand-alone signal processors in large systems. The second trend, however, will probably result in a continued increase of the demand for SC anti-aliasing and smoothing filters, A/D and D/A converters, loss and delay equalizers, and so on. On balance the volume of SC circuits in commercial applications should remain at least on its current level in the foreseeable future.

Many new application areas will open up for SC filters when their speed will be sufficient for radio and video communication systems. It is already possible to use them to replace discrete circuits used in AM and FM communication receivers. These represent a very high volume market, and if the cost per chip can be lowered sufficiently, they will offer a new, large, and lucrative application area for SC filters.

REFERENCES

1 R. N. Battista, C. G. Morrison and D. H. Nosh (1963) Signaling system and receiver for touch-tone calling. *I.E.E.E. Trans. Commun. Electron.* **82**, 9–17.

2 W. C. Black, Jr., D. J. Allstot and P. A. Reed (1980) A high-performance low-power CMOS channel filter. *I.E.E.E. J. Solid-State Circuits* **15**, 929–939.

3 M. J. Callahan Jr. (1979) Integrated DTMF receiver. *I.E.E.E. J. Solid-State Circuits* **14**, 85–90.

4 J. T. Caves, C. H. Chan, S. D. Rosenbaum, L. P. Sellars and J. B. Terry (1979) A PCM voice codec with on-chip filters. *I.E.E.E. J. Solid-State Circuits* **14**, 65–73.

5 R. B. Datar and A. S. Sedra (1983) Exact design of strays-insensitive switched-capacitor ladder filters. *I.E.E.E. Trans. Circuits Systems* **30**, 888–898.

6 T. Foxall, R. Whitbread, L. Sellars, A. Aitken and J. Morris (1980) A switched-capacitor band split filter using double polysilicon oxide isolated CMOS. *I.S.S.C.C. Tech. Digest* **Feb.**, 90–91.

7 P. R. Gray, D. Senderowicz, H. Ohara and B. M. Warren (1979) A single-chip NMOS dual-channel filter for PCM telephony applications. *I.E.E.E. J. Solid-State Circuits* **14**, 981–990.

8 R. Gregorian and W. E. Nicholson (1979) CMOS switched-capacitor filters for a PCM voice codec. *I.E.E.E. J. Solid-State Circuits* **14**, 970–980.

9 R. Gregorian, W. E. Nicholson and G. C. Temes (1980) Integrated band-split filter system for a dual-tone multifrequency receiver. *Microelectron. J.* **11**, 5–12.

10 R. Gregorian, G. A. Wegner and W. E. Nicholson (1981) An integrated single-chip PCM voice codec with filters. *I.E.E.E. J. Solid-State Circuits* **16**, 322–333.

11 R. Gregorian, K. W. Martin and G. C. Temes (1983) Switched-capacitor circuit design. *Proc. I.E.E.E.* **71**, 941–966.

12 R. Gregorian and G. C. Temes (1986) *Analog MOS Integrated Circuits for Signal Processing*: Wiley Interscience, New York.

13 K. Haug, G. C. Temes and K. W. Martin (1984) Improved offset compensation schemes for SC circuits. *Proc. I.E.E.E. Internat. Symp. Circuits and Systems*, 1054–1057.

14 Y. A. Haque, R. Gregorian, R. W. Blasco, R. A. Mao and W. E. Nicholson Jr. (1979) A two-chip PCM voice codec with filters. *I.E.E.E. J. Solid-State Circuits* **14**, 961–969.

15 M. E. Hoff Jr., J. Huggins and B. M. Warren (1979) An NMOS telephone codec for transmission and switching applications. *I.E.E.E. J. Solid-State Circuits* **14**, 47–53.

16 A. Iwata, H. Kikuchi, K. Uchimura, A. Morino and M. Nakajima (1981) A single-chip CMOS codec with switched-capacitor filters. *I.E.E.E. J. Solid-State Circuits* **16**, 315–321.

17 L. T. Lin, H. F. Tseng and L. Querry (1982) Monolithic filters for 1200 baud modems. *I.S.S.C.C. Dig. Tech. Papers* **Feb.**, 148–149.

18 D. G. Marsh, B. K. Ahuja, T. Misawa, M. R. Dwarakanath, P. E. Fleischer and V. R. Saari (1981) A single-chip CMOS PCM codec with filters, *I.E.E.E. J. Solid-State Circuits* **16**, 308–315.

19 J. L. McCreary and P. R. Gray (1975) All-MOS charge redistribution analog-to-digital conversion technique—Part I. *I.E.E.E. J. Solid-State Circuits* **10**, 371–379.

20 H. J. Orchard (1966) Inductorless filters. *Electron. Lett.* **2**, 224–225.

21 H. J. Orchard, G. C. Temes and T. Cataltepe (1975) Sensitivity formulas for terminated lossless 2-ports. *I.E.E.E. Trans. Circuits Systems* **32**, 459–466.

22 J-B. Shyu, G. C. Temes and F. Krummenacher (1984) Random error effects in matched MOS capacitors and current sources. *I.E.E.E. J. Solid-State Circuits* **19**, 948–955.

23 G. C. Temes (1986) Simple formula for estimation of minimum clock-feed-through error voltage. *Electron. Lett.* **22**, 1069–1070.

24 Y. Tsividis, Mihai Banu and J. Khoury (1986) Continuous-time MOSFET-C filters in VLSI. *I.E.E.E. J. Solid-State Circuits* **21**, 15–30.

25 B. J. White, G. M. Jacobs and G. F. Landsburg (1979) A monolithic dual-tone multifrequency receiver. *I.E.E.E. J. Solid-State Circuits* **14**, 991–997.

26 K. Yamakido, T. Suzuki, H. Shirasu, M. Tanaka, K. Yasunari, J. Sakaguchi and S. Hagiwara (1981). A single-chip CMOS filter/codec. *I.E.E.E. J. Solid-State Circuits* **16**, 302–307.

6 Charge-Transfer Device Filters

JOHN MAVOR
University of Edinburgh, Scotland, UK

The origins of charge-transfer device (CTD) filters are traced in this chapter, and their key operational features and performance bounds are described. Practical design aspects are treated and significant CTD filtering modules are introduced and practical examples given. The general impact of CTD filtering is discussed, and future trends in this versatile filter family are also outlined.

6.1. THE ORIGINS OF CHARGE-TRANSFER DEVICES

The potential for realizing electrical filters in miniature form has been evident since silicon integrated circuit (IC) technology matured in production in the late 1960s. Early attempts at integrated, active RC filters were not very successful because of the inability at that time to define critical circuit components to close tolerance, and improvements in this area are detailed in Chapter 3. Although it has always been evident that filter realizations employing digital methods should not be as susceptible to component tolerances and variation, the then low integration potential of silicon technology implied that only the most primitive digital filter structures could be implemented.

The introduction of the CTD family in 1969 [3, 17] enabled high performance time- and frequency-domain filters to be achieved in solid-state form, by teaming up discrete-time analog signal processing concepts with MOS LSI (metal-oxide-silicon technology at large-scale integration density). Further, charge-transfer filtering became a reality because the processing is performed directly on sampled-data *analog* signals, rather than by adopting more complicated digital techniques. Thereby, the considerable arithmetical computation required to achieve a particular filtering function in real time can be performed efficiently in hardware form, and the need for analog-to-digital conversion is obviated.

The CTD family is a grouping of structures that are basically high density, analog signal, serial stores [19]. The signal voltage applied at the CTD input is first sampled in time, and then stored within the device as a time sequence of isolated packets of charge. Each packet represents an analog sample of the signal, and is stored individually in capacitor sites formed within the CTD. Filtering is achieved by introducing, delaying, adding, dividing and sensing the individual charge packets (electrons or holes within the semiconductor body of the device) in a controlled manner.

The main CTDs are the *bucket brigade device* (BBD) [17], the *charge-coupled device* (CCD) [3] and the more recently introduced *charge-domain device* (CDD) [21]. First introduced in 1969, the BBD realizes a highly dense delay-line function which has not been developed extensively for FIR and IIR filtering. When integrated with MOS transistors, the CCD approach is far more versatile. Both time-domain and frequency-domain filters can be achieved in compact form. Whilst the CCD is good for FIR structures, it took a further ten years for a high performance IIR filter to be produced in semiconductor technology with the introduction of the CDD.

CTDs have made a substantial impact in signal processing by implementing miniature, low power, transversal, and recursive filters at low cost by taking advantage of high volume, MOS semiconductor production processes. They offer essentially signal bandwidths up to 1 MHz or occasionally to video frequencies at baseband. CTD delay lines and simple filters have been demonstrated at signal bandwidths exceeding 100 MHz, although the vast majority of charge-transfer filters that have been produced commercially have been designed for under 10 MHz signal processing applications. These applications include speech processing, sonar, and radar. As CTD filter structures are integrated monolithically, normally in silicon MOS technology, they can be combined with other functions, such as digital circuits and silicon sensors, to produce a full system-on-a-chip. Currently, such an integrated system implementation has power, speed, weight, and size advantages over alternative technologies. For this reason CTD solutions have been favored for mobile, military signal processing requirements. As such systems are often custom designed or application specific, it is not surprising that CTD catalog parts have not become widely available commercially. Indeed, many available CTD products were originally custom designed for specific applications and then released onto the open market (The design example presented in §6.6.2 is a case in point.).

The focus in this chapter is very largely on the CCD, which exemplifies the general characteristics and limitations of the CTD family. In addition, performance examples of the CDD are included to illustrate its complementary attributes. Both time-domain and frequency-domain applications for a range of signal processing requirements are discussed, following an initial introduction to CCD operating principles. The reader is also given a basic appreciation of the main design parameters and performance limiting factors. Some technological features enabling FIR, IIR, and adaptive filter architectures to be achieved are also presented. The reader is referred to several textbooks for a full exposition on CTDs and their applications [1, 2, 9].

6.2. CHARGE-COUPLED DEVICE: OPERATING PRINCIPLES AND PERFORMANCE LIMITATIONS

The objective of this section is to present a general view of the physical operation of the CCD and the mechanisms limiting its performance.

6.2.1. The Storage Concept and Charge Coupling

The charge-transfer concept covers the periodic spacial transfer, followed by storage, of a pattern of discrete charge packets, each representing the amplitude of an analog input signal that has been sampled at regular time instances from the past to the present. The pulsed input structure of any CCD suitably provides a sampling mechanism at the delay line input. The physical process by which the collective transfer of the charge packets is achieved is known as *charge-coupling*, and gives this device its name.

Let us commence this examination of a CCD by considering the storage operation of an MOS capacitor, fundamental to its understanding. In MOS technology, a dynamic storage site or *potential well* for charge packets may be created locally at (or below) the surface of the silicon substrate, and beneath a conducting electrode (which forms part of an elemental MOS capacitor) that has a voltage pulse applied to it. The storage capacity of this local storage site or the *depth* of the potential well is a function of the oxide capacitance (which is fixed after fabrication) and the applied potential, as shown in Figure 6.1. The well depth reduces due to minority carrier

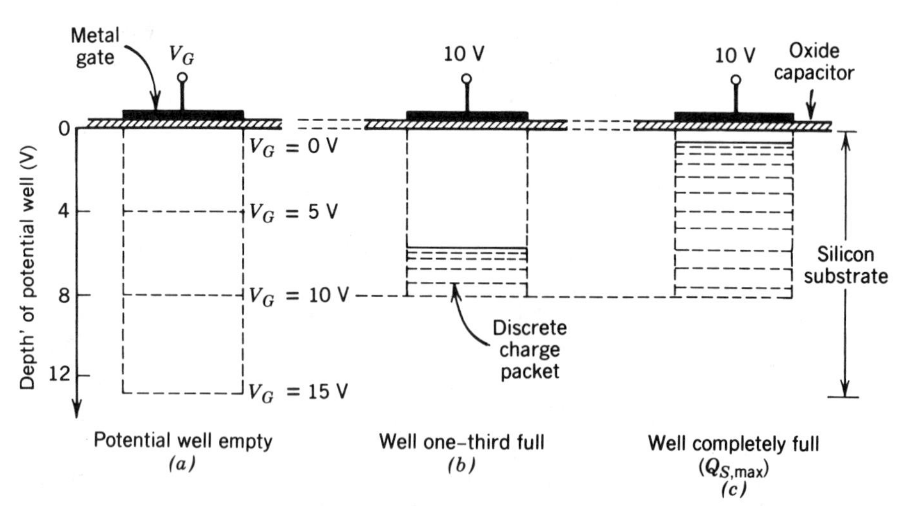

Figure 6.1 The potential well concept. (a) The depth of an empty well as approximately proportional to the gate voltage; (b), (c) for a given gate voltage, the well depth decreases linearly with increasing charge. Reprinted with permission from Beynon and Lamb, *Charge-Coupled Devices and their Applications*, © 1980, McGraw-Hill Book Co. (UK) Ltd.

generation processes within the semiconductor which try continually to fill this, and any other, storage wells. However, at room temperature the carrier generation rates are such that for many milliseconds storage wells appear *empty*, and any deliberately introduced (signal) charge packets are essentially uncorrupted in original size. For a typical MOS process, the voltage swing needs to exceed about 5 V for the creation of a deep well that can store useful charge levels. With no applied voltage on the top electrode or *gate* of the MOS capacitor no storage capability exists; and indeed, under this condition a *potential barrier* exists barring minority carrier accumulation at the semiconductor surface — a natural means for charge packet isolation. Thus it follows that the simplest CCD delay line is realized by a serial array of MOS capacitors, with alternating potentials on their electrodes to either create or collapse wells and provide isolation barriers between the individual wells.

Signal charge can be injected into the input of the CCD delay line by several different techniques: a simple means is described here with reference to Figure 6.2. The usual input structure, shown here, appears similar to a MOS transistor* [13] with the familiar source diode and a gate electrode, but the drain diode is replaced by a potential well region (virtual drain) into which injected charge may flow and be stored temporarily. The continuous-time analog input signal, V_{IG} is applied to the input gate. When the first transfer gate electrode is pulsed with ϕ_1, and the input diode held momentarily at a voltage, V_{ID}, near to 0 V, the source diode fills or *sources* the potential well (under ϕ_1) with a full charge packet, $Q_{S,\mathrm{MAX}}$. The diode input voltage is then increased to a higher positive value, as shown in Figure 6.2, causing the full charge packet to tend to spill back out to the diode source. However, when the surface potential under the first transfer electrode (with ϕ_1 applied) is equal to the surface potential below the input gate (to which the input signal V_{IG} is applied) no more charge leaves the well (because there is no surface drift field caused by a difference in surface potential). Thus the surface potential under the storage electrode is set to a potential the same as the input gate; consequently, a charge level, Q_S, is deposited that is essentially proportional to the instantaneous input signal sample, V_{IG}. This sequence is repeated every clock period, ϕ_1, enabling charge packets, $Q_S(t)$, to be injected into the CCD delay line proportional to the instantaneous value of V_{IG}. Clearly, a linear correspondence between V_{IG} and Q_S is crucial for good delay line linearity. Input/output methods are treated in more detail in §6.3.2.

Charge transfer between two neighboring storage wells is achieved by successively creating and collapsing the potential wells. By collapsing a well containing charge (a negative going ϕ_1 edge) and simultaneously creating an empty neighboring well (a positive going ϕ_2 edge), charge transfers by

* Note also that the following CCD description is consistent with an *n*-channel MOS transistor formed on a *p*-type silicon substrate, in which positive voltages on the electrodes cause potential well formation.

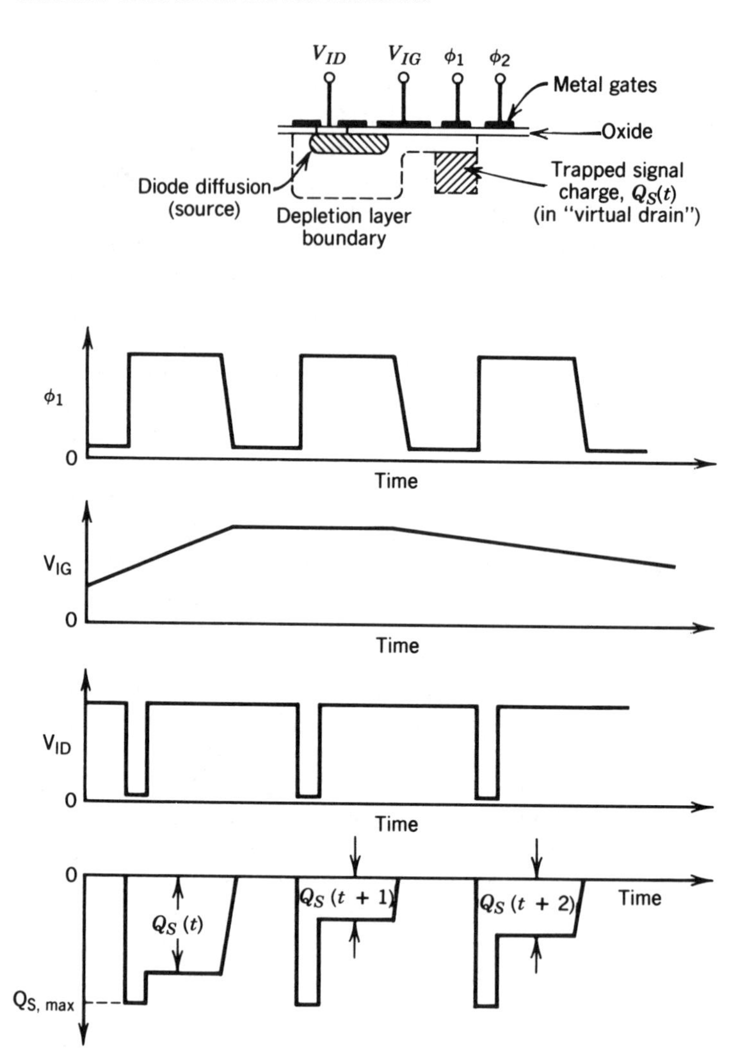

Figure 6.2 Technique for the linear injection of charge into a CCD. This material first appeared in *Wireless World* (now entitled *Electronics and Wireless World*), February 1975, as part of a four-part series on charge-coupled devices.

charge coupling into the newly created well, as shown in Figure 6.3. Clearly, a new charge packet must not be introduced into a well that is in the process of depositing charge into a neighboring well. Therefore it is crucial in delay line design that each storage well be isolated from neighboring storage sites. This is achieved by each delay line storage site or storage *bit* being formed by at least two wells, one for holding charge, the other for isolation. To constitute a serial delay line, the first, second, third etc., gate of each MOS capacitor per storage bit must be electrically connected and driven by

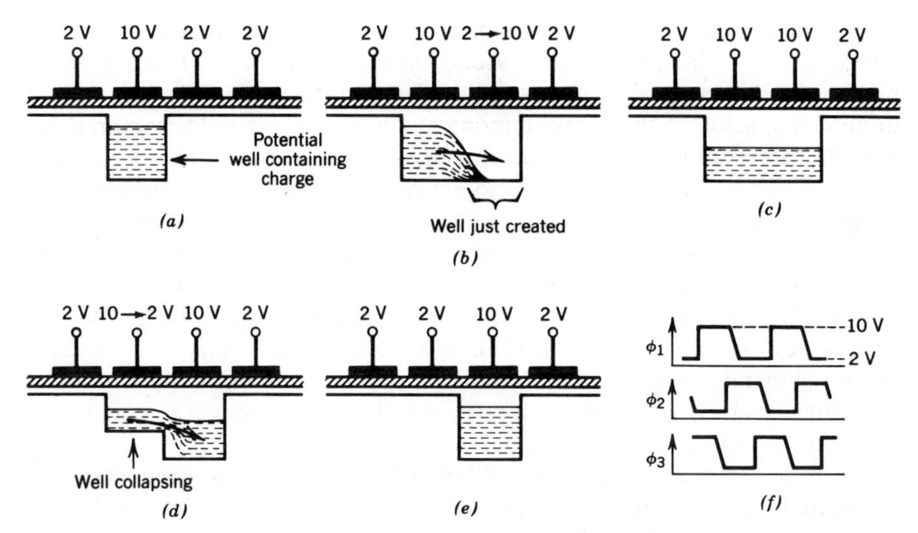

Figure 6.3 Movement of potential well and associated charge packet motion by clocking of gate voltages, (a)–(e): clocking waveforms. Reprinted with permission from Beynon and Lamb, *Charge-Coupled Devices and their Applications*, © 1980, McGraw-Hill Book Co. (UK) Ltd.

separate voltage supplies. This enables charge packets to be transferred in isolation along the device (see Figure 6.5(a)) without combining them. Therefore, in practice, transfer gate electrodes are connected appropriately to each phase ($\phi_1, \phi_2, \ldots, \phi_N$) of a multiphase clock generator, supplied via conductor *busbars* often along both sides of the CCD. These multiphase clocks are usually derived from an on-chip generator driven by an external single-phase clock supplied to the CCD integrated circuit. The number of phases required equals the number of MOS capacitors, M, employed per storage bit in the delay line, which is dependent on the technology. Typically, between two to four MOS capacitors are required per bit. Thus the number of transfers, N, required to pass an injected charge packet down the entire length of a delay line of P bits is the product PM. The period of the clocking, T, (common for each phase) determines the overall device delay, D, and for a P-bit design

$$D = NT = PMT \tag{6.1}$$

which must not exceed a few milliseconds at room temperature or else the charge packets are successively corrupted owing to the background minority carrier generation. The latter process is known as the *dark current* mechanism.

Charge packet propagation in the CCD is caused by charge coupling, in which drift (fringing) fields and diffusion processes cause carrier transport during the successive creation/collapsing of neighboring potential wells. The charge-transfer efficiency (CTE), usually given the symbol α, needs to be as near to 100% as possible, but is limited in practice primarily by: (a) the minimum spacing of the gates, limiting the high fringing fields required for good transfer; (b) the imperfect characteristics of the oxide-silicon interface, causing trapping of the otherwise mobile charges; and (c) the practical pulse shapes of the clocking voltages used to move the charges, which must provide time for adequate carrier diffusion. Several other physical mechanisms also limit the CTE. However, for practical devices, it is fortunately sufficiently high per transfer (99.99% being typical) that charge-transfer inefficiency (CTI), rather than CTE is usually quoted. The CTI value is given the symbol ϵ ($\epsilon = 1 - \alpha$, where α is per unit). Later it is pointed out how imperfect CTE adversely affects the device performance and is the major limitation in CCD operation.

There are several techniques available for converting the signal charge at the output, or, alternatively, outputs along the side, of a CCD delay line to an electrical signal. The floating-diffusion reset (FDR) technique is commonly employed for this purpose and is described here with reference to Figure 6.4. By collapsing the voltage on the last storage well under the final ϕ_3 transfer gate, the charge is dumped onto an output diode that has been previously reverse-biased to a voltage, V_{GG}, via the low channel *on* resistance of a reset MOS transistor, T_1 (under a high positive voltage on its gate). The charge transferred from the collapsing well on the back edge of ϕ_3 causes a change of voltage, V_{od}, on the output diode. This signal transition is detected by the sensing transistor, T_2, and appears as a signal at the output, which is superimposed on the dc level. For an n-channel CCD fabricated on a p-channel substrate*, the output electron charge packets cause negative signal transitions of usually a few hundred millivolts on a positive V_{GG} supply level. Notice in Figure 6.4 that the signals are basically rectangular in shape, indicating the low leakage of the reverse-biased diode and, also, the fast response of the reset transistor used to restore the output to V_{GG} after each charge transfer. The output signal waveform can be restored back to a continuous-time equivalent by subsequent sample-and-hold and low-pass filtering. These operations can be performed on-chip, although there is usually an external requirement for some post-processing of the output signal.

As can be appreciated from this discussion and the diagrammatical cross-section through a complete CCD given in Figure 6.5(a), the whole device may be regarded as a multigate MOS transistor or perhaps, more precisely, as a *virtual drain* input transistor supplying current to a *virtual source* output transistor. However, this analogy is of limited importance

* This is the preferred polarity arrangement, because n-type electron charge packets have higher mobilities than p-type holes enabling faster speeds to be obtained for the former embodiment.

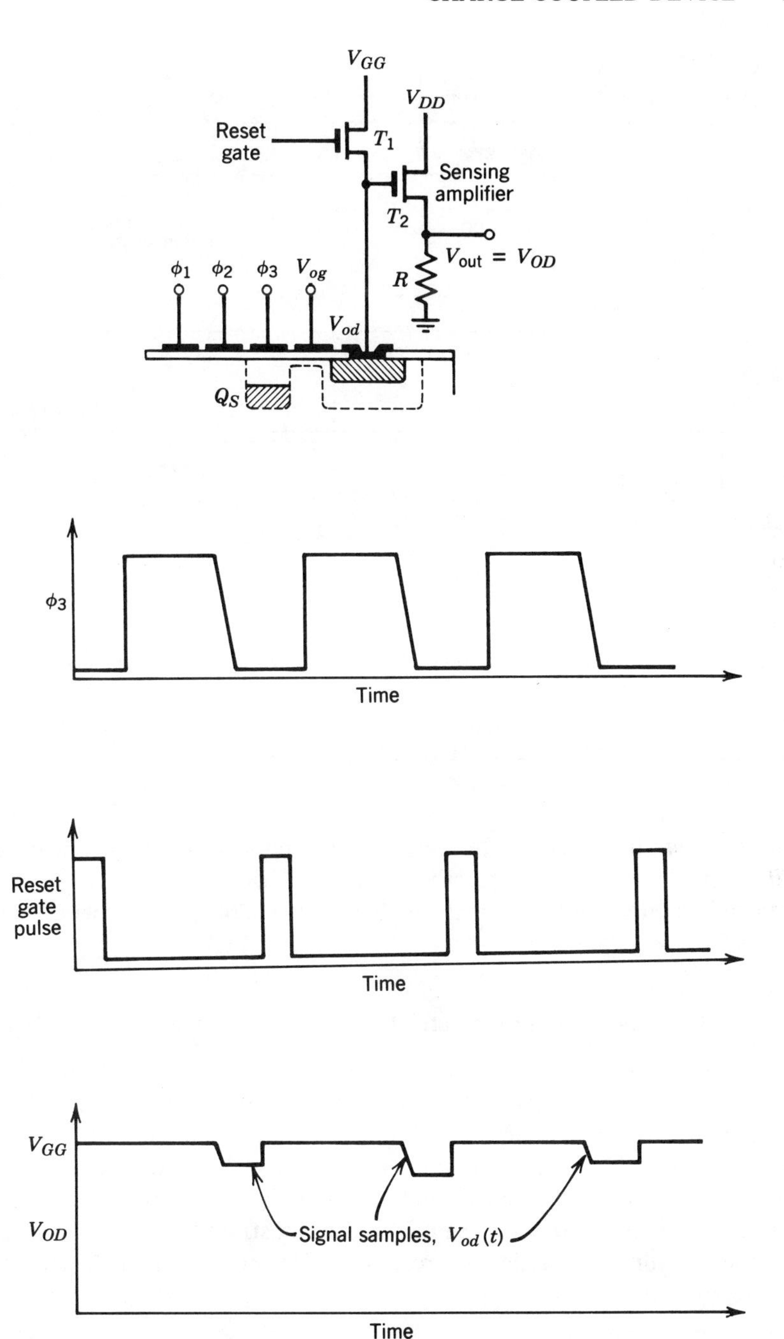

Figure 6.4 Floating-diffusion reset method of converting the output charge in a CCD to an electrical voltage. This material first appeared in *Wireless World* (now entitled *Electronics and Wireless World*), February 1975, as part of a four-part series on charge-coupled devices.

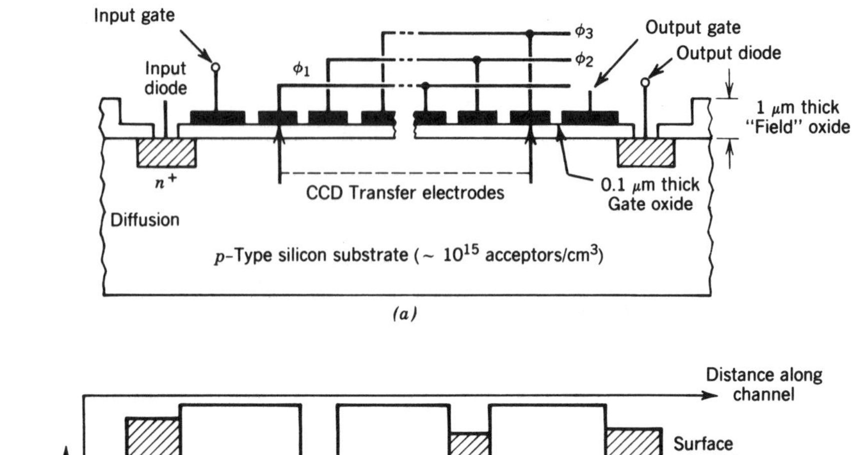

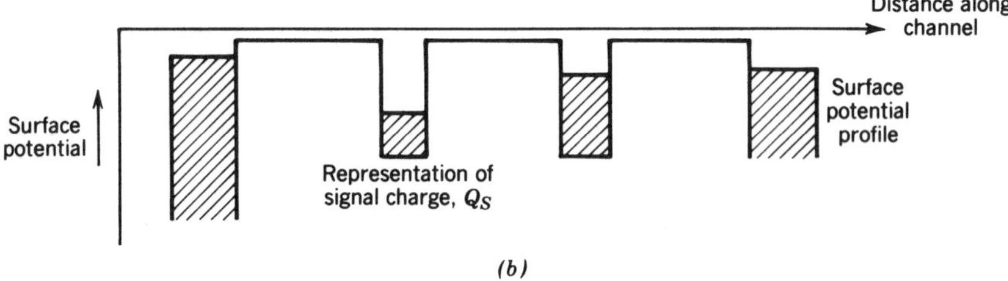

Figure 6.5 Three-phase surface-channel CCD. (a) Cross-section through structure (diagrammatic); (b) surface-potential model showing potential wells. Reprinted with permission from Beynon and Lamb, *Charge-Coupled Devices and their Applications*, © 1980, McGraw-Hill Book Co. (UK) Ltd.

because rather than being a transistor that conducts signal current continuously in its conducting state, the CCD is a dynamic shift register that propagates signal information through variable amplitude, discrete charge packets.

6.2.2. Performance Limitations and Governing Equations

We now provide a quantitative basis for CTD design by introducing a number of key equations. Generally, the mathematical expressions employed here are in simplest form, but more detailed derivations can be found elsewhere [2, 9]. Throughout this section and in the remainder of the chapter it is assumed, unless stated otherwise, that the device under discussion is fabricated on a *p*-type silicon substrate, with *n*-type diffused input/output junction diodes (corresponding to an *n*-channel, MOS transistor configuration).

6.2.2.1. Charge Packet Size. To a good approximation the maximum charge handling capability, $Q_{S,\mathrm{MAX}}$, of a CTD can be found [19] by considering the structure as an array of elemental MOS capacitors. For each capacitor structure, the absolute maximum stored charge in the well is given by

$$Q_{S,\mathrm{MAX}} = kC_{OX}V_GA \tag{6.2}$$

where k is a constant of value less than unity, C_{OX} is the oxide capacitance per unit surface area, V_G is the potential applied to the gate, and A is the area of an individual transfer gate (of length L in the transfer direction and of width W). For $V_G = 5$ V, $L = 10\ \mu$m, $W = 50\ \mu$m with $C_{OX} = 4 \times 10^{-4}$ F/m^2, then $Q_{S,\mathrm{MAX}} = 1$ pC for $k = 1$ (strictly a negative sign should be associated with this charge value because electrons are the minority carriers within a p-type substrate), corresponding to a well charge of $N_W = 6.25 \times 10^6$ electrons. Normally practical CTDs do not employ a maximum well capacity much in excess of a few million electrons to minimize certain undesirable nonlinear effects and because of reduced barrier heights in certain CTD structures. Further, the above calculation for $Q_{S,\mathrm{MAX}}$ strictly applies only to so-called *surface channel* CCDs having uniform doping profiles in their substrates (as indicated in the figures so far in this chapter).

It has been mentioned earlier that the imperfect characteristics of the oxide-silicon boundary adversely affects the CTE of a surface channel CCD (SCCD), as the signal charge is in intimate contact with the interface traps. The *bulk* or *buried channel* CCD (BCCD) was introduced [6, 10, 22] to provide a potential minimum beneath the silicon surface whereby the signal charge is confined within the volume of a diffused or implanted layer—the n-type layer in Figure 6.6. As charge packets in the bulk device are transferred within the silicon wafer, but still near to the silicon surface, much higher CTEs and operating bandwidths are obtained. This is due to the near elimination of surface-state effects and improved bulk carrier mobilities and higher fringing fields. However, these improvements in the BCCD are gained at the expense of lower charge handling capability and poorer charge storage times. Typically, the value of k in Eq. (6.2) is about 0.8 for SCCDs whereas for BCCDs $k = 0.3$–0.5. This arises because charge

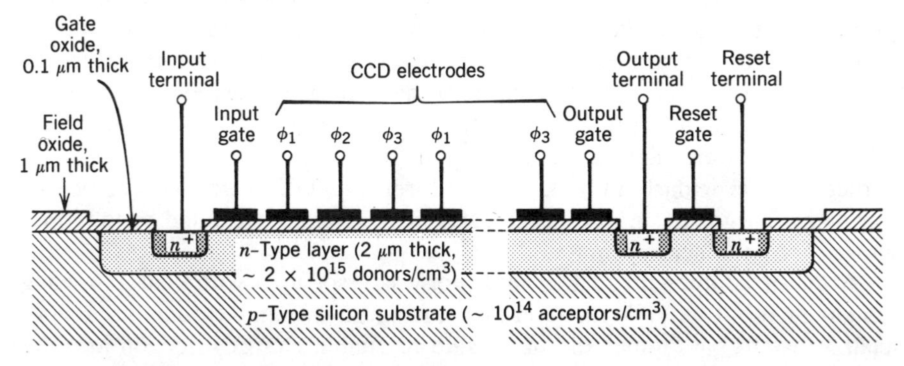

Figure 6.6 Three-phase buried (bulk)-channel CCD. Reprinted with permission from Beynon and Lamb, *Charge-Coupled Devices and their Applications*, 1980, McGraw-Hill Book Co. (UK) Ltd.

stored near or at the semiconductor surface (SCCD) experiences higher supporting fields than charge confined within the semiconductor (BCCD). Nevertheless, both surface- and bulk-channel CCDs are in common use, with SCCDs being generally preferred for low-frequency high-linearity applications whereas BCCDs are favored for high-bandwidth applications and where long devices (greater than say 100 bits) are needed.

A further factor that can restrict the charge handling capability of a CCD is the choice of clocking waveshapes. Generally near-rectangular pulses are used (see Figure 6.3(f)), however, it is difficult to preserve fast edges when driving CCDs at high frequencies owing to the capacitive load presented to each clock phase, associated with the total capacitance of all the MOS capacitors being driven. For some CCD designs this can easily exceed hundreds of picofarads per phase. Alternately, a sinusoidal clocking waveform that is easy to generate can be adopted for certain high-frequency designs, but the charge handling capacity may be reduced. A further clocking variant is often employed in certain tapped delay line designs (refer to §6.4.2) in which one of the three-phase clocks is replaced by a dc level, half the value of the clock amplitude. With this so-called $2\frac{1}{2}$-phase scheme, the charge handling will be reduced by nearly 50%. However, this clear disadvantage, which impairs the CCD signal-to-noise ratio (S/N), is outweighed for certain designs by the advantages of a particular charge sensing arrangement.

6.2.2.2. Noise Mechanisms. For good analog operation, the CTD S/N must be optimized, implying a high $Q_{S,\mathrm{MAX}}$ value and low internal noise. Although the $Q_{S,\mathrm{MAX}}$ value for an SCCD may exceed the BCCD value, the latter device has a much reduced noise. In fact S/N ratios for both CTD variants usually lie in the range 70–90 dB.

The noise analysis of any semiconductor device is involved, but the treatment for a CTD is particularly complex because it involves many separate physical processes associated with the injection, transfer, and sensing of charge packets, all of which add to the overall device output noise. Further, its sampled-data nature and low signal levels in the presence of high voltage clock pulses make noise measurements exceedingly difficult to perform and interpret. Fortunately, however, it was established soon after CTDs were first introduced that they were inherently low-noise structures, particularly in bulk-channel form. Indeed, as we shall see below, for many designs the true CTD transfer noise is often exceeded by the input noise or by the sensing noise.

A major component of the input noise is associated with charge flowing under an input gate. With reference to Figure 6.2 and earlier discussion, charge has to pass under the input gate to enable charge injection into the virtual-drain storage site. When the diode potential rises to isolate Q_S, charge under the input gate will momentarily divide or partition (adding to Q_S), and also be removed through the diode circuit. This causes a partitioning noise by corrupting the magnitude of Q_S. Fortunately, input techniques have been developed that do not inherently possess a partition noise mechanism (see §6.3.2).

Transfer noise has its origins in imperfect charge transfer and charge-carrier trapping processes (particularly for SCCDs in which the signal charge is in intimate contact with the imperfect oxide-silicon interface). The mean deviation of the CTI induced noise is given by

$$N_I = (2\epsilon N N_W)^{1/2} \tag{6.3}$$

where ϵ is the CTI, N is the total number of transfers, and N_W is the well capacity in electrons. For the SCCD discussed above; $\epsilon = 10^{-4}$, $N_W = 6.25 \times 10^6$, and if a 2-ϕ, 100-bit device is considered, $N_I = 500$ electrons. Additionally, the trapping contribution may exceed N_I, although usually it is of the same order. In a BCCD, the trapping contribution should be below half of the SCCD value because charge is held away from the surface. A typical SCCD could have a transfer noise of about 1,000 electrons, whereas a BCCD could be as low as 100 electrons.

As CCDs are invariably fabricated in MOS technology, it is very natural to employ MOS transistors (a) as amplifying devices to increase the milli-volt-level output signals and (b) as switches to reset various circuit potentials between clocking periods. Function (a) causes thermal noise associated with the transistor channel resistance with a superimposed $1/f$ frequency characteristic, whose corner frequency is usually about several kilohertz. Function (b) implies a reset noise contribution because of the uncertainty of resetting the potential across a capacitive circuit to a precise value. This follows because of the thermal noise generated by the finite channel resistance of the *closed* MOS transistor switch (typically, thermal noise associated with an *on* resistance of perhaps 1,000 Ω). This reset or so-called *kTC* white noise component in electrons, N_R, at room temperature is of value

$$N_R \approx 400 C_N^{1/2} \tag{6.4}$$

equivalent to $6.45 \times 10^{-5}/C_N^{1/2}$ V r.m.s. at room temperature; where C_N is the capacitance (in pF) of the circuit whose voltage is to be reset. For a typical CCD with $C_N \approx 1$ pF, $N_R \approx 400$ electrons; a value that often also applies at the device input. At the device output the noise may be larger, because of a corresponding increase in the C_N value. This is due to the appreciable additional loading of the gate capacitance presented by the output sensing transistor (T_2 in Figure 6.4). Although the thermal noise of the output MOS transistor may be small because the transistor is usually designed to be high gain, the large gate area required for high gain presents a high capacitance to the output node and introduces a high reset noise contribution. Therefore a trade-off exists between the device noise and the output amplifier gain.

Several other noise sources may be identified, including that due to the dark current [19] caused by the continuous thermal generation of minority carriers. Even in the dark internal state of an encapsulated device such minority carriers try to fill a storage well, whether it contains signal charge or not. This mechanism exhibits itself as a noise source. A dark current density of 10^{-4} A/m^2 is typical for a good substrate although, occasionally, due to poor device processing and/or substandard silicon wafers, highly localized dark current spikes can occur. The background dark current generation only contributes a few hundred electrons of noise; however, when they exist, dark current spikes can be disastrous to device performance, particularly in a stop clock mode which is occasionally employed to produce long delays in short devices.

From the above brief review of noise in CTDs, it is clear that for signal capacities exceeding 10^6 electrons and with noise equivalences around 10^3 electrons, S/N ratios exceeding 60 dB are easily achievable at room temperature. As indicated earlier, improved S/N ratios exceeding 90 dB are obtainable for a carefully designed CTD produced on a low dark current process having a low density of interface traps. However, to achieve the lowest possible noise in operation, clever circuit techniques need to be employed in addition to the use of a good fabrication technology. A well-established circuit technique for low noise operation is correlated double-sampling, and the reader is referred to Beynon and Lamb [2] for a detailed treatment.

6.2.2.3. Charge-Transfer Effects. The physics of charge transfer in CTDs is extremely involved, and numerical techniques are usually required to solve the semiconductor continuity equations for the current density in the device. In many technical papers treating this subject, the qualitative dependencies of CTI with device geometry, bias conditions, and so on, have been well established. However, the prediction of absolute values of CTI using such approaches has been relatively unproductive. The need exists to supply a range of parameters and explicit boundary conditions, such as the density and time dependence of interface traps, which are extremely difficult to measure in a completed device. Furthermore, all semiconductor device parameters suffer considerable spreads in value ($\simeq$20%) and temperature dependencies that defeat the benefit of detailed device modeling. In a design laboratory, therefore, there is usually a continuing device evaluation/simulation program to resolve empirical differences between existing device models and experimental results.

The practical effect of CTI is to smear an injected charge packet across a number of trailing cells, increasingly more severely as it propagates along the delay line. This process reflects itself in a reduced bandwidth due to the distortion of the original charge packet in the time-domain, as shown in Figure 6.7. Actually the CTI value is not constant with operating conditions and device geometries, and varies for different devices. However, for

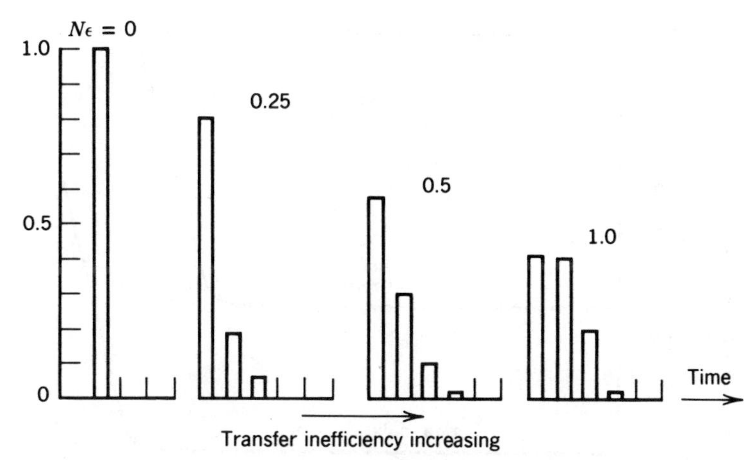

Figure 6.7 CCD output for single "one" input followed by a stream of "zeroes". Note effect of increasing charge-transfer inefficiency, ϵ, through nonzero $N\epsilon$ product. Reprinted with permission from Beynon and Lamb, *Charge-Coupled Devices and their Applications*, © 1980, McGraw-Hill Book Co. (UK) Ltd.

first-order design purposes, an average value is normally taken for ϵ, with an associated tolerance spread usually obtained from device measurements. In practice, the product $N\epsilon$ rarely exceeds 0.1 for reasonable device performance.

Surface-state traps located at the oxide-silicon interface have a profound effect on the CTI in SCCDs, as has previously been discussed. Consequently, it is common practice to inject a constant, background charge (at about 10–20% of $Q_{S,\mathrm{MAX}}$) into the device at all times to ensure that on average the traps will always be occupied. This procedure prevents the severe distortion that a freshly injected charge packet would otherwise suffer, in the absence of the background charge. As the background charge produces a modified zero signal charge level, the technique is known as *fat zero*, and is widely adopted for SCCD operation.

BCCDs have improved CTI values because the signal charge is largely constrained from interacting with surface traps but interacts instead with the much lower density of bulk traps. For this reason, also, they have a superior frequency performance to SCCDs with $10^{-5} < \epsilon < 10^{-4}$ to 10 MHz or above, whereas SCCDs have $\epsilon \approx 10^{-4}$ with a break point in the several MHz range, as depicted in Figure 6.8.

Inter-electrode drift fields, charge repulsion and diffusion mechanisms during charge transfer set an optimum value of gate length, L, for a given CTI. Experience indicates that the CTI versus L dimension relationship is lowest and not too sensitive, in the range $3\ \mu\text{m} < L < 5\ \mu\text{m}$. This dimensional range coincides with a minimum feature size for many current analog processes. When the CCD was first introduced, however, the minimum

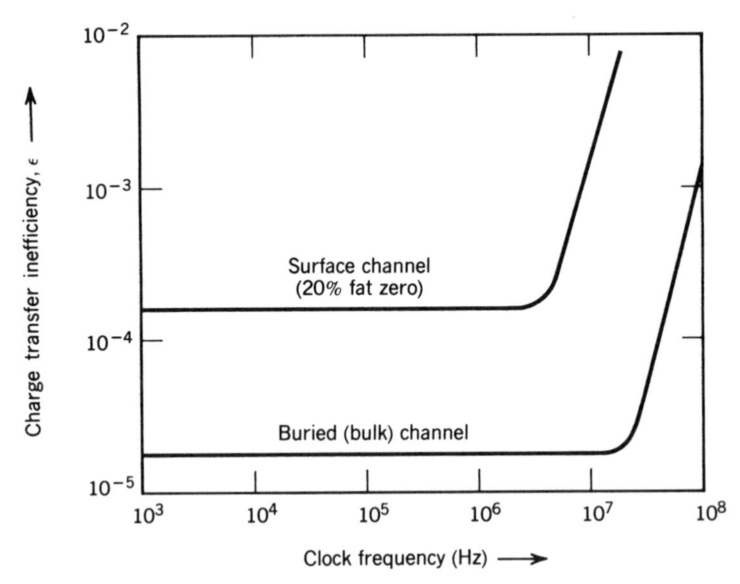

Figure 6.8 Typical charge-transfer inefficiency dependence on clocking frequency for both surface- and buried-channel CCDs.

feature size in MOS technology was 10–15 μm, which has now shrunk in production currently to 1–3 μm for digital processes. However, future advances in sub-micron VLSI will not benefit the CCD for analog signal processing applications — a situation that is shared generally with all analog MOS transistor circuits — where S/N ratios reduce drastically below 60 dB for under 2-μm feature sizes. For examples where the performance advantages of CCD filters are to be combined with a VLSI digital circuit, CCD gate lengths should be maintained around 3 μm to preserve CTI and S/N values. Digital structures, however, can take full advantage of the minimum feature size offered because of their far superior noise tolerance.

6.3. PRACTICAL DEVICE DESIGN

In §6.1 and §6.2 we discussed the main physical processes in CTDs and the basic limitations on the device performance. Here we extend these concepts into design, within the framework of the sampled-data nature of these devices. Thereafter, a number of basic performance trade-offs are discussed which the designer has to address before any CTD filter can be realized. Section 6.4 treats architectural design aspects and actual charge-transfer filtering examples for a number of different applications and device specification is covered.

6.3.1. Time-Bandwidth Product and Insertion Loss

Due to the sampled-data nature of CTDs, Nyquist's sampling theorem applies and limits their signal bandwidth, f_s, to less than or equal to half the sampling frequency, f_c/M, to avoid aliasing effects. It is thus mandatory to bandlimit the input signal frequencies, f_s, to $f_c/2M$ often by adding a low-order, continuous-time, anti-aliasing prefilter at the CTD input.

The maximum time-bandwidth product, TB, is a prime figure of merit for any signal processing device; it may be derived for a CTD in terms of the total number of transfer bits, P (typically, for $M = 2$ or 3 storage sites per bit) and f_c

$$TB = Df_s = Df_c/2M \tag{6.5}$$

From Eq. (6.1) $D = PMT$, and $T = 1/f_c$. Hence the maximum $TB = P/2$. Practical CTDs rarely exceed a TB product of 1,000, because of CTE limitations, indicating a usual maximum device length of 2,000 bits of delay.

As the output of a CTD is a succession of flat topped, amplitude-modulated spikes caused by charge packets being dumped on the output diode (see Figure 6.2), a stage of sample-and-hold is always employed at the output to help reconstitute the original, but time-delayed, continuous-time signal. This sample-and-hold effect causes the frequency-domain amplitude response to be weighted by a $(\sin x)/x$ function, in common with any sampled-data processor. When required, compensation for this weighting function can be designed into the output circuit.

A further frequency-domain distortion effect arises owing to imperfect CTE, as shown in Figure 6.9. Charge-smearing effects reduce bandwidth and cause phase distortion in addition, as these "bell" (shaped) curves show. For $N\epsilon$ products as high as 0.1, tolerable amplitude and phase distortion will be experienced by signals; corresponding, say, to a 256-bit, two-phase SCCD (total transfers $N = 512$) with $\epsilon = 2 \times 10^{-4}$, or a longer bit-length BCCD. This consideration limits practical CTDs to having around 1,000 bits of delay, for high performance analog operation. Note that a short device can have a long delay (up to the dark current limit) by momentarily stopping or reducing the clock frequency, following the signal being fed in at a higher frequency which satisfies the $f_c > 2Mf_s$ requirement. The maximum device bandwidth is usually dictated by the frequency dependence of the CTI; this has already been discussed in the last section with reference to Figure 6.8.

Occasionally, also, a device designer will arrange for signal charge packets to be transferred between several CTD delay lines, operating at different clocking speeds, to produce a long overall delay. A common example of this approach is to be found with the serial-parallel-serial (SPS) architecture. With this technique, signal charge packets introduced into a serial input, n-bit, CCD delay line at rate f_c are transferred through parallel

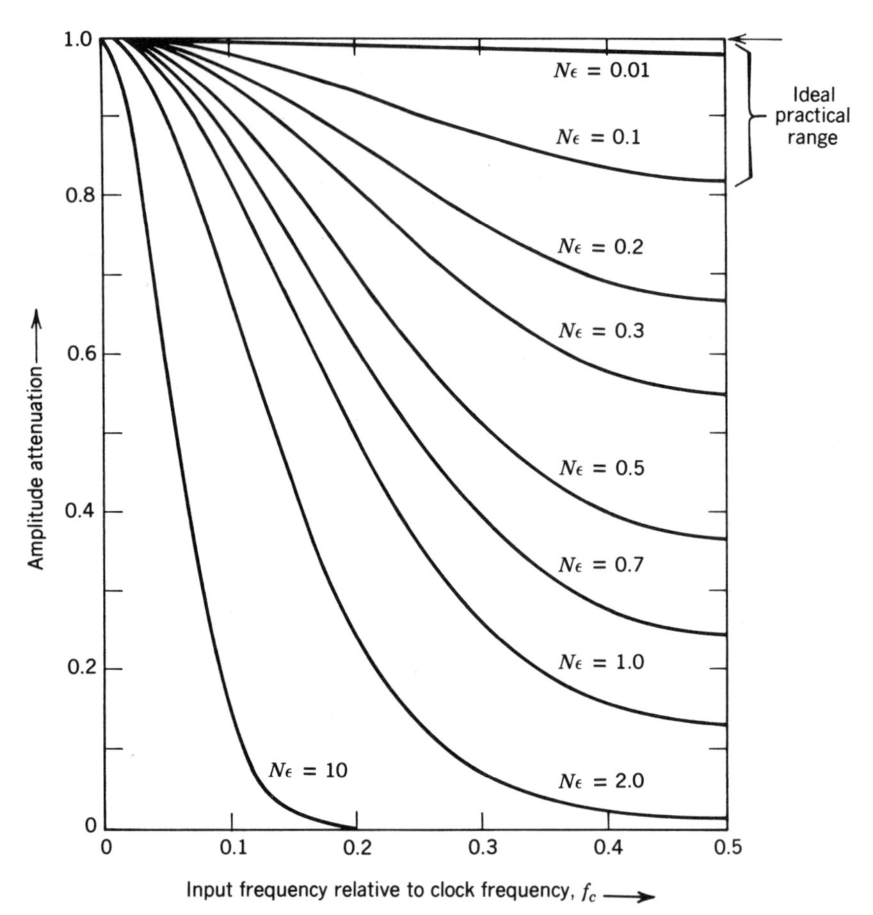

Figure 6.9 Amplitude attenuation of a CTD transfer function for different values of $N\epsilon$ product. Note $N\epsilon = 0.1$ is usually a practical upper operating limit. Reprinted with permission from Séquin and Tompsett, *Charge Transfer Devices*, © 1975, Academic Press, New York.

outputs at rate f_c/n (Figure 6.18(b) indicates the serial-parallel translation). The charge packets are then clocked at rate f_c/n down n-parallel CCD delay lines each of m-bits in length. Following a subsequent parallel-serial conversion, the charge packets are then clocked out at the higher (input) rate, f_c. Thus from the external terminals the SPS structure appears as a serial delay line clocked throughout at rate f_c. Actually, charge packets have only experienced $(m + n)$ transfers and have incurred a total delay of $n(m + 1)/f_c$. To achieve this overall delay in an equivalent, serial, CCD at the f_c rate, charge packets would have to experience $(m + 1)$-times the transfers. Bearing in mind that the m-parallel transfers for the SPS configuration are made at only the f_c/n rate, the $m\epsilon$ product will be low (see Figure 6.8).

Therefore, the overall $N\epsilon$ of the SPS device is much improved on the corresponding serial device value. This is particularly so in device designs where the f_c rate has to be near to the *knee* of the CTI-clock frequency curve, in which case the advantage of the SPS design can be extremely signficant.

As the CTD is an active structure that can possess gain by suitable design of the MOS sensing amplifier, there is no fundamental reason why there should be an insertion loss. Practical devices often have modest insertion losses of several decibels but, in general, this could be compensated for by increasing the gain of the amplifying output stage. There remains, nonetheless, the difficult design issue of being able to predict the absolute value of CTD input/output gain, purely from a theoretical basis, that is, without knowledge from previous device measurement. The problem stems from the fact that the CTD is a complicated distributed device that is fundamentally dynamic in nature; it also requires three-dimensional modeling programs to predict even the most basic internal dependencies. To the author's knowledge, only one technique, called feedback linearization, exists [2] that ensures specified device gain (and also linearity improvement). It functions by launching a charge packet in a CCD, which has previously been monitored by an initial sensing amplifier. The nondelayed signal obtained before charge launch is compared with the input—within a feedback loop—and charge packets within the CCD are sensed by exactly similar sensing amplifiers along its length. Thus the outputs should be closely related in magnitude to the input signal, within the gain tolerances of all the matched amplifiers employed (say, $\approx 2\%$). Because of the settling time of the feedback loop, practical signal bandwidths for this technique do not exceed 50 kHz.

6.3.2. Charge-Coupled Device Linearity Design

That CTDs should exhibit good input/output voltage transfer function linearity to achieve satisfactory filtering performance is clearly a necessity. This covers the charge generation and its internal transfer through the CTD and subsequent sensing. In addition, the overall gain of the device must be predictable to close limits across the filter production volume, without the need for individual gain adjustment. These strict gain and linearity requirements are very difficult to achieve and maintain in practice, because of the complexity and design freedom of CTD filters, which often employ many different output and sensing structures. The problem stems very largely from the spreads in key MOS parameters, particularly tolerances associated with the threshold voltage, V_T, and the gain factor, K'. Indeed, in any analog MOS circuit design the V_T and K' parameters are absolutely crucial to performance and techniques must be employed to reduce circuit sensitivity to semiconductor parameter variation.

There are two distinctly different approaches to the design of linear CTD filters: one relies on setting the precise size of the signal charge packets, Q_S, which are generated, launched, and then sensed; the other relies on the precise control at the input and detection at the output of the surface potential, ψ_S, which is set by supplying an appropriate amount of injected charge (as a secondary variable). Although the difference between the two techniques may at first appear rather subtle to the reader, they are quite distinct in approach. Their difference stems from the fact that the surface potential (set locally at the oxide-silicon interface under any gate electrode) is not linearly related to the quantity of charge in a chosen potential well; in fact, the $\psi_S - Q_S$ relationship involves a square root term in addition to a proportional term (refer to Eq. (6.6)). For a well-designed SCCD with a lightly doped silicon substrate ($X \approx 0$), the linear term can be made to dominate permitting to a good approximation $\psi_S \propto Q_S$. Under this condition, there is little difference between the two approaches to linear design, giving the designer a wide choice of appropriate input and output circuits based on criteria other than linearity alone, such as layout area.

The complex structure of BCCDs causes their $\psi_S - Q_S$ relationship to be quite nonlinear under all circumstances. For this reason, BCCDs are often designed to have an SCCD input, but with a buried-channel transfer section to preserve optimum CTI. (In principle, this can be achieved simply by reducing the length of the *n*-type layer in Figure 6.6 so that it only extends to under the first ϕ_1 electrode thereby leaving the input as a surface channel structure.) Thus to preserve good linearity it is essential for the designer to distinguish between the two approaches, Q_S or ψ_S setting, and ensure that these are not mixed in a particular CTD design or else severe nonlinearities may occur, probably resulting in poor filtering performance.

Table 6.1 gives a basic comparison of three popular input and output techniques and indicates in each case which signal setting technique is invoked. For good linearity it may be seen that either diode cut-off and floating-gate reset (FGR) sensing should be employed or any combination of the charge packet-size sensing techniques. However, if an SCCD is to be designed in which a low substrate doping level is acceptable, as discussed above, then both techniques are equally applicable making the input/output choice unrestricted. Full descriptions of each of the six techniques mentioned in Table 6.1 are outside the scope of this book. For further details of these and other charge inputting and sensing methods the reader is referred to a more specialized text [2, 9].

In order to give the reader an appreciation of a popular input technique, a brief description of the potential equilibration principle is given with reference to Figure 6.10. With this technique, the objective is to make the input signal charge, Q_S, a direct function of the input voltage, V_{SIG}, by eliminating the nonlinearity in ψ_S which appears in the following relevant equation

$$Q_S = C_{OX}[V_G - \psi_S - (2X\psi_S)^{1/2}] \tag{6.6}$$

TABLE 6.1 Comparison of Three CCD Input and Output Techniques Indicating the Signal-Setting Technique

	INPUT			OUTPUT		
Description of Technique	Dynamic Current Injection or Current Integration	Diode Cut-off	Potential Equilibration or Fill-and-Spill	Current Output Sensing	Floating Diffusion	Floating-Gate Reset (FGR)
Charge packet size setting, Q_S, or surface potential setting, ψ_S	Q_S	ψ_S	Q_S	Q_S	Q_S	ψ_S
Comments	Performance depends on gating period, which needs resetting if clocking changes	Susceptible to charge partitioning	Popular input technique which in modified form is very elegant	High quality operational amplifier required	Popular output technique	Popular for multitapped delay lines; reduced dynamic range

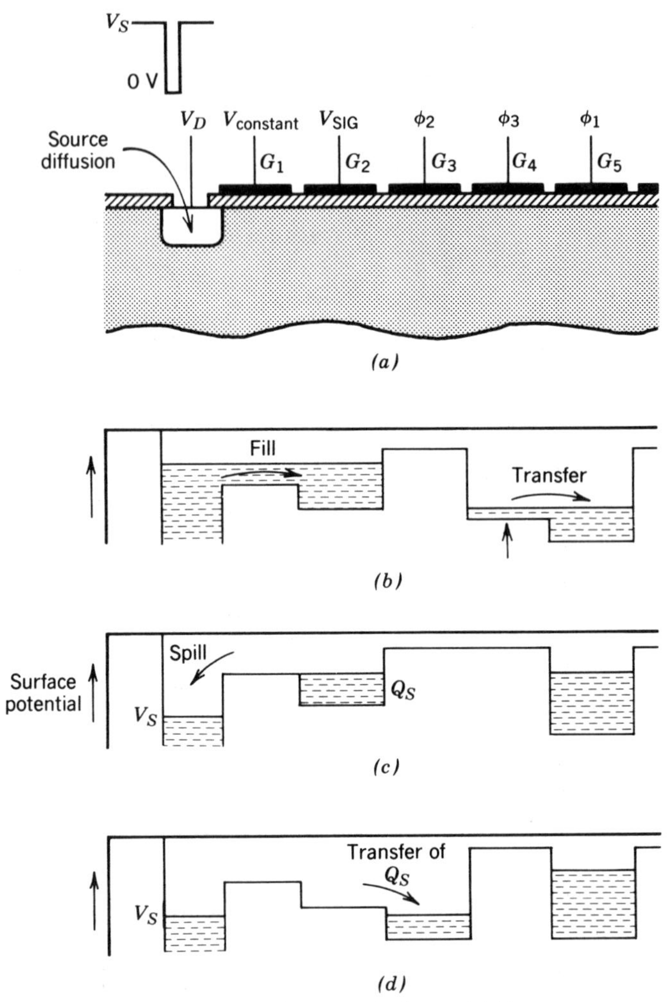

Figure 6.10 Potential equilibration input technique. (a) Input gate structure; (b), (c) potential profiles showing fill-and-spill operations; (d) potential profile showing transfer of input signal charge, Q_S, from input well to CCD register. Reprinted with permission from Tompsett (1975) Surface potential equilibration method of setting charge in CCDs. *I.E.E.E. Trans. Electron. Devices* **22**, 305–309 © 1975 I.E.E.E.

where X and C_{OX} are constants for a particular technology. ψ_S is fixed by applying a constant potential V_{constant} to the first gate, G_1, of the CCD, with the signal V_{SIG} applied to G_2 making $V_{G2} = V_G = V_{SIG}$. This arrangement yields $Q_S = C_{OX}V_{SIG}$ and ensures a linear relationship between the input voltage and the charge packet size setting, as given in Table 6.1. Practically, separate *fill and spill* operations are applied for each charge packet generated to set ψ_S constant, by setting a charge threshold above which charge flows. Eventually the level settles to provide a constant surface potential after each such operation. The provision (sourcing) and removal (draining) of charge required for the separate fill and spill operations are performed by a source (drain) diffusion to which a periodic pulse, V_S, is applied. All input and sensing techniques employ a different interpretation of Eq. (6.6), and a correspondingly different arrangement of the potentials.

A second crucial advantage of the fill-and-spill input technique is that it does not inherently suffer from partitioning noise (refer to discussion on noise mechanisms). The magnitude of Q_S is not going to be affected by charge partitioning from under G_1 because any such charge will spill towards the input diode, over the constant threshold, and not contribute. As we have seen above, Q_S is solely determined by the depth of the completely-filled storage well set by V_{SIG}.

6.4. TOPOLOGICAL DESIGN ASPECTS

In the previous sections, the electrical characteristics of CTDs have been established and key concepts have been introduced relating to delay line design. In this section, the physical features of CTD implementation are surveyed and important technological considerations are discussed. For simplicity of explanation the emphasis remains on delay line design, although in §6.5 advanced filtering architectures and applications are treated.

6.4.1. Charge-Transfer Device Technology

Charge-transfer filters are usually fabricated in silicon integrated circuit technology, although significant research results aimed at very high bandwidth delay line performance have been reported on gallium arsenide processes. CTD structures are invariably implemented in MOS form, although BBDs can be produced to advantage in higher speed bipolar junction transistor (BJT) processes [19] as well as in MOS technology. As the vast majority of CTDs are currently fabricated on MOS processing lines, the discussion here will focus exclusively on MOS implementation.

Although CTDs are members of the MOS device family, they cannot in general be produced on standard MOS production lines. This is due to the fact that certain performance factors and constructional features in the CTD require modifications to commonly available MOS (digital) integrated circuit processes. Strictly speaking there is no standard MOS process, as each process has minor procedural and material differences. For example, the multilayer structure usually required for CTD implementation requires two levels of polysilicon conductor to formulate closely spaced gate electrodes, whereas, often a logic MOS process employs a single layer of polysilicon. A key CTD processing requirement for good performance is a low level of so-called interface states or fast surface states (N_{SS}). These states [8] are associated with the presence of charge trapping centers at the semiconductor-oxide interface. Their characteristics are involved and give rise to statistical fluctuations in signal charge, present at the semiconductor surface. This mechanism causes CTE deterioration and introduces a dominant source of noise in the device. Several other important process variations are required to accommodate CTD design rule peculiarities and technological differences. Therefore, a fully optimised CTD process is really different to widely available digital MOS processes: a fact that has adversely affected CTD availability; although, as explained earlier, a further factor is that CTDs are usually application specific in design and are, therefore, not widely available as catalog parts.

6.4.2. Clock Phase Formation

In §6.2, a basic description of a three-phase (3-ϕ) CCD was given as this was the first device demonstrated. Although 3-ϕ CCDs are probably the easiest to fabricate, they suffer from the obvious layout problem that at least one cross-over is required to enable the three clock phase conductors to supply each bit of the CCD. This applies even when a phase is passed down each side of the CCD channel region (see Figure 6.11(a)). Although cross-overs are designed routinely in silicon technology, by employing an underpass diffusion or a conductor over a thick (or field) oxide, they are inconvenient and also lead to capacitive feedthrough effects between overlapping phases. This may be a critical layout consideration when millivolt signal levels are passed across 5 V or larger clock phases. However, 3-ϕ CCDs have a larger signal handling capacity and can, in principle, provide a higher packing density than alternative 2-ϕ or 4-ϕ devices.

The strict requirements on the leading/falling edges of 3-ϕ CCDs, and the need for three-phase clock generation led device engineers to investigate structures having simpler drive demands. This resulted in many new structures being introduced based on 2-ϕ CCDs. The basic principle of a 2-ϕ CCD is to incorporate asymmetry in well-depth potential profiles, under every gate to which a phase voltage is applied. The most popular approach to producing a dual well depth under each gate is to: either (a) selectively thin the gate oxide; or (b) partially ion-implant the silicon substrate under each gate (see Figure 6.12). If the maximum level of signal charge injection, $Q_{S,\mathrm{MAX}}$ corresponds under all conditions to the difference in the dual well depth, as illustrated in Figure 6.12(a), then each gate cell acts as a storage site plus a potential barrier (when a phase voltage is applied). Now, when the gate potential on one cell collapses, under which charge has been stored, and then a neighboring gate simultaneously receives a phase voltage, charge will be transferred between cells because the barrier has been removed through the collapsing phase. It can be readily appreciated that for each gate cell, the shallow well part should be about half the depth of the deeper well portion.

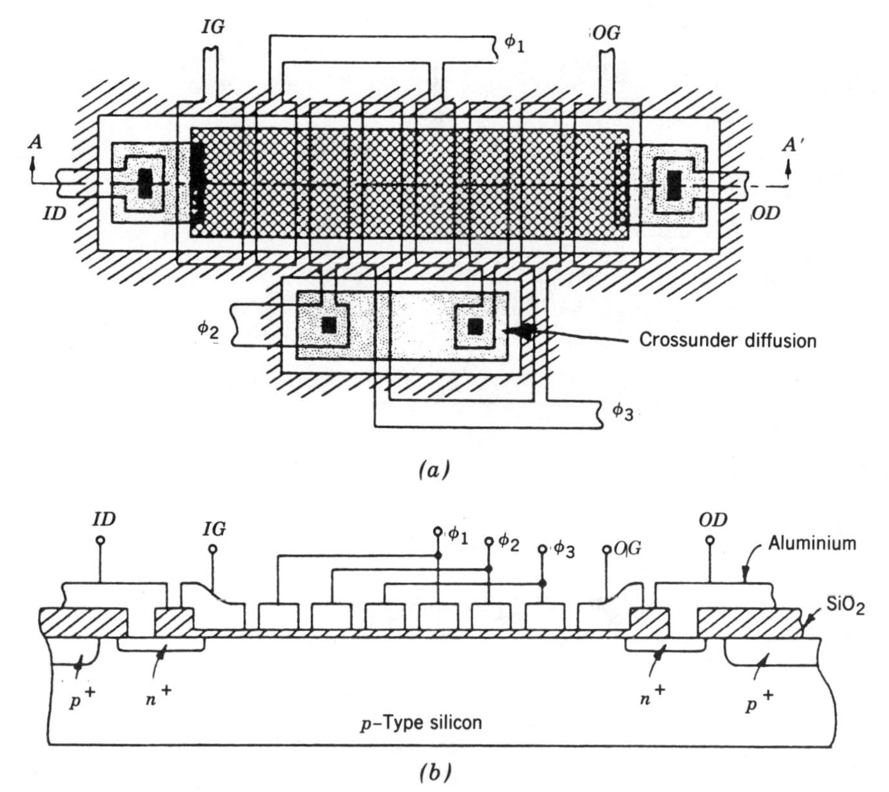

Figure 6.11 Three-phase, two-bit, *n*-channel CCD. (a) Layout of the device; (b) cross-section diagram along *A-A'*. From Howes and Morgan, *Charge-Coupled Devices and Systems*, © 1979. Reprinted by permission of John Wiley & Sons Ltd., UK.

Whilst the charge handling capacity of a 2-ϕ CCD is about half that of a deep well 3-ϕ device (under similar conditions), it also requires a more specialized technology. However, the 2-ϕ device has several key advantages: easier topological design, simpler clocking and charge transfer per delay-bit requires two, rather than three, inter-cell transfers. As these advantages outway the disadvantages in practice, 2-ϕ CCDs are more popular, mainly due to the simpler driving requirements.

Two other clocking schemes, $2\frac{1}{2}$-ϕ and 4-ϕ, are commonly employed in CTD design. $2\frac{1}{2}$-ϕ clocking is an arrangement frequently employed in charge sensing (typically, FGR tapping in Table 6.1), in which charge packets are dumped from a full amplitude clock phase into a half depth well, previously created by a dc level of half the clock amplitude. Although the charge handling is halved and the CTE is slightly impaired, thc technique is very convenient for charge sensing in CTD filters.

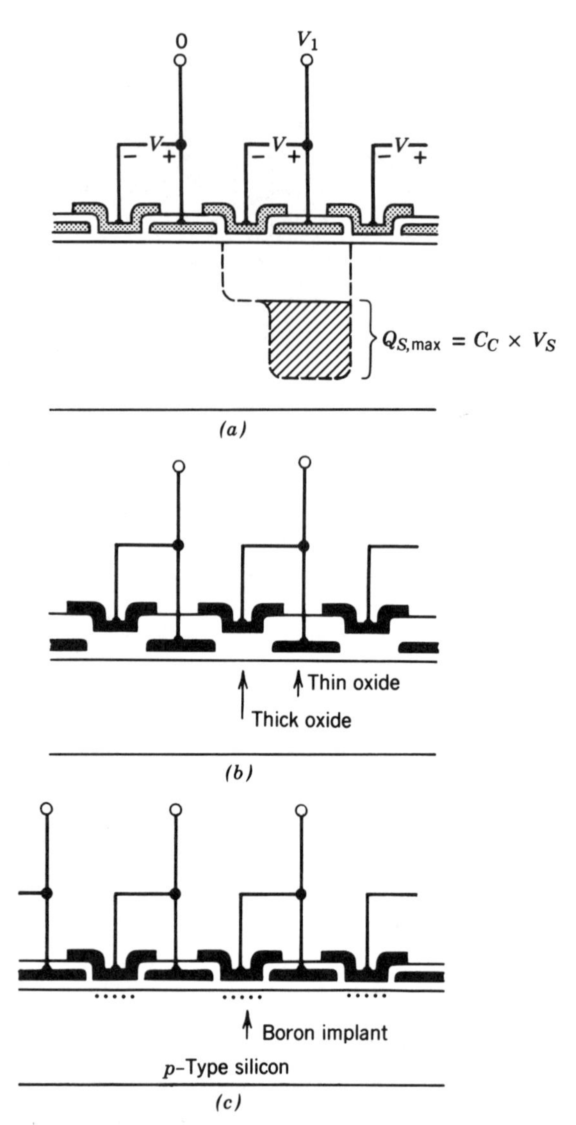

Figure 6.12 Two-phase CCD gate structure. (a) Voltage offset; (b) stepped (profiled) oxide; (c) implanted barrier. From Howes and Morgan, *Charge-Coupled Devices and Systems*, © 1979. Reprinted by permission of John Wiley & Sons Ltd., UK.

So-called 4-ϕ CCDs are essentially 2-ϕ in operation, but the two pairs of different well depths are produced individually and, thereafter, coupled together electrically by joining each pair of cell electrodes to two phases. As in a 2-ϕ CCD, the different well depths are usually produced by methods (a) or (b), above, as illustrated in Figure 6.12.

6.4.3. Layout Design

The design engineer has to be able to produce a prototype filter to a given specification. Decisions in this respect that relate to device layout are crucial, and many man-months of design effort are usually devoted to this study. A number of basic design constraints emerge in this connection which link layout geometries to S/N ratio and linearity considerations (raised in §6.2).

6.4.3.1. Input/Output Diodes. These structures must be as small as possible because kTC reset noise is a function of junction diode capacitance $C_N^{1/2}$ (refer to Eq. (6.4)). They must also be small since any transistor switch associated with a capacitance, C_N, requires the lowest possible channel resistance. This enables large charging currents for high speed resetting of capacitive-node voltages. Small C_N values also permit small gate area transistors (e.g., T_1 in Figure 6.4) to be employed, as their channel *on* resistances are inversely proportional to gate width, W (with L at the minimum feature size). Invariably, therefore, small area low-capacitance diodes are designed to preserve speed and area.

6.4.3.2. Gate Lengths. The CTI is a function of the gate length, L, and in practice $L \approx 3\ \mu m$ is generally acceptable. For any given process, the CTI versus L dependence may be available. Often L is determined more by layout considerations than by electrical requirements, about a nominal 3 μm value.

Occasionally, the designer is faced unavoidably with a situation where charge is to transfer under a long gate dimension (say, $W = 10\ \mu m$ but $10\ \mu m < L < 50\ \mu m$). This occurs in advanced CTD filtering architectures mentioned in §6.4: a common example is the corner turn application in long delay lines where, to preserve an approximately square integrated circuit area preferred for high yields and easy packaging, the device channel is periodically folded around. The general effect of making $L > 10\ \mu m$ is to reduce internal drift fields and increase required carrier diffusion time, causing a deleterious effect on CTI that can only be partially alleviated at greatly reduced clocking rates.

6.4.3.3. Gate Widths. The charge handling capacity of a CCD is proportional to the gate area, A_G, and so the area should be adequately large. However, as the reset and certain other noise origins have a $A_G^{1/2}$ dependence, the best S/N ratio is achieved for a large gate area CTD ($100 \times 5\ \mu m$ being suitable for many analog signal processing applications). However, the upper limit to the gate area is set by the drive capability of the clock generators and MOS reset transistors, which have to drive a capacitance of value proportional to A_G. Furthermore, the size of the CTD is essentially determined by the total gate area and as production cost is related through device yield to active area, very large area structures are in general discouraged wherever possible. Normally, gate widths lying in the range $20\,\mu\text{m} < W < 800\,\mu\text{m}$ are suitable for analog filtering applications: the broadest gates are required for precision split-gate transversal filters (discussed in §6.5) in which the gate area has to be bifurcated with precision.

6.4.3.4. Conductor Tracks. In silicon integrated circuit practice, it is usual to run high current supplies in aluminium, rather than in diffusion or polysilicon, to minimize series voltage drops. In a typical analog filtering situation, the total gate capacitance could easily be 100 pF/phase. At modest clocking speeds the clock edges could easily exceed 10^8 V/s, indicating peak currents that exceed 10 mA. Such a current level would normally require phase conductor widths that exceed the minimum specified value given in the process design rules. This is particularly true of complicated CTD filter design in which dc bias currents to the sense amplifiers also have to be considered. Often two, or more, sets of busbars are used to avoid coupling clock pulse edges to the signal lines.

6.5. CHARGE-TRANSFER DEVICE FILTERING MODULES

The CTD concept permits a compact realization of a sampled-data, serial-in/serial-out delay line that can operate on either analog or digital signals (by coding the binary or multiple levels in terms of appropriate fractions of an empty to full charge packet size). Further, by exploiting the natural availability of MOS analog circuits, the basic delay line can easily be extended to provide parallel-in/serial-out or serial-in/parallel-out architecture by introducing additional charge injectors and/or sensors. In some CCD filters, charge packets are detected in the form of a voltage from a charge/voltage sensor and then re-introduced into a new delay line by a further voltage/charge conversion. In an alternative charge-domain (CDD) technique, charge packets are divided, recirculated and re-introduced all in the charge domain. The extreme versatility of the CTD concept, together with the option of mixed analog/digital circuit functions, enables the designer to produce custom-specific filtering modules that can form part of a monolithic signal processing system. This versatility has meant that virtually every filtering configuration required for general signal processing has been realized since CTDs were first introduced, with varying degrees of success.

In this section, we review a number of established approaches to recursive, transversal, and adaptive filter modules which have been introduced for a range of applications including sonar, radar and communications. It is to be remembered here that the CTD is a baseband filtering technique that requires an anti-aliasing prefilter, and output sample-and-hold and low-pass filter stages. In general, bandwidths to 1 MHz are readily obtainable for advanced filter modules, whereas bulk-channel delay lines easily exceed full video bandwidth capability.

6.5.1. Recursive Filters

In recursive filter implementations, a suitable means has to be provided for allowing signal feedback and/or feedforward from various points along a tapped delay line. The tapped signals must also be multiplied, individually, by the filter coefficients (in voltage or charge form) before being reintroduced into the CTD input. This operation can be achieved in either of two ways. Firstly, by multiple charge/voltage and voltage/charge conversions with intermediate voltage multiplication. Secondly, by recirculating a predetermined proportion of the delay line propagating charge, without the need for such conversion (to voltage), as employed in CDDs. Charge/voltage conversion may be achieved conveniently, for example, by employing the FDR technique described in §6.2.1. The voltage/charge conversion corresponds to a CCD input structure which again is covered in §6.2.1. The voltage amplification operation may be achieved using an MOS transistor, as discussed later in §6.5.2. Results for CCD recursive filters employing multiple charge/voltage conversions have been disappointing in respect of their accuracy, sensitivity, and linearity; mainly because of the spread in gain and linearity of the sensing and input amplifiers, and poor precision associated with integrated analog voltage multipliers that set the filter coefficients. Nevertheless, comb filters and low-order filtering sections have been reported [2] in prototype CCD form.

Probably the best recursive filter implementations in CCD technology have employed [1] time-multiplexed, multiplying digital-analog converters (MDACs) which greatly alleviate the problem of accurately controlled gain. Such MDAC circuits are widely available commercially, and may be realized in integrated form using MOS technology. They condition, or multiply, an analog input signal with a digital input word (of typically up to 12 bits in monolithic form) and produce a corresponding analog output signal. The provision of MDACs adds two new facilities to CCD filter realization: electronic (digital) control of the filter characteristics or *programmability* is possible leading to improved filter stability; and, further, a means is provided to enable CTE and gain effects to be compensated. The former advantage is provided by digitally programming filter coefficients, stored in memory, into the filter through each MDAC. The tapped CCD analog signals applied to the MDAC input emerge after multiplication, and are then introduced into the CCD channel. Filter banks have been realized by this approach in which the coefficients have been stored in ROMs. In practice, the whole filter can be integrated monolithically in MOS technology, the CCD, MDACs, and ROM. Filter banks based on this approach can provide resonator-like characteristics with a maximum center frequency of about 20–50 kHz, limited by the MDAC conversion times. It is unlikely that Q-values greater than 100 can be achieved by this technique, with reasonable stability.

Compensation for certain deficiencies in such a CCD recursive filter can be achieved by applying a test pulse to its input and then storing in the

ROM the digital error words necessary to adjust for imperfect CTE and gain. Such error correction must cancel residual effects (see Figure 6.7) and also restore the desired output pulse to its original input amplitude.

With the alternative CDD technique [21], all the filter operations are performed by charge manipulation, without the need for charge/voltage conversion (apart from the output). The coefficient (tap) values for a particular filter are preset by splits in the transfer gates, which are predefined by interceding potential barriers created by narrow implantations or thick-oxide regions. Charge packet splitting using the latter technique is illustrated in Figure 6.13. Here, a thick-oxide barrier region divides a single thin-oxide region (necessary for charge coupling action) into two channels of width a_1 and a_2. With this approach the split ratio a_1/a_2 can only achieve coefficients less than unity. However, techniques exist [21] to implement ratios greater than unity having, where desired, negative sign. Divided

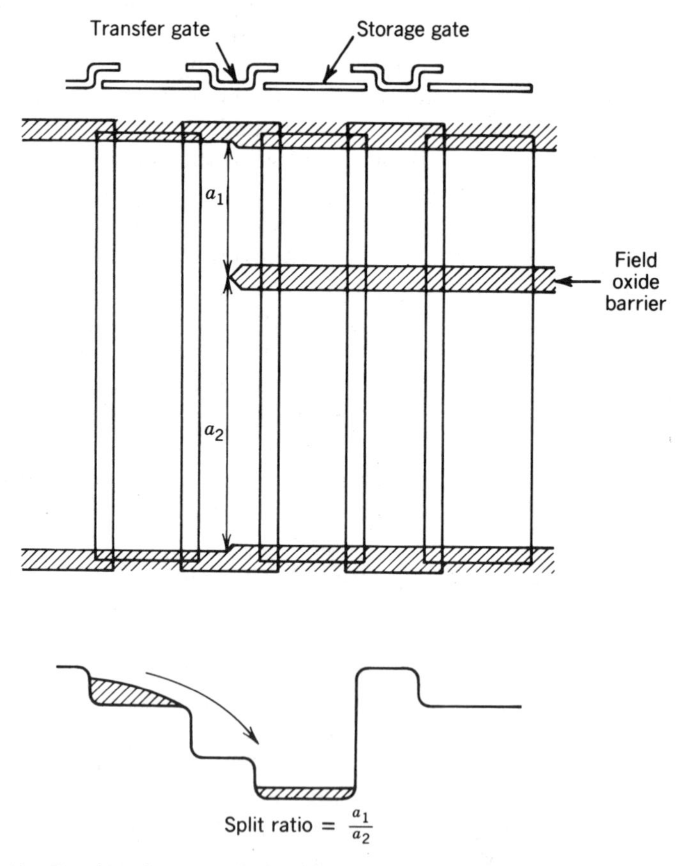

Figure 6.13 Barrier charge splitting. Reprinted with permission from Vogelsong, Tiemann and Steckl (1985) Charge-domain integrated circuits for signal processing. *I.E.E.E. J. Solid-State Circuits* **20**, 562–570 © 1985 I.E.E.E.

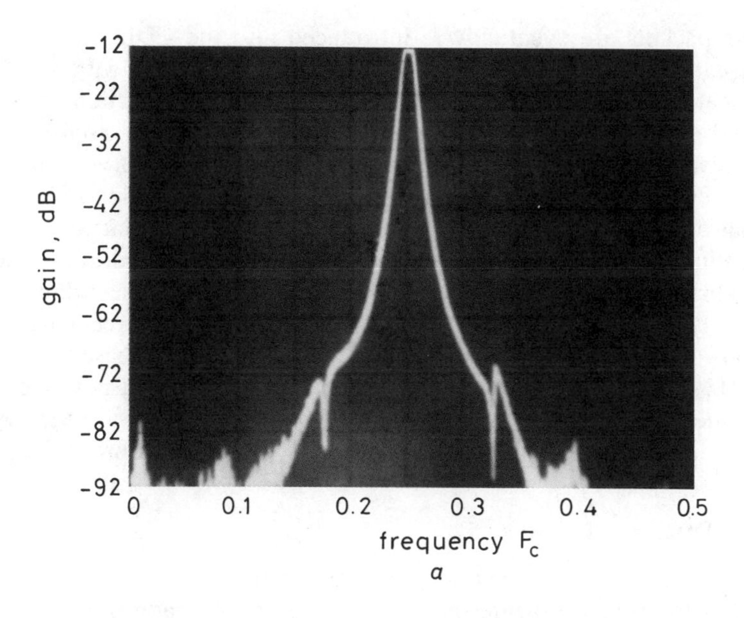

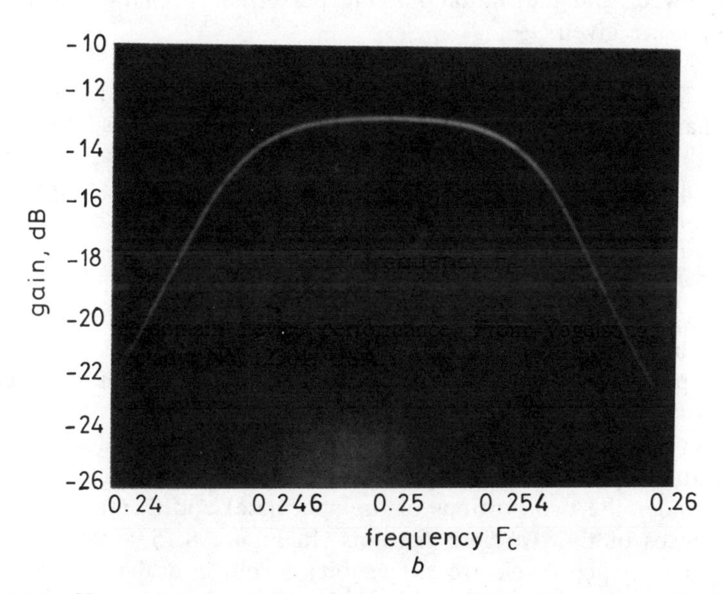

Figure 6.14 Charge-domain device performance. From Vogelsong, © General Electric Co., Schenectady, NY 12301, USA.

charge packets are eventually re-introduced into the CDD, after a number of separate transfers, by merging the re-entrant packets with the forward propagating signal charge packets. The whole filter function is therefore defined exclusively in the charge domain, and coefficient accuracy, set by gate division, can be maintained to three decimal places. Figure 6.14 illustrates the performance of an eight-pole, CDD bandpass filter design having a stop-band attenuation of around 100 dB (the latter value being determined by device noise rather than by coefficient inaccuracy). The CDD was fabricated on a 4 μm, surface channel, double-polysilicon layer process originally developed for CCDs. The CDD approach to recursive filtering appears extremely promising, and in advance of CCD-based techniques. 20 MHz clocking has been demonstrated and device size, power efficiency and wide dynamic range are also in its favor. Such devices appear useful for IF filtering, FM communications, radar and video signal processing.

6.5.2. Transversal Filters

Probably the most widespread application of CTD technology has been directed to transversal filtering. A fixed response, finite impulse response filter can be produced very elegantly in hardly more complexity than required for producing the basic analog CCD delay line. Alternatively, a programmable or variable response transversal filter can be implemented in CCD/MOS technology, but at considerably more complexity. In this section, both fixed and programmable response transversal filter realizations are reviewed, and the limitations and performance characteristics of each technique are given.

6.5.2.1. Fixed-Response Filters. The most popular approach for achieving a fixed-characteristic transversal filter is referred to as the split-electrode or split-gate technique [9]. With this efficient approach, the sensed charge packets are multiplied by the filter tap weights, set by splitting transfer gates in proportion to the desired impulse response.

The principle of operation is that as charge moves into the region under a gate, an equal and opposite image charge must flow onto the gate electrode from the clock phase to which it is attached. Therefore, if the current flow in a clock line is monitored for a single CCD cell, then its time-integrated value is proportional to the gate area and the size of the charge packet that has just flowed underneath the gate. With the gate split into two portions, the difference between the current flow in both supplying clock lines will be a quantity proportional to the product of the charge packet size (distributed evenly under the two portions of the split gate) and the ratio between the capacitances of the two gate portions. In Figure 6.15, which outlines this arrangement, each ϕ_2 electrode (per bit) is split in design according to the desired tap weight: a portion $\frac{1}{2}(1 + h_k)$ connected to a common positive clock phase, ϕ_2^+; and a portion $\frac{1}{2}(1 + h_k)$ connected to a common negative

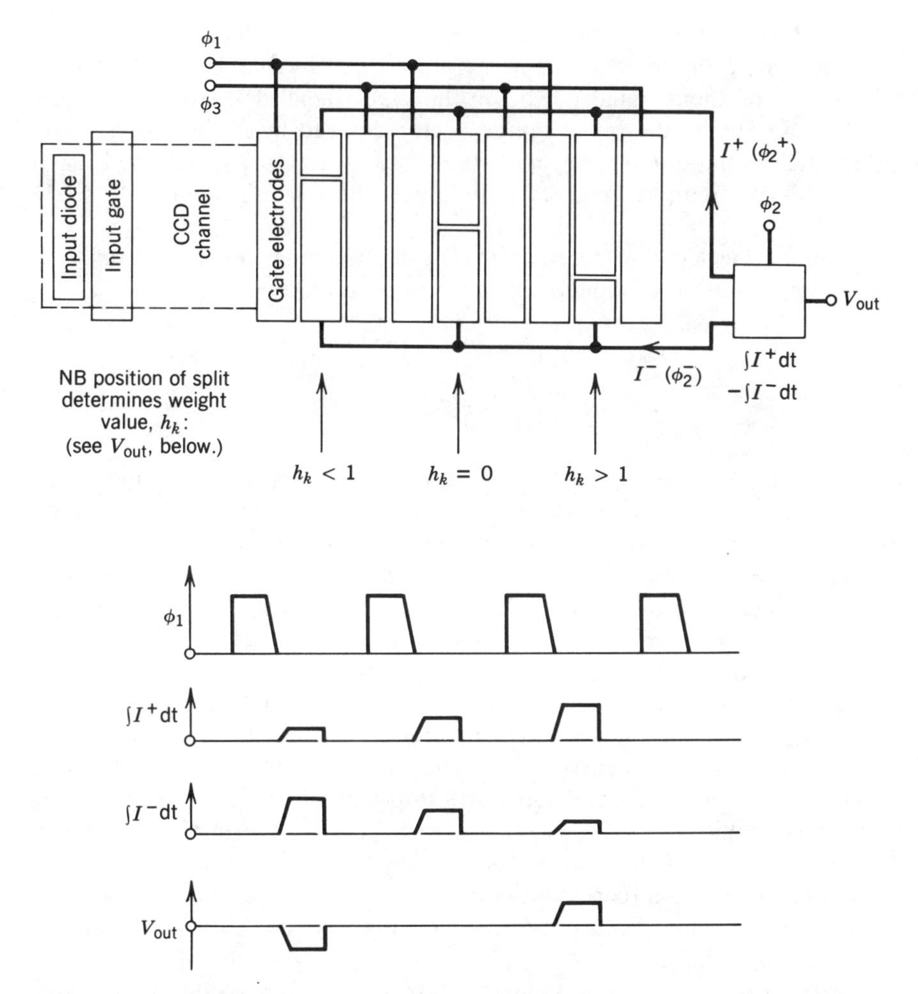

Figure 6.15 CCD transversal filter employing split-gate technique. Note clock current integration performed during ϕ_2 clock time. This material first appeared in *Wireless World* (now entitled *Electronics and Wireless World*) February 1975, as part of a four-part series on charge-coupled devices.

clock phase, ϕ_2^-. When the entire gate is connected to the positive clock phase, a weight $h_k = +1$ is generated, or $h_k = -1$ is produced when the entire gate is connected to the negative clock phase. This approach would produce binary weight filter coefficients. By choosing the split to lie within these two extremes, weighting values between $+1$ and -1 can be obtained. A differential, integrating charge amplifier is used to determine the difference between the two integrated current values, for the whole delay line. The desired transversal filtering operation, $\Sigma_1^N h_k q_k$, is produced at its output whose value is determined by the functions of tapping, multiplication by a fixed number and summation. Here, h_k is an individual tap weight and $\mathbf{H}_k$ is the whole tap weight vector. q_k represents (in charge value) delayed samples of the input signal (q_1 being the input signal delayed by one time unit, and $\mathbf{Q}_N$ being the input charge vector delayed by N time units). The differential amplifier, together with the clock generator, is designed using analog MOS circuit techniques and integrated with the split-gate CCD array.

The filter specification determines [16] the minimum number of taps, N, for an optimum, finite impulse response low-pass filter according to

$$N \geq \frac{-10 \log_{10}(\delta_1 \delta_2) - 15}{\frac{14}{f_c}(f_B - f_A)} + 1 \tag{6.7}$$

where, δ_1 and δ_2 define the peak-to-peak pass-band ripple, and f_A and f_B are the pass-band and stop-band edge frequencies, respectively. The clock frequency, f_c, is set by the anti-aliasing constraint. As an example, a low-pass telephone filter with an out-of-band rejection of 40 dB, pass-band ripple of 0.5 dB, pass-band, stop-band and clock frequencies of 3.4 kHz, 4.6 kHz and 64 kHz, respectively, would require at least 78 taps. In practice, however, factors such as tap weight tolerances dictate that about 100 taps would be required for this application. Above this number, the additional small-value tap weights serve only to introduce noise into the output.

Generally, for a nonrecursive, FIR filter: (a) the pass-band roll-off rate is a function of the number of taps; (b) stop-band rejection and pass-band ripple are functions of tap weight accuracy; and (c) the impulse response duration is given by the number of taps, and is inversely proportional to clock frequency. From these considerations, it can be deduced that the filter response may be improved either by reducing the clock frequency and/or increasing the number of taps.

The tap weights may be found from standard design procedures [16], with the effects of tolerance and truncation minimized by applying appropriate window functions, for example, Hamming or raised cosine. Alternatively, interactive design programs based on the McClellan–Parks algorithm [14] may be adopted for tap weight design procedures.

CCD transversal filters suffer from a number of inherent limitations, which are detailed here:

- Weighting coefficient inaccuracy in a split-gate filter provides a practical limit to pass-band ripple and to the available stop-band attenuation of typically 40–50 dB. This inaccuracy arises from quantized split placing, caused by processing inconsistencies and in photomask production. For example, in an electron-beam exposure system a step size of 0.5 μm is achievable for a gate width of about 200 μm. By designing wider gates this quantization effect can be minimized; however, wider gates result in a larger clock phase capacitance limiting bandwidth. Tap weight errors from all sources in practical split-gate devices can usually be maintained to 0.5%, but in some designs 1% error may be unavoidable; through really careful designs, 0.1% tap weight accuracy is possible and is necessary for 50 dB stop-band rejection.

- Photomask misalignment: A fixed shift of all the split locations can occur in fabrication due to one mask layer being displaced relative to another (say, gate metal to the channel-definition diffusion). This has an adverse effect on the pass-band ripple in low-pass filters of almost 1 dB, in the worst case.
- CCD filters suffer inevitably from CTI effects, the extent of the problem depending on the particular application. For large $N\epsilon$ products effects of CTI are evident and cause a perceptive tapering in the impulse response. In frequency-selective filter realization, CTI effects cause a reduction in the apparent clock frequency, with a proportional scaling to lower frequencies of the filter response.
- General processing errors may be correlated or uncorrelated according to their origins. Such errors contribute to the overall device noise and limit the weight precision. Increasing the number of taps in a filter does not necessarily improve the response, because random errors contribute to the frequency response as $N^{1/2}$-correlated errors follow an N dependence. Very long delay lines usually induce only noise rather than signal from the low value tap weights, near the device extremity.
- Non-linearities between the charge packet sizes being sensed and the integrated currents induced on the clock lines by the image charges cause errors in a filter transfer function. As discussed in §6.3.2, the linearity issue is involved and is determined by the combined effects of the input and sensing techniques. For BCCD split-gate filters the linearity is somewhat inferior to SCCDs, because the depletion layer charge is much greater and is more sensitive to charge packet size. By employing surface-channel input stages in a bulk channel CCD, reasonable linearity can be achieved, although it is not always possible to compensate this nonlinearity at the input.

In spite of the above performance limitations, split-gate CCD filters are employed as a powerful approach to the realization of time-domain and frequency-domain filters. The CCD FIR approach offers linear phase, high transition (150 dB/octave has been achieved, roughly equivalent to a 25-pole RC filter), large time-constants without large values of resistors and capacitors, and a dynamic range exceeding 60 dB.

Once the designer has specified the tap weights for a particular filter, the mask defining the split positions in the gate electrodes can be produced. Should further modification to the device coefficients be required, then the mask-making costs for a single layer (out of a set of 7–9 for a typical CCD process) will be incurred again. Thereafter, only a single metal-etch processing step is necessary to complete the device fabrication sequence. Thus the split-gate CCD is an applications-specific component that may be customized rapidly and fabricated from a stock of near-finished blank filters.

A refinement of the split-gate technique is achieved [7] by dividing the CCD into two separate channels, for positive and negative valued weights, by providing two splits in a given tap electrode. Consequently, a reduction of 80% in the inter-electrode capacitance may be obtained leading to a greatly improved common-mode rejection ratio (CMRR) figure. As well as providing improved CMRR results, this dual-split electrode technique enables better tap weight definition leading to a possible 3 dB increase in stop-band attenuation over single-split electrode devices. Improved dynamic range, lower noise and insertion loss are other advantages that make the dual-split electrode transversal filter appealing to the designer for a variety of applications, including voice band CODECs.

A 100-bit split-gate CCD transversal filter with tap weights defining a low-pass filter is shown in Figure 6.16(a), with the corresponding metal-mask pattern defining the split positions (just discernible) in Figure 6.16(b). In addition to the CCD, the chip contains MOS operational amplifiers to perform the charge sensing and differencing functions and output buffering. This device, when clocked at 64 kHz, exhibits a 3.4 kHz bandwidth and frequency sidelobes at about −35 dB. The latter are extendible to −40 dB by increasing the filter length, but improvements beyond this are limited by inaccuracies in tap weight coefficients. This type of filter is relatively insensitive to CCD charge-transfer inefficiency effects, because the number of transfers is only several hundred, making $N\epsilon \approx 0.05$ (refer to Figure 6.9). The filter characteristic, shown in Figure 6.16(c), is broadly similar to that produced by a five-pole elliptic filter, commonly employed in telephony applications.

One of the most successful matched filtering applications of split-electrode, CCD transversal filters is the realization of the convolution function involved in the chirp-z transform (CZT). The CZT algorithm [19] enables a compact, real-time, implementation of the discrete Fourier transform (DFT), commonly used to perform spectrum analysis. The CZT algorithm is realized by: (a) premultiplying the input signal with a chirp (a sinusoid, modulated by a linear FM law); (b) filtering with a chirp convolution filter; and (c) post-multiplying with a chirp waveform. When only amplitude information is required at the output (rather than, more generally, amplitude and phase) the third operation, (c), can be changed to that of square-and-add (as shown in Figure 6.17).

Commercially, CCD/MOS LSI technology has been used to produce fully integrated versions of the CZT: chirp prefiltering being performed usually by an MDAC, in MOS analog circuitry, whose coefficients (representing the sampled chirp waveform) are supplied from ROM; and convolution being performed by four, split-gate CCD filters weighted as two sine and two

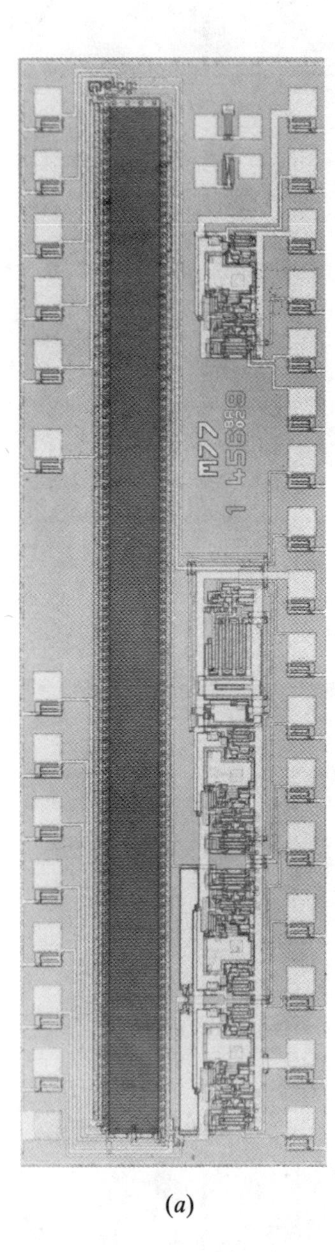

(*a*)

Figure 6.16 One-hundred bit, split-gate CCD transversal filter. (a) Photomicrograph of device; (b) metal-mask split electrode pattern defining impulse response; (c) low-pass frequency response of this filter. The M77 split-gate CCD filter chip photomicrograph and frequency response were supplied by Plessey Research (Caswell) Ltd., UK.

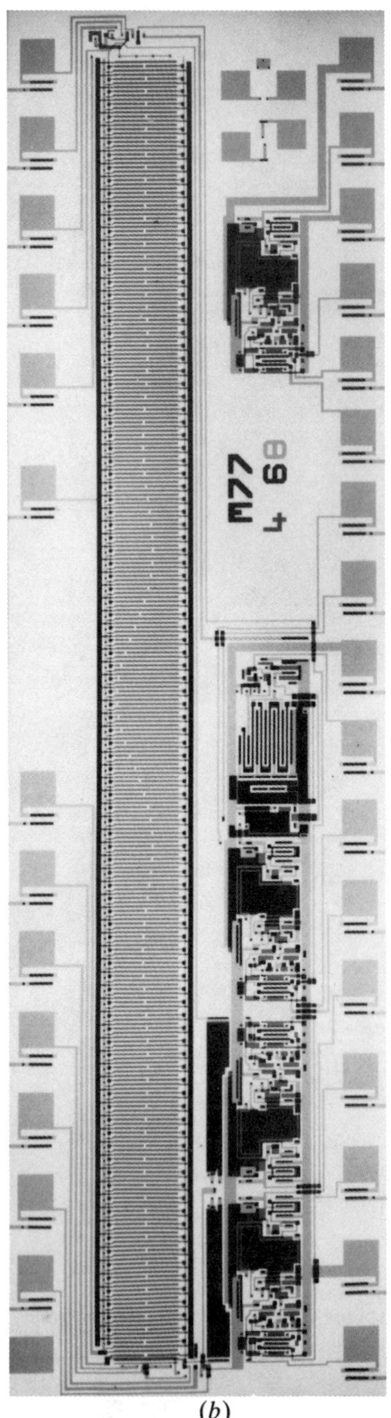

(*b*)

Figure 6.16 (*Continued*)

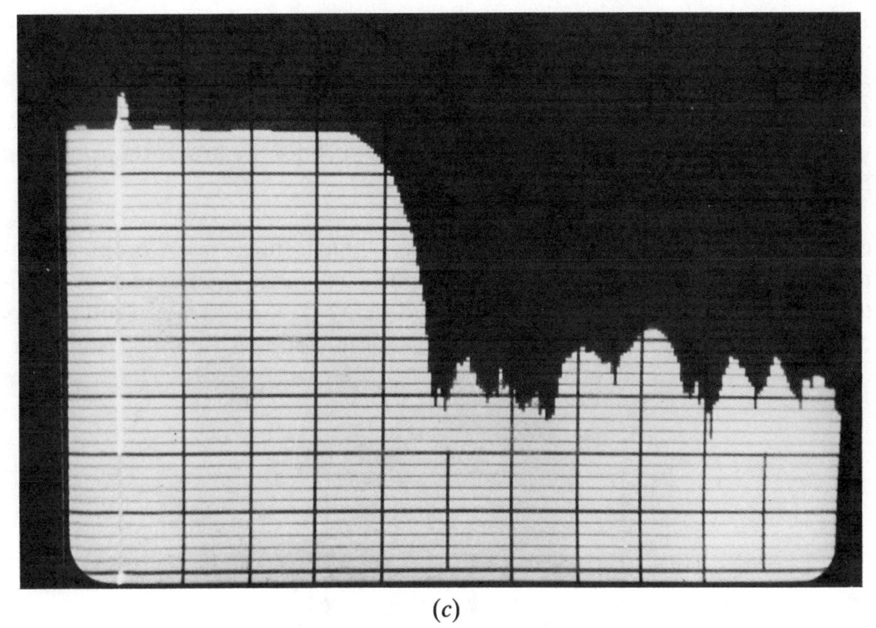

(*c*)

Figure 6.16 (*Continued*)

cosine chirps. A monolithic version would typically offer a 100-point CZT implementation with a 1 MHz signal capability, at baseband. Alternatively, quadruple filter banks are available commercially that contain 512-tap,

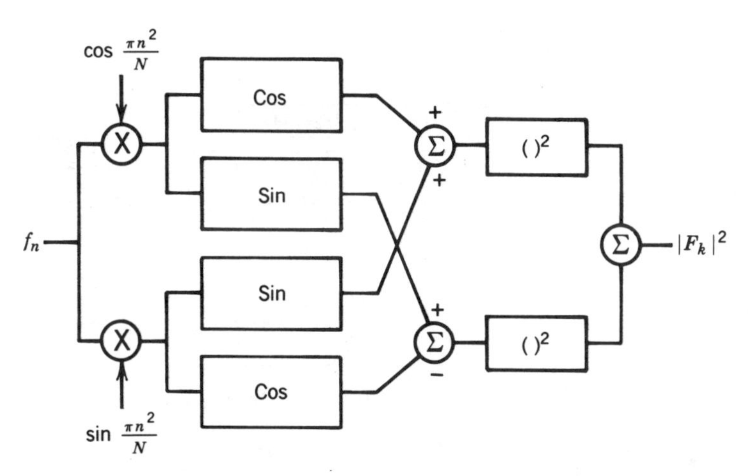

Figure 6.17 Block diagram of the complex arithmetic of the CZT algorithm for computing power-density spectrum. Note the CCD transversal filters are designated by COS and SIN. From Howes and Morgan, *Charge-Coupled Devices and Systems*, © 1979. Reprinted by permission of John Wiley & Sons Ltd., UK.

split-gate CCD transversal filters. These can be teamed up with additional integrated circuits that perform the pre- and post-convolution operations, to obtain a 256-point CZT spectrum analyzer. Usually the output accuracy of such analyzers is about equivalent to 7 bits digital. Probably the cost, weight, size, and power advantages of CCD technology are at their greatest in this application.

6.5.2.2. Variable-Response Filters. Many schemes have been proposed for realizing variable or programmable impulse response CCD transversal filters. The elements required are a tapped delay line, an array of multipliers or a single time-multiplexed multiplier, and an optional summing amplifier.

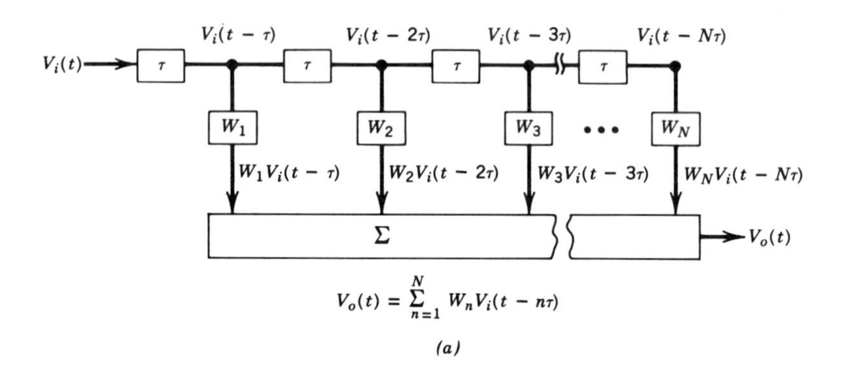

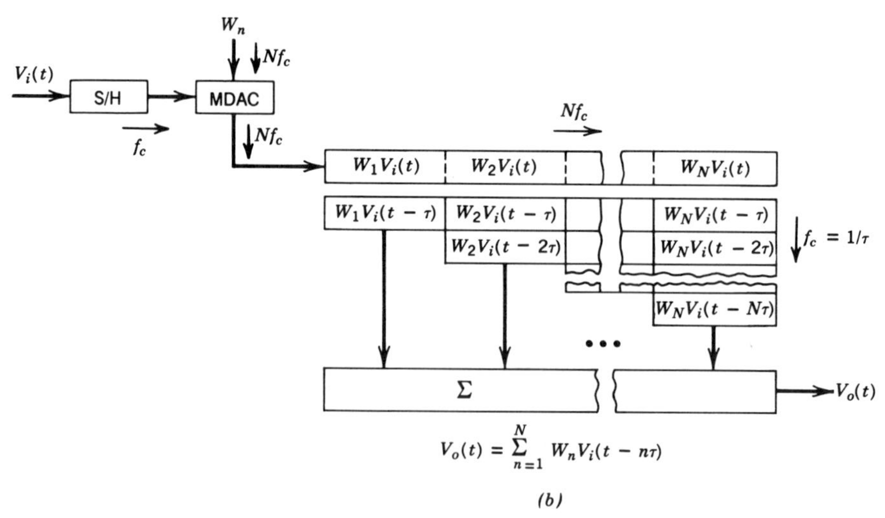

Figure 6.18 Alternative implementations of transversal filter. (a) Using a tapped and weighted SI/PO CTD delay line; (b) using a PI/SO CTD delay line (time-delay and integrated (TDI) configuration). From Barbe, *Charge-Coupled Devices*, Vol. 38, *Topics in Applied Physics*, © 1980, Springer-Verlag, West Germany.

However, the order in which these elements are used and the consequent filter architecture leaves great scope to the designer, according to the performance required. The individual filter coefficients are programmed into the filter by multiplication of signal levels; and all-binary, binary-analog and analog-analog filter formations have been reported each placing different requirements on the multiplier array. The usual approach, given in Figure 6.18(a), is to post-multiply each of the delayed signal samples, after they have passed down a serial-in/parallel-out CCD and, thereafter, to sum them in a separate operation. An attractive alternative shown diagrammatically in Figure 6.18(b), is to premultiply the input signal using a high-speed, time-multiplexed MDAC whose weights are stored digitally in a ROM. The weighted outputs can then be fed to a parallel-in/serial-out CCD which conveniently sums, by charge packet addition, the individual delayed and preweighted signals within the serial section. The CCD channel configuration employed here is referred to as (signal charge) time delay and integrate or TDI; a device example of which is described in §6.6.2. Both approaches have merit and subtle design trade-offs exist between each of the realizations.

We concentrate here on the former alternative which has been employed [5] for sonar signal processing in monolithic form with 256 taps. In this approach, the tapped CCD delay line poses no particular problems, but the task arises of producing an array of several hundred multipliers, particularly analog-analog forms (multiplication of two analog signals). The usual technique for achieving this is to employ an array of matched, MOS transistors each biased into their triode region of operation. In Figure 6.19, which

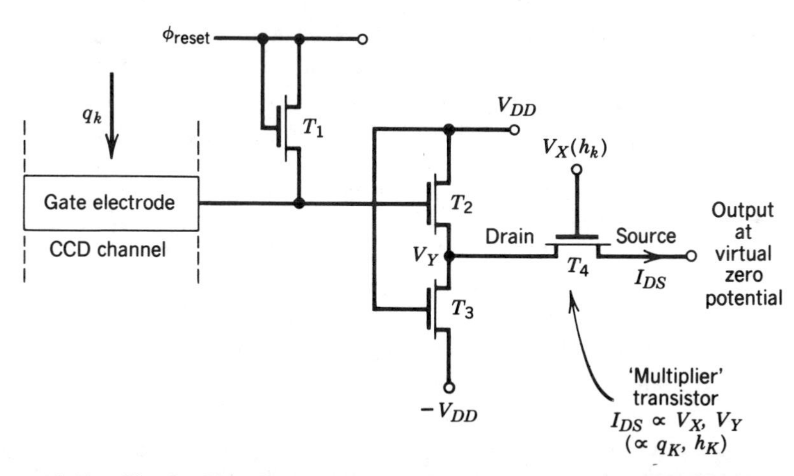

Figure 6.19 Circuit diagram for floating-gate sensing amplifier (T_2 and T_3) with variable transconductance signal weighting (through T_4), and voltage reset (via T_1). Reprinted with permission from Beynon and Lamb, *Charge-Coupled Devices and their Applications*, © 1980, McGraw-Hill Book Co. (UK) Ltd.

illustrates the general arrangement, the MOS transistor T_4 performs the analog, transconductance multiplication. For each multiplier transistor (T_4) under this condition, the output (source-drain) current, I_{DS}, is proportional to the product of the gate-source (V_x) and drain-source (V_y) voltages. Thus, by applying a tap voltage (following a reset pulse, ϕ) from the CCD (proportional to the signal charge, q_k) to the T_4 transistor drain, V_y, and a voltage corresponding to an individual tap weight (h_k) value, V_x, to the transistor gate-source terminals, the output current of an individual multiplier is proportional to the product $q_k h_k$. By summing simultaneously all these outputs in a current amplifier (at a virtual zero potential), the desired convolution $\Sigma_1^N q_k h_k$ can be obtained. Nonlinearities in the MOS transistor characteristics can be minimized by injecting a signal zero into the CCD input between each signal bit and, thereafter, subtracting the stored nonlinear current term from the signal. A higher performance multiplier arrangement can be achieved by employing four MOS transistors in a bridge [18], rather than the single-transistor scheme just described. With this approach, the inherent properties of a bridge configuration make the multiplier cell highly tolerant of errors not only in the transistors but also in the signal path, including the CCD input and output devices. However, the price to be paid for the performance increase is the extra complexity of the CCDs and their interconnections to the multipliers.

A parallel correlator or parallel convolver, shown schematically in Figure 6.18(a), can be implemented by temporarily storing the weight voltage vector in an additional CCD delay line, whose contents can be updated when required by serial input from either end of this register. An experimental parallel correlator based on these principles and employing one MOS transistor per multiplier cell has been designed [5], and a typical chip microphotograph is given in Figure 6.20. This has a signal bandwidth of 500 kHz, a processing accuracy of 7–8 binary bits and is 256 points in length (corresponding to a $TB = 256/2$, according to Eq. (6.5)). With this correlator structure, the weight vector delay line is replaced by an MOS sample-and-hold array that does not suffer from CTI effects. Eight of these devices have been cascaded in an active sonar experiment to yield a maximum time-bandwidth product of 1024. Matched-filter results for this arrangement are given in Figure 6.21 for: (a) a reduced TB product of 256 (two chips in cascade); and, (b) a full TB product of 1024. The characteristic $(\sin x)/x$ response of this transversal filter is clearly observed in both cases.

Such fully programmable structures are extremely flexible, as we shall see in §6.5.3 on adaptive filtering, especially when teamed-up with a microprocessor that provides a stable (digital) reference for periodic compensation against CTI effects and other imperfections. Other applications for such CCD programmable transversal filters are potentially numerous and include, for example, the realization of a DFT using the prime transform [11] for spectrum analysis.

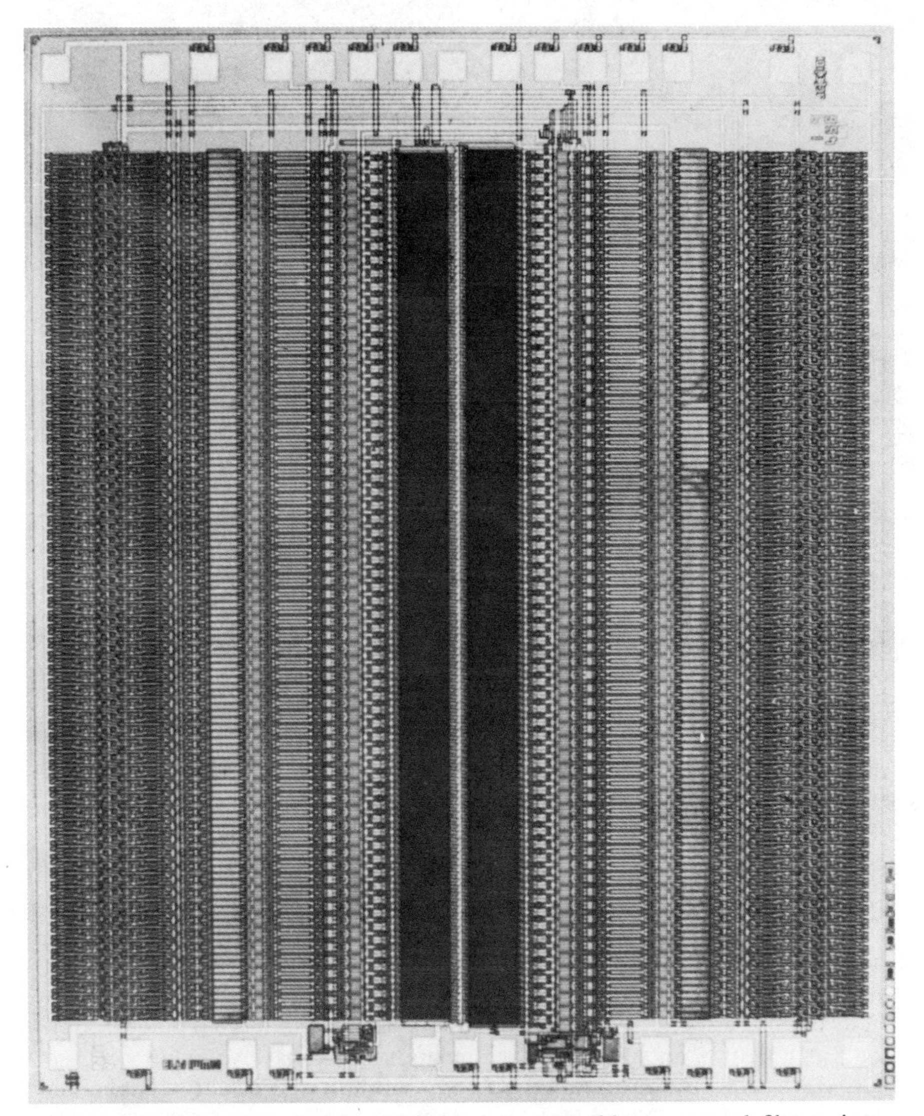

Figure 6.20 Photomicrograph of 256-bit programmable transversal filter using an MOS transistor transconductance array (as in Figure 6.19). From Denyer and Mavor (1979) Monolithic 256-Point Programmable Transversal Filter. *Electron. Lett.* **15**, 710–712 © 1979 I.E.E.

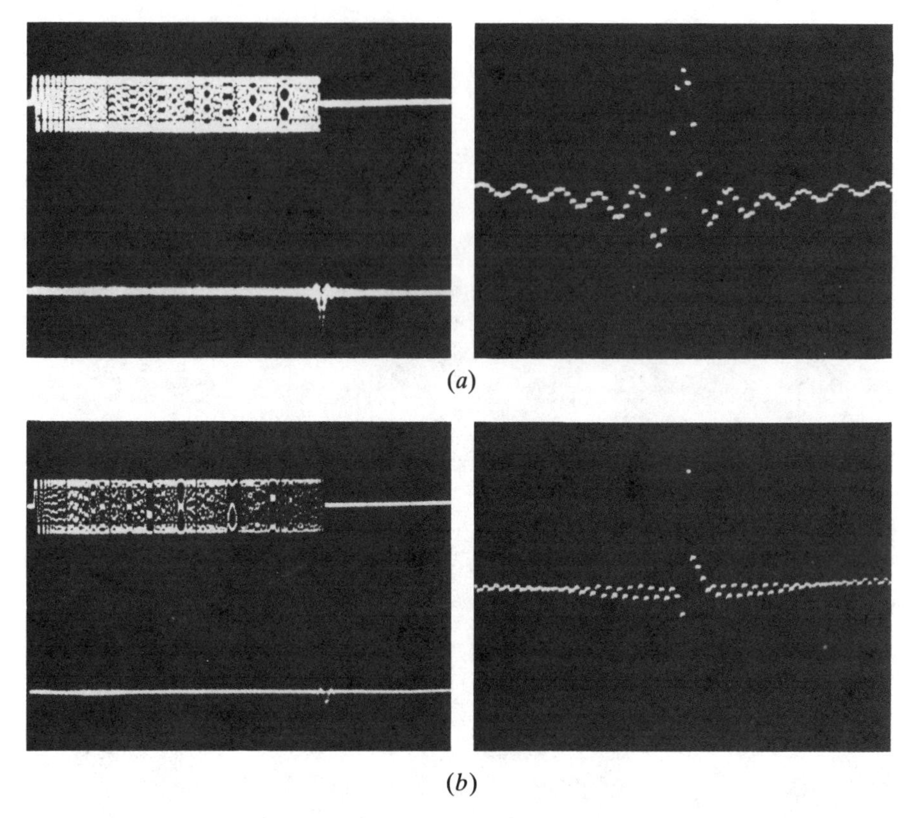

Figure 6.21 Results for experimental FIR filter module employing Figure 6.20 devices: (a) $TB = 256$, two devices in cascade showing peak detail; (b) $TB = 1024$, eight devices in cascade showing peak detail. With permission, Wolfson Microelectronics Ltd, Edinburgh, UK.

6.5.3. Adaptive Filtering

It has been recognized for many years that the so-called adaptive filter is a key filtering concept that can be used for tasks such as echo cancellation (and its corollaries, e.g., noise and multipath cancellation) in communication links and adaptive equalization. The adaptive filter is invaluable because these tasks can be achieved potentially in real-time within an unknown and changing environment, such as a switched telephone line. It is inherently suited to this type of application because it is designed to learn from incoming signal statistics about the environment in which those signals exists. By modeling the impulse response of the transmission medium or its inverse (in the case of a communication link), effective echo cancellation or channel equalization may be achieved. Most practical adaptive filter realizations, based in the past on digital technology, have been very complex [12] with generally low bandwidths (few kilohertz). The CCD programmable transversal filter approach (described in §6.5.2), however, lends itself very elegantly to compact, high performance implementation and this has been achieved for some of the above applications [4].

The adaptive filter tunes itself automatically to the operating environment by a control signal supplied from an internal adaptive algorithm, which adjusts the transversal filter tap weight values. Preferably, the tap weight vector would be changed every clock period (as each new signal sample is fed into the adaptive filter); however, in the particular CCD-based processor described here it is adjusted less frequently leading to a slower convergence time. Many forms of adaptive algorithm exist, but in selecting a simple algorithm, such as the Widrow least-mean-square (LMS) adaption algorithm [23], monolithic implementation may be achieved because analog circuits can be employed throughout. Figure 6.22 shows a block diagram of an adaptive system, in which the filter is required to estimate and cancel components of a reference input x, in a primary channel, d. The resultant system output, e, becomes an error to be minimized by the filter via the controlling algorithm. The adaptive filter performance is governed by the set of tap weight values or vector, $\mathbf{H}(n)$, in the time period, n, presented to the embedded transversal filter (refer to Fig. 6.17(a)). Its output in the $(n+1)$th period is determined by the updated, $\mathbf{H}(n+1)$, vector found from a knowledge of the previous input (or reference) set or vector, $\mathbf{X}(n)$, stored in the filter and the current value of the error signal, $e(n)$. With the LMS algorithm the tap weight vector, $\mathbf{H}$, is adjusted according to

$$\mathbf{H}(n+1) = \mathbf{H}(n) + 2\mu e(n)\mathbf{X}(n) \tag{6.8}$$

The term, 2μ, determines how quickly the adaptive filter system converges

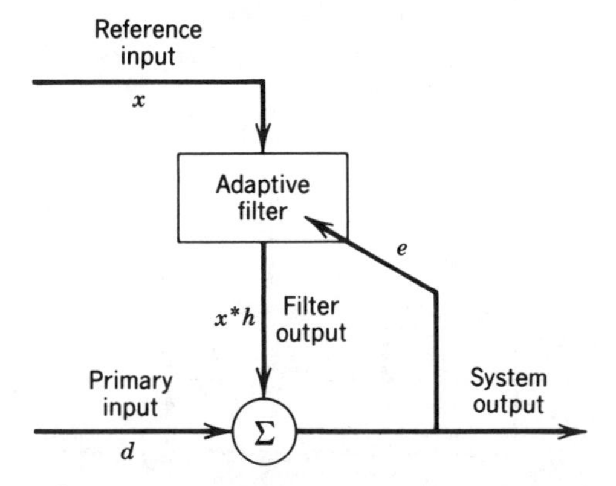

Figure 6.22 Block diagram of an adaptive system. From Denyer, Cowan, Mavor, Clayton and Pennock (1983) A monolithic adaptive filter. *I.E.E.E. J. Solid-State Circuits* **18**, 291–296 © 1983 I.E.E.E.

to minimize the derived error signal to zero. For this reason, 2μ is called the convergence factor. The error signal is derived as follows

$$e(n) = d(n) - \mathbf{X}(n) \cdot \mathbf{H}(n) \qquad (6.9)$$

where $\mathbf{X}(n) \cdot \mathbf{H}(n)$ is the transversal filter output.

A fully concurrent implementation of this algorithm requires the tap weights within a transversal filter to be updated via a multiply/accumulate circuit that forms, and integrates, the product $2\mu e(n)x(n-i)$. A schematic diagram of the necessary filter architecture is shown in Figure 6.23. In a particular realization described here, a CCD/MOS circuit approach using entirely analog techniques is advocated, which includes the multiply/accumulate operation realized in a novel MOS analog circuit. A photomicrograph of the prototype, 65-bit adaptive filter is given in Figure 6.24, in which the signal delay line is in the center of the device about which the multiplier/accumulators are located. The chip area is 4.75×3.5 mm and the pitch between filter sections (see Figure 6.23) is only 28 μm, indicating the compact packing density achieved in this approach. This experimental filter performs correctly up to sample rates in excess of 500 kHz, and Figure 6.25 is indicative of its filtering capability. This two-order bandwidth advantage, coupled with a power requirement for this 65-point adaptive filter of only 250 mW, compares extremely favorably with alternative all-digital implementations. An illustration of adaptive tone cancellation is given in

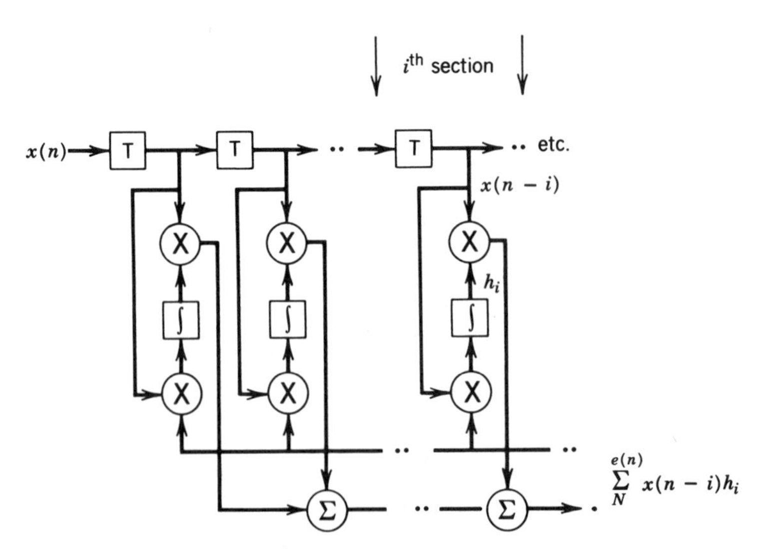

Figure 6.23 Adaptive filter architecture based on a tapped and weighted transversal filter. From Denyer, Cowan, Mavor, Clayton and Pennock (1983) A monolithic adaptive filter. *I.E.E.E. J. Solid-State Circuits* **18**, 291–296 © 1983 I.E.E.E.

Figure 6.24 Photomicrograph of a prototype 65-bit adaptive filter based on Figure 6.23 architecture. From Denyer, Cowan, Mavor, Clayton and Pennock (1983) A monolithic adaptive filter. *I.E.E.E. J. Solid-State Circuits* **18**, 291–296 © 1983 I.E.E.E.

Figure 6.25, whereby an unwanted component of two sinusoids at the filter primary input (a) is adaptively cancelled by coherent subtraction to produce trace (b). The adaptivity or rejection figure of 20 dB compares favorably with other approaches.

This adaptive filter implementation illustrates the power of CCD/MOS techniques applied to solid-state filtering. The designer may achieve arbitrarily long, high-bandwidth adaptive filters either in a single integrated circuit or by cascading a number of such circuits for higher time-bandwidth products. Further, it indicates that by changing the architecture and by designing other novel supporting circuitry, more complex, monolithic, adaptive filtering systems can be implemented.

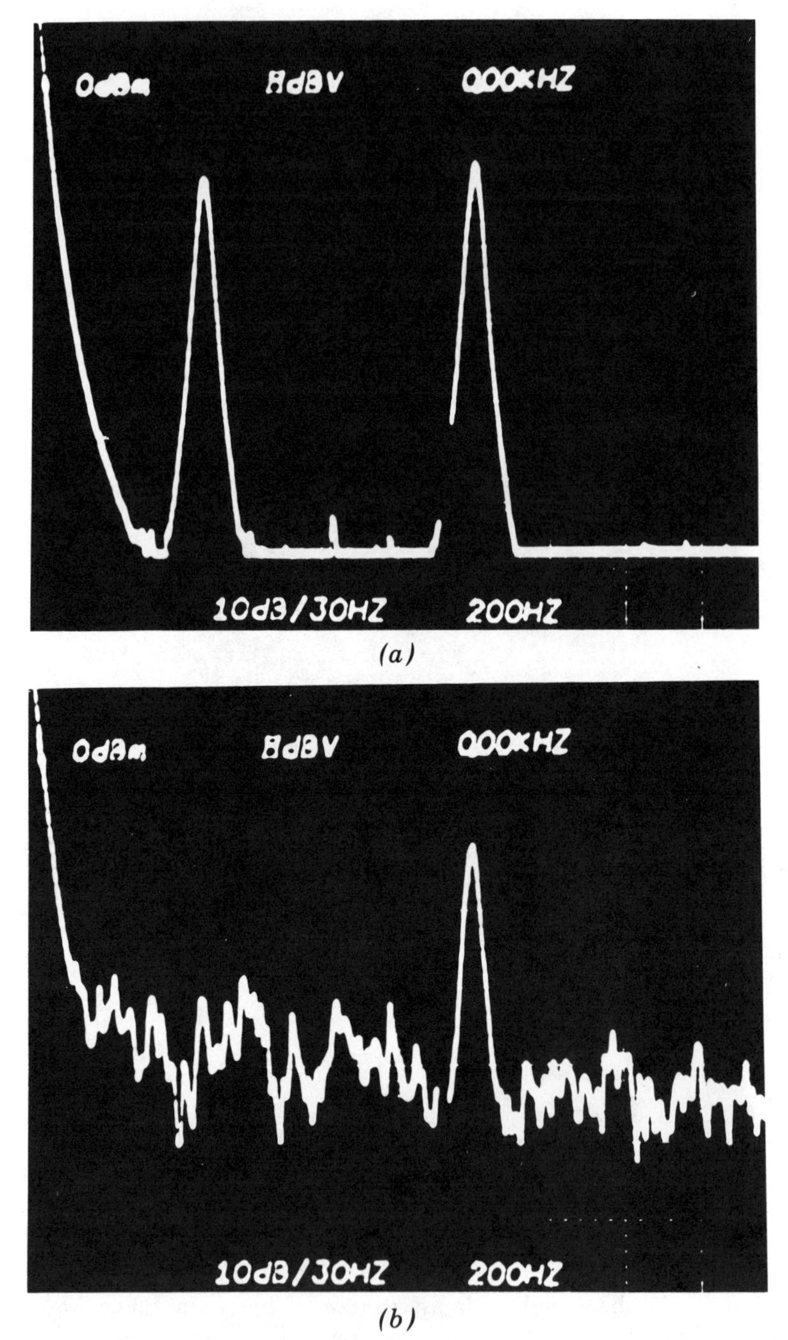

Figure 6.25 Demonstration of adaptive tone cancellation for the filter shown in Figure 6.24. (a) Input signal, $d(t)$, containing unwanted tone; (b) output signal, $e(t)$ showing unwanted tone cancelled by 20 dB. From Cowan, Denyer, Clayton, Mavor and Pennock (1982) A monolithic 65-point CCD Adaptive filter. *I.E.E. Saraga Colloquium*, © 1982 I.E.E.

6.6. THE IMPACT OF CHARGE-TRANSFER FILTERING

The CTD concept has now matured and made a valuable impact on solid-state filtering applications. It has re-inforced the view that we are now in the age of the system-on-a-chip, as CTD solutions for a host of applications have already been evaluated; including sonar, radar, and communications. The charge-transfer device approach offers compact, low-power filtering functions to be achieved at megahertz bandwidths with medium precision (7–8 bits digital equivalent). However, the CTD has made a broader impact than would normally be associated with a newly introduced technology, as it truly offers the prospect of monolithic signal processing and imaging systems (not treated here, but the reader is referred to Howes and Morgan [9] or Barbe [1] for a treatment of CTD imaging). Therefore, in situations that demand this broad requirement, including on-chip filtering, the unique CTD approach to systems integration can yield the advantages of a compact, low-cost, high performance realization.

6.6.1. Commercial Availability of Charge-Transfer Device Products

The CTD contribution to filtering is essentially the charge-transfer concept which can be applied to a variety of signal processing tasks and produced in silicon, integrated circuit form. As CTD technology can be applied to a host of applications, their design is normally user specific. Indeed, the CTD is one of the first truly application-specific integrated circuits (ASICs). For this reason, catalog CTD parts have never been widely available although some ubiquitous structures are marketed. These are invariably a spin-off from a custom design program for a particular requirement. The typical performance figures and applications areas of some commercially available CCD/BBD products are listed in Table 6.2. The various functions are usually available at signal bandwidths exceeding 1 MHz (bearing in mind that $f_s \leq f_c/2M$), although custom designed parts can easily cover video frequencies (especially in CDD form).

Although the available products are of relatively modest performance, the variety of functions makes them of interest for a wide range of applications. Virtually every conceivable signal processing task has been realized in prototype CTD form. However, their widespread adoption for many applications is limited by their requirement for external support circuitry (bias supplies, sample-and-hold circuits, post-filtering, etc.). In addition, the set-up procedures that are often involved, typical of any analog device, deter design engineers from incorporating them into systems.

A number of commercial organizations have been active for many years in CTD research and development and, currently engage in their custom design and/or supply a range of catalog parts. Such companies include: EG&G Reticon (BBD and CCD) and GE Semiconductors (CDD) in the USA; GEC Semiconductors (CCD), Plessey Semiconductors (CCD),

TABLE 6.2 Summary of Performance and Applications of Commercially Available BBD and CCD Signal Processing Products

Function	Stages	Dynamic Range (dB)	Total Harmonic Distortion (dB)	Gain (dB)	Clock Frequency Range	Power Dissipation (mW)	Application
Bucket-Brigade Devices							
Analog delay lines	256 512 1024	60	−40	0.5	500 Hz to 1.6 MHz	100	Electronic music Voice scrambling systems Telephone time compression
Charge-Coupled Devices							
Analog delay lines	851 910 1024	—	—	±2	to 5 MHz	200	Full PAL/NTSC TV line store Timebase correction Time compression/expansion
Tapped delay lines	32 64	60	−40	—	10 Hz to 5 MHz	200	Frequency filtering Phase equalizers/shifters Function generators
Time-delay and integrate	15	60	—	0.1	to 2 MHz	500	Sonar beam steering General purpose filtering
Quad. chirped transversal filters	4 × 512	60	−40	—	4 kHz to 2 MHz	500	Convolution filters for CZT
Programmable transveral filters	64 256	74	−45	0.1	100 Hz to 5 MHz	250	Sonar matched filtering Adaptive filtering Channel equalization

Walmsley (Microelectronics) Ltd (CCD) and Wolfson Microelectronics Ltd (CCD) in the UK. Philips in Europe, and several Japanese companies are also suppliers of CTD products. Due to their application-specific nature, however, the variety of CTD parts available in catalog lists is a poor reflection of the design activity for this device type. This applies particularly to the military market where CTDs have already established an important niche.

6.6.2. Design Example

A design example of a commercially available CCD is summarized here. This particular device was originally designed* to a customer's specification and, later, made available as a catalog part (WA 1001). This example, and the accompanying description, has been chosen to give the reader an insight into the design sequence and the major factors influencing the design.

The example chosen is the ubiquitous TDI structure, first introduced in §6.5. It is versatile in application as a signal processing element, especially for infrared imaging, sonar beam-forming, as well as transversal filtering (see Figure 6.18(b)). Functionally, the TDI is comprised of a group of delay lines of different lengths, $n\tau$, together with an output summer (see Figure 6.26). Its output, V_{out}, can be expressed mathematically as

$$V_{out}(t) = G \sum_{j=i}^{n} V_j[t - M(n+1-j)T] \qquad (6.10)$$

where G is the device through gain, T is the clock period, n is the number of

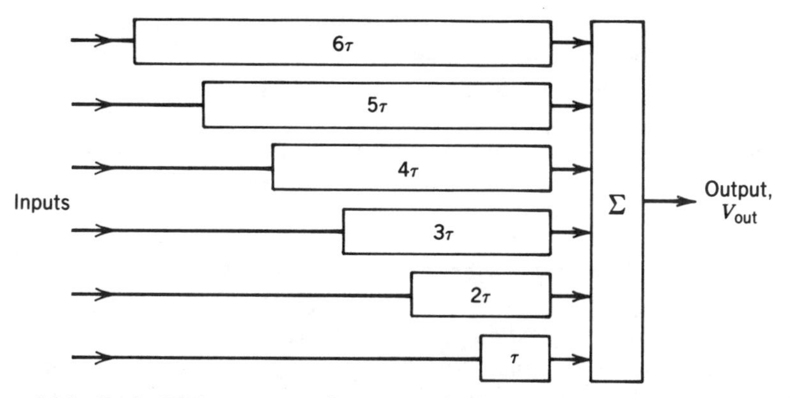

Figure 6.26 Basic TDI structure. © 1988, Walmsley (Microelectronics) Ltd, Edinburgh, UK.

* by Walmsley (Microelectronics) Ltd, Edinburgh, EH14 4AF, Scotland, UK.

TABLE 6.3 Outline Specification of TDI

CCD type	Two-phase, buried N-channel
Clocks	Single, TTL level input
Package	18-pin DIL ceramic
Operating temperature	−40 to +70°C
Storage temperature	−55 to +125°C
Power supplies	16 ± 1 V, 0 V, −5 V
Power dissipation	≤175 mW at 25°C
	≤300 mW at −40°C
Input channels	10, with six delays between each input

Courtesy © 1988 Walmsley (Microelectronics) Ltd.

parallel input channels and j is the input number, M is the delay between each input (in clock periods), and V_j is the jth input signal. As the delay τ $(=1/T)$ can be varied at will, the TDI is eminently suitable for use in variable or swept frequency systems.

6.6.2.1. Target Specification. The customer provided an outline specification for the TDI that is summarized in Table 6.3. The specification for the original device required six delays between each channel ($M = 6$), with ten input channels ($j = 10$). From Eq. (6.10), the transfer function for this TDI structure becomes

$$V_{\text{out}}(\tau) = G \sum_{j=1}^{10} V_j[\tau - 6(11 - j)] \tag{6.11}$$

The customer also specified that the clock breakthrough on the output signal (of 4 V peak-to-peak maximum) should not exceed 100 mV. Further, that the dynamic range should exceed 80 dB over 1 MHz signal bandwidth. Differential channel matching for any given TDI was not to exceed 2%.

Other less critical parameters included the clock and signal input capacitances, both of 5 pF, and the output buffer resistance not to exceed 1 kΩ.

6.6.2.2. Design Decisions/Performance Estimates. A two-phase, buried N-channel technology was available to the designer. This offered high performance and, for design flexibility, two levels of polysilicon-conductor interconnect. Previous experience with this process provided extra confidence valuable for such an analog design. The buried-channel technology affords very low CTE values (see Figure 6.8) necessary to achieve acceptable $N\epsilon$ products, even for the low transfer numbers associated with the longest (10τ) delay line $(10 \times 6 \times 2$ transfers/bit $= 120$ transfers).

For good linearity and low (partitioning) noise, a fill-and-spill input technique (see §6.2.2) was chosen. The input structure was arranged to be surface-channel (the buried-channel implant being excluded from this region) to ensure better linearity.

The restriction on a single, external clock required that the internal multiphase clocks were derived from this, through a specially designed generator. This is a nontrivial task, even for the relatively low capacitances associated with this design. At 25°C the generator was predicted to dissipate about 100 mW, being about half the estimated power for the whole circuit.

As a first step, the designer estimated the relative capacitances (and thereby the relative gate areas) of the following structures: (a) the input gate, C_I; (b) the first (smallest) transfer gate, C_C; and (c) the last (largest) output gate, C_O. For the process selected, the thin-oxide gate capacitance is 3.6×10^{-4} pF/μm^2. Within the CCD buried channel, however, the effective gate capacitance per unit area, C'_C, is reduced to about one-half to one-third of this value (corresponding to the k factor in Eq. (6.2)). Experimental evidence indicates that the per-unit capacitance in the surface-channel, input region, $C'_I \approx 2C'_C$. If each input signal has, say, a maximum amplitude of 1 V, then equating charges in C_I and C_O yields, $C_O \approx 2C_I$. Appropriate scaling of the gate areas must be made to account for the relationship that $C'_I \approx 2C'_C$.

Now the maximum charge in a transfer well is given by $Q_S = C_C V_S$, in which the equivalent barrier height in voltage for this 2-ϕ CCD process $V_S \approx 4.5$ V (refer to Figure 6.12(a)). Therefore, Q_S represents the maximum charge value that is injected by the input gate, $Q_I = C_I V_I = C_I$ coulombs. By again equating charge values we find that $C_I = 4.5 C_C$. In terms of gate dimensions defined in Figure 6.27, it follows that if the CCD is not to saturate with a 1 V input signal, then

$$W_I L_I C'_I < 4.5 W_C L_C C'_C \tag{6.12}$$

where $C'_I \approx 2C'_C$ and, thus

$$W_I L_I < 2.25 W_C L_C \tag{6.13}$$

or the area of the input gate must be less than twice the area of a transfer gate. Equation (6.12) must not be achieved by increasing L_I much above the nominal L_C design value of 9 μm, or else the CTE will suffer. Making W_I large will also deteriorate the CTE value, because newly introduced charge packets have to travel in the W-direction before being clocked as usual (in the L-direction) under the CCD transfer gates (see Figure 6.27). In this design, the following gate dimensions were chosen as a result of Eq. (6.12); $W_I = 37$ μm, $L_I = 13$ μm, $W_C = 19$ μm, and $L_C = 12$ μm. However, in order

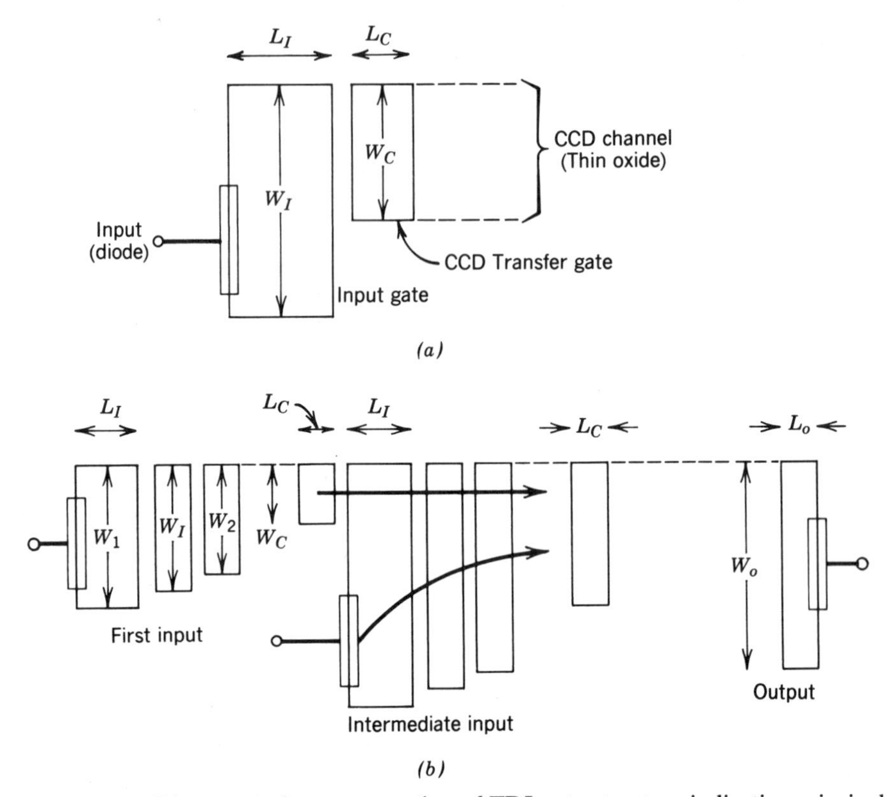

Figure 6.27 Diagrammatic representation of TDI gate structure indicating principal dimensions. (a) Simple input; (b) input taper design.

TABLE 6.4 Noise Analysis for TDI

Noise Source	Equivalent Magnitude (electrons)
Input kTC noise	540
Transfer noise	116
Bulk trapping noise	88
Dark current shot noise	40
Partitioning noise (output reset gate)	150
Output kTC noise	212

Courtesy © 1988 Walmsley (Microelectronics) Ltd.

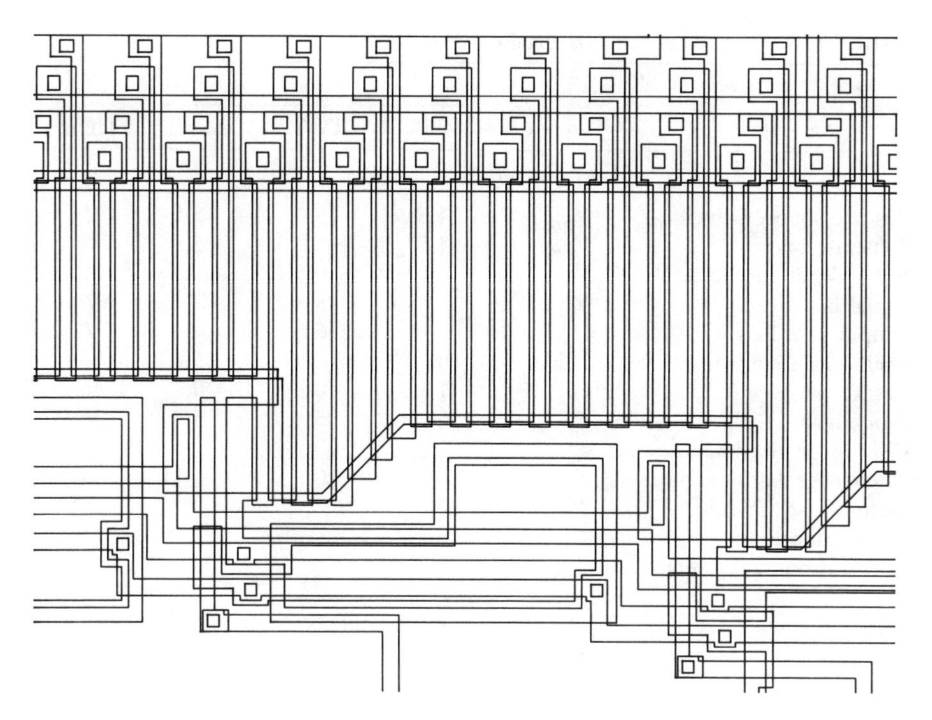

Figure 6.28 Computer plot of WA 1001 TDI gate structure. © 1988, Walmsley (Microelectronics) Ltd, Edinburgh, UK.

to minimize the otherwise abrupt transition in effective gate lengths associated with the inputs, a taper is associated with the input electrode widths (see Fig. 6.28 for a computer plot showing this particular feature). This reduces the loss in CTE that would otherwise be experienced. The gate width taper, $W_I > W_1 > W_2 > \cdots > W_C$ occurs over the next five electrodes.

A noise analysis based on the TDI structure, which emerges from the above calculations, indicates that input kTC noise (see §6.2.2) dominates the device performance. Table 6.4 indicates the predicted noise levels for each source. From these data, the dynamic range was estimated to be 83 dB, for all inputs active at a 3.9 V output signal.

6.6.2.3. Final Specification. Following device processing, sample devices were characterized against the parameters specified, over the desired temperature range. From this experimental information a list of recommended operating conditions and electrical characteristics were prepared. This information forms the basis of the Data Sheet, an extract from which is detailed in Table 6.5.

TABLE 6.5 Data Sheet for TDI Product (WA 1001)

Parameter	Symbol	Value Min	Value Typ.	Value Max.	Units
	(a) Operating Conditions				
Positive supply voltage	V_{DD}	14.5	15.0	15.5	V
Substrate bias	V_{BB}	−4.5	−5.0	−5.5	V
Input signal amplitude	V_{IN}		0.5		V
Signal bias offset	V_{BIAS}		8.8		V
Output gate bias	V_{OG}		6.5		V
Digital gain control	DGC	0		15	V
Clock low	$\mathrm{CK_L}$	−1	0	0.5	V
Clock high	$\mathrm{CK_H}$	2.7	5	5.5	V
Clock edge risetime	t_r		20		ns
Clock edge fall time	t_f		20		ns
Output load: resistance	R_L	10			kΩ
capacitance	C_L			20	pF
Ambient operating temp.	T	0		70	°C
	(b) Electrical Characteristics				
Supply current	I_{DD}	10	12	14	mA
Substrate current	I_{BB}		10	50	μA
Reference voltage output	V_{REF}		9.8		V
Buffer bias output	V_L		4.2		V
Gain, DGC = 15 V	G_{H}	−7.1	−6.6	−6.1	dB
Gain, DGC = 0 V	G_{L}	−9.6	−9.1	−8.6	dB
Operating clock frequency	f_c	0.01		10	MHz
Output signal range	V_{out}			4.0	$\mathrm{V_{PP}}$
Random noise (rms)	RN		0.2	0.3	mV
Dynamic range	DR	85	87		dB
Differential gain	DG		1.5	2	%
Sample and hold clock breakthrough	V_{BT}		70	100	mV
Clock capacitance	C_c	4	5	7	pF
Signal input capacitance	C_I	4	5	7	pF
Signal input resistance	R_I		0.4		MΩ
Buffer output resistance	R_{out}		150	250	Ω

Courtesy © 1988 Walmsley (Microelectronics) Ltd.

A photomicrograph of a device is shown in Figure 6.29(a), and in (b) an input structure with electrode tapering is clearly visible. Other features include the clock generator and the output buffer amplifier. Results for the TDI with a square-wave input signal are given in Figure 6.29(c).

In all respects the design met the original specification and set the basis for other TDI designs with enhanced features.

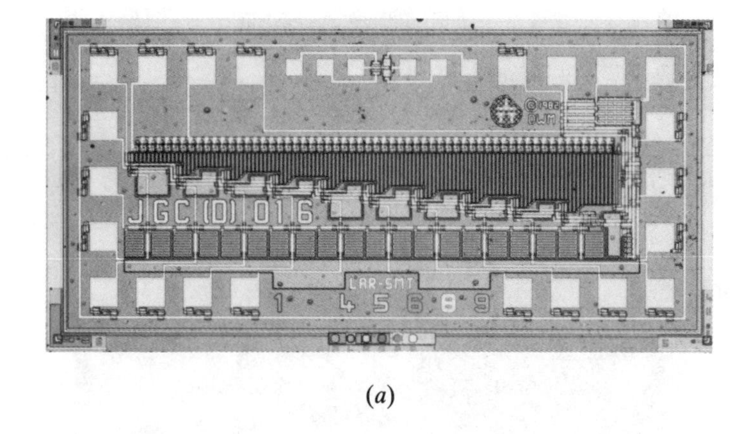

(*a*)

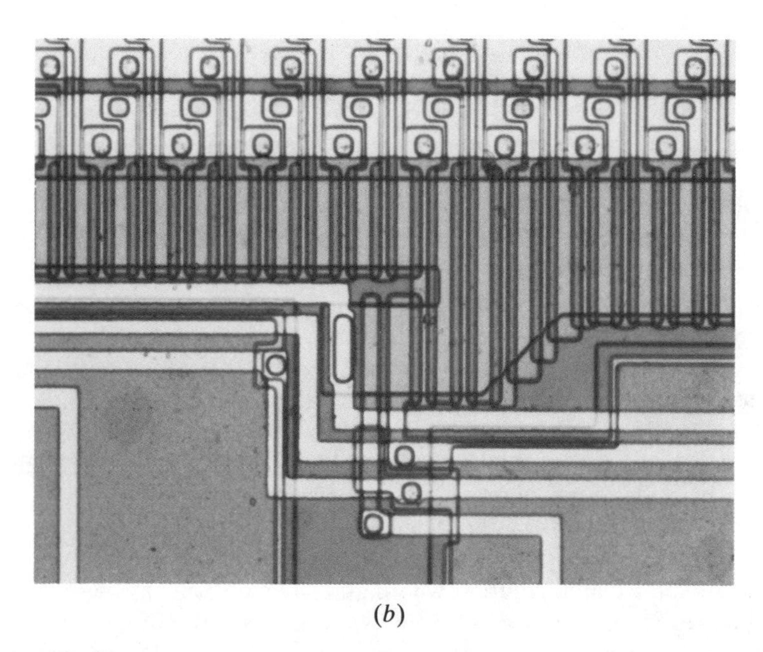

(*b*)

Figure 6.29 Photomicrograph and operation of WA 1001 TDI circuit. (a) Complete circuit; (b) enlarged view of input structure taper; (c) performance of WA 1001 showing integration of square-wave input (upper trace) to give triangular waveform output (lower trace). Clock frequency, 13 MHz; signal frequency 110 kHz. Scale: 0.5 V per division (vertical), 2 μs per division (horizontal). © 1988, Walmsley (Microelectronics) Ltd, Edinburgh, UK.

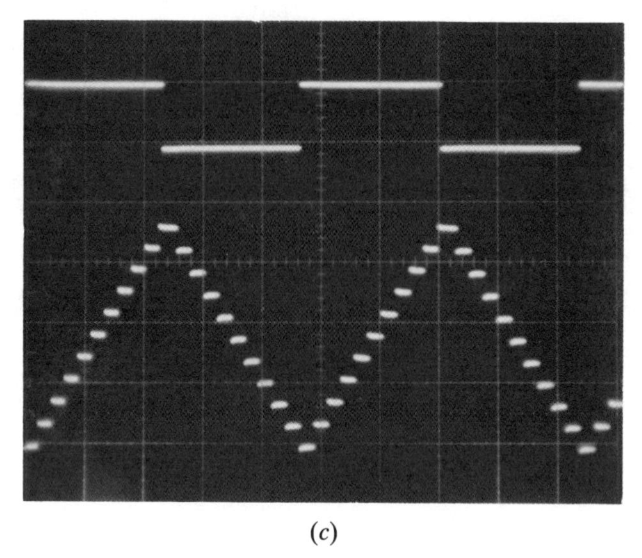

(*c*)

Figure 6.29 (*Continued*)

6.6.2.4. TDI Applications. The WA 1001 TDI featured here can realize a transversal filter (considered in §6.5.2) for frequency-domain requirements. Another major use is in the preprocessing of analog signals from transducer arrays as required in, for example, infrared or ultrasonic detection and imaging systems.

IR imaging systems often employ a sequential scanning technique whereby a linear array of detectors is scanned across an image. The basic scheme, shown in Figure 6.30, provides a set of signals that are essentially identical but staggered in the time-domain. To remove this stagger and combine the resultant signals, a TDI function is necessary [20]. The coherent signals will be integrated, but the noise from each will be uncorrelated and give an r.m.s. summation. Thus the S/N ratio of an N-detector, scanned array will show an improvement of $\sqrt{N}$ over that of a single detector. The TDI function also yields an improvement in display uniformity and, thus, the uniformity requirements of the detectors can be reduced.

The monolithic CCD approach has the advantage in size, speed and simplicity over a digital solution where an A-D converter for each detector, plus shift registers and adders, would be required. Real-time operation is possible with much reduced system complexity. With a TDI, changes in the IR mirror scan rate can be compensated for simply by changing the clock rate, something that would be impossible using fixed, passive delays.

The signal-to-noise performance of a sonar detector array can be enhanced by filtering in the frequency-domain to limit the system response to the probable frequency range of the target. Further gains can be realized [15] by spatial-domain filtering or beam-forming, where the response to the

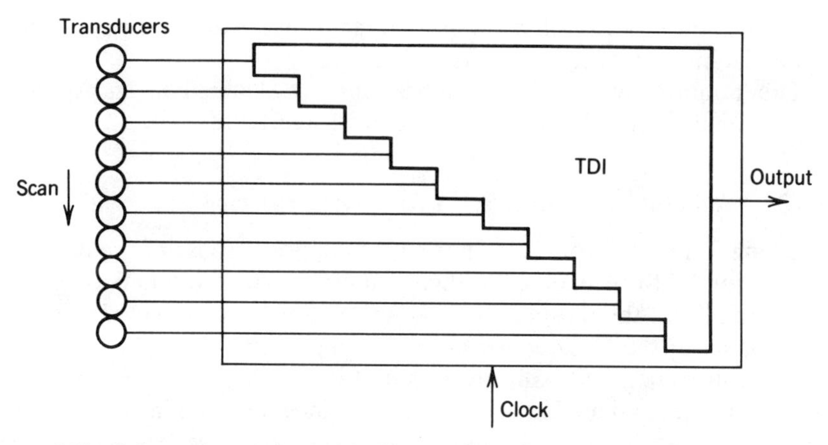

Figure 6.30 Infra-red scanning with TDI circuit. © 1988, Walmsley (Microelectronics) Ltd, Edinburgh, UK.

detector array can be limited to a particular direction. In order to steer the array in this manner, a delay must be applied to each transducer output in such a way as to cancel the propagation delay of a wave incident on the array from the required direction. Signals from other directions will be out of phase and will not add coherently. The technique is illustrated in Figure 6.31. The delay, τ, required to steer a beam at an angle θ degrees to the plane of the detectors is given by

$$\tau = \frac{d \sin \theta}{v} \tag{6.14}$$

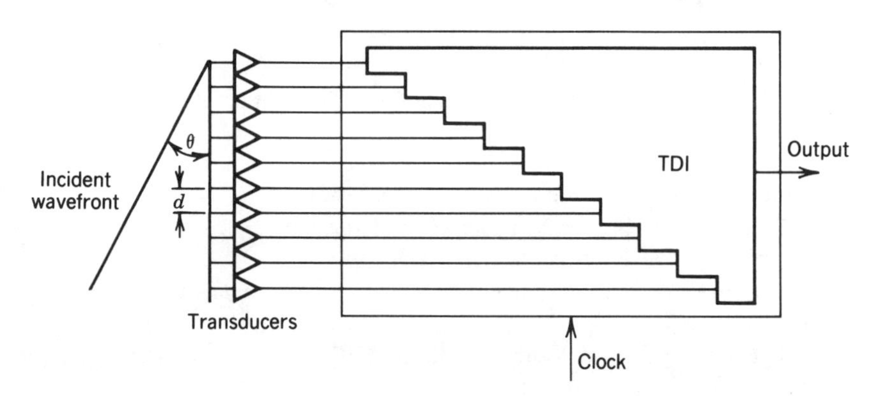

Figure 6.31 Basic TDI beamforming. © 1988, Walmsley (Microelectronics) Ltd, Edinburgh, UK.

where v is the velocity of sound in the operating medium (e.g., water), and d the distance between detectors.

Other more involved TDI applications are detailed in an Application Note available from Walmsley (Microelectronics) Ltd.

6.6.3. The Future of Charge-Transfer Device Filtering

The long term future of CTD filtering is in doubt because of the potential of sub-micron VLSI technology to digital filtering, which will undoubtedly oust CTDs in all but the most specialized applications (which includes imaging). The impact of the CTD approach to filtering has not been fully realized to date because specialist skills are required in the design of this applications-specific device; and analog MOS design engineers will be in short supply for the foreseeable future. In addition, as CTD structures tend to be custom designed for a particular requirement, and as these are often for low volume military applications, then the high design costs will rarely be offset by the production phase. These restrictions are unfortunate because of the capability of the CTD for implementing compact, low-power, high-performance processors which should remain competitive with digital approaches for some years to come. Undoubtedly, the versatile CTD family will retain its place as a significant filtering technology into the 1990s.

REFERENCES

1 D. F. Barbe (Editor) (1980) *Charge-Coupled Devices*. Springer-Verlag, Berlin.

2 J. D. E. Beynon and D. R. Lamb (1980) *Charge-Coupled Devices and their Applications*. McGraw-Hill, London.

3 W. S. Boyle and G. E. Smith (1970) Charge coupled semiconductor devices. *Bell Systems Tech. J.* **49**, 587–593.

4 C. F. N. Cowan and P. M. Grant (Editors) (1985) *Adaptive Filters*. Prentice-Hall, Englewood Cliffs, NJ.

5 P. B. Denyer and J. Mavor (1979) Monolithic 256-point programmable transversal filter. *Electron. Lett.* **15**, 710–712.

6 L. J. M. Esser (1972) The peristaltic charge coupled device: a new type of charge-transfer device. *Electron. Lett.* **8**, 620–621.

7 T. G. Foxall, A. A. Ibrahim and G. J. Hupe (1977) Double-split electrode transversal filters. *Electron. Lett.* **13**, 223–234.

8 A. S. Grove (1967) *Physics and Technology of Semiconductor Devices*. Wiley, New York.

9 M. J. Howes and D. V. Morgan (Editors) (1979) *Charge-Coupled Devices and Systems*. Wiley, Chichester, UK.

10 D. Kahng (1972) Charge coupled devices. US Patent No. 3, 651, 349.

11 N. Kapur, J. Mavor and M. A. Jack (1980) Convolutional architectures for spectrum analysis employing CCD programmable transversal filters. *Int. J. Electron.* **49**, 131–146.

12 J. Mavor and P. M. Grant (1987) Operating principles and recent developments in analogue and digital signal processing hardware. *Proc. I.E.E.*, Part F, **134**, 305–334.

13 J. Mavor, M. A. Jack and P. B. Denyer (1983) *Introduction to MOS LSI Design*. Addison-Wesley, London.

14 J. H. McClelland, T. W. Parks and L. R. Rabiner (1973) A computer program for designing optimum FIR linear phase filters. *I.E.E.E. Trans. Audio Electroacoust.* **21**, 506–526.

15 K. J. Petrosky and M. H. White (1978) CCD sonar beamforming. *Proc. Int. Conf. on the Application of CCDs*, San Diego, pp. 3B–63/71.

16 L. R. Rabiner and B. Gold (1975) *Theory and Application of Digital Signal Processing*, Prentice-Hall, Englewood Cliffs, NJ.

17 F. L. J. Sangster and K. Teer (1969) Bucket-brigade electronics — new possibilities for delay, time-axis conversion and scanning. *I.E.E.E. J. Solid-State Circuits* **4**, 131–136.

18 J. P. Sage and A. M. Cappon (1980) CCD analog-analog correlator with four-FET bridge multipliers. *Jpn. J. Appl. Phys.* **19**, 265–268.

19 C. H. Séquin and M. F. Tompsett (1975) *Charge Transfer Devices*. Academic Press, New York.

20 G. F. Vanstone, J. G. Harp, J. M. Keen, D. V. McCaughan and D. B. Webb (1976) A time delay and integration CCD for a serial scanned IR imager. *Proc. Int. Conf. on the Technology and Applications of CCDs*, Edinburgh, pp. 315–325.

21 T. L. Vogelsong and J. J. Tiemann (1984) Charge domain circuits for signal processing. CRD, GE Schenectady, NY, Report No. 84CRD117.

22 R. H. Walden, R. H. Krambeck, R. J. Strain, J. McKenna, N. L. Schryer and G. E. Smith (1972) The buried channel charge coupled device. *Bell Systems Tech. J.* **51**, 1635–1640.

23 B. Widrow, J. R. Glover Jr., J. M. McCool, J. Kaunithy, C. S. Williams, R. H. Hearn, J. R. Ziedler, E. Dong Jr. and R. C. Goodlin (1975) Adaptive noise cancelling: principles and applications. *Proc. I.E.E.E.* **63**, 1692–1716.

7 Crystal Filters

ROBERT C. SMYTHE
Piezo Technology Inc., Orlando, Florida

This chapter describes the design and performance of bandpass and band-stop crystal filters. Primary emphasis is given to monolithic crystal filters as being of greatest current and anticipated future importance. Discrete-resonator crystal filters, however, continue to be the best choice for a number of applications and are given adequate coverage. Indeed, the circuit aspects of monolithic and discrete-resonator crystal filters are closely related, although device aspects of the two may be quite different. A unified treatment is therefore appropriate.

During the 1930s, the need for highly selective voice channel filters for telephone FDM transmission systems in the US and abroad led to the development of bandpass filters using quartz crystal resonators [25]. These early filters were in the frequency range from 60 to 108 kHz, where it was possible to obtain high Q, stable resonators in a form suitable for incorporation into elegantly simple (in contemporary terms) filter networks.

From the early low-frequency applications in telephony, the use of bandpass crystal filters has expanded many-fold and today encompasses a broad range of communications, electronic navigation, and position-locating radar, as well as instrumentation systems. Beginning in the late 1960s, much of this growth was associated with the development of monolithic crystal filters. A range of typical present-day crystal filters is shown in Figure 7.1.

Before going further, we need to clarify the terminology to be used. For purposes of this chapter, an acoustically coupled resonator (ACR) is a monolithic device having two or more resonators that are mechanically, that is, acoustically, coupled. Its most common form is the monolithic dual resonator (MDR) which has exactly two resonators. In agreement with common usage, a monolithic crystal filter, or simply monolithic filter, is any crystal filter using ACRs. Thus a monolithic filter, need not be monolithic; it merely must use MDRs or other ACRs.

It is also possible to introduce names for specific monolithic filter configurations. Some that have been used in the past are bilithic to denote a filter having two ACRs, polylithic to denote more than two ACRs and/or single resonators, and tandem monolithic to denote a monolithic filter

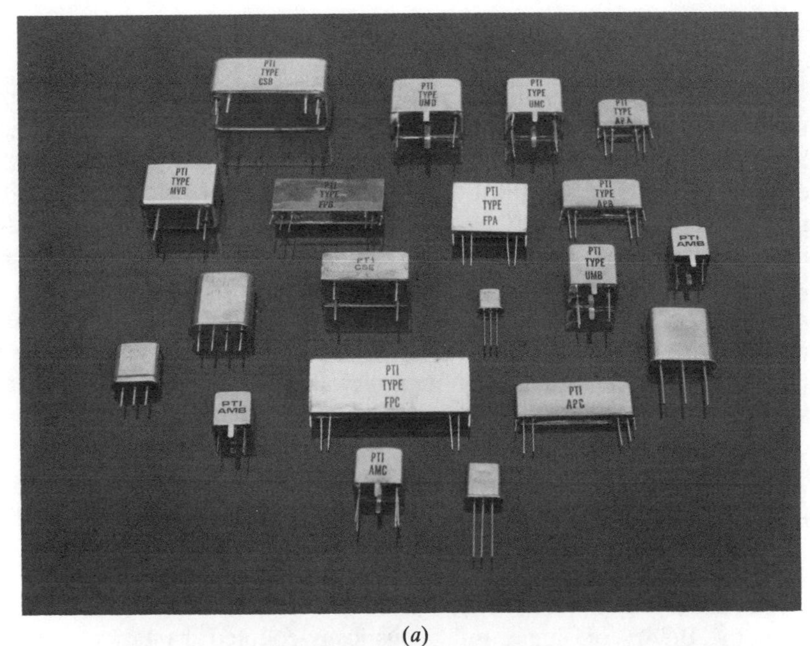

(*a*)

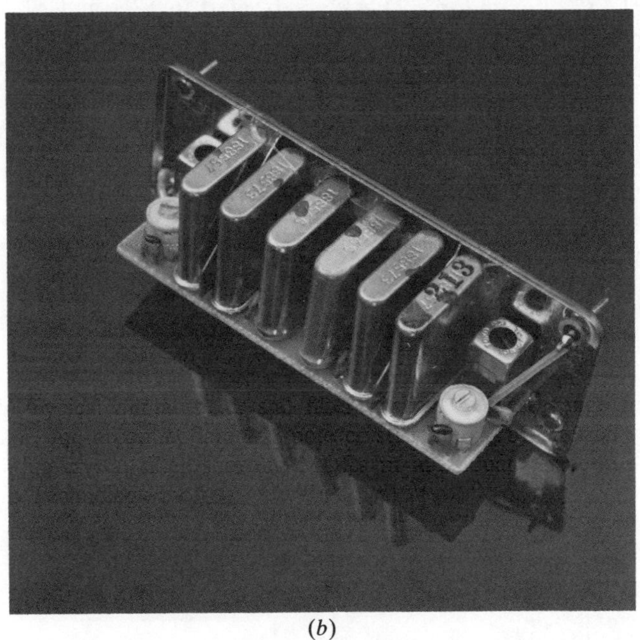

(*b*)

Figure 7.1 (a) Typical miniature crystal filter packages. These range from 0.2–11 cm^3 in volume and accommodate 2–8-pole crystal filters; (b) Internal view of a 12-pole monolithic crystal filter showing MDRs in individual hermetic packages. (Courtesy Piezo Technology Inc.)

having ACRs, and especially MDRs, connected in tandem, together with other circuit elements. Except for the last term, we will not use any of these.

Quartz crystal filters are essentially narrow-band networks. The combination of high Q (typically 1×10^4 to 1×10^6), high stability, small size, and low aging found in quartz bulk-wave resonators is unequalled in any other electrical technology. Typical quartz crystal filter bandwidths are less than a few tenths of one percent of center frequency. By incorporating inductors into the filter network, maximum bandwidths of several percent can be achieved. Alternatively, by using a crystalline material that has a larger piezoelectric effect than quartz, wider bandwidths can be obtained. At present, the only suitable commercially available material is lithium tantalate, which is in limited use. Lithium niobate, commonly used for wide-band surface acoustic wave filters (see Chapter 8) possesses no temperature-compensated bulk wave orientations.

Figure 7.2 shows approximate bandwidth and frequency limits for band-pass quartz crystal filters. This chart is intended only as a general guide and does not take into account trade-offs involving selectivity, phase or group delay requirements, size, and so on.

An understanding of the design of crystal filters requires knowledge of both the theory of single and acoustically-coupled crystal resonators and classical LC filter theory. These are treated in §7.1 and §7.2, respectively. Section 7.3 then gives examples of discrete-resonator and monolithic band-pass crystal filters.

It was recognized at an early date that quartz crystals could be advantageously incorporated into low-pass and high-pass LC filters to realize critical

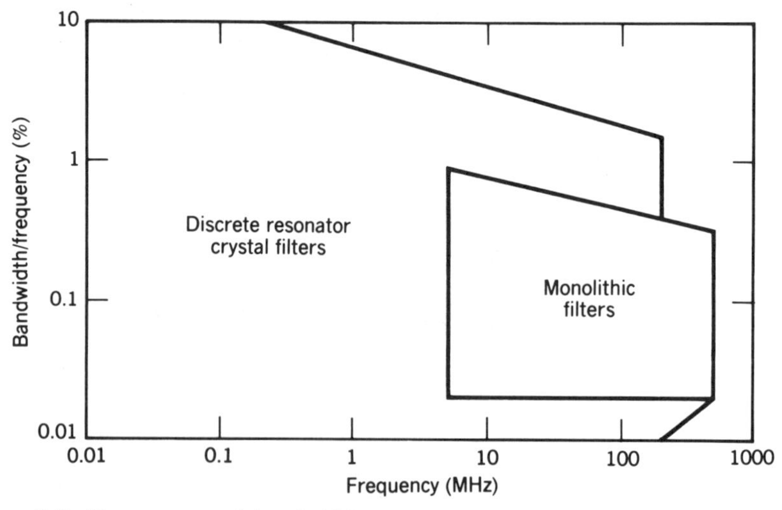

Figure 7.2 Frequency and bandwidth range for discrete-resonator and monolithic crystal filters.

low-loss network branches required for very high selectivity; that is, branches producing transmission zeroes at stop-band frequencies close to the cut-off frequency. As such filters are quite limited in use and because they cannot be considered miniature filters, they will not be discussed here.

Bandstop crystal filters also date from the early days. Although the majority of system applications call for bandpass filters, bandstop requirements are sometimes encountered. Accordingly, bandstop crystal filters will be briefly treated in §7.4.

In §7.5, some mechanical design aspects of crystal filters are considered. Nonlinear behavior, important for a number of applications, is covered in §7.6. Finally, §7.7 discusses possible future trends.

For the reader who wishes more detailed information, the book by Kinsman [15] is highly recommended. For further background, the volume edited by Sheahan and Johnson [23] reprints a number of important papers in the field up to about 1975. An extensive bibliography can be found in Vol. I of Gerber and Ballato [10].

7.1. CRYSTAL RESONATORS AND ACOUSTICALLY-COUPLED RESONATORS

7.1.1. Piezoelectric Materials

In piezoelectric materials, an electric field in certain directions will induce a mechanical strain, not necessarily in the same direction. Conversely, a mechanical strain induces an electric displacement. The strain may be in flexure, extension or shear. Crystal resonators and related devices utilize the piezoelectric effect and an alternating electrical field to excite mechanical vibrations in plates, rods, and the like which are designed to have prescribed mechanical resonances.

While many materials are piezoelectric, only a few also possess the low internal friction required for low mechanical loss and hence for a high Q resonance. Still fewer combine these qualities with good stability as a function of temperature and time.

By far the most widely used material is crystalline quartz. Although the piezoelectric effect in quartz is moderately small, limiting its application in filters to relatively narrow bandwidths, quartz is an extremely stable material with very low mechanical loss.

Accordingly, most of the examples in this chapter will be quartz crystal filters. Originally, quartz crystal devices were made from natural quartz, mined chiefly in Brazil. Nowadays, cultured crystalline quartz, grown hydrothermally, is used almost exclusively, for both economic and technical reasons.

Another piezoelectric material of some importance for filters is lithium tantalate ($LiTaO_3$), a synthetic single-crystal material not found in nature.

Lithium tantalate is strongly piezoelectric, making it useful for wide bandwidths. Its temperature stability is inferior to that of quartz, however, limiting its use in narrow-band filters. An example of a lithium tantalate filter will be given in §7.3.3. Other piezoelectric materials which may prove useful for crystal resonators are under development and will be discussed briefly in §7.7.

7.1.2. Crystal Resonators

In a crystal resonator [4], one or more modes of motion of a piezoelectric body (typically a thin quartz plate) are piezoelectrically excited by an alternating electric field product by electrodes on or near its surface. A portion of the applied electric energy is thus converted to mechanical energy and stored. For each mode of motion the body possesses a theoretically infinite number of mechanical resonances, not all of which are piezoelectrically excited. In the resonator equivalent circuit (Figure 7.3(a)) each resonance is represented by an LCR branch, while C_0 represents the static capacitance of the electrodes.

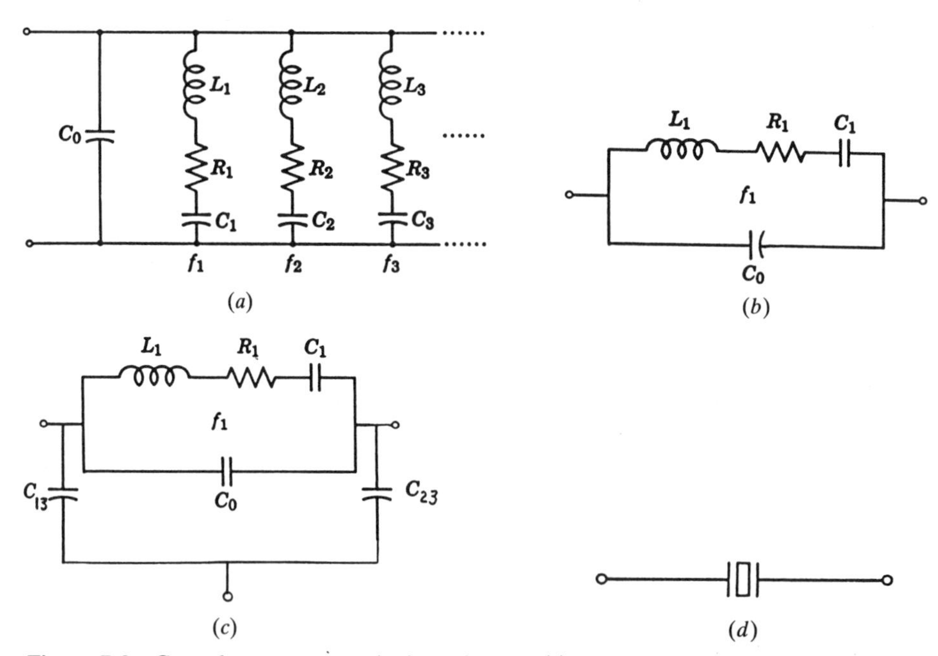

Figure 7.3 Crystal resonator equivalent circuits. (a) Each mode is represented by a series resonant branch whose resistance represents its electrical and mechanical loss. C_0 represents the electrode static capacitance. (b) and (c) An isolated resonance can be represented over a narrow frequency range by this simplified equivalent circuit. C_0 now includes the susceptances due to other resonances; (c) includes package capacitances; (d) drafting symbol.

For an isolated resonance, this equivalent circuit reduces to the familiar form of Figure 7.3(b). In using this circuit, the existence of the neglected resonances must always be borne in mind. At present most resonators are enclosed in hermetically sealed metal packages, adding two capacitors to the equivalent circuit (Figure 7.3(c)). The standard drafting symbol for a crystal unit is shown in Figure 7.3(d).

Parameters of the resonator equivalent circuit can be related to crystal filter properties. Thus, the filter designer must have a working knowledge of the range of resonator parameters. Of primary importance are the motional parameters, C_1, L_1, and R_1 (or, equivalently, $Q = 2\pi f L_1/R_1$) and the ratio of static to motional capacitance, r

$$r = C_0/C_1$$

Briefly, C_1 and L_1 set the impedance level of the filter. L_1 may range from many Henries at very low frequencies to 1 mH or less in the VHF range, but in general it will turn out that crystal filters are high impedance circuits. The capacitance ratio sets the maximum fractional bandwidth. Minimum bandwidth is set by Q and the stability of frequency as a function of temperature.

All piezoelectric materials are anisotropic; their mechanical and electrical properties are different in different directions. Consequently, a plate cut from such material will have properties that depend on its orientation with respect to the crystal axes. Table 7.1 lists commonly used types of quartz resonators. At frequencies below approximately 1 MHz, modes of motion are employed for which the resonance frequencies are determined primarily

TABLE 7.1 Commonly Used Types of Quartz Resonators

Cut	Frequency Range (MHz)	Capacitance Ratio	Mode of Motion
		Low Frequency Resonators	
NT	0.01–0.10	900	Flexure
5° X	0.05–0.20	130	Length-extension
DT	0.20–0.50	400	Length-width shear
CT	0.20–0.70	350	Length-width shear
SL	0.35–0.70	400	Length-width shear
		*High Frequency Resonators**	
AT	0.80–500	$180n^2$	Thickness-shear
BT	1.0 –50	$450n^2$	Thickness-shear
SC	1.0 –100	$575n^2$	Thickness-shear

* Capacitance ratio excludes holder capacitance; n is overtone: 1, 3, 5, . . .

by the lateral dimensions of the device; these include, in order of increasing frequency range, flexure, length extension, and length-width shear. At higher frequencies, the lateral dimensions of such resonators become impractically small, and thickness modes are used instead.

The thickness modes include thickness-extension and thickness-shear. For quartz, only thickness-shear is of practical importance. The frequencies of thickness modes are determined primarily by wafer thickness. The lowest, or fundamental, mode corresponds to a wafer thickness of half a wavelength. (Bear in mind that the acoustic wave velocity in quartz is about 10^{-5} times the electromagnetic wave velocity. At 10 MHz, an AT-cut quartz plate is 0.166 mm in thickness.) Higher order modes correspond to multiple half-wavelength thicknesses; thus a third overtone resonator has a wafer thickness of three half-wavelengths, a fifth overtone five half-wavelengths, and so on. Only the odd orders are excited piezoelectrically.

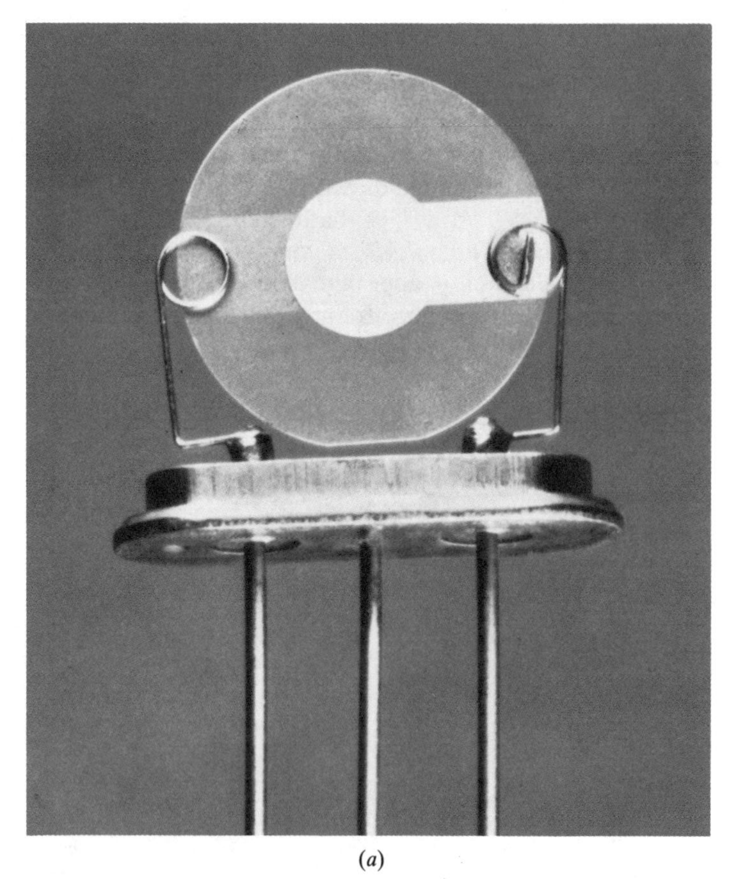

(a)

Figure 7.4 (a) An AT-cut filter crystal, typical of 5–30 MHz range; (b) mode plot for a 21.4 MHz filter crystal. Modes below the blank frequency, f_p, are trapped; above, untrapped.

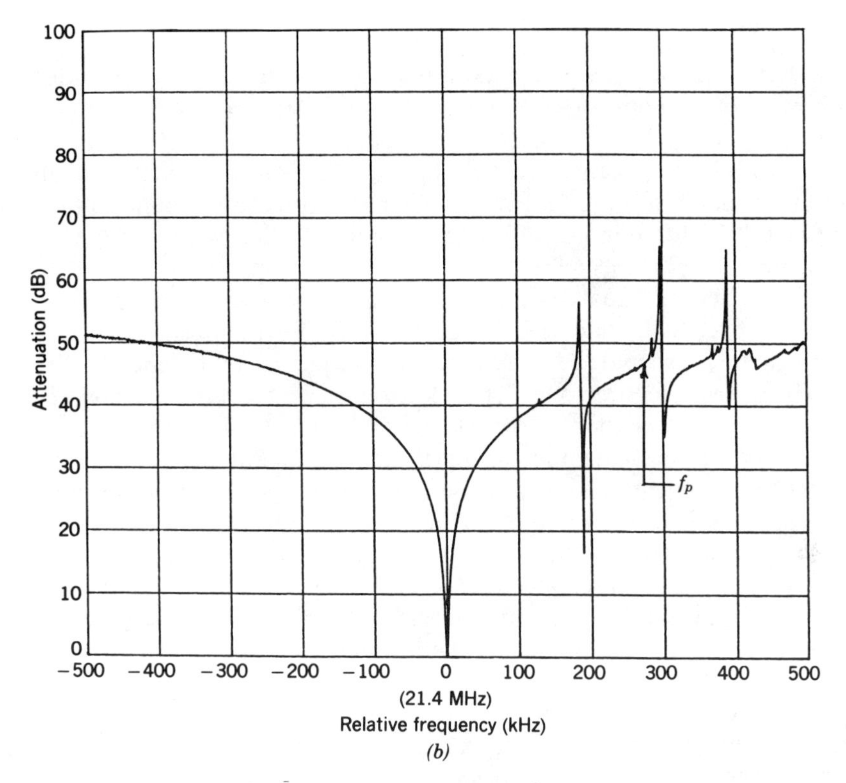

Figure 7.4 (*Continued*)

Of the several cuts that possess useful thickness-shear modes, the AT-cut is by far the most important for filter applications because of its superior temperature stability and impedance level. Nonlinear effects are smaller in BT- and SC-cut resonators than in AT-cuts; consequently, these resonators are to be considered in achieving low intermodulation and low amplitude nonlinearity. The frequency of a thickness-shear resonator is inversely related to its thickness, the minimum practical thickness being determined by the method of fabrication. Since the thickness of a BT-cut resonator is about one and a half times that of an AT-cut resonator of the same frequency, fundamental mode BT-cut resonators, although considerably inferior in temperature stability, are occasionally used in place of third overtone ATs to realize VHF filters.

For frequencies of the range of roughly 1–5 MHz, thickness-shear resonators are often made as plano-convex or bi-convex plates in order to minimize their size. At frequencies above a few MHz, however, they are made as plates of uniform thickness with thin-film metal electrodes (Figure 7.4(a)). The design of these resonators is based on a very elegant theory, the energy-trapping theory.

7.1.3. Energy trapping

The following is a simplified treatment of energy-trapping analysis. The energy-trapping point of view is useful in understanding not only thickness-shear resonators but also acoustically-coupled resonators. Our intent is to provide a physical picture of the trapping mechanism as well as analytical results in usable form [29]. The underlying theory can be found in the work of Onoe [21], Mindlin [19, 20], and Tiersten [33, 35], as well as the tutorial presentation by Spencer [31].

Trapped-energy analysis makes use of acoustic wave propagation concepts. The crystal plate is an acoustic analogue of a parallel-plate electromagnetic waveguide. It has a cut-off frequency that depends upon its thickness and the overtone. The electrodes form a region whose cut-off frequency is below that of the plate. At frequencies above the electrode region cut-off but below the cut-off of the surrounding plate, acoustic shear waves can propagate laterally in the electroded region, but not in the plate. Hence, there is at least one frequency between the two cut-offs at which a standing wave exists in the electrode region, corresponding to a resonance.

This is the fundamental concept of energy trapping. Additional resonances, if they exist, are anharmonically related to the lowest, or principal, resonance. These anharmonics, called unwanted modes or spurs, are accurately predicted by trapping analysis. All of these resonances are trapped modes; that is, they are confined in varying degree to the electroded region by the difference in cut-off frequencies. At frequencies above the plate cut-off, shear waves can propagate in both regions, and are said to be untrapped. Anharmonics also exist in this frequency range due to the finite lateral dimensions of the plate; their frequencies can be calculated. Figure 7.4(b) shows the response of a resonator used as a one-pole filter, showing both trapped and untrapped anharmonic modes. For purposes of illustration, a resonator having relatively strong anharmonics has been selected.

The analysis begins by considering a simplified model. Figure 7.5(a) depicts an infinite piezoelectric plate of uniform thickness, t, with strip electrodes. Let $\bar{c}$ be the appropriate piezoelectrically stiffened thickness-shear elastic constant; for the AT-cut, this is $\bar{c}_{66}$. Then if ρ is the mass density of the quartz plate, the thickness-shear wave velocity in the x_2-direction is

$$v = \bar{c}/\rho \tag{7.1}$$

In the absence of an electric field, the plate possesses thickness-shear mode resonances at frequencies, f_p, for which the plate thickness is an integral number, n, of half-wavelengths; n is the overtone of the plate.

$$f_p = nN/t$$

where $N = v/2$.

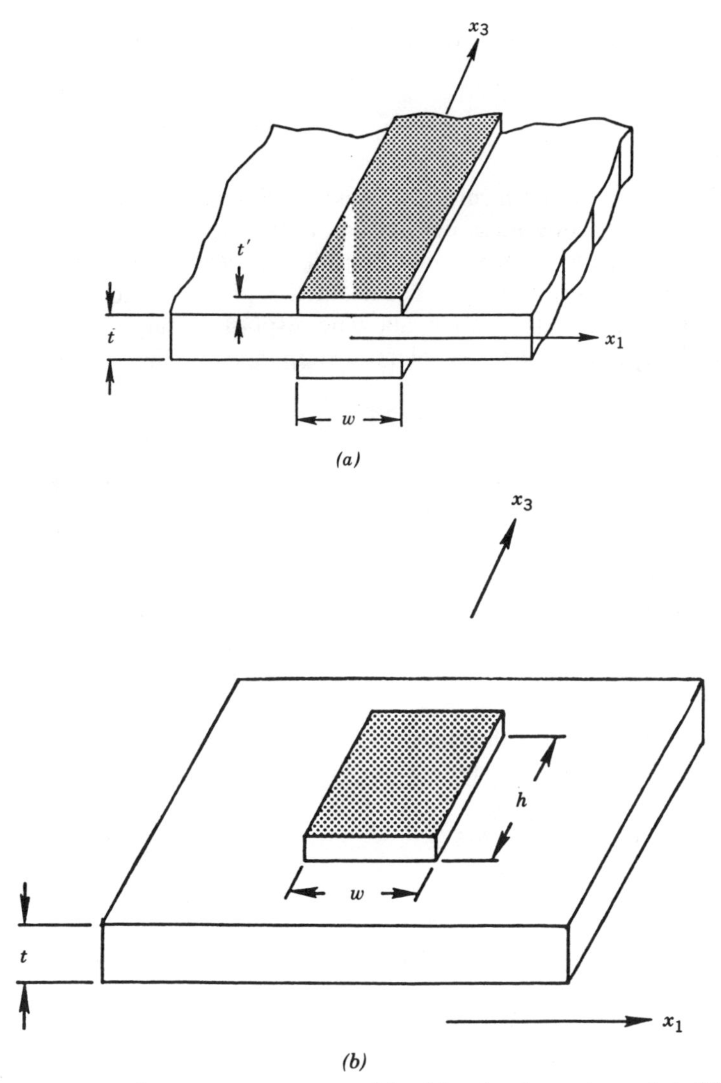

Figure 7.5 Trapped-energy resonator models: (a) strip electrode model; (b) rectangular electrode model. From Smythe (1985) In: *Precision Frequency Control*, edited by A. Gerber and A. D. Ballato. © 1985 Academic Press, Orlando.

The f_p are the waveguide cut-off frequencies for the unelectroded plate. N is called the frequency-thickness constant. For AT-cut quartz, the wave velocity is approximately 3320 m/s, so that $N = 1660$ Hz-m. For a plate 0.1 mm thick, for instance, the lowest thickness-shear frequency ($n = 1$) is 16.6 MHz. It should be noted that, for reasons of symmetry, the modes for even values of n are not piezoelectrically excited, and thus not electrically observable.

In the electroded region, the cut-off frequency, f_e, will be less than f_p by an amount Δf

$$\Delta f = f_p - f_e = (2\rho' t'/\rho t + 4k^2/n^2\pi^2) f_p \tag{7.2}$$

In Eq. (7.2) the first term is the mass loading due to the electrode mass, ρ' being the electrode mass density and t' the electrode film thickness. By analogy, the second term is called the piezoelectric loading and represents the reduction in stiffness that occurs in the piezoelectric plate when the electrodes are shorted together, allowing displaced charge to equalize. The quantity k represents the piezoelectric coupling. For the AT-cut

$$k^2 = 7.752 \times 10^{-8}$$

We now have a region defined by the strip electrodes whose cut-off frequency, f_e, is less than that of the unelectroded region. As in the previous discussion, at frequencies above f_e but below f_p, shear waves can propagate in the electroded region but not in the adjacent unelectroded region. Consequently, for at least one frequency between f_e and f_p, a standing wave, that is, a resonance, exists in the electroded region, while in the unelectroded region the shear wave decays exponentially with distance.

This model can be described by a set of linear partial differential equations [20, 33]. A solution is obtained in the form of the following transcendental equation, whose roots give the resonance frequencies.

$$\tan(\xi_e w - m\pi/2) = \xi_p', \quad m = 0, 1, 2, \ldots, M \tag{7.3}$$

where

$$\left.\begin{aligned} \xi_e &= (4\pi^2\rho/C)(f^2 - f_e^2) \\ \xi_p &= (4\pi^2\rho/C)(f^2 - f_p^2) \\ \xi_p' &= -j\xi_p \end{aligned}\right\} \tag{7.4}$$

The quantities ξ_e and ξ_p are the wave numbers, or propagation constants, in the electroded and plate regions, respectively. C is an effective elastic constant whose value, for the AT-cut, is obtained from Table 7.2. In the first column, z' and x are crystallographic axes [4].

Equation (7.3) is readily solved for f, using Eq. (7.4). The lowest root, corresponding to $m = 0$, is the principal thickness-shear resonance for the particular overtone, n, being considered. The solutions for $m > 0$ correspond to the trapped anharmonic overtones (spurs or unwanted modes) of that resonance. If m is even, the mode is symmetric about x_3. If m is odd, the mode is antisymmetric, and is not excited by the symmetric electrode

TABLE 7.2 Effective Elastic Constant, C, for AT-Cut Quartz

x_1-axis	n	C (10^9 N/m)
z'	all	68.81
x	1	109.94
	3	75.80
	5	90.09

structure shown; however, any variation from perfect structural symmetry will result in some degree of excitation. In Figure 7.4(b), a symmetric spur at approximately +190 kHz is strongly excited; at +125 kHz an antisymmetric spur can just be detected. In designing the resonator, the number, M, and spacing of trapped anharmonic modes can be traded off against other parameters. Also, by making the plate width finite, the frequencies of the untrapped anharmonics can also be calculated; however, it is usually unnecessary to carry out this exercise.

A three-dimensional trapped-energy model is shown in Figure 7.5(b). If we write the strip electrode frequencies resulting from Eq. (7.3) in the form

$$f_{nm} = f_e + \delta_{nm} \tag{7.5}$$

then it can be shown that the resonance frequencies for Figure 7.5(b) are

$$f_{npq} = f_e + \delta_{np} + \delta_{nq}$$

where p and q are the x_1- and x_3-direction mode indices. Each of these frequencies is obtained by solving the strip electrode equation twice, once for each direction of propagation. The frequency offsets thus obtained are added together to get the total offset from f_e.

The motional capacitance, C_1 in Figure 7.3(c), can also be obtained using trapping analysis. For most purposes, however, the capacitance ratio, r, of the principal mode (but not the spurs) is essentially constant for a given n and a given cut, so that an exact analysis is not necessary.

First, the static capacitance is given by the parallel plate equation

$$C_0 = \varepsilon hw/t \tag{7.6}$$

in which ε represents the dielectric constant.

For an AT-cut plate, the ratio of static to motional capacitance is approximately

$$r = 180n^2 \tag{7.7}$$

(It is important to note that these values are exclusive of package capacitance.)

Then C_1 is just

$$C_1 = C_0/r \tag{7.8}$$

As a typical example

$$f = 21.4\,\text{MHz}$$
$$n = 1$$
$$w = h = 1.5\,\text{mm}$$

For the AT-cut

$$\varepsilon = 39.82\,\text{pF/m}$$
$$t = 1660/21400 = 0.0776\,\text{mm}$$
$$C_1 = 39.82 \times 10^{-3} \times 1.5 \times 1.5/0.0776 = \underline{1.155\,\text{pF}}$$
$$C_1 = 1.55/180 = \underline{6.42\,\text{fF}}$$

7.1.4. Acoustic Coupling

7.1.4.1. The Monolithic Dual Resonator. Figure 7.6 shows two trapped-energy thickness-shear resonators arranged in close proximity on a single

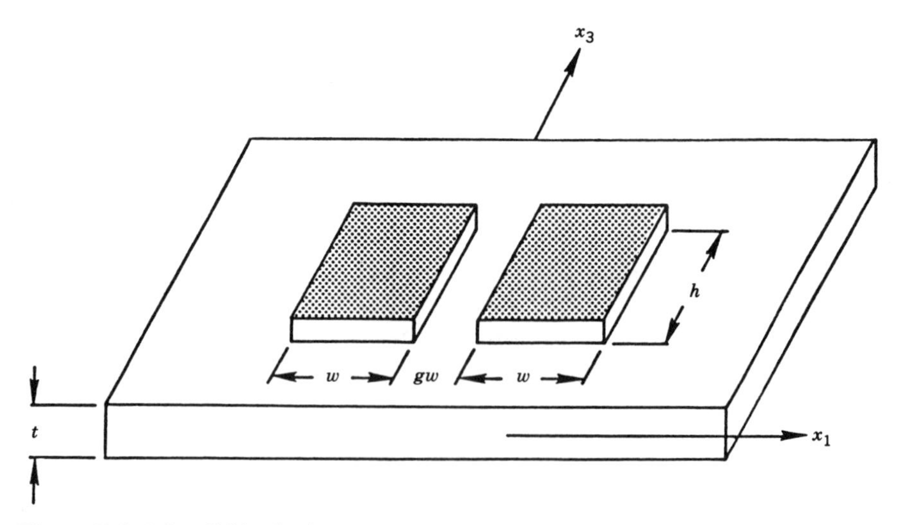

Figure 7.6 Monolithic dual resonator rectangular electrode model. From Smythe (1985) In: *Precision Frequency Control*, edited by A. Gerber and A. D. Ballato. © 1985 Academic Press, Orlando.

piezoelectric plate to form an MDR. For configurations of this type, two points of view are useful. On the one hand, we can think of the two resonators as being coupled via the waveguide-below-cutoff region separating them. Alternatively, we can consider the MDR as a single structure having two resonant modes.

The latter point of view is used to analyze the device. First consider the strip electrode case. In Figure 7.6, let h become infinite. Since the device is symmetrical about the x_3-axis, its resonant modes are either symmetric or antisymmetric. The frequencies of the symmetric modes are given by the solutions of

$$\xi_e w = \tan^{-1}(\xi_p'/\xi_e) + \tan^{-1}[(\xi_p'/\xi_e)\tanh(\xi_p' g w/2)] + m\pi\,, \qquad m = 0, 1, 2, \ldots, M \tag{7.9}$$

and for the antisymmetric modes by

$$\xi_e w = \tan^{-1}(\xi_p'/\xi_e) + \tan^{-1}[(\xi_p'/\xi_e)\coth(\xi_p' g w/2)] + m\pi\,, \qquad m = 0, 1, 2, \ldots, M \tag{7.10}$$

where ξ is defined above (Eq. 7.4).

These modes can be represented by the symmetrical lattice equivalent circuit of Figure 7.7(a). The modes corresponding to $m = 0$ are the principal symmetric and antisymmetric modes, whose frequencies we will call f_S and f_A, while the modes corresponding to $m > 0$ are unwanted anharmonics of the principal modes. Omitting unwanted modes, the lattice reduces to Figure 7.7(b). C_0 represents the symmetric static capacitance, that is, the capacitance between each pair of electrodes; $2C_g$ represents the difference between the antisymmetric and symmetric capacitance between adjacent electrodes.

In most instances C_{1S} and C_{1A} are nearly equal. Assuming equality, and applying a network equivalence given in §7.2.1 (Figure 7.12), we obtain Figure 7.7(c), which will be recognized as a coupled-resonator ladder, modified by C_g.

The lattice and ladder element values are related by

$$\begin{aligned} C_1 &= C_{1S} = C_{1A} \\ L_1 &= L_{1A} \\ L_{12} &= k_{12} L_1 \\ k_{12} &= S/f_m \\ f_m &= (f_S + f_A)/2 \\ S &= f_A - f_S \end{aligned}$$

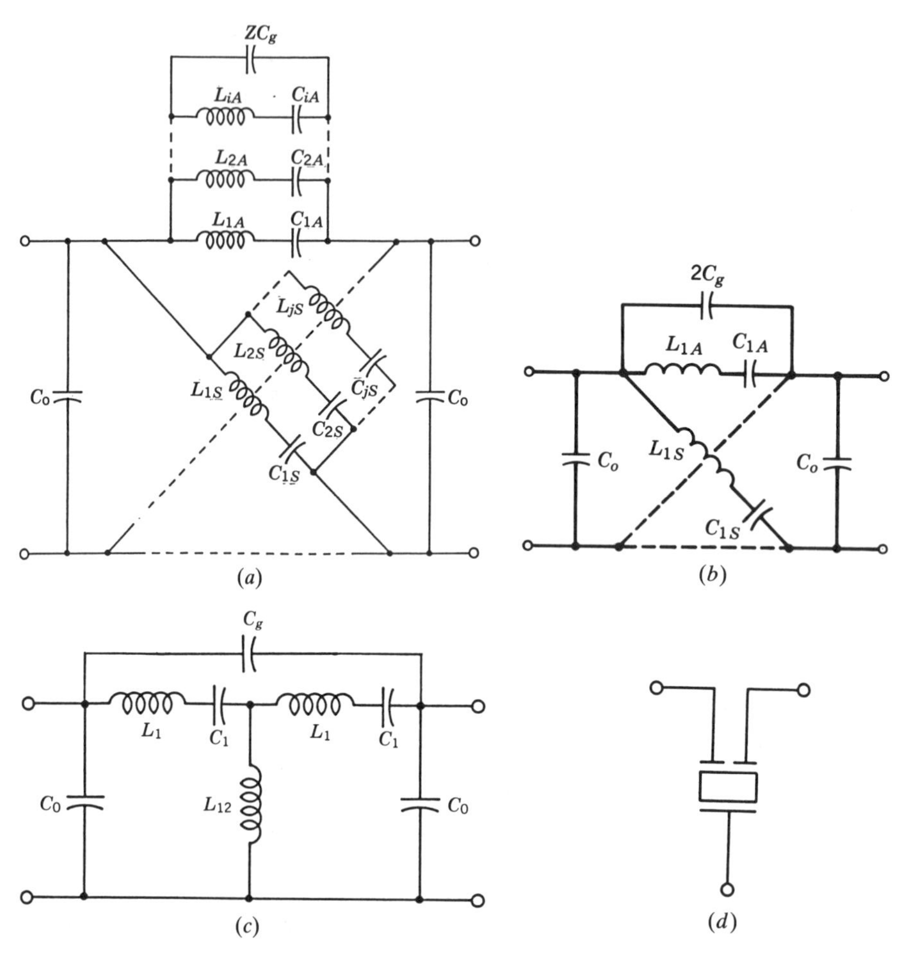

Figure 7.7 Monolithic dual resonator equivalent circuits. (a) Lattice equivalent circuit, including unwanted modes. From Smythe (1985) In: *Precision Frequency Control*, edited by A. Gerber and A. D. Ballato. © 1985 Academic Press, Orlando. (b) Lattice representation of principal modes. (c) Bridged ladder network equivalent to (b) if $C_{1S} = C_{1A}$; (d) drafting symbol.

S is the spacing between symmetric and antisymmetric modes; f_m is the mean of the mode frequencies. Their ratio, S/f_m is the coupling coefficient, k_{12}, between the two resonators. The mean of the mode frequencies, f_m, is also the frequency of each resonator (considered as if the other were not present).

Now let h in Figure 7.6 be finite. The mode spacing is unchanged. The frequency offset due to finite h, calculated by applying Eq. (7.5) in the x_3-direction, is added to the solutions of Eqs. (7.9) and (7.10) to obtain the mode frequencies. The equivalent circuits of Figure 7.7 remain valid. C_0 and C_1 can be calculated with good accuracy from the single-resonator relations.

It should again be pointed out that because C_{1S} and C_{1A} may not be exactly equal, the ladder representation shown is somewhat approximate, although extremely useful. More accurate ladder representations are possible, but are beyond our present scope.

The restriction to a symmetrical device, although convenient and representative of common practice, is not necessary. In the most common departure from symmetry, the two resonators are allowed to have slightly different values of mass loading and hence slightly different frequencies, f_1 and f_2. With obvious element value changes, the ladder representations still apply. Since the difference in mass loading is small, the coupling may be

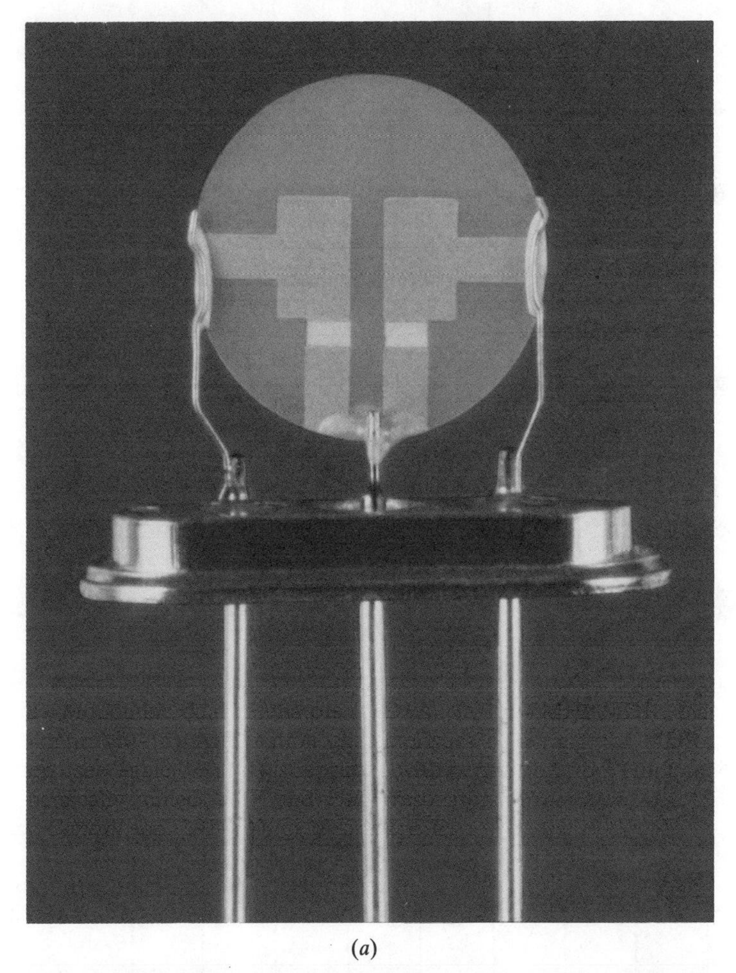

(*a*)

Figure 7.8 Monolithic dual resonators (MDRs): (a) 21.4 MHz MDR, blank diameter is 6.4 mm; (b)–(d) Attenuation characteristics of a range of MDRs used as 2-pole filters (see Table 7.3). (d) is reprinted with permission from Hunt and Smythe (1985) Chemically milled VHF and UHF resonators. *Proc 39th Ann. Symp. on Frequency Control*, pp. 292–300 © 1985 I.E.E.E.

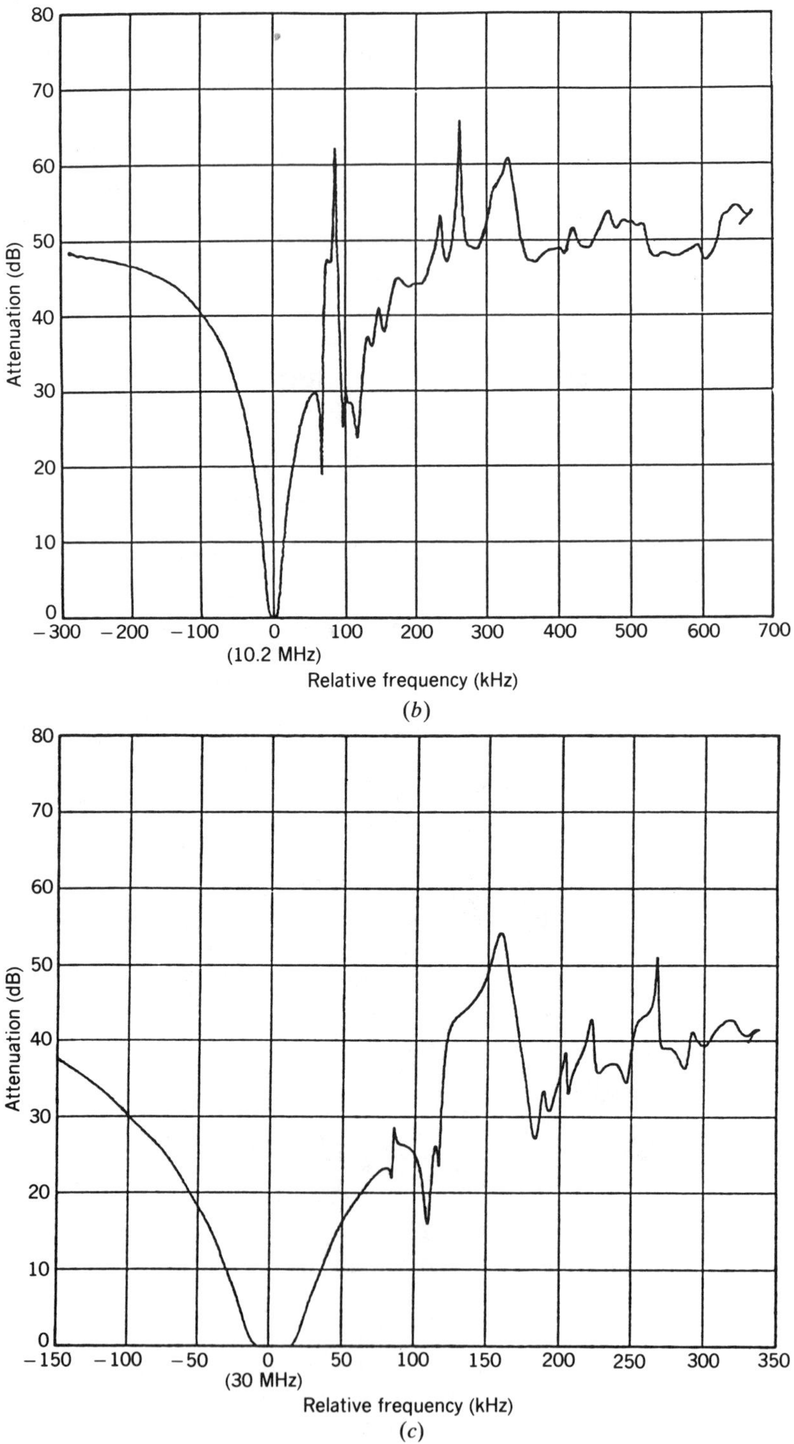

(*b*)

(*c*)

Figure 7.8 (*Continued*)

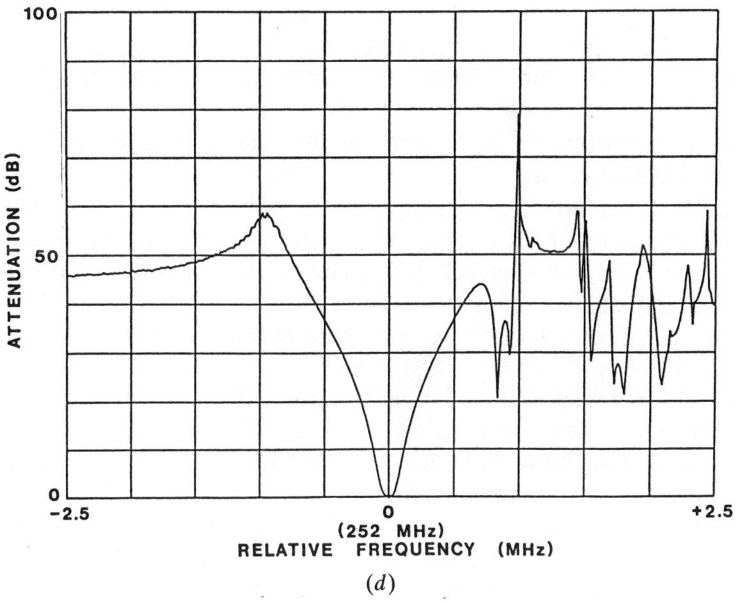

(*d*)

Figure 7.8 (*Continued*)

calculated using the average value. As the resonators are at different frequencies, the mode spacing, S', is related to both the coupling and the frequency unbalance, U, according to

$$S' = (S^2 + U^2)^{1/2}$$

where

$$U = f_2 - f_1$$

TABLE 7.3 Monolithic Dual Resonators

Frequency (MHz)	n	Electrode Dimensions w (mm)	gw (mm)	h (mm)	S (kHz)	L_1 (mH)	Two-Pole Filter Parameters Bandwidth (kHz)	Term. Res (Ω)	Case Style	Figure
5.3	1	3.9	1.2	5.1	7.8	65.0	11.0	2,700	HC-6	
10.2	1	1.9	0.6	1.9	14.5	50.0	18.0	3,600	HC-18	7.8(b)
21.4	1	1.0	0.6	1.5	12.0	11.7	18.3	1,500	HC-45	
30.0	1	1.0	0.3	1.3	26.0	4.8	36.5	700	HC-18	7.8(c)
45.0	1	0.6	0.3	0.6	28.0	4.1	40.0	1,200	HC-45	
85.0	3	0.75	0.2	0.75	8.5	13.3	12.0	600	TO-5	
125.0	1	0.25	0.10	0.30	72.5	1.2	102.0	1,000	HC-18	
150.0	5	0.40	0.20	0.60	20.0	39.0	28.0	2,600	HC-18	
252.3	1	0.13	0.08	0.13	95.0	0.7	11.1	2,700	HC-18	7.8(d)

Going farther, h and w (Figure 7.6) may be different for the two resonators. We will not treat this case, except to say that, as before, the coupling is obtained from a modal analysis of the complete structure. The device is no longer represented by a symmetrical lattice, but by a circuit using ideal transformers [16], which can be placed in a somewhat more complicated ladder form.

Examples of monolithic dual resonators and their filter responses are given in Table 7.3 and Figure 7.8. These illustrate the range of bandwidth and frequency obtainable using MDRs. The MDR design is typically carried out using the MDR frequency equations in an interactive computer program written so that design variables can be readily manipulated.

7.1.4.2. Multiresonator ACRs. Taking the coupled-resonator viewpoint, it is easily seen that a third resonator can be coupled to the second, a fourth to the third, and so on, to form higher order ACRs. To a first approximation, the coupling between adjacent resonators can be calculated as just outlined. An example of a four-resonator VHF ACR developed by Motorola Inc. for use in a highly miniaturized paging receiver is described in §7.3.3 [8].

Another example is the AT&T eight-resonator filter [22]. This unit, with a bandwidth of 3.2 kHz at a frequency around 8 MHz, is used as a single-sideband voice channel filter for telephone FDM transmission, an application that requires large numbers of highly selective filters. The filter is packaged in a coldweld-seal flatpack measuring approximately $45 \times 25 \times 8$ mm. The quartz plate itself measures $34.4 \times 11.2 \times 0.2$ mm.

The elegant simplicity of the multiresonator ACR has its price. For the MDR, three parameters are highly sensitive to process variables: the two resonator frequencies and the coupling coefficient. For each additional resonator, we add another coupling coefficient and another resonator frequency. Hence, a four-resonator ACR has seven critical parameters. In order that manufacturing yield be high, extremely tight process control is necessary, and for many applications this is not economic. Consequently, the MDR is the basic building block of monolithic filter technology.

A further, less obvious, reason for favoring MDRs over multiresonator versions is that many filters can be realized using tandem symmetrical two-pole sections, which are then realized by MDRs. The symmetry is advantageous in manufacturing. Division into higher order sections realized with multipole ACRs is also possible, but the requirement of symmetry must usually be dropped.

7.1.5. Ring-Supported Resonators and MDRs

Quartz wafers for thickness-shear resonators are made by a series of specialized machining operations. The machining process sets a minimum on the fundamental shear frequency, since it is inversely related to the wafer thickness. For the AT-cut, the limit is typically 30–50 MHz, corresponding

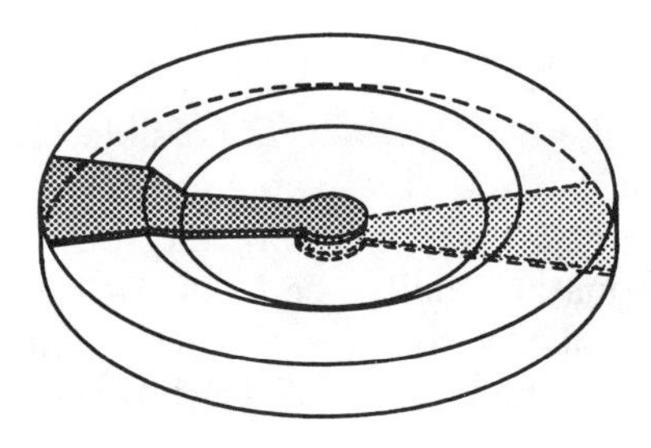

Figure 7.9 Ring-supported VHF resonator. Wafers of this type may also be used for MDRs. From Hunt and Smythe (1985) Chemically milled VHF and UHF resonators. *Proc. 39th Ann. Symp. on Frequency Control*, pp. 292–300 © 1985 I.E.E.E.

to thicknesses of 55–33 μm. Above this limit, overtones must be used; however, for a given frequency, the motional inductance, and hence the filter impedance level, is proportional to the cube of the overtone, while the capacitance ratio is proportional to the square. Thus, a third-overtone resonator has a motional inductance and a capacitance ratio which are, respectively, 27 times and 9 times those of a fundamental mode resonator with the same frequency and electrode dimensions. This presents serious difficulties in circuit construction and severely limits the maximum achievable bandwidth. Consequently, there is considerable advantage to being able to make the fundamental frequency as high as possible.

The frequency limit imposed by conventional machining can be overcome by using other processes, especially ion milling [2] and chemical etching [14, 36]. With these processes, fundamental frequencies of several hundred MHz are practical, and 1.6 GHz has been achieved experimentally using chemical etching [14]. In the manufacturing process, wafers are first lapped and polished to a convenient thickness, say 60 μm. Then the central portion of the wafer is thinned (Figure 7.9), leaving a supporting ring. This configuration alleviates the problems associated with processing and mounting an extremely thin wafer. Plates made in this manner may be used in either discrete-resonator or monolithic filters [30]. Figure 7.8(d) shows the response of a 252 MHz fundamental mode MDR. An example of a 6-pole monolithic filter using ring-supported MDRs is given in §7.3.3 (Figure 7.23).

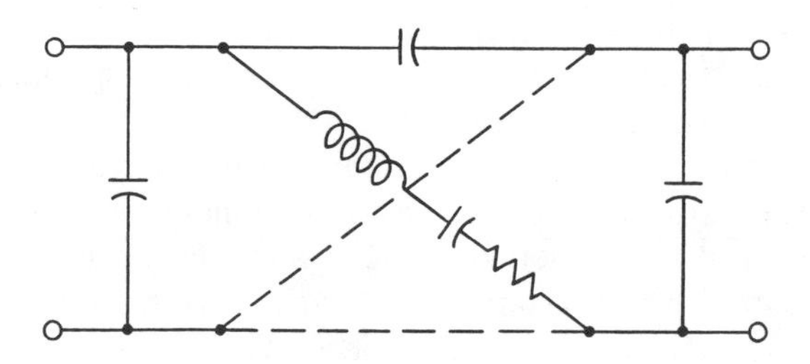

Figure 7.10 Low frequency divided-electrode resonator equivalent circuit.

7.1.6. Low-Frequency, Divided-Electrode Resonators

For some low-frequency cuts, in particular the +5° X-cut, it is possible to make a two-port resonator by splitting each electrode into two and providing connections to each half [17]. A resonator of this type is thus referred to as a split-electrode or divided-electrode resonator. Unlike the MDR, it has only a single resonance. Usually the electrode is split symmetrically, in which case the equivalent circuit is the lattice network (§7.2.1) shown in Figure 7.10.

7.2. CIRCUIT CONSIDERATIONS

Crystal filter circuit design is essentially an application of classical filter theory (see Chapter 2). However, because the crystal resonator equivalent circuit (Figure 7.3) unavoidably includes the static capacitance C_0, as well as package capacitances, and because of the limited range of resonator parameters at a given frequency, the detailed design differs considerably from that for LC filters. Similar remarks apply to monolithic filters. In general, the circuit aspects of monolithic and discrete-resonator crystal filters are more alike than different. In this section we will attempt to present the most salient circuit considerations for both. Filter circuit design will not, however, be treated in detail. For those interested, an excellent introduction can be found in Chapters 1–3 of Temes and Mitra [32]. Chapter 4, by Szentirmai, supplements this to provide a good overview of crystal filter circuit design. A number of worked examples are given by Kinsman [15], who treats a number of practical problems.

7.2.1. Lattice Networks and Equivalents

Most LC bandpass filters are essentially ladder structures. If we attempt to realize crystal bandpass filters as ladders, we quickly realize that the antiresonance introduced by C_0 severely limits the maximum bandwidth and ultimate attenuation. Further limitations arise from the range of motional parameters required. One reason for this is that ladder networks attenuate by voltage divider action; hence, for high stop-band attenuation, the ratio of series and shunt branch impedances must be high. On the other hand, lattice networks achieve attenuation by bridge-balancing: for high attenuation, the impedance of one arm must nearly equal that of its opposite number.

As a consequence, most crystal filters using discrete resonators are symmetrical lattice networks or, more often, tandem lattices. As we shall see, the symmetrical lattice network allows static capacitance in one lattice arm to be neutralized by static capacitance in the opposing arm. In monolithic filters, the lattice are realized as MDRs or multiresonator devices. In discrete-resonator filters, at low frequencies, divided-electrode

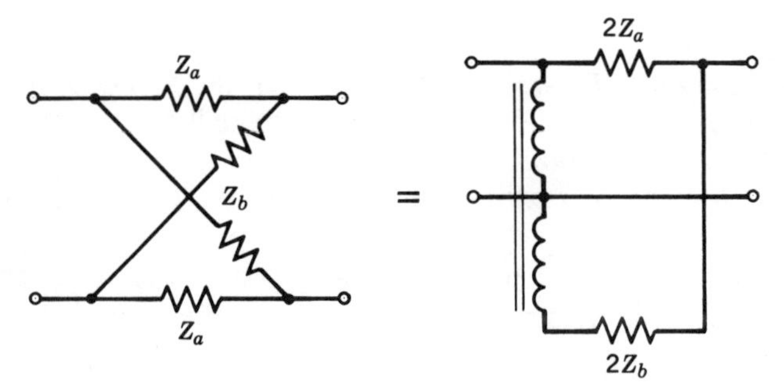

Figure 7.11 *Left*, symmetrical lattice network; *right*, its unbalanced network half-lattice equivalent, using an ideal 1:1 transformer.

crystal units can be used; otherwise, each lattice is realized as a half-lattice network.

Figure 7.11 shows, on the left, a symmetrical lattice network, which, when unfolded, takes the form of a bridge. When $Z_a = Z_b$, the bridge is balanced and there is no transmission through the network. The number of bridge arms is reduced from four to two by the equivalent half-lattice, or transformer bridge, network on the right, which incorporates an ideal 1:1 transformer.

In Figure 7.12, at the left is shown a symmetrical lattice with a crystal resonator in each arm. An unbalanced, bridged ladder equivalent is shown on the right. (Equivalences of this sort are readily obtained by application of Bartlett's bisection theorem [3, 38].) At frequencies sufficiently far removed from the crystal resonances, the crystal motional reactances are very large, and may be neglected: in this region the transmission is then determined by $(C_{0a} - C_{0b})$.

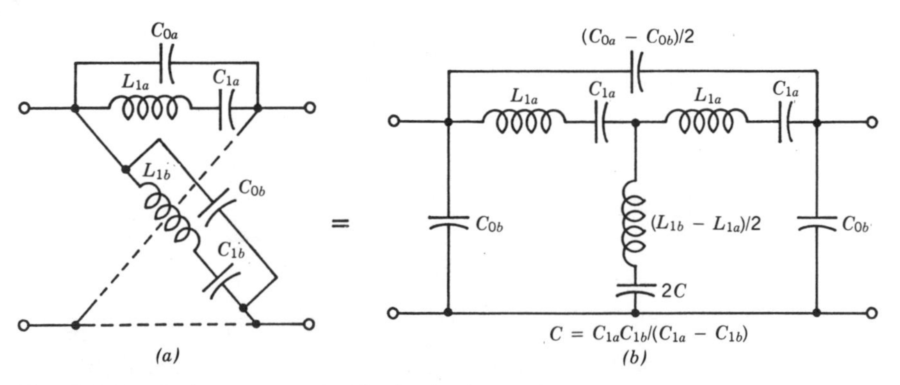

Figure 7.12 *Left*, a symmetrical lattice having a crystal resonator in each arm; *right*, its bridged ladder equivalent.

7.2.2. Narrow-, Intermediate-, and Wide-Band Filters

7.2.2.1. Narrow-Band Filters. Bandpass crystal filters may usefully be classified according to fractional bandwidth. For small fractional bandwidths, crystal filters may be realized without inductors, using only capacitors and crystal units, MDRs, or multiresonator ACRs. We call these narrow-band (NB) crystal filters.

The maximum fractional bandwidth that crystal filters can achieve without incorporating inductors is determined by the capacitance ratio, r,

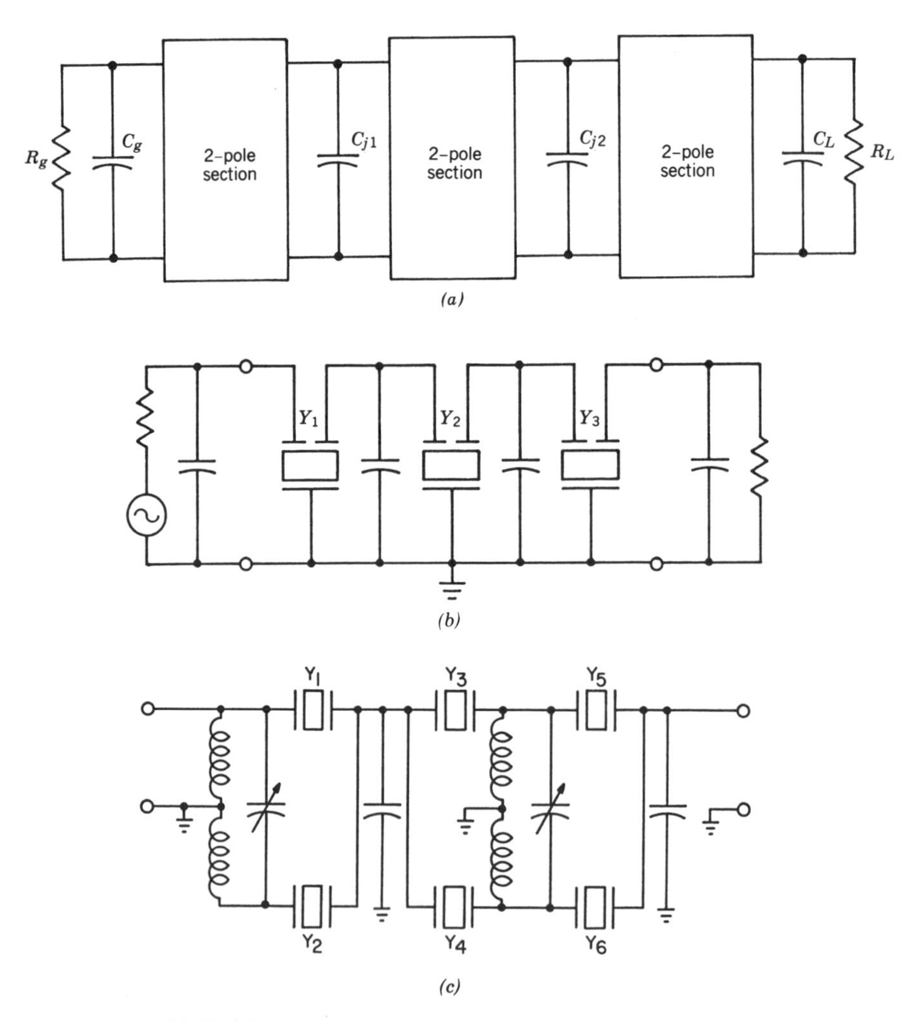

Figure 7.13 (a) Tandem 2-pole arrangement typical of NB crystal filters. This configuration may be realized using MDRs (b), or discrete resonators in half-lattice sections (c).

(§7.1.2) of the devices, including circuit strays. Two different inductorless limits may be recognized. In the first, and more restrictive, inductors at the filter input and output, as well as other internal inductors, are excluded. Since it is frequently necessary to transform from the filter natural impedance to specified source and load impedances, and since the transformation invariably requires inductors, this limit is of restricted practical significance.

A more meaningful definition of the maximum inductorless bandwidth excludes only internal inductors. (Half-lattice balanced transformers are ignored.) The limit depends upon the circuit configuration and the type of response; for lattice-equivalent forms, including MDRs, the limit is never greater than $1/r$, and is typically $0.7/r$ to $0.9/r$. For AT-cut devices the inductorless bandwidth limit is typically $0.2/n^2$ to $0.3/n^2$ percent, the exact value depending upon a number of factors, especially response type and crystal package configuration. If parasitic capacitances could be completely eliminated, the limit would be very nearly $0.5/n^2$ for a 0.1 dB ripple Chebyshev design.

Referring to Figure 7.13(a), the inductorless bandwidth limit can readily be appreciated. As bandwidth is increased, the nodal capacitances, C_j, decrease until, at the NB inductorless limit, they vanish. C_j appears in parallel with the static capacitances of the two-pole sections; reduction of the latter, therefore, increases the maximum inductorless bandwidth. Due to the high impedance level, a few tenths of 1 pF may change the maximum bandwidth significantly.

For ladder crystal filters the maximum bandwidth is much lower. Since resonances of the shunt-arm crystals and antiresonances of the series-arm ones produce transmission zeroes, the maximum fractional bandwidth is much less than the resonant-to-antiresonant frequency spacing, $1/2r$.

7.2.2.2. Wide-Band Filters. Fractional bandwidths up to $\sqrt{2}/r$ are realized by the wide-band (WB) class, in which inductors are used in combination with the static capacitance of the crystal units or MDRs (plus additional capacitance as dictated by the circuit design) to create additional resonators. In this type of realization, typically about one-half of the resonators are piezoelectric, the other half being LC. A four-pole NB filter, for example, employs four discrete crystal units or two MDRs, while a four-pole half-lattice WB filter requires just two crystal units (Figure 7.14), or one MDR, but also uses two inductors, including the balanced transformer, as resonators.

WB filters are most common at low frequencies. At 100 kHz, fractional bandwidths up to 8% or 10% can be realized. At 20 MHz, 4% or 5% is about the maximum, based on capacitance considerations, but unwanted modes usually reduce this to 3% or less. The minimum bandwidth is determined by the requirement that inductor Q be at least several times f/bandwidth. This is most easily satisfied at low frequencies, where high-permeability core materials are available. For this reason, and because the

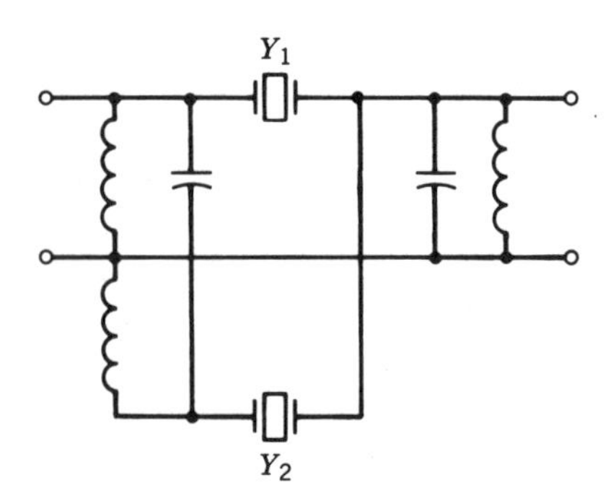

Figure 7.14 A wide-band 4-pole half-lattice filter.

capacitance ratio and unwanted mode spacing are most favorable for certain low-frequency cuts, especially the +5° X-cut, WB filters are most common in the low frequency range.

7.2.2.3. Intermediate-Band Filters. In WB filters, the minimum bandwidth is set by inductor Q. It turns out that this bandwidth is always less than the maximum NB bandwidth. To fill this gap, the intermediate band class (IB) was developed. In the most common type of IB design, part or all of the device static capacitance and associated circuit strays are *tuned out* with inductance at the filter center frequency. The LC combination is treated as if the capacitance had been reduced or eliminated (net negative values are also possible), and frequency dependence is ignored. This allows fractional bandwidths greater than the NB limit, the maximum being set either by inductor Q or by the limitations of the constant reactance assumption. Typically, the maximum IB fractional bandwidth is between $2/r$ and $4/r$. Circuit configurations are similar to Figure 7.13(b) and (c), except that there is an inductor at each node.

7.2.3. Minimum Fractional Bandwidth

For WB filters, as noted, the minimum bandwidth is set by inductor Q. For NB crystal filters, the minimum fractional bandwidth is determined by the Q, stability, and manufacturing tolerances of the crystal devices. If necessary, stability can be achieved by use of an oven; manufacturing tolerances generally reduce to questions of cost; Q, however, is an inherent limitation.

The Q limitation is the same as for other resonator-type filters, and is easily calculated. For a filter having resonators of equal Q, a first-order expression for the dissipation loss in dB, is [3]

$$L(f) = 8.686 \cdot T_g(f) \cdot f/Q$$

$T_g(f)$ is the group delay of the lossless filter, and is inversely proportional to bandwidth; hence, the loss is also. Normalized delay plots for common filter functions may be found, for example, in Zverev [38]. Group delay increases with the order of the filter; an eight-pole filter has roughly twice the mid-band delay of a four-pole, and hence twice the mid-band loss. When we also take into account the decrease in manufacturing tolerances required for higher-order filters, it is clear that the narrowest bandwidths can be achieved only by low-order filters.

7.2.4. Circuit Design Methods

The techniques of crystal filter circuit design are those of classical network synthesis. The advent of high-speed digital computers in the 1950s and 1960s made it possible to apply insertion loss design techniques economically, and most crystal filters are now designed on an insertion loss basis.

The insertion loss method proceeds in two stages. First, given a specification, a realizable network function is found to satisfy it; this process is called approximation [7]. Second, from the network function, a network is realized. Neither stage has a unique outcome. In particular, it is usually necessary to manipulate the form of the realized network to obtain the desired configuration and to obtain a practical set of element values. Computer programs for carrying out both stages of the insertion loss design process are commercially available (see also Chapter 2).

Many filters can be realized from low-pass prototype networks by frequency transformation, impedance scaling, and manipulation using network equivalences such as the Norton transformation [38], thus eliminating both the approximation stage and the first part of the realization stage. When these procedures are computerized they can be carried out very quickly indeed. To facilitate this, closed-form expressions are available giving Butterworth and Chebyshev low-pass element values [37]. In addition, element values for these as well as elliptic-function filters and many linear-phase types are widely tabulated [1, 5, 38].

7.2.5. Transfer Functions

For most applications, the primary electrical requirement is selectivity. In general, all-pole functions are most easily realized, and many selectivity requirements are economically satisfied by Chebyshev designs. The Butterworth approximation is relatively inefficient, and is used less often.

By introducing transmission zeroes, for instance by using elliptic function designs, fewer resonators may be required, but some simplicity is lost, especially for monolithic realizations. When the selectivity requirements are not symmetrical with respect to center frequency, as for sideband filtering requirements, symmetrical designs are inefficient, and the response function is tailored to the specification.

In many applications, the signals being filtered are angle-modulated. Voice-modulated analog FM is the most common type of angle modulation, but various types of PSK and QAM modulation are of rapidly increasing importance as more use is made of digitally-coded signals. For these applications, attention must be paid to the filter phase response as well as its attenuation characteristic. In general, what is desired is an approximately linear phase response; that is, a nearly constant group delay.

One difficulty that arises is determining phase or delay requirements from given system specifications. The analytical relations between analog signal distortion or digital bit error rate and filter phase nonlinearity or group delay variation are not simple. Requirements are often determined empirically.

Selectivity is realized most efficiently by minimum-phase transfer functions [3]. For functions of this class, however, the attenuation and phase characteristics are related by an integral transform, and cannot be specified independently. We can obtain linear phase or constant delay only by sacrificing selectivity. An example is given in §7.3.3 (Figure 7.24).

When both a high degree of selectivity and constant group delay are specified, a nonminimum-phase transfer function is required. In some cases, the requirement is met by first designing a minimum-phase filter that meets the attenuation requirements and then designing an all-pass phase correction network to be used in tandem. While this can give a combined phase characteristic that is linear, the total delay is increased. Alternatively, a nonminimum-phase function can be realized directly [11, 29] (see also Chapter 2).

Since phase linearity has a relatively high price, the system designer should not assume that for transmission of phase- or frequency-modulated signals, a linear phase or constant delay filter is required. If the available spectrum permits, it may be more economical to use a Chebyshev or elliptic-function filter, with the pass bandwidth perhaps 20% or 30% greater than the signal bandwidth. Since the group delay peaks sharply near the band edge, in this way the signal occupies the central portion of the pass-band where the delay variation is smaller.

7.3. BANDPASS FILTERS

In this section, we sample the very wide variety of bandpass crystal filters currently in use, both discrete and monolithic. Crystal filter center frequencies range from about 10 kHz to over 500 MHz, with the great majority of filters lying between 100 kHz and 200 MHz. Bandwidths may be from below 0.01% to above 10% of center frequency. The number of poles may be from 1 to 12, and occasionally even higher.

It will be convenient to divide our examples into low frequency (below 1 MHz), high frequency (1–30 MHz), and VHF/UHF (above 30 MHz) ranges. These will illustrate both typical practice and state-of-the-art techniques.

7.3.1. Low-Frequency Bandpass Filters

The earliest crystal filters were low-frequency ones; today, low-frequency crystal filters are still of importance. Their frequency-bandwidth range overlaps that of mechanical filters (see Chapter 4), but each type of filter has its own strengths [23].

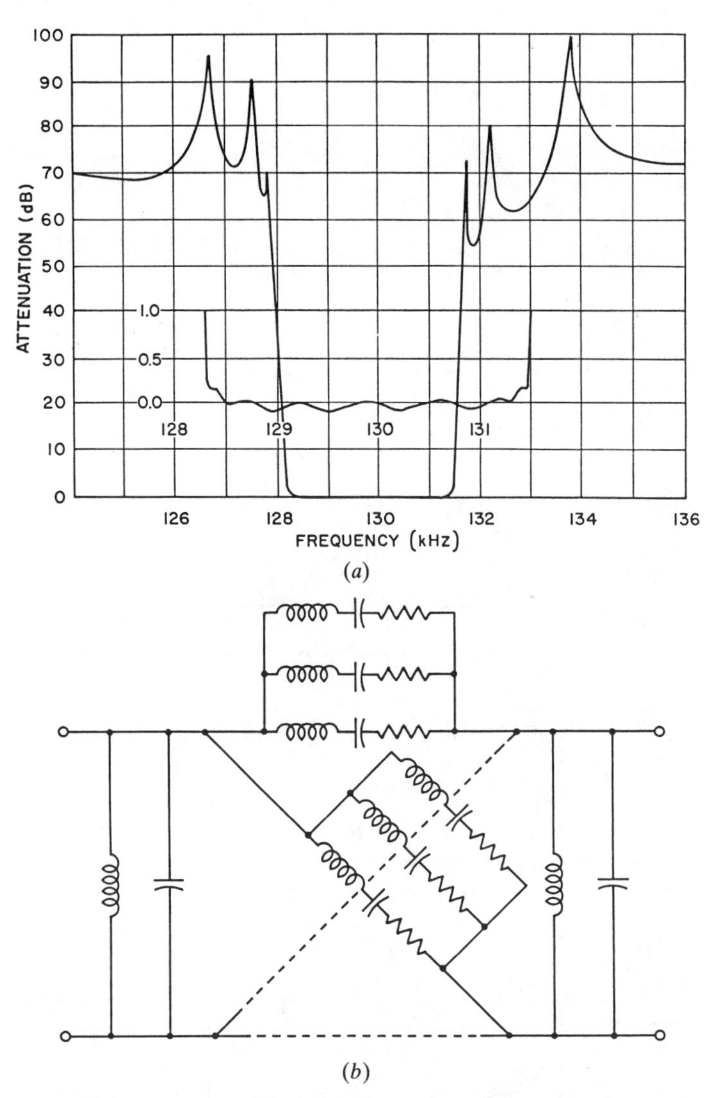

Figure 7.15 AT&T Technologies' FDM voice-channel filter. (a) Attenuation characteristic. From McLean et al. (1979) New discrete crystal filters for Bell system analog channel banks. *Proc. 33rd Ann. Symp. on Frequency Control*, pp. 166–172 © 1979 Electronics Industries Association. (b) Circuit diagram.

A modern, low-frequency voice-channel filter for FDM telephone transmission equipment is shown in Figure 7.15 [18]. Its novel mechanical design is described in §7.5. The same circuit configuration was used for this application as early as 1944, but the present filter is two orders of magnitude smaller than the 1944 version. This 8-pole, 128 kHz upper sideband filter uses six split-electrode, +5° X-cut crystals in a single WB lattice section and meets international (CCITT) standards of performance. High ultimate

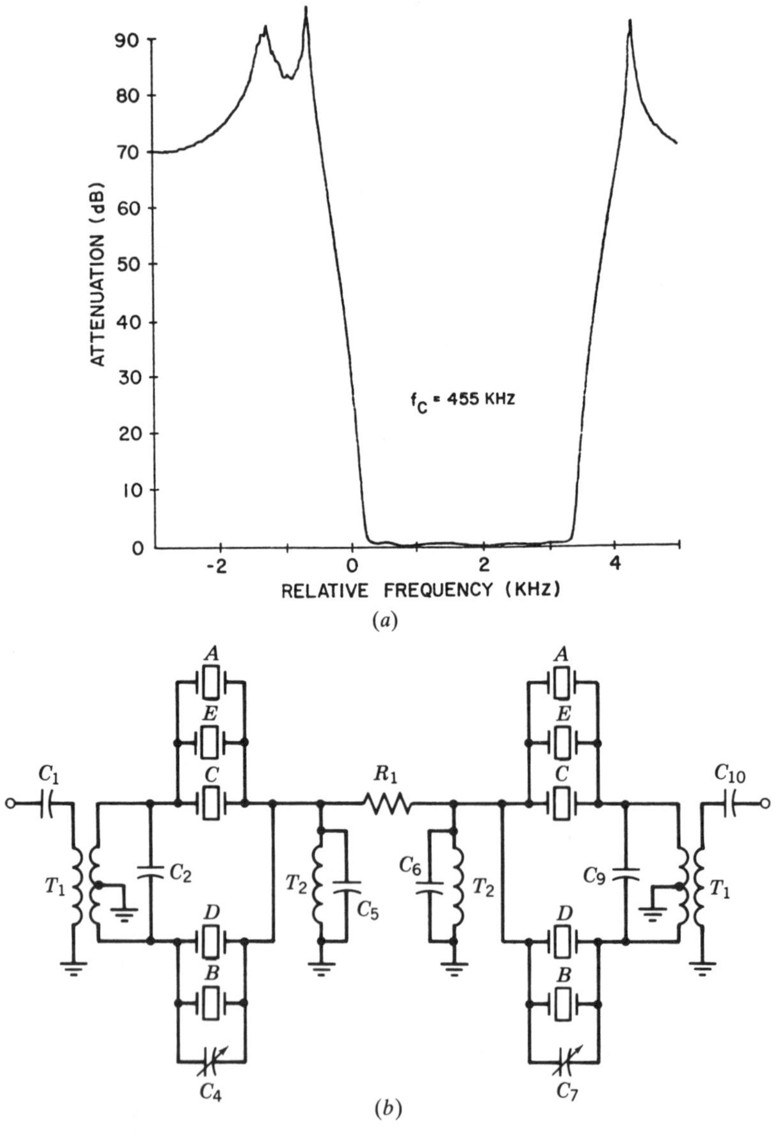

Figure 7.16 455 kHz upper sideband filter: (a) attenuation characteristic; (b) circuit diagram.

attenuation is achieved using a differential trimmer capacitor (not shown) to obtain a wide-band balance of the lattice arms.

Figure 7.16 describes a 455 kHz upper sideband filter for point-to-point military HF communications. It consists of two, 5-pole IB half-lattice sections. All transformers and inductors are slug-tuned ferrite pot cores, typical of construction at frequencies below a few MHz. DT-cut crystals are used.

This filter illustrates one of the short-cuts that are sometimes possible in filter circuit design. The two sections of the filter are identical, an advantage in manufacturing, and are designed on an insertion-loss basis. The insertion loss design of the individual sections can be carried out much more easily than could the insertion loss design of the complete filter. While there is some interaction between the two sections, so that the overall response is not merely the product of the individual section responses, analysis shows that it is acceptable. Further improvement is obtained by the small, empirically determined series resistor, R_1, which adds about 0.5 dB of flat loss but reduces the pass-band ripple.

7.3.2. High-Frequency Bandpass Filters

While BT-cut and SC-cut quartz resonators as well as lithium tantalate resonators are used for a few special applications, the great majority of high-frequency filters use AT-cut elements. Above about 5 MHz, these may be either single resonators or ACRs; below 5 MHz, discrete resonators are used. Below 1.5 MHz, acceptable levels of miniaturization of AT-cut resonators are difficult to achieve.

A good discrete-resonator example is the 1.75 MHz upper sideband filter of Figure 7.17. The application is again point-to-point HF communications. The circuit diagram of this NB filter illustrates two common tricks. First, the two center sections share a balanced transformer. When this arrangement is used it is important that the transformer coupling coefficient be very close to unity; its leakage inductance appears in the common lead from center tap to ground, changing the stop-band attenuation. Second, in order that all crystal units may be made according to a single design, differing only in frequency, capacitors are placed in series with some crystals to give the effect of a reduced motional capacitance. Another use of such a capacitor is in a very narrow-band filter, where it may be used to adjust the resonance frequency of the network branch, allowing a looser manufacturing tolerance for the resonators.

Typical of monolithic filters in the HF range are the two 8-pole NB filters of Figure 7.18, which use tandem MDRs. The filter whose response is shown in Figure 7.18(a) is a custom design used in a UHF communications transceiver for military aircraft. Its 3 dB bandwidth is approximately ± 40 kHz; the center frequency is 30 MHz.

Standardized monolithic filters are available from a number of manufacturers, particularly at center frequencies of 10.7 and 21.4 MHz. Figure

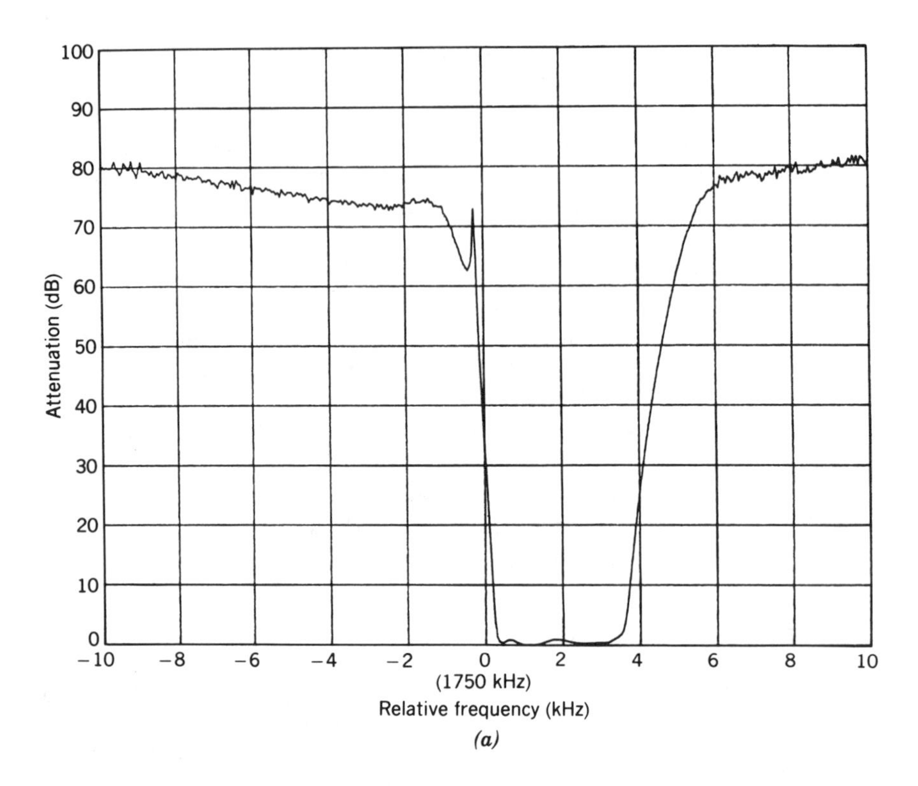

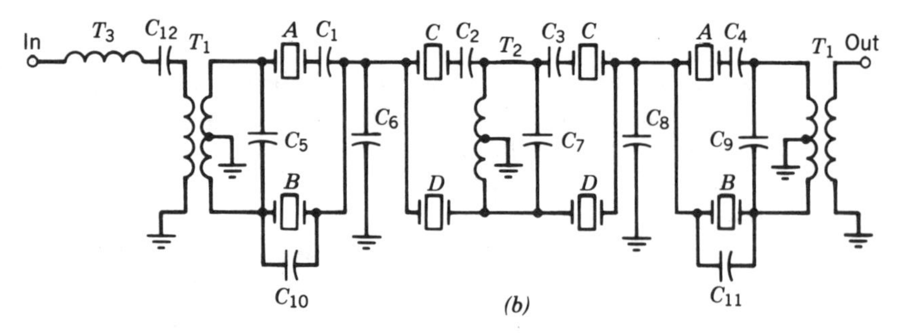

Figure 7.17 1.75 MHz upper sideband filter: (a) attenuation characteristic; (b) circuit diagram.

7.18(b) shows the response of one such 21.4 MHz filter. Most manufacturers offer 2- to 10-pole filters and several different bandwidths. Many of these filters were originally designed for use as IF filters in VHF and UHF mobile radios, but because of their wide availability and low cost, they are now used in many other applications as well.

A comparison of the tandem MDR circuit diagram (Figure 7.18(c)) with Figure 7.17(b) illustrates the simplicity which is a very significant characteristic of monolithic filters. For the monolithic filter, the total number of components is greatly reduced, saving assembly time and allowing smaller size and increased reliability.

Of even greater significance, the balanced transformers used for discrete-resonator filters are eliminated. Besides occupying considerable space, transformers impose electrical performance limitations. As frequency increases, both high coupling and good balance become more and more difficult to achieve. Additionally, stability and Q decrease. Transformer losses increase the filter flat loss, change the shape of the passband, and degrade return loss. It is probably safe to say that the elimination of balanced transformers, and the consequent economic and technical benefits, is the dominant reason for the success of monolithic filter technology.

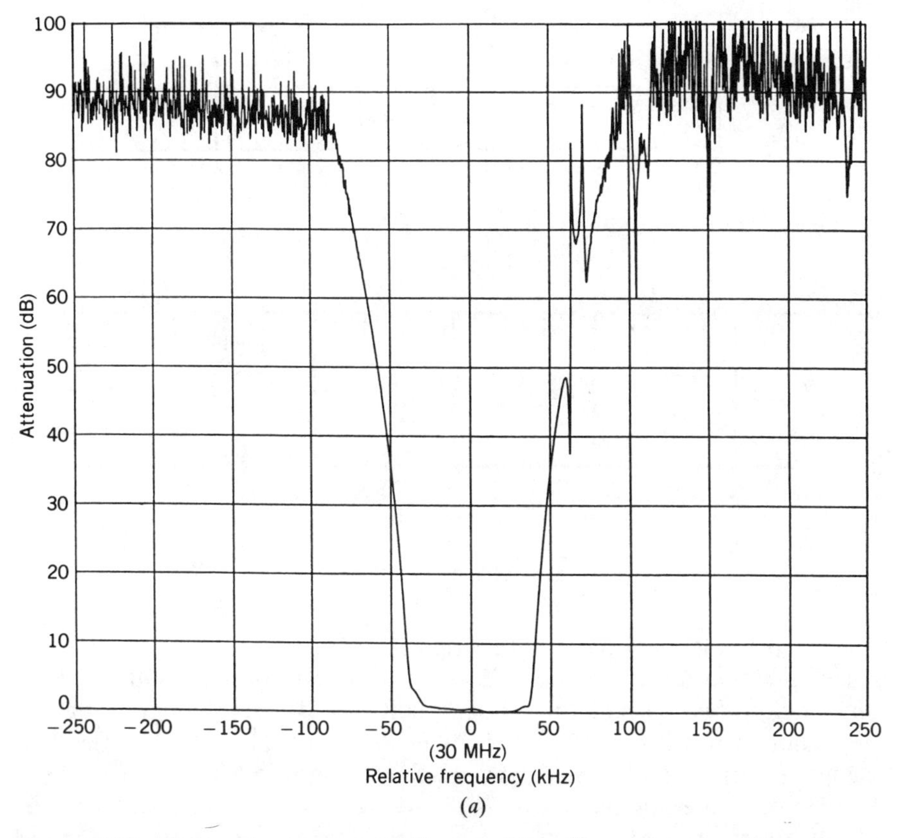

(*a*)

Figure 7.18 Eight-pole, tandem MDR monolithic filters: (a) attenuation characteristic for a filter with a center frequency at 30 MHz with a ±37.5 kHz 3-dB bandwidth; (b) attenuation characteristic for a filter with a center frequency at 21.4 MHz with a ±6 kHz 3-dB bandwidth; (c) circuit diagram.

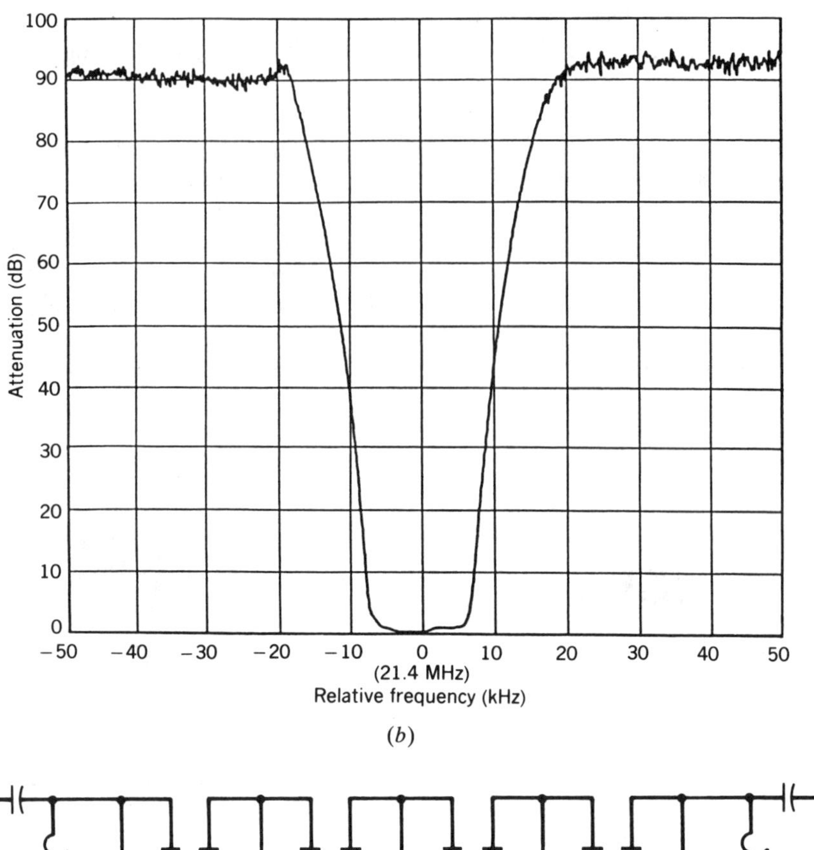

(*b*)

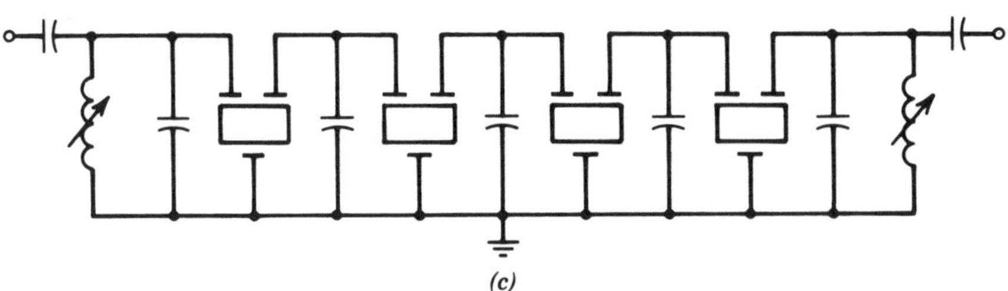

(*c*)

Figure 7.18 (*Continued*)

When the fractional bandwidth is much smaller than $1/2r$, one way of achieving simplicity and eliminating balanced transformers is the ladder configuration. Because of the ladder's voltage divider topology, ultimate attenuation varies inversely with bandwidth, as noted in §7.2.2. For the example shown in Figure 7.19, the fractional bandwidth of about 0.005% allows an ultimate attenuation of 70 dB with only three fundamental mode AT-cut resonators.

WB filters are fairly common at low frequencies, where the crystal resonances tend to be widely separated. At high frequencies, unwanted anharmonic modes may lie quite close to the desired resonance. While the

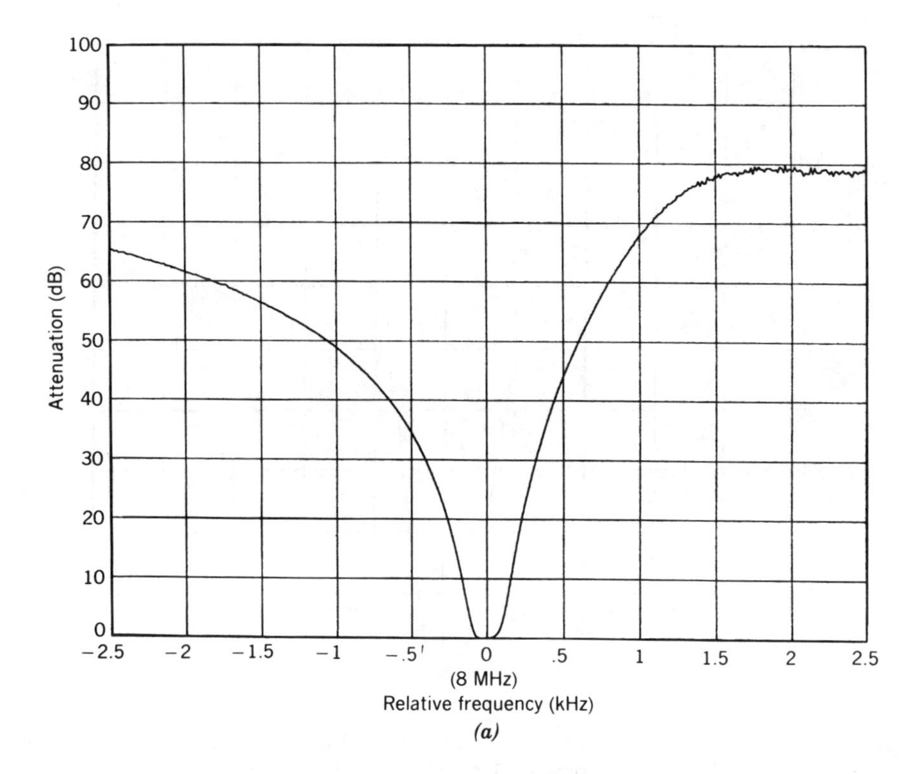

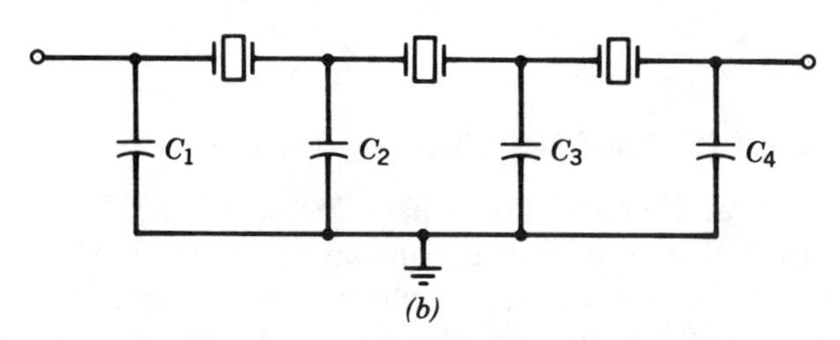

Figure 7.19 Three-pole ladder crystal filter: (a) attenuation characteristic; center frequency is at 8 MHz and 3-dB bandwidth is 250 Hz; (b) circuit diagram.

principles determining unwanted mode frequencies are well understood, the design trade-offs become increasingly less favorable as frequency increases. Nevertheless, WB filters are still manufacturable in the high-frequency range, although with a more restricted range of fractional bandwidths than at lower frequencies. A typical 4-pole WB filter at 21.4 MHz has the half-lattice configuration of Figure 7.14; its attenuation characteristic is shown in Figure 7.20.

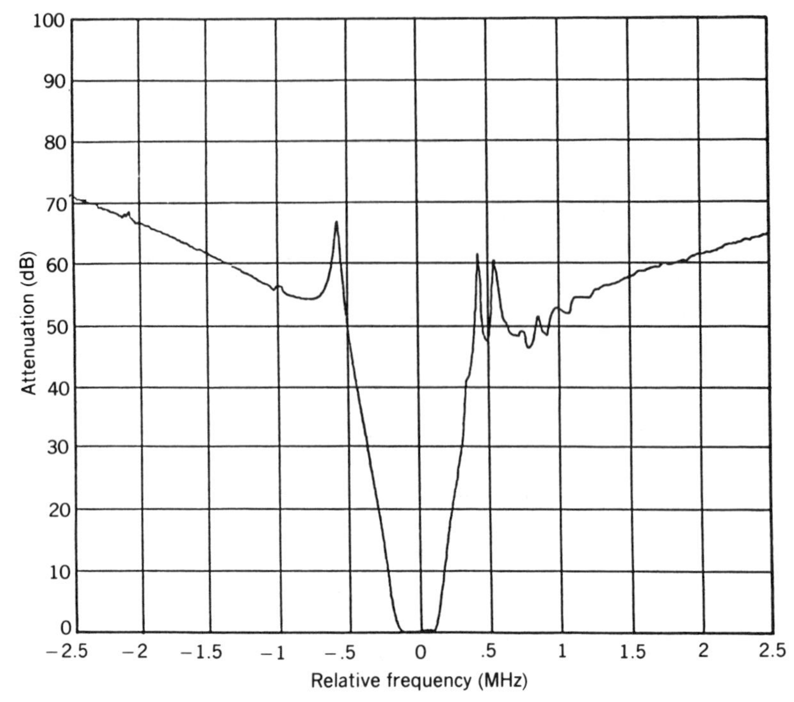

Figure 7.20 Attenuation characteristic of a 4-pole WB filter with a 3 dB bandwidth of 300 kHz at a center frequency of 21.4 MHz. The circuit configuration is shown in Figure 7.14.

7.3.3. VHF and UHF Bandpass Filters

Applications for VHF crystal filters have increased rapidly in recent years, consistent with both the general expansion of communications and navigation systems and the trend toward high operating frequencies. While UHF crystal filters are still uncommon, there is no doubt a latent demand.

As for HF filters, the most common configuration uses tandem MDRs. Half-lattice discrete-resonator sections and multiresonator ACRs are also used. Until recently, most VHF crystal filters used overtone devices; consequently, they were IB designs, like Figure 7.21. The increased availability of VHF fundamental mode resonators and MDRs has now made inductorless filters possible for many applications.

Figure 7.22 shows a 4-pole tandem MDR filter used in the first IF stage of an 800 MHz cellular mobile telephone or base station receiver. Another fundamental mode filter is shown in Figure 7.23. It uses three ring-supported MDRs, as described in §7.1.5. Both of these filters are inductorless.

In Figure 7.24 is shown the pass-band attenuation and group delay of a 4-pole tandem MDR filter for use as an IF filter in the NTT (Japan) car

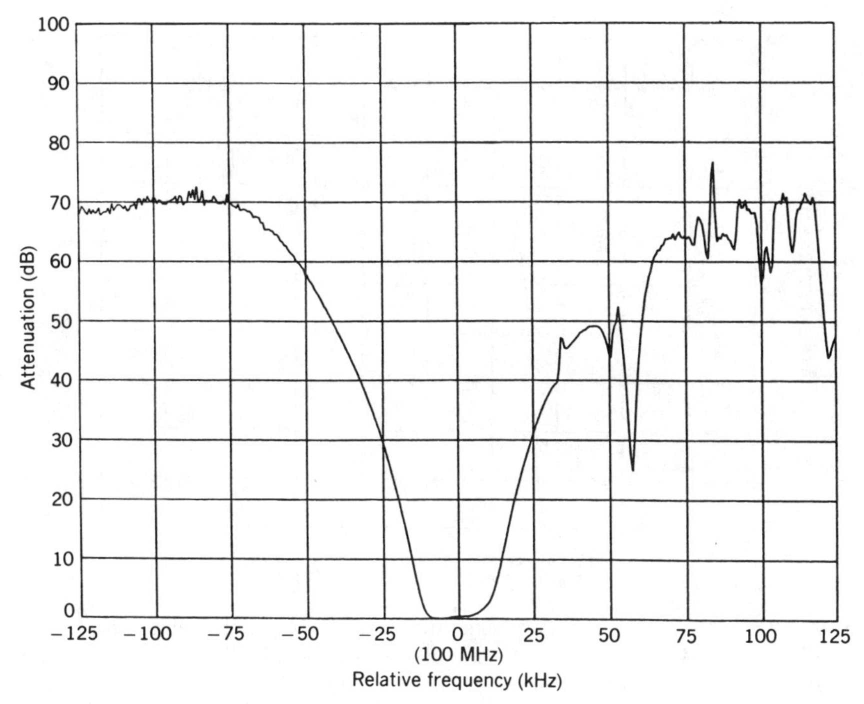

Figure 7.21 Attenuation characteristic of a 100 MHz 4-pole monolithic filter using two-third overtone MDRs. 3 dB bandwidth is 25.6 kHz.

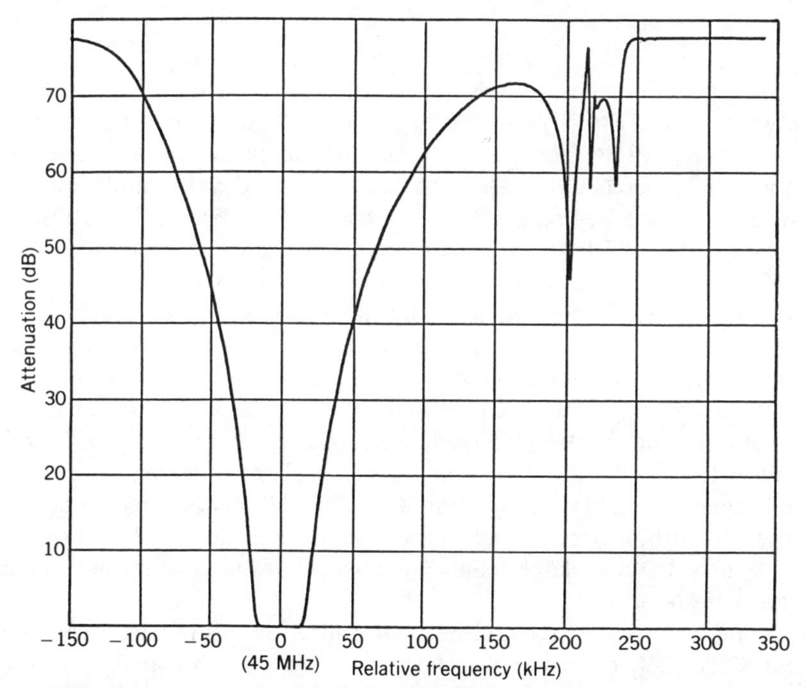

Figure 7.22 Attenuation characteristic of a 45 MHz 4-pole inductorless monolithic filter using fundamental mode MDRs. The 3 dB bandwidth is 30 kHz.

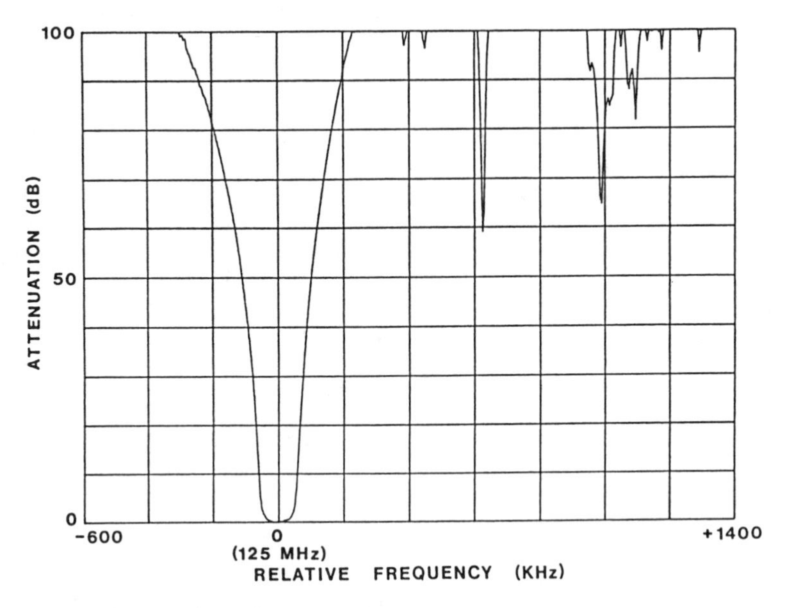

Figure 7.23 Attenuation characteristic of a 125 MHz 6-pole filter with a 3 dB bandwidth of 100 kHz and ultimate attenuation of 120 dB. This inductorless filter uses ring-supported fundamental mode MDRs. From Smythe et al. (1985) VHF monolithic filters fabricated by chemical milling. *Proc. 39th Ann. Symp. on Frequency Control*, pp. 481–485 © 1985 I.E.E.E.

telephone. This filter has a transitional Butterworth–Thomson characteristic, in which selectivity is sacrificed to obtain constant group delay.

As overtone VHF filters are almost invariably IB designs requiring inductors, there has been considerable incentive to develop multiresonator devices in this frequency range. Two approaches have been followed. In the first, described in §7.1.4, resonators are acoustically coupled in tandem. Figure 7.25 shows a four-resonator ACR developed by Motorola, Inc. [8]. The version shown is used in a highly miniaturized paging receiver as a front-end preselector. For this application, a fifth resonator, formed on the same quartz wafer but acoustically isolated from the filter array is used to control the frequency of the local oscillator, whose output is mixed with that of the filter to convert the received signal to the IF frequency of 35 kHz. Because of the low IF, the filter must have excellent rejection at the mixer image frequency, 70 kHz away from the 175 MHz center frequency. The filter uses the fifth overtone of an 8 mm AT-cut wafer mounted on a ceramic substrate, in a TO-8 package sealed by resistance welding. Internal connections are wire bonded.

A second way to realize a multiresonator filter on a single wafer is shown in Figure 7.26 [26]. In this approach, two MDRs are fabricated on a single

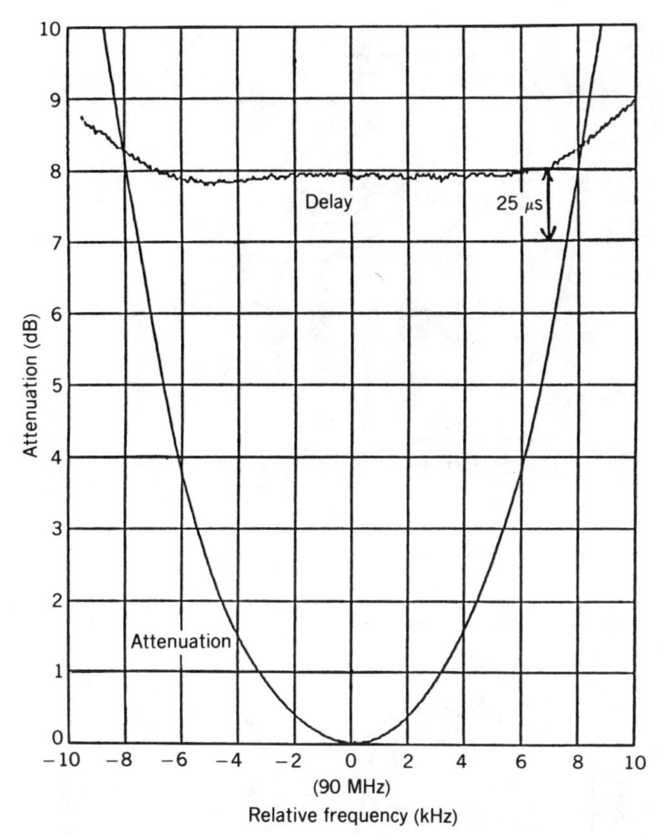

Figure 7.24 Passband attenuation and group delay of a 90 MHz tandem monolithic filter with a 4-pole transitional Butterworth–Thomson response characteristic. Stopband attenuation is 60 dB at ±50 kHz. (Courtesy Toyo Communication Equipment Co., Ltd.)

wafer. The MDRs are far enough apart that very little energy is transmitted between them acoustically. The two MDRs are connected in tandem as shown, to form a 4-pole filter. Filters of this type are used in a number of applications as front-end filters, IF filters, and spectrum clean-up filters.

For either approach, the cost of a single 4-pole device may be greater than two MDRs. However, total filter cost is generally less. In addition, great savings in size and reliability are achieved by elimination of the spreading coil required with separate MDRs.

Much greater bandwidths can be obtained by using lithium tantalate instead of quartz. A 6-pole, fundamental mode, discrete-resonator filter is shown in Figure 7.27 [6]. This filter, which is used in an L-band IFF transponder, has a bandwidth of 1 MHz, or 1.7% of the 60 MHz center frequency.

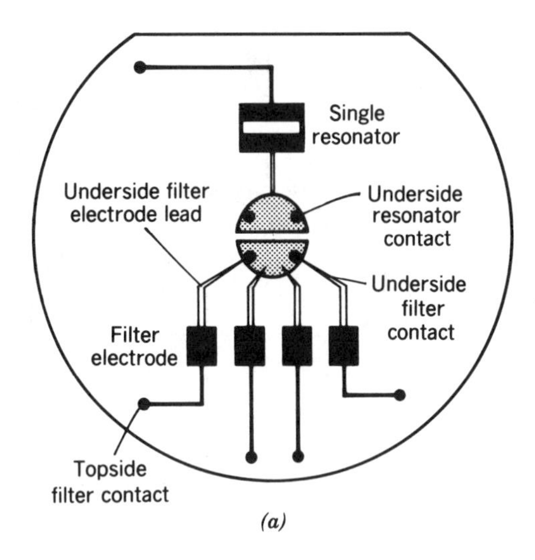

(a)

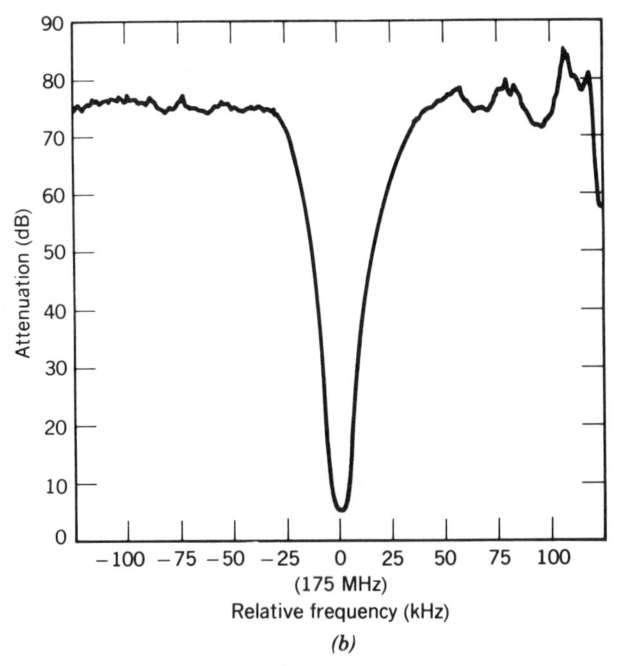

(b)

Figure 7.25 Four-pole ACR. Center frequency is 175 MHz, using the fifth overtone; 3 dB bandwidth is 12 kHz: (a) Electrode arrangement. Grounding is via a novel central mounting pad. From Dworsky and Shanley (1985) The Motorola multipole monolithic filter project. *Proc. 39th Ann. Symp. on Frequency Control*, pp. 486–490 © 1985 I.E.E.E. (b) Attenuation characteristic showing 70 dB ultimate attenuation. (Courtesy Motorola Inc.)

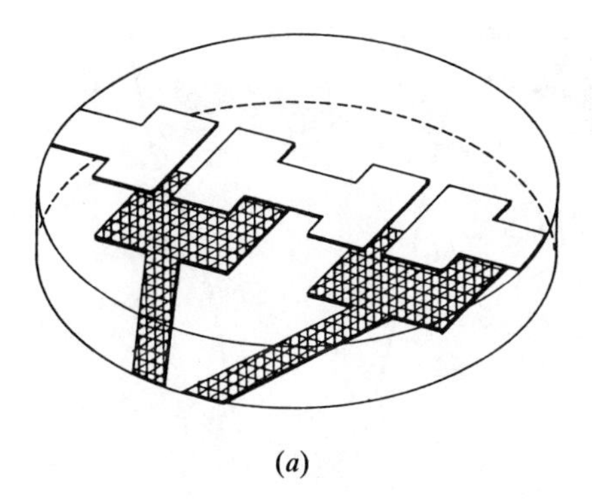

(*a*)

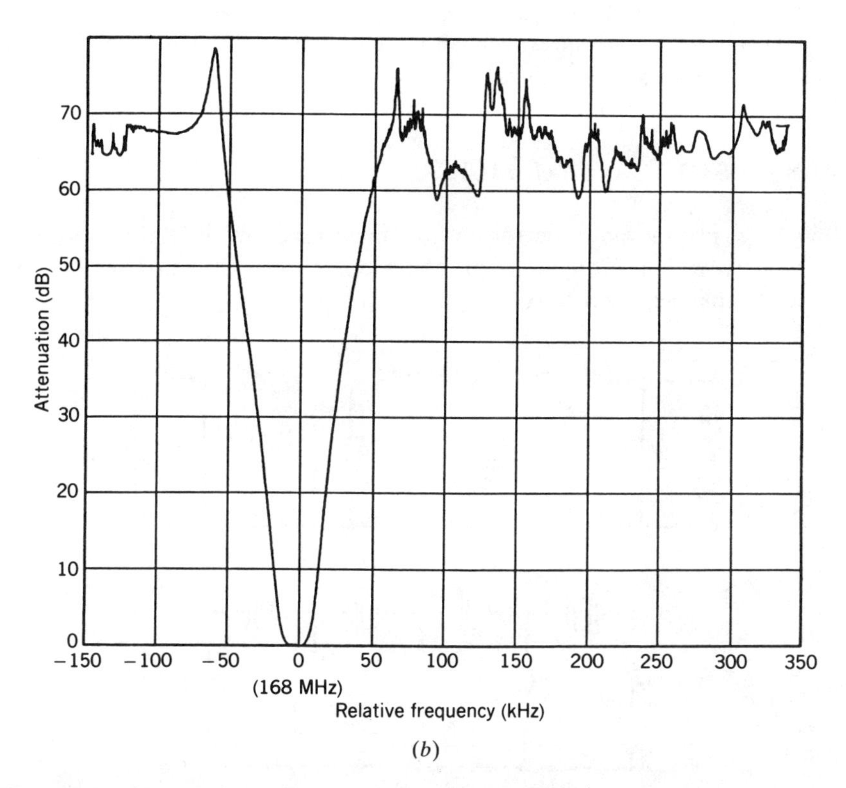

(*b*)

Figure 7.26 Four-pole monolithic filter comprised of two MDRs fabricated on a single 6.4 mm wafer: (a) electrode arrangement; (b) attenuation characteristic for a fifth overtone device with a 3 dB bandwidth of 25 kHz at 168 MHz.

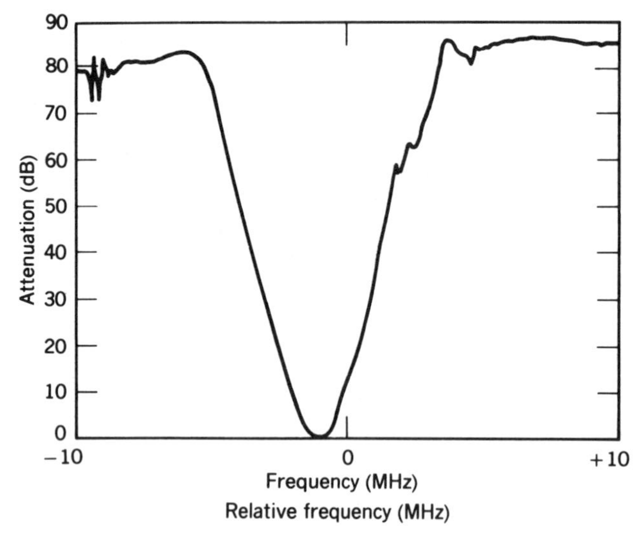

Figure 7.27 Attenuation characteristic of a six-resonator lithium tantalate filter with a center frequency of 60 MHz and a 3 dB bandwidth of 1 MHz. From d'Albaret and Siffert (1982) Recent advances in UHF crystal filters. *Proc. 36th Ann. Symp. on Frequency Control*, pp. 405–418 © 1982 I.E.E.E.

7.4. BANDSTOP CRYSTAL FILTERS

While most crystal filters perform a bandpass function, it is also possible to realize narrow band-rejection, or bandstop, crystal filters. The most important forms use discrete resonators.

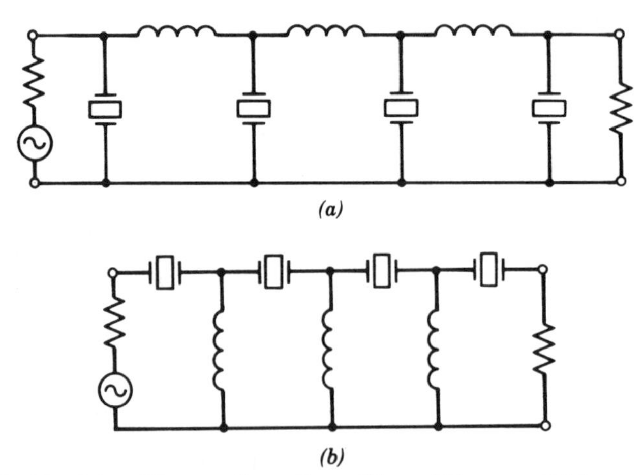

Figure 7.28 Bandstop filter configurations. These filters are (a) low-pass/bandstop and (b) high-pass/bandstop networks.

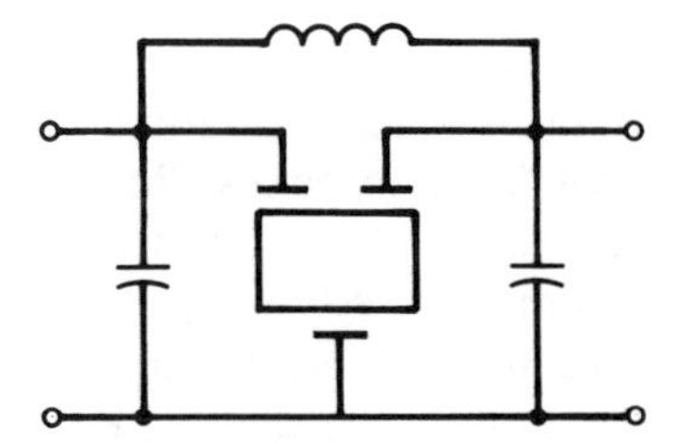

Figure 7.29 Low-pass/bandstop filter using an MDR.

The nature of the passband requirement affects the choice of network structure. For low-pass or high-pass forms, particularly simple networks are possible (Figure 7.28) [1]. Figure 7.29 shows a low-pass structure in which the notch is realized by an MDR [9, 24]. This configuration has been used as a 42.88 MHz pilot rejection filter for telephone FDM systems.

7.5. MECHANICAL DESIGN

Crystal filters are essentially high impedance RF circuits. Their construction reflects this: attention must be paid to controlling stray capacitances and to shielding. Except for this, most crystal filters follow conventional electronic assembly practice. Crystal devices are in individual, hermetically-sealed enclosures (see §7.5.1.) The complete filter circuit is enclosed in a metal case, that may or may not be hermetic. A printed circuit board may be used, but point-to-point wiring is not uncommon and may allow better use of available space. Further miniaturization may be obtained by the use of surface-mount components, such as chip capacitors. However, the size of the filter is essentially determined by the crystal devices and inductive components.

For some low-cost applications, the filter may be furnished as a printed circuit assembly, without an outer enclosure. In this case, the user must give careful consideration to shielding requirements. For many high-volume, low-cost monolithic filter applications, the MDRs or ACRs are assembled directly on the equipment circuit board. Again, shielding must be considered.

As the crystal devices are supplied in packaged form, they may be handled like other components; the filter manufacturing process then divides naturally into device fabrication and filter assembly. Since many crystal filters are custom designs made in lots of several thousand or less this type of construction is the norm.

A greater degree of miniaturization may be possible with an integrated type of construction, in which crystal units and other circuit components occupy a single hermetic package. In this case, assembly must be carried out with the same scrupulous attention to cleanliness and contamination as in crystal manufacture. At higher frequencies, for example, inductive components might not be allowed, because of outgassing problems. In addition, special tooling is generally required. This approach tends to be less flexible than conventional construction. Since initial costs are higher, it has been used only where justified by large quantity requirements.

Integration is more easily achieved at lower frequencies, where crystals are less sensitive to contamination. An excellent example is the 128 KHz FDM channel filter described in §7.3.1. This filter uses six, split-electrode, +5° X-cut crystal resonators, each about 20 mm long by 5 mm wide. All six are mounted in a special plastic holder that also provides interconnections. Together with other components this assembly is mounted on a circuit board and hermetically packaged in a conventional drawn metal can. The novel crystal configuration, plus miniaturization of the two cup-core inductors used, allowed the size to be reduced by a factor of 12. Package volume is 36 cm^3.

7.5.1. Crystal Device Packages

For filter use, crystal units (MDRs and ACRs) are packaged in hermetic metal enclosures that are either evacuated or filled with a dry gas. Table 7.4 gives approximate dimensions of some commonly used packages. Nomenclature varies among manufacturers, and a number of proprietary variations of these are in use. The last two entries are adaptations of metal transistor packages; the remainder have been specifically developed for crystal devices.

TABLE 7.4 Crystal Unit Packages

	Dimensions (mm)			
Designation*	Length	Width	Height†	Seal‡
HC-6	19.0	9.0	20.0	S, R, C
HC-6 Slim	19.0	6.4	20.0	S, R
HC-12	15.0	6.4	15.7	S, R
HC-12 Slim	15.0	4.7	15.7	S, R
HC-18	11.0	4.7	11.4	S, R, C
HC-45	7.9	3.2	8.0	R
TO-5	10.7§		6.6	S, R, C
TO-8	15.5§		6.6	S, R, C

* Nomenclature varies among suppliers.
† Other values may be available.
‡ S, solder seal; R, resistance welded; C, coldweld.
§ Diameter.

7.6. NONLINEAR EFFECTS

While quartz crystal filters are, by their nature, linear networks, for a number of high-performance applications, nonlinear effects must be considered. The limitations of inductors and transformers with ferrite or powdered-iron cores are too well known to need further mention. The nonlinear properties of quartz resonators and ACRs are somewhat less familiar.

The most important nonlinear effects in quartz crystal filters include intermodulation [12, 13, 27, 34], variation of attenuation and phase shift with signal level, and excess phase noise. These in turn can be related to nonlinear behavior of the quartz devices themselves.

Nonlinear effects may occur in all types of crystal filters, but have been studied primarily in high-frequency filters using AT-cut crystals. At low currents the resistance of a crystal resonator may vary as a function of current, and may exhibit hysteresis. Such behavior in a filter circuit may give rise to both attenuation and phase shift variation with signal level and intermodulation of two or more signals. The intermodulating signals may lie in either the pass-band or the stop-band, provided that their frequencies are such as to produce an in-band product.

To illustrate the case in which the signals are in the stop-band, two 0 dBm test tones, 20 kHz and 40 kHz below center frequency, are applied to the input of the filter whose attenuation characteristic was shown in Figure 7.18(b). Figure 7.30 shows the output spectrum. The test tones are attenuated by approximately 90 dB. A −80 dBm third-order intermodulation product is seen at band center. As the test tones lie in the stop-band, the resonator nearest the filter input sees the largest current at test tone frequencies. Each succeeding resonator sees a successively smaller test tone current. Consequently, the first resonator in the chain is the one primarily responsible for the intermodulation. It follows that, if filter input and output are reversed, the intermodulation may change. When the test tones lie in the pass-band, each resonator is a potential source of intermodulation; nevertheless, reversing input and output may change the intermodulation produced.

Nonlinearity at low current levels is generally due to the surface condition of the quartz devices, and is controlled through careful attention to wafer preparation, electrode deposition, and cleaning. An example of the amplitude linearity that can be achieved is shown in Figure 7.31, which displays the variation in attenuation of center frequency as a function of input signal level for a 4-pole monolithic filter composed of two MDRs. Over a 40 dB signal range, the filter loss varies by less than 0.005 dB.

At somewhat higher currents, elastic nonlinearity of the quartz becomes significant, causing intermodulation of in-band signals. Tiersten [34] has analyzed this cause of intermodulation for thickness-shear resonators, and a design relation based on his analysis has been developed [29]. BT- and SC-cut resonators exhibit smaller anelastic effects than AT-cut, and some

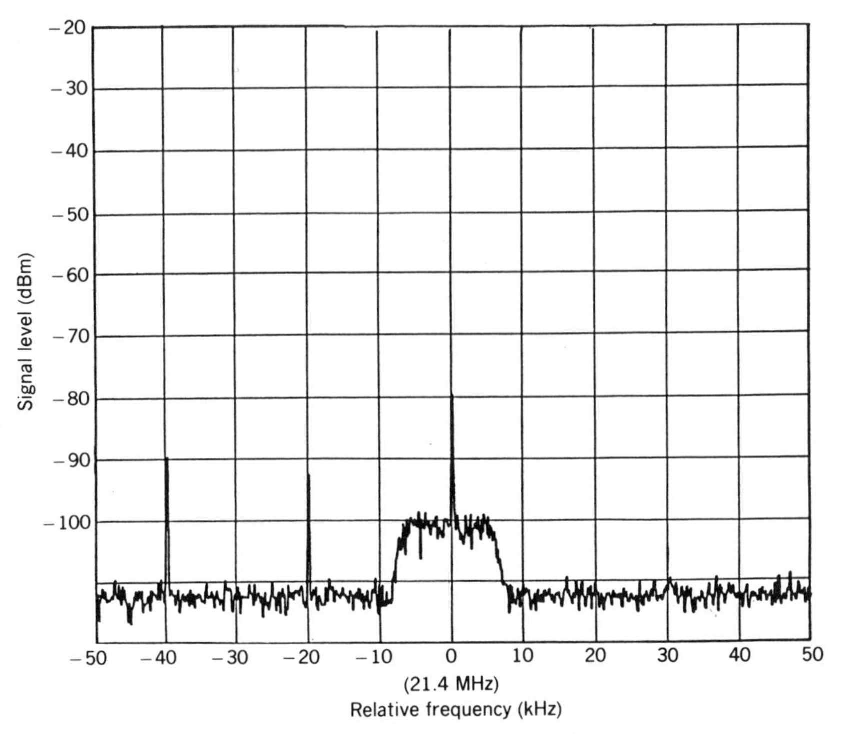

Figure 7.30 Output spectrum of the filter of Figure 7.18(b) showing third order intermodulation at 21.4 MHz induced by 0 dBm test tones in the lower stop-band at 21.36 and 21.38 MHz.

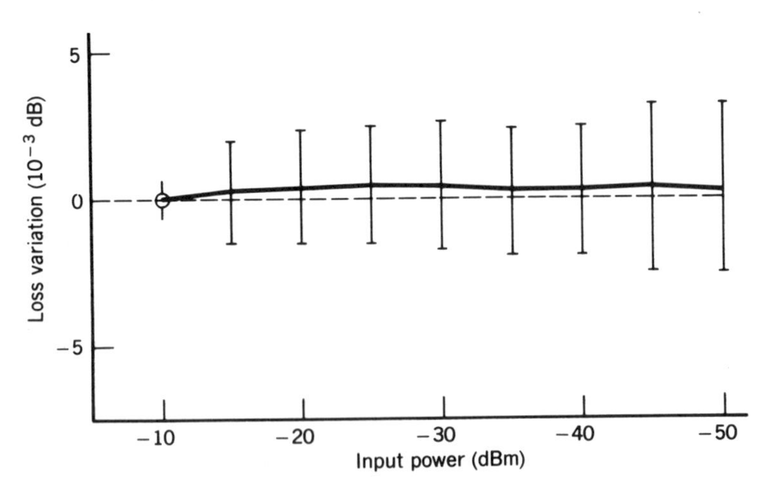

Figure 7.31 Variation of loss at center frequency with signal level for a 4-pole, 20 MHz monolithic filter. Data shown are for a lot of 60 filters; bars denote 2 sigma limits.

improvement in intermodulation performance can be obtained through their use [13].

For still higher currents, heating due to dissipation in the resonators becomes significant. The resulting thermal gradients can cause changes in resonator Q as well as frequency, and the filter is no longer well-described by its attenuation and phase response, since these depend upon the duration of the signal. Operation in this current regime is to be avoided.

Excess phase noise in crystal filters is of interest primarily for spectrum clean-up filters used in frequency synthesizers and other frequency sources. The causes of excess phase noise are not completely understood. At low to moderate current level, however, an important source of excess noise is again related to surface conditions, and is controlled as indicated above.

Finally, for all nonlinear mechanisms, the possibility exists of coupling to other resonances via the nonlinearity. When this occurs, the various nonlinear effects, such as intermodulation, are affected.

7.7. CURRENT TRENDS

Crystal filters have been in commercial use for over 50 years. The technology must be considered a mature one, but it is by no means stagnant. The use of crystal filters in military and commercial communications and navigation systems and in electronic instrumentation continues to grow, and the past decade has seen, for the first time, extensive use of crystal filters in consumer products.

Throughout this period, there has been a steady trend toward higher filter frequencies. The recent introduction of ring-supported resonators and MDRs should allow this trend to continue.

Less easy to predict is the future use of resonator materials other than quartz. For filter use, the outstanding limitation of quartz is its weak piezoelectric effect. This results, for resonators, in a high capacitance ratio, and consequently limits the maximum filter bandwidth to a few tenths of one percent for inductorless (NB) filters or a few percent when inductors are used. With higher piezoelectric coupling, wider fractional bandwidths are possible; moreover, for wider bandwidths the stability and Q of the material need not be as good as those of quartz.

The prime materials under consideration at present are lithium tantalate and Berlinite (aluminum metaphosphate, $AlPO_4$). As noted earlier, lithium niobate, widely used for SAW devices, is unsuitable for crystal filters, since it does not have a bulk-acoustic-wave cut with good frequency-temperature characteristics.

Lithium tantalate, on the other hand, possesses a number of plate orientations that are potentially useful for filters and have higher piezoelectric coupling than quartz. It is available from a number of sources. The X-cut is already used commercially to some extent; future use of this material is expected to increase. Berlinite has not yet emerged from the laboratory, but if growth problems can be resolved, it may find considerable use. Useful cuts are known to exist; the piezoelectric coupling, while not as high as lithium tantalate is considerably greater than for quartz.

Another historical trend is miniaturization. The demand for yet smaller crystal filters is expected to continue due to anticipated system requirements, and will probably be satisfied by a combination of techniques. In addition, if the current trend toward surface-mount technology continues, compatible filter packages will no doubt emerge.

REFERENCES

1 Anonymous (1975) Reference Data for Radio Engineers, 6th edn. Howard W. Sams & Co., Inc., Indianapolis, Indiana, Chap. 8.

2 M. Berte (1977) Acoustic bulk wave resonators and filters operating in the fundamental mode at frequencies greater than 100 MHz. *Proc. 31st Ann. Symp. on Frequency Control*, pp. 122–125.

3 H. W. Bode (1945) *Network Analysis and Feedback Amplifier Design*. D. Van Nostrand, Princeton, NJ, pp. 242–244.

4 V. E. Bottom (1982) *Introduction to Quartz Crystal Unit Design*. Van Nostrand Reinhold, New York.

5 E. Christian and E. Eisenmann (1966) *Filter Design Tables and Graphs*. Wiley, New York.

6 B. d'Albaret and P. Siffert (1982) Recent advances in UHF crystal filters. *Proc. 36th Ann. Symp. on Frequency Control*, pp. 405–418.

7 R. W. Daniels (1974) *Approximation Methods for Electronic Filter Design*. McGraw-Hill, New York.

8 L. N. Dworsky and C. S. Shanley (1985) The Motorola multi-pole monolithic filter project. *Proc. 39th Ann. Symp. on Frequency Control*, pp. 486–490.

9 J. L. Garrison, A. N. Georgiades and H. A. Simpson (1970) The application of monolithic crystal filters to frequency selective networks. *Proc. I.E.E.E. Int. Symp. on Circuit Theory*, pp. 177–178.

10 E. A. Gerber and A. D. Ballato (Editors) (1985) *Precision Frequency Control*, Vol. I: *Acoustic Resonators and Filters*. Academic Press, Orlando.

11 P. A. Herzig and T. W. Swanson (1978) A polylithic crystal filter employing a Rhodes transfer function. *Proc. 32nd Ann. Symp. on Frequency Control*, pp. 233–242.

12 W. H. Horton and R. C. Smythe (1973) Experimental investigations of intermodulation in monolithic crystal filters. *Proc. 27th Ann. Symp. on Frequency Control*, pp. 243–245.

13 M. D. Howard, R. C. Smythe and P. E. Morley (1985) Monolithic Crystal Filters Having Improved Intermodulation and Power Handling Capability. *Proc. 39th Ann. Symp. on Frequency Control*, pp. 491–503.

14 J. R. Hunt and R. C. Smythe (1985) Chemically milled VHF and UHF resonators. *Proc. 39th Ann. Symp. on Frequency Control*, pp. 292–300.

15 R. G. Kinsman (1987) *Crystal Filters: Design, Manufacture, and Application*. Wiley, New York.

16 P. Lloyd (1967) Equations governing the electrical behavior of an arbitrary piezoelectric resonator having N electrodes. *Bell Systems Tech. J.* **46**, 1881–1900.

17 W. P. Mason (1948) *Electromechanical Transducers and Wave Filters*, 2nd edn. D. Van Nostrand, New York, Chap. 8.

18 D. I. McLean, A. F. Graziani and J. J. Royer (1979) New discrete crystal filters for Bell system analog channel banks, *Proc. 33rd Ann. Symp. on Frequency Control*, pp. 166–172.

19 R. D. Mindlin and P. Y. Lee (1966) Thickness-shear and flexural vibrations of partially-plated crystal plates. *Int. J. Solids Struct.* **2**, 125–139.

20 R. D. Mindlin (1967) Bechmann's number for harmonic overtones of thickness-twist vibrations of rotated Y-cut plates. *J. Acoust. Soc. Am.* **41**, 969–973.

21 M. Onoe and H. Jumonji (1965) Analysis of piezoelectric resonators vibrating in trapped-energy modes. *Electron. Commun. Jpn.* **48**, 84–93.

22 G. T. Pearman and R. C. Rennick (1977) Unwanted modes in monolithic filters. *Proc. 31st Ann. Symp. on Frequency Control*, 191–196.

23 D. F. Sheahan and R. A. Johnson (Editors) (1977) *Modern Crystal and Mechanical Filters*. I.E.E.E. Press, New York.

24 H. A. Simpson, E. D. Finch, Jr. and R. K. Weeman (1971) Composite filter structures incorporating monolithic crystal filters and LC networks. *Proc. 25th Ann. Symp. on Frequency Control*, pp. 287–296.

25 T. H. Simmonds, Jr. (1979) The evolution of the discrete crystal single-sideband selection filter in the Bell system. *Proc. I.E.E.E.* **67**, 109–115.

26 R. C. Smythe (1972) Communications systems benefit from monolithic crystal filters. *Electronics* **45**, 48–51.

27 R. C. Smythe (1974) Intermodulation in thickness-shear resonators. *Proc. 28th Ann. Symp. on Frequency Control*, pp. 5–7.

28 R. C. Smythe and M. D. Howard (1983) Current Trends in crystal filters. *Proc. 37th Ann. Symp. on Frequency Control*, pp. 349–353.

29 R. C. Smythe (1985) Bulk acoustic wave filters. In: *Precision Frequency Control*, Vol. I, edited by E. A. Gerber and A. D. Ballato. Academic Press, Orlando, pp. 188–228.

30 R. C. Smythe, M. D. Howard and J. R. Hunt (1985) VHF monolithic filters fabricated by chemical milling. *Proc. 39th Ann. Symp. on Frequency Control*, pp. 481–485.

31 W. J. Spencer (1972) Monolithic crystal filters. In: *Physical Acoustics*, Vol. IX, edited by W. P. Mason and R. N. Thurston. Academic Press, New York, pp. 167–220.

32 G. C. Temes and S. K. Mitra (Editors) (1973) *Modern Filter Theory and Design*. Wiley, New York.

33 H. F. Tiersten (1974) Analysis of trapped energy resonators operating in overtones of thickness-shear. *Proc. 28th Ann. Symp. on Frequency Control*, pp. 44–48.

34 H. F. Tiersten (1975) An analysis of intermodulation in thickness-shear and trapped-energy resonators. *J. Acoust. Soc. Am.* **57**, 667–681.

35 H. F. Tiersten (1977) An analysis of overtone modes in monolithic crystal filters. *J. Acoust. Soc. Am.* **62**, 1424–1430.

36 J. R. Vig, J. W. LeBus and R. L. Filler (1977) Chemically polished quartz. *Proc. 31st Ann. Symp. on Frequency Control*, pp. 131–143.

37 L. Weinberg (1962) *Network Analysis and Synthesis*. McGraw-Hill, New York, pp. 651–669.

38 A. I. Zverev (1967) *Handbook of Filter Design*. Wiley, New York.

8 Surface Acoustic Wave Filters

ROBERT L. ROSENBERG
AT&T Bell Laboratories, North Andover, Massachusetts

Surface acoustic wave (SAW) technology has grown into a field of great sophistication and versatility since the introduction in 1965 [63] of the interdigital surface wave transducer, the first efficient electro-acoustic generating and receiving structure consistent with planar photolithographic fabrication methods. The appearance of a single-surface acoustic technology brought a new flexibility to the design of bandpass filters, access to a much higher range of filtering frequencies than previously attainable with crystal filters, and the potential for very low cost in applications requiring large quantities of a fixed design. Today, SAW bandpass filters are used for such varied purposes as TV IF filtering; channel selection in satellite, paging, and mobile telephone systems; timing recovery in optical-fiber digital transmission systems; frequency control of high-frequency oscillators; pulse compression in chirp radar systems; broadband signal processing and matched filtering; and a great many other passive and dynamic filtering applications.

The practical center frequency range of SAW technology runs from about 30 MHz to a few GHz. Below 30 MHz, SAW devices tend to be inconveniently large, while above two GHz or so, where the crystal chip size tends to be inconsequential, the technology is limited by the decreasing (submicrometer) feature size of the transducer structures, and also by the emerging dominance of material losses, which increase as the square of frequency. Achievable ratios of 3 dB passband width to center frequency span a range from one part in ten or twenty thousand in high-Q SAW resonator filters, to values approaching unity in broadband octave-spanning transversal filters. Aging rates can be extremely low. Through procedures long used with bulkwave filters, stability to a few parts per million (ppm) over 25 years has been projected [36] for SAW timing recovery filters manufactured according to strict standards for use in undersea fiber cables. Extreme ruggedness is also available. Artillery deployment of modules with simply-packaged quartz bulkwave filters in a military application [18] also establishes the ruggedness of SAW filters.

Despite the broad capabilities of the technology, acceptance has been rather gradual, in part because the fundamentals seem far removed from the usual background of the electrical engineer. This review is primarily aimed at reaching the larger world of system designers still unfamiliar with the SAW art. Subjects addressed include the basic questions of a new user: What is it? What can it do? What are the limitations? How does it relate to more familiar things? How does it interface with established technology? How big is it? What must be specified to a vendor? How much is it likely to cost? Answers are developed from the relevant physical concepts, structural properties, practical considerations, and representative examples from the literature and commerce.

8.1. PHENOMENOLOGY OF A SIMPLE FILTER

8.2.1. Nature of Surface Acoustic Waves

A surface acoustic wave, or Rayleigh wave [41], is a special propagating mode of material deformation in a solid. The wave is bound to the free surface of the medium (substrate) by a waveguiding effect caused by the abrupt termination of material restraints at the surface. The material motions and energy transport are concentrated within a wavelength of the surface, and tail off exponentially beyond that, as shown schematically in Figure 8.1. The wave amplitudes in Figure 8.1(a) are highly exaggerated for clarity. Surface displacements seldom exceed atomic dimensions. The wave displaces material points along elliptical paths, illustrated for various depths in Figure 8.1(b). In an isotropic material, the motion has a longitudinal component along the propagation direction **p** and a transverse component along the surface normal **n**. The depth dependence of the components along **n** and **p** is shown in Figure 8.1(c). The normal component is the stronger. A tangential third component is common in single-crystal substrates.

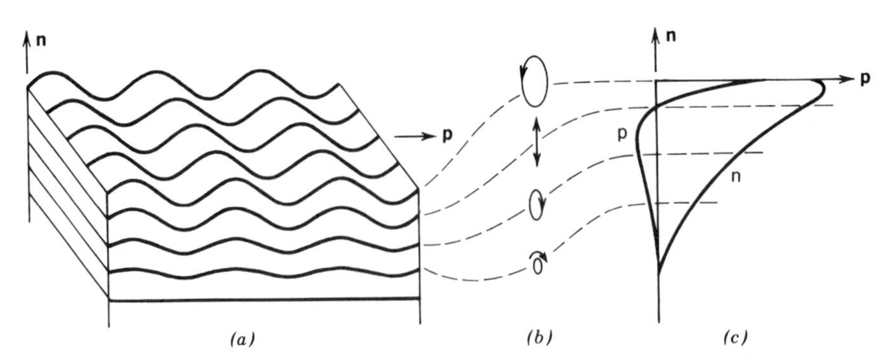

Figure 8.1 Representation of motions in an isotropic surface region excited by a Rayleigh wave: (a) tapering wave amplitude and constant propagation phase along depth coordinate; (b) local mass trajectories; (c) component displacements versus depth.

In a homogeneous substrate, the Rayleigh wave is non-dispersive, so that the phase velocity v_{ph} is independent of frequency. As a result, the acoustic wavelength λ scales with velocity according to a light-like law

$$\lambda = v_{ph}/f \tag{8.1}$$

where f is frequency. The value of v_{ph} varies with the material properties of the substrate, but the variations are ordinarily quite limited. In commonly used crystals and glasses, v_{ph} tends to fall within 20% of 3 km/s. By comparison with electromagnetic waves, then, the velocities and wavelengths of SAWs are five orders of magnitude smaller. At all frequencies of practical interest, however, the wavelengths are still orders of magnitude larger than the interatomic distances in solids, so that the substrate is effectively a continuous medium. The reciprocal of the phase velocity gives a phase delay of roughly 3 μs/cm. Even though SAWs remain within a thin material layer, the wavefronts are effectively planar when diffraction in the surface is ignorable.

8.1.2. Basic Transversal Filter Structure and Operation

A prototypical SAW transversal filter is illustrated in Figure 8.2(a). The wave generating and receiving transducers consist of interleaved metallic-film combs deposited on the flat surface of a piezoelectric crystal. The metal is usually aluminum, which provides a fairly good acoustic match to common substrate materials. As shown in Figure 8.2(b), a voltage applied across the pads of the transmitting transducer appears with alternating sign across successive gaps between fingers. Correspondingly, an alternating pattern of material deformations is piezoelectrically induced in the substrate just under the finger pattern. A sinusoidal applied voltage thus produces a standing wave of material deformation, which decomposes into counter-propagating acoustic waves running orthogonally to the long finger edges. The wave

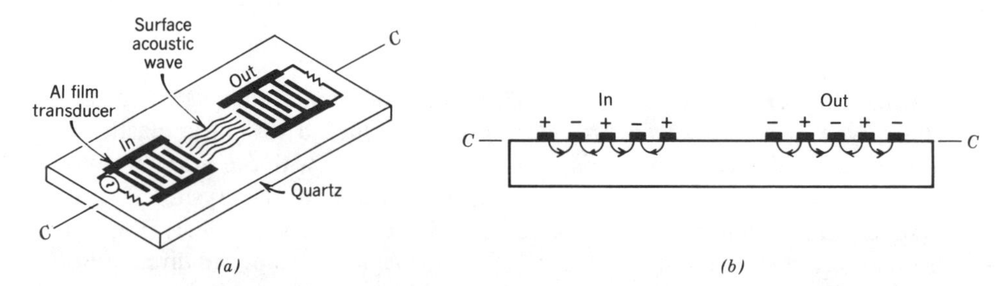

Figure 8.2 (a) Prototype SAW transversal filter, with metal interdigital transducers on a piezoelectric substrate. (b) Alternating substrate deformations induced by alternating potentials on transducer fingers.

passing under the receiving transducer excites a sinusoidal voltage across the receiver pads through the inverse of the generation process. (The electrode patterns by themselves are usually referred to as the transducers, even though the transduction process also requires the piezoelectric coupling provided by the substrate.)

In its dependence on frequency, a transducer functions like an endfire antenna array. The highest acoustic generation efficiency occurs at the frequency where the surface wavelength matches the voltage periodicity of the finger pattern; in that case, continuous deformation waves generated by successive finger periods show maximum coherent reinforcement along the propagation direction. Generation efficiency rolls off as frequency moves away from this *synchronous* frequency f_{syn}, as discussed further in §8.2.1.

In filters utilizing the simple transducers considered here, the generated wave traveling away from the receiver is discarded. The generation process thus introduces a 3 dB loss, and reception produces another 3 dB loss in accord with reciprocity. Various ways to reduce the 6 dB bidirectionality penalty of single-pass transmission have been developed and will be discussed below as appropriate. For many applications, the penalty is simply accepted to avoid the added complexity and cost of loss reduction. Indeed, much higher insertion loss is often accepted in exchange for desired performance characteristics.

The development of a second major type of SAW filter, the resonator filter, was stimulated in part by a desire to overcome the bidirectionality loss of transversal filters. Resonator filters are configured to trap the surface waves moving in both directions inside one or more resonant cavities. This SAW filter class will be described in §8.3. The resonator filters are completely analogous to conventional LC filters across the 3 dB passband and skirts, and can be designed with the same pole-zero methods [44] or other standard procedures [35]. The transversal filters, on the other hand, function without feedback, and are operationally quite different as we shall see below.

8.1.3. Role of Materials

Material properties control the characteristics of SAW filters in a number of crucial ways, reviewed in this section.

8.1.3.1. Importance of Anisotropy. The most common choice of filter substrate type is a piezoelectric single crystal. As such a crystal is elastically anisotropic, *all* SAW characteristics vary substantially with the orientation of the crystal cut relative to the principal symmetry axes of the crystal. The cut is specified in terms of two special vectors, illustrated in Figure 8.1(a): **n**, the normal to the surface on which the SAWs propagate, and **p**, the direction of propagation on that surface. Figure 8.3 illustrates the variation of the inverse phase velocity, or slowness, on a given surface (fixed **n**) as a function of the propagation (wavefront normal) direction **p**. The anisotropy of the

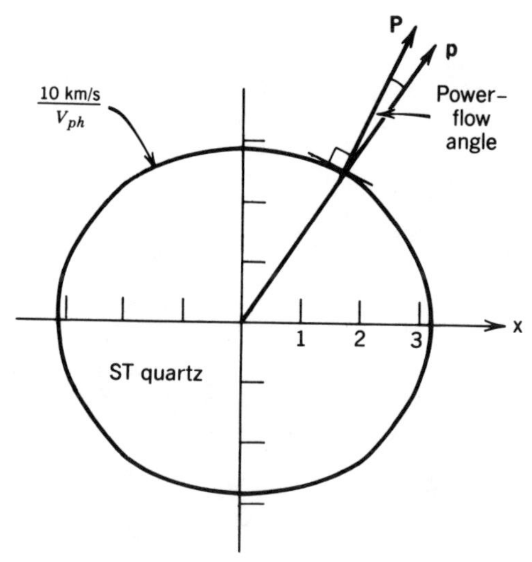

Figure 8.3 Anisotropic slowness polar diagrams for Rayleigh waves on ST quartz. Energy flow and wave propagation are generally not parallel.

medium generally results in noncollinearity between **p** and the power-flow direction **P**, just as in the case of electromagnetic waves. **P** is normal to the slowness curve at each point, as shown. The noncollinearity is one of several reasons why designers prefer to use propagation along a symmetry axis, where **P** and **p** are parallel, despite the associated loss of design flexibility. (It is also worth noting that for some crystal cuts, the material parameters cannot support a wave trapped to the surface.)

8.1.3.2. Electroacoustic Coupling Strength. A second material parameter of fundamental importance is k^2, the coefficient of coupling between electric fields and material deformations. Broadly speaking, a piezoelectric crystal belongs to one of two categories, distinguished by whether the crystal is simply piezoelectric or is also ferroelectric, possessing a permanent electric dipole moment. Ferroelectrics (e.g., lithium niobate) tend to be strong-coupling substrates, while simple piezoelectrics (e.g., quartz) tend to be weak-coupling substrates. Although many other crystalline materials spanning a range of k^2 values have been put to use, quartz and lithium niobate have been the most highly developed for commercial purposes, and consequently tend to have the best purity and homogeneity. Of the two materials, quartz is the more rugged mechanically, while lithium niobate is somewhat susceptible to damage by thermal or mechanical shock. Despite this difference, lithium niobate remains attractive for its strong coupling. Reasonable care is enough to avoid damage.

The dependence of k^2 on propagation direction for commonly used crystal surfaces of the two materials is shown in Figure 8.4. The disparity in

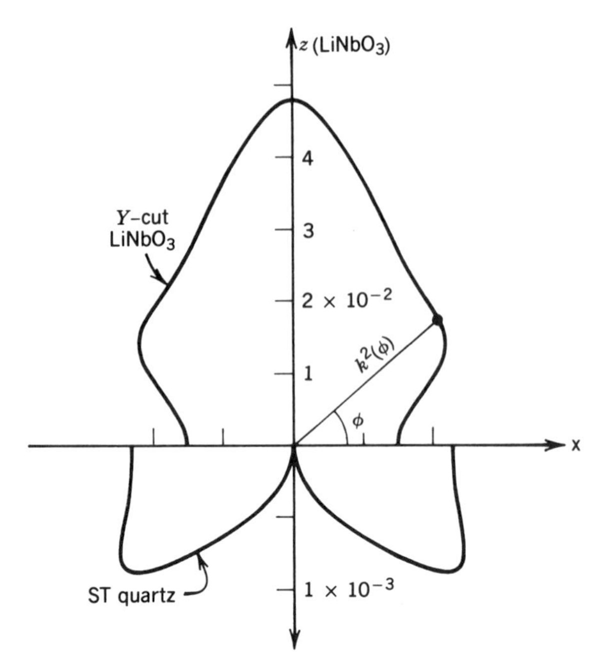

Figure 8.4 Piezoelectric coupling versus propagation angle ψ for Rayleigh waves on special cuts of two favored materials.

the general sizes of k^2 has profound effects on filter design. Many finger periods are needed on quartz to generate SAW amplitudes that require only a few finger periods on lithium niobate. Fewer finger periods result in electroacoustic coupling over a broader bandwidth (§8.2.4), so that lithium niobate is the choice for broadband applications. In fact, its high k^2 allows design options that are simply not practical on quartz.

At this point, we should note that TV IF filters, the largest-volume SAW filters in production, are sometimes made with (nonpiezoelectric) glass substrates. Electroacoustic coupling can be accomplished as shown in Figure 8.5. Zinc oxide microcrystallite films, sputtered onto the transducer regions

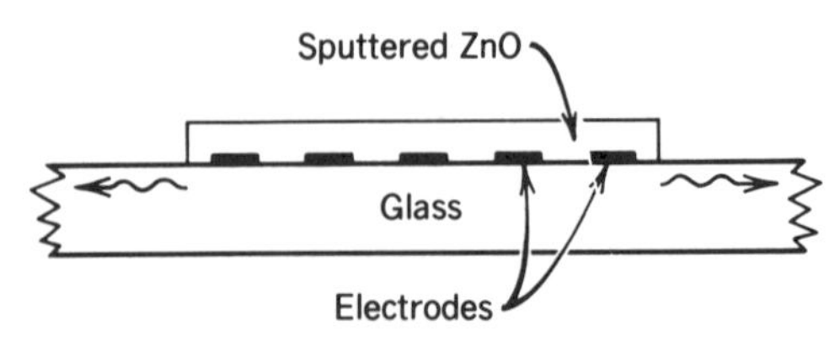

Figure 8.5 Common arrangement for transduction of surface waves in a nonpiezoelectric substrate.

so that the crystallite c axes are oriented normal to the surface, provide a high value of k^2 in a structure whose unit cost is very low.

8.1.3.3. Thermal Characteristics. For most applications, filters must have stable characteristics over a considerable range of operating temperatures T, such as 0–30°C in undersea fiber transmission systems, or 0–65°C in sheltered terrestrial environments. An appreciation of SAW thermal characteristics is therefore essential.

The principal thermal effect is that the entire passband of a transducer or filter can be frequency-shifted by a change in temperature. The shift is often described as a fractional thermal shift of the synchronous frequency f_{syn}. SAW devices may exhibit one of three types of temperature variation of f_{syn}. A linear variation is the most common at normal operating temperatures. Typical linear shifts in lithium niobate are generally 70–90 ppm/°C [51]. In quartz and some other crystalline materials, certain combinations of surface normal **n** and propagation direction **p** have been discovered for which the linear component of thermal shift vanishes at some temperature. Thermal shifts then become parabolic or even cubic [7] about that temperature, as shown in Figure 8.6(a). The rotated-Y cuts of quartz, with wave propagation along the X axis, provide a family of parabolas with a range of turnover temperatures, T_t, as illustrated in Figure 8.6(b) [15]. The best-known member of the family is the ST-X cut [49], with T_t at room temperature and a downward bandshift amounting to $-32(T - T_t)^2$ ppb (parts per billion) at T°C. An appropriate choice of surface normal can place T_t at any temperature of ordinary interest. Although a given cut tends to have a shifted T_t when transducer metallization is added [38], the family of cuts includes another member that can restore T_t to its value before metallization.

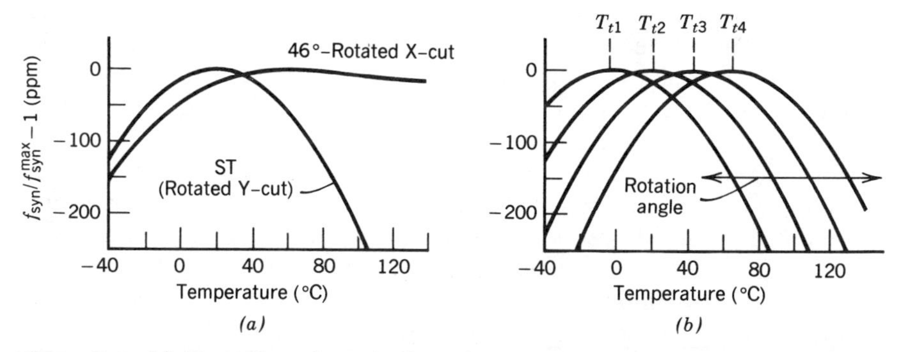

Figure 8.6 (a) Examples of parabolic and cubic temperature dependence of synchronous frequency on special cuts of quartz. (Adapted by permission of Electronic Industries Association.) (b) Turnover temperature dependence on surface-normal orientation in the YZ plane of quartz.

The cubic type of frequency-temperature characteristic can be fairly flat over some finite temperature range. The occurrence of a cubic characteristic implies the vanishing or near-vanishing of both the linear and quadratic components of thermal variation, and is consequently much rarer than a parabolic behavior. As indicated in Figure 8.6(a), quartz allows cubic behavior at somewhat elevated temperatures, but only on doubly-rotated cuts, which have a number of drawbacks. For most practical purposes, parabolic cuts are the choice.

Before leaving this subject, we should mention two useful circuit approaches to the stabilization of narrow-band filter circuits against temperature changes. In both cases, the filter operates in series with an electronic circuit whose phase-temperature characteristic is also (roughly) parabolic, but with curvature of the opposite sense. As a result, the transmission phase of the series combination shows a much smaller variation with temperature than either the filter or the electronics alone. In the first approach, the circuit is a varactor diode controlled by a temperature-sensing transistor [24, 30]. In the second approach, the circuit is the normal electronic part of the timing loop of a digital regenerator in a fiber cable [1, 48]; the degree of compensation is controlled by an appropriate choice of filter turnover temperature. These approaches show that valuable improvements in performance are available at modest or no cost. Ovening will be mandatory only for applications with stringent stability requirements.

8.2. CHARACTERISTICS OF TRANSDUCERS AND TRANSVERSAL FILTERS

8.2.1. First-Order Operational Features

8.2.1.1. General Features. Some fundamental differences between transversal filters and conventional resonator filters can be identified with the help of Figure 8.2. The single-pass nature of the acoustic-wave transmission indicated in the figure accounts for several important characteristics of transversal filters. They have a finite impulse response (FIR), which vanishes after a wave train spanning the input transducer (generated there by a voltage impulse) has propagated past the output transducer. Energy storage exists only during the transit time of the wave train. Internal oscillations in the conventional resonator sense are absent. The transmission phase has a major contribution from simple acoustic propagation along the path between the effective input and output ports, which are commonly located about midway along the axial lengths of the transducer patterns. Since the propagation path often spans a great many wavelengths, the filter transmission phase tends to be linear with frequency, and the associated transmission delay more or less independent of frequency. Phase-frequency linearity may

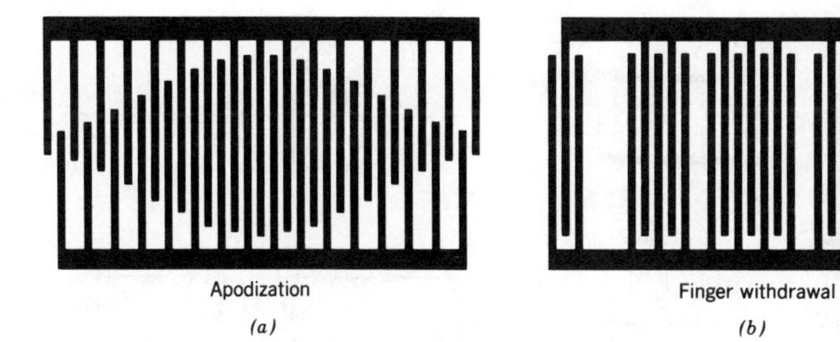

Figure 8.7 Methods of weighting the array impulse response of a transducer: (a) amplitude weighting, and (b) phase weighting.

be modified significantly, however, through a position (phase) dependence of amplitude weighting in asymmetrically apodized transducers [17]. (Apodization weighting is illustrated in Figure 8.7(a).) In addition, the single-pass character of transmission isolates the transducers electrically, so that each can be terminated without reference to the other, in contrast to the interrelated input and output terminations of LC resonator filters.

The widely diverse responses achievable with SAW transversal filters are all based on modifications of the unweighted periodic finger structures shown in Figure 8.2. To see how the more elaborate designs function, we begin by examining the first-order behavior of the prototype.

8.2.1.2. Time Behavior. The first-order impulse response of a simple transducer can be grasped almost literally at a glance. If we ignore circuit effects by imagining that a unit voltage impulse applied across the finger pairs translates instantaneously into an acoustic excitation pattern, then the acoustic wave train traveling in either direction will be an *rf* pulse with an envelope that replicates the finger overlap pattern. This circuit-independent response is sometimes called the *array response*, since it depends only on the finger array. The simplest method of weighting the impulse response is to apodize the overlap pattern as shown in Figure 8.7(a). Another common weighting method is finger withdrawal [20], illustrated in Figure 8.7(b).

For the impulse response of a complete filter, we must also consider what happens at the receiving transducer. There the arriving acoustic wave train slides across the finger pairs and generates currents that are summed at the bus bars. If we again ignore circuit effects, the first-order response is as shown in Figure 8.8. The output waveform is $N_{Tr1} + N_{Tr2}$ cycles in length, and has a trapezoidal envelope in the case of uniform finger overlaps (Figure 8.8(a)). If one of the transducers is apodized, the output envelope has the shape of the convolution of the apodization function with a rectangular window, as depicted in Figure 8.8(b). The envelope becomes much more

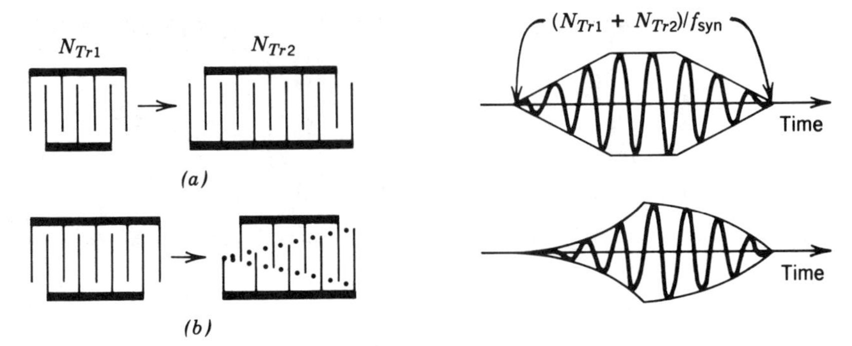

Figure 8.8 Dependence of filter transmission impulse response on weighting: (a) both transducers unweighted, and (b) one transducer apodized.

complicated if one attempts to use a filter with two apodized transducers. As indicated by the dotted lines in Figure 8.9, when a generated pulse with a non-uniform envelope slides past a receiver with a non-uniform finger-overlap pattern, the receiver periods that span the full aperture will detect all the incoming signals, but periods with shorter finger overlaps will miss the part of the wave pattern passing beyond the overlaps. As a result, the laterally weighted incoming wave is re-weighted at each receiver finger pair. Under these conditions, design of the individual transducers to provide a prescribed filter transfer function leads to computational costs that are rarely justified.

It should be mentioned that this difficulty is curable in hardware with the help of a multistrip coupler [34], an ingenious and versatile SAW component to be illustrated later. Use of a multistrip coupler is practical only on high-coupling substrate materials, and requires a significantly larger crystal area than would otherwise be needed. More commonly, the problems of double apodization are avoided by confining apodization weighting to one transducer of a pair. The other transducer is either unweighted or else weighted by a non-apodizing technique, such as finger withdrawal, or phase weighting either by repositioning fingers or finger segments [22] or by reversing the polarity of some bus-bar connections. Regardless of how the transducers are weighted, the filter time response is always further compli-

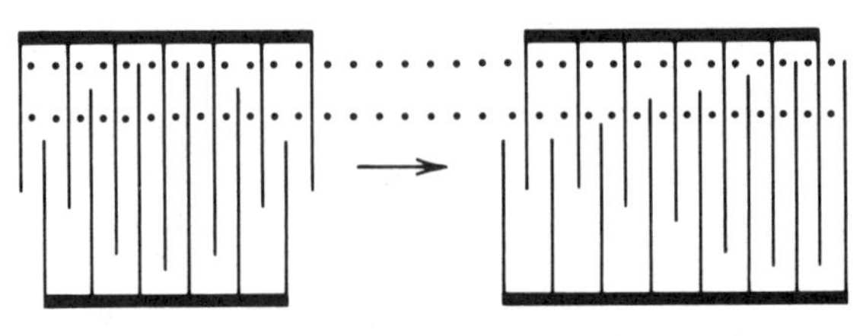

Figure 8.9 Illustration of design problem when both transducers are apodized.

cated by the transducer equivalent circuits and the electrical terminations, which alter the simple array phenomena just described.

The time response of SAW filters is used primarily in broadband signal processors such as code-matching filters and pulse compressors. Matched filtering, illustrated in simple form in Figure 8.10, uses N_{Tr2} isolated taps in the receiving transducer. Each tap is separated from the next by a chip length $N_{Tr1}\lambda_{\mathrm{syn}}$ (λ_{syn} = wavelength at the synchronous frequency) of the transmitting array, that is, the length of the transmitter impulse response. A code pattern such as that shown in Figure 8.10(b) is generated by a sequence of N_{Tr2} voltage impulses applied to the transmitter at intervals of N_{Tr1} synchronous periods. In the simplest case, the coding is embodied in the signs of the voltage impulses. To select a particular code sequence out of a large number of possibilities, one connects the receiving pairs to match the signs in the selected sequence. With each receiver tap sampling a distinct chip in the pattern, the sum of tap outputs will be large only when there is a match to the whole sequence; coding constraints are used to avoid large outputs from mismatched sequences. A characteristic waveform is illustrated in Figure 8.10(c). Although active devices are beyond the scope of this chapter, it should be noted that with the help of electronics and ingenuity, device flexibility can be greatly enhanced through switchable taps and other arrangements [29, 39]. Sequences of more than 1,000 chips have been implemented in reasonably compact and power-efficient form [50].

While pulse compressors also merit description, they involve both time and frequency filter responses, and are more easily described after the next section.

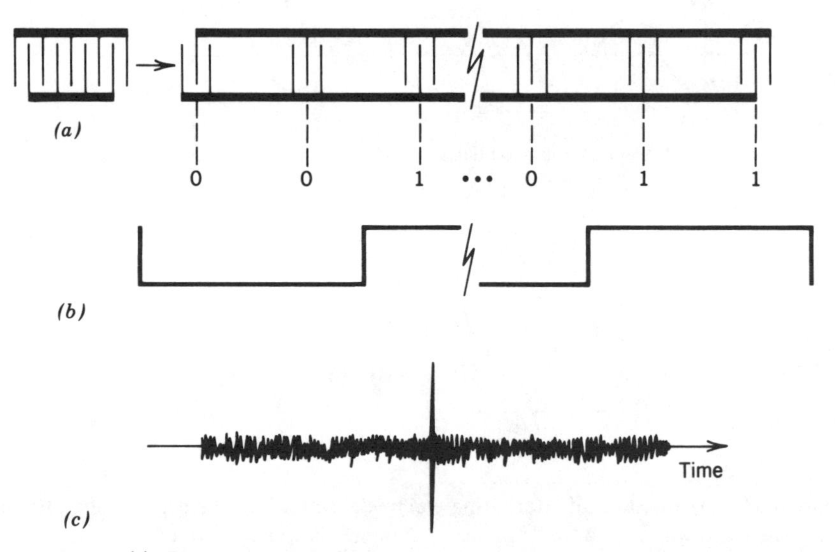

Figure 8.10 (a) Prototype transducer pair for code-matched filtering. (b) Code matched by receiver tap phases. (c) Typical filter response to a matched code word.

8.2.1.3. Frequency Behavior. The nature of the filter frequency response can be understood by first considering a continuous acoustic wave propagating past a periodic receiving transducer with uniform finger overlaps. At the synchronous frequency, the piezoelectrically generated voltage pattern will be the same at all finger periods; output currents from all the periods will thus add in phase at the bus bars to produce the largest possible output current through the electrical termination. Figure 8.11(a) illustrates what happens at a different frequency, not synchronous with the pattern period. The acoustic stresses across different receiver periods vary in both magnitude and sign, in accord with the mismatch between acoustic wavelength and finger period. Since the two combs of the transducer are effectively equipotential surfaces, the potential difference generated piezoelectrically is clamped to the instantaneous average over all periods, and the net output current through the termination will consequently be smaller than the value at synchronism. The response vanishes altogether in a case like that illustrated, where the average potential difference over all periods vanishes. Output nulls will occur at any frequency where the receiver spans a number of wavelengths differing by an integer from N_{Tr}, the number of receiving periods. A local output maximum will of course occur between each pair of nulls. For the unweighted periodic prototype receiving transducer, the complete transducer array frequency response is the Fourier transform of

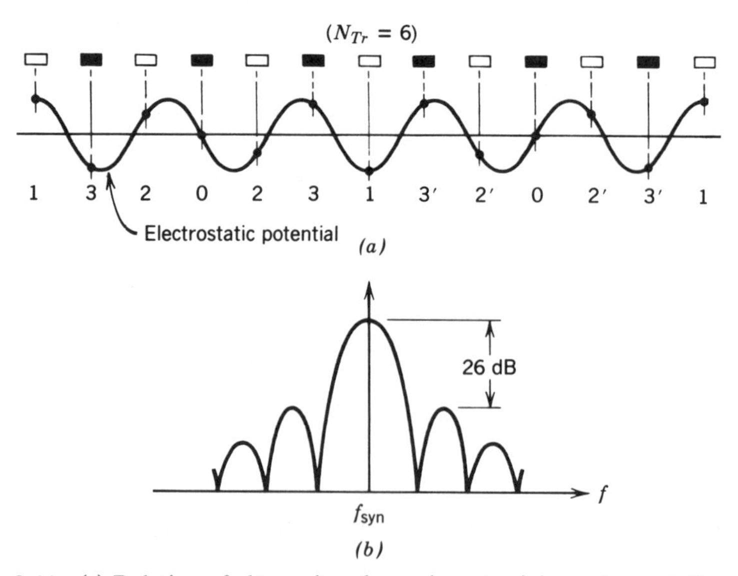

Figure 8.11 (a) Relation of alternating electrode potentials to electrostatic potential of a Rayleigh wave at a null in the frequency response (b) of an unapodized transducer. The numerals identify pairs of antiphase points on the wave associated with in-phase finger connections.

the rectangular-array impulse response. The array transfer function is thus [4]

$$F(x) = \sin x / x \tag{8.2}$$

where

$$x = \pi N_{Tr} \frac{(f - f_{\text{syn}})}{f_{\text{syn}}} = \pi \left(\frac{L_{Tr}}{\lambda} - \frac{L_{Tr}}{\lambda_{\text{syn}}} \right) \tag{8.3}$$

and L_{Tr} is the axial length of the N_{Tr} receiver periods. By reciprocity, the transmitting transducer will have a similar transfer function. If the two transducers are identical, the array transfer function of the *filter* will have magnitude $(\sin x/x)^2$, and the power transmission will go as the square of this expression, as illustrated in Figure 8.11(b). The first sideband maxima are 26 dB below the passband peak.

SAW transversal filtering derives much of its great versatility from modifications of the transducer finger patterns to narrow, broaden, or otherwise shape the central peak and to suppress the sideband levels to meet specific requirements. As in the case of time response, the transfer functions of the individual arrays and the filter may be distorted by circuit effects and by a number of higher-order effects, all of which can be partly or wholly compensated through transducer pattern modifications. The qualitative impact of overlap apodization on frequency response follows directly from the above description of the reception process. Since the relative tap weights of an apodized transducer are different, the output currents from the various taps make proportionate rather than equal contributions to the total output current. The result must be a shift in the null frequencies and related changes in bandwidth and sideband structure relative to the unapodized case. Similar statements can be made about transducers weighted by finger omission or by finger phase modifications. Weighting techniques allow an enormous variation in filter transfer characteristics, as a few examples will illustrate.

Figure 8.12(a) shows the passband response of a TV IF filter [14] with a 50 dB attenuation requirement for audio traps near 41 and 47 MHz. Figure 8.12(b) shows the mask pattern designed to meet these requirements with particular choices of the lithium niobate substrate orientation and electrical terminations. One transducer is finger-overlap weighted, with detached finger ends retained to control SAW diffraction [58]. The multistrip coupler is a periodic array of vertical metal strips with a period substantially smaller than the transducer finger and gap width. In this application, the coupler functions as a piezoelectric track changer. Pairs of strips sense differences of electrostatic potential along the incident Rayleigh wave, and regenerate the wave in a parallel acoustic channel with high efficiency. The arrangement shown is effective in suppressing spurious bulk-wave transmission, since bulk waves are not coupled across by the surface strips.

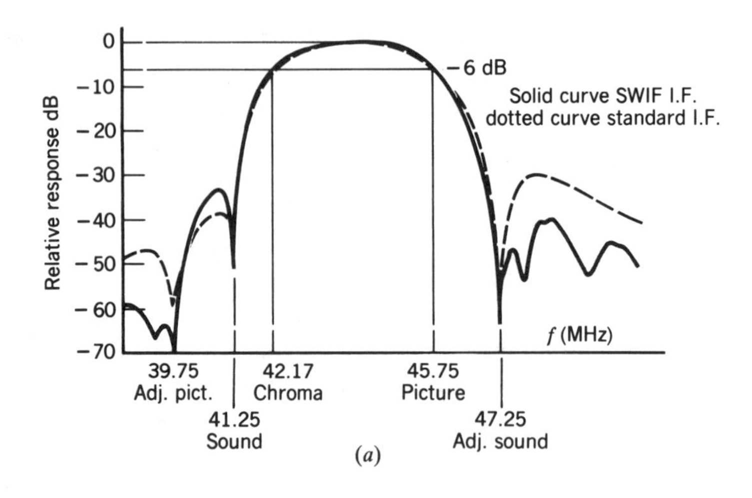

(*a*)

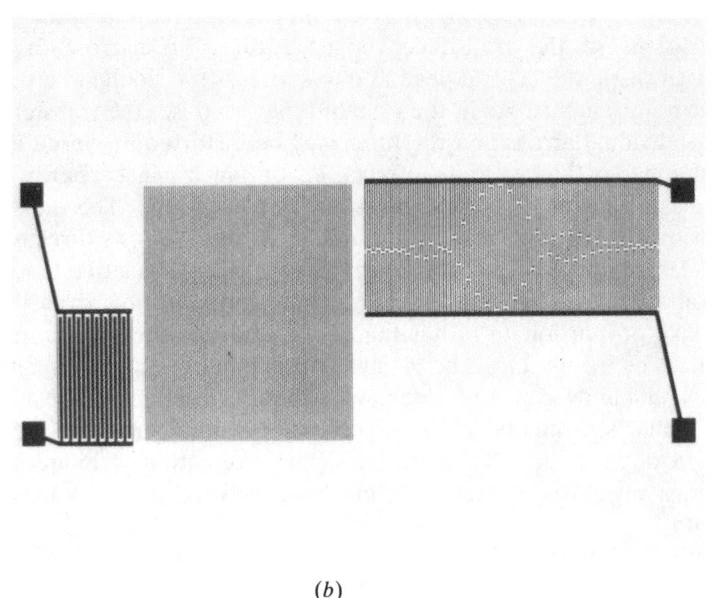

(*b*)

Figure 8.12 (a) TV IF filter response (solid curve) produced by the structure shown in (b). From DeVries and Adler (1976) Case history of a surface-wave TV IF filter for color television receivers. *Proc. I.E.E.E.* **64**, 671–676 © 1976 I.E.E.E. (photo courtesy of Zenith Electronics Corp.).

The discussion to this point has focused on filters based on spatially periodic or nearly periodic transducer finger placements. Achievement of flat transmission over the broadest bandwidths, however, requires dispersive transducer designs, in which the finger period changes monotonically from one end of the transducer to the other. Such transducers are illustrated in Figure 8.13(a) [59], which shows a pair of dispersive transducers forming a

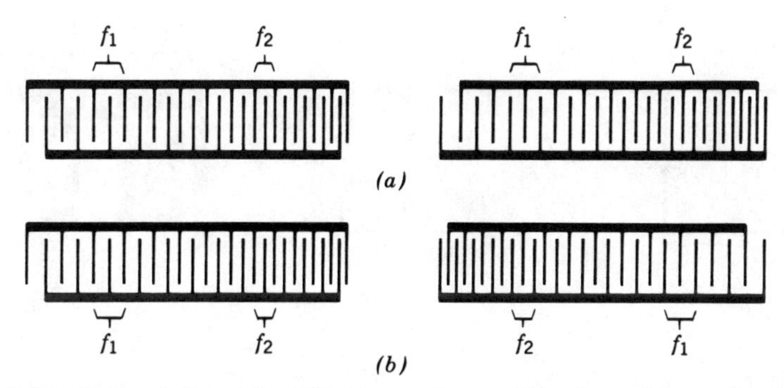

Figure 8.13 Pairs of dispersive transducers arranged to form (a) nondispersive and (b) dispersive filters. f_1 and f_2 identify finger groups synchronous with unique frequencies.

nondispersive filter. This arrangement is equivalent to a cascade of periodic filters, all with the same effective propagation path length. Reversal of one of the transducers of Figure 8.13(a) yields the configuration of Figure 8.13(b), which can be used for time compression of an up-chirped radar return signal. The time compression is made possible by the frequency dependence of the propagation delay between transducer segments of like synchronous frequency. A variant of this concept, the reflective array compressor (RAC), has found more favor, and will be described at an appropriate point.

8.2.2. Transducer Equivalent Circuit

As noted earlier, the array responses of the two transducers do not provide a complete description of filter behavior; the circuit characteristics of each transducer with its electrical termination can modify its array response in important ways. In this section, we briefly examine a transducer equivalent circuit developed by Smith et al. [52] to account for both the capacitance of the finger pattern and the equivalent electrical effects of electroacoustic conversion. This simple crossed-field model, represented by the transducer part of Figure 8.14(a), is the key to understanding a number of fundamental transducer electrical characteristics, as we shall see in this and later sections.

The input admittance Y_{Tr} consists of three shunt elements described by the following relations [52] and Figure 8.14(b)

$$C_{Tr} = c_{Tr} N_{Tr} W_{Tr} \tag{8.4}$$

$$G_e(f) = G_e(f_{\text{syn}})(\sin x/x)^2 \tag{8.5}$$

$$B_e(f) = G_e(f_{\text{syn}})[\sin(2x) - 2x]/(2x^2) \tag{8.6}$$

$$G_e(f_{\text{syn}}) = 8k^2 f_{\text{syn}} C_{Tr} N_{Tr} \tag{8.7}$$

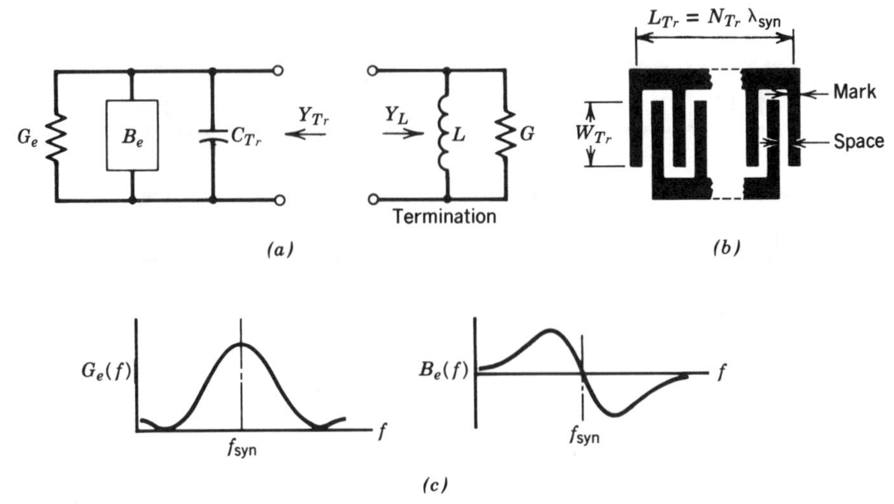

Figure 8.14 (a) Crossed-field equivalent circuit model of a transducer, shown with a conjugate matchable termination. (b) Transducer parameter definitions. (c) Frequency dependence of transducer conductance and susceptance near synchronism.

Equation (8.4) expresses the static capacitance C_{Tr} of the transducer pattern in terms of c_{Tr}, the capacitance/finger period/unit overlap length. The value of c_{Tr} is independent of scale, but depends on the *relative* widths of fingers and gaps (the mark/space ratio) and on the dielectric constants of the substrate. The finger overlap length W_{Tr} shows the explicit dependence of C_{Tr} on the radiating aperture. The dynamic conductance $G_e(f)$ in Eq. (8.5) scales the electrical equivalent of the power radiated bidirectionally into acoustic waves at the transmitter, where x is the relative frequency variable defined in Eq. (8.3). The dynamic susceptance $B_e(f)$ in Eq. (8.6) measures the quadrature component of the circuit effect of electro-acoustic generation. The dependence of G_e and B_e on frequency in the vicinity of the major response peak is sketched in Figure 8.14(c). Both dynamic elements are scaled by the value of $G_e(f_{syn})$ (Eq. (8.7)), which is linear in C_{Tr}, and hence linear in the aperture W_{Tr}. W_{Tr} can be used in some designs to match $G_e(f_{syn})$ to a desired termination conductance G (e.g. $(50\,\Omega)^{-1}$), for optimum power delivery and minimal VSWR. (For that purpose, it would also be necessary to tune out the static capacitance with an inductor as shown. More general termination considerations are discussed in the next section.) For a fixed pattern length, weighted patterns require larger apertures W_{Tr} than unweighted patterns to obtain the same sum of finger overlap lengths and hence the same static capacitance and radiated acoustic power at band center. For a fixed aperture, weighted transducers would have to be longer than unweighted transducers to obtain the same radiated power at f_{syn}. These comments relate to a discussion of filter size in §8.2.4.

8.2.3. Filter Termination Considerations

8.2.3.1. Triple-Transit Ripple. As for all other passive filter types, the passband shape of a SAW filter is sensitive to the terminations. Unlike other filter types, however, transversal SAW filters have a unique source of sensitivity related to the three-port character of a bidirectional transducer. The scattering consequences are illustrated in Figure 8.15. When an acoustic wave a_1 comes into a receiving transducer, the outgoing waves include the transmitted electrical signal, $b_3 = S_{31}(f)a_1$, a transmitted acoustic wave, $b_2 = S_{21}(f)a_1$, and a reflected acoustic wave, $b_1 = S_{11}(f)a_1$. This scattering description applies with an arbitrary termination. It is characteristic of the scattering [47] that the magnitude $|S_{11}|$ of acoustic reflection increases when the acousto-electric transmission $|S_{31}|$ is improved through impedance matching. The result is that a filter with relatively low loss typically exhibits significant acoustic-wave reflections from bidirectional transducers. The most important reflections produce the triple-transit signal illustrated in Figure 8.15(b). This signal contributes to filter output after a delay of one round trip between effective acoustic ports separated by a distance D, and thereby distorts the array time response and adds a ripple to the frequency response. The triple-transit signal is weaker than the desired first-transit signal by a factor of $|S_{11}|^2$. Figure 8.16 illustrates passband amplitude and phase ripple effects in a very narrow-band ($Q = 800$) prototype transversal filter for selected values of D differing by intervals of $\lambda_{syn}/8$. The greatest amplitude distortions typically occur in the uppermost part of the passband, where $S_{31}(f)$, and $S_{11}(f)$ have their largest magnitudes. Although $|S_{11}|$ and $|S_{31}|$ are not uniquely related, their ratio is bounded. Instructive limits on the ratio may be obtained, for example, from a scattering analysis of a

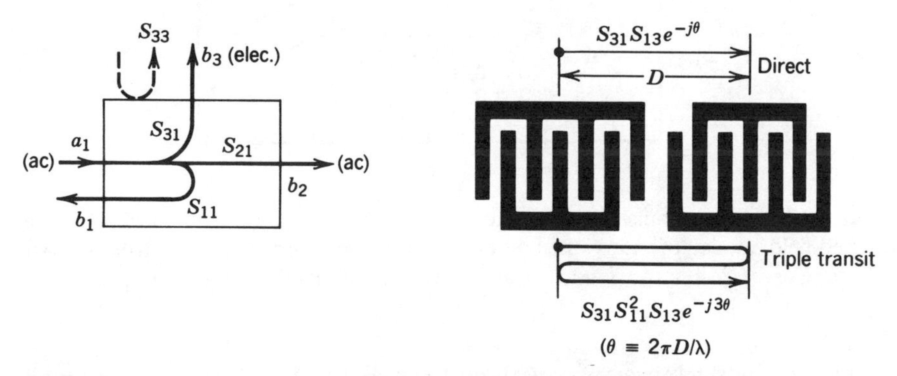

Figure 8.15 (a) Scattering representation of a bidirectional transducer irradiated by a Rayleigh wave at one acoustic port. Waves are scattered to the electrical port and both acoustic ports. Also illustrated is the reflection coefficient S_{33} at the electrical port. (b) Scattering description of a directly transmitted signal and the triple-transit (twice reflected) signal.

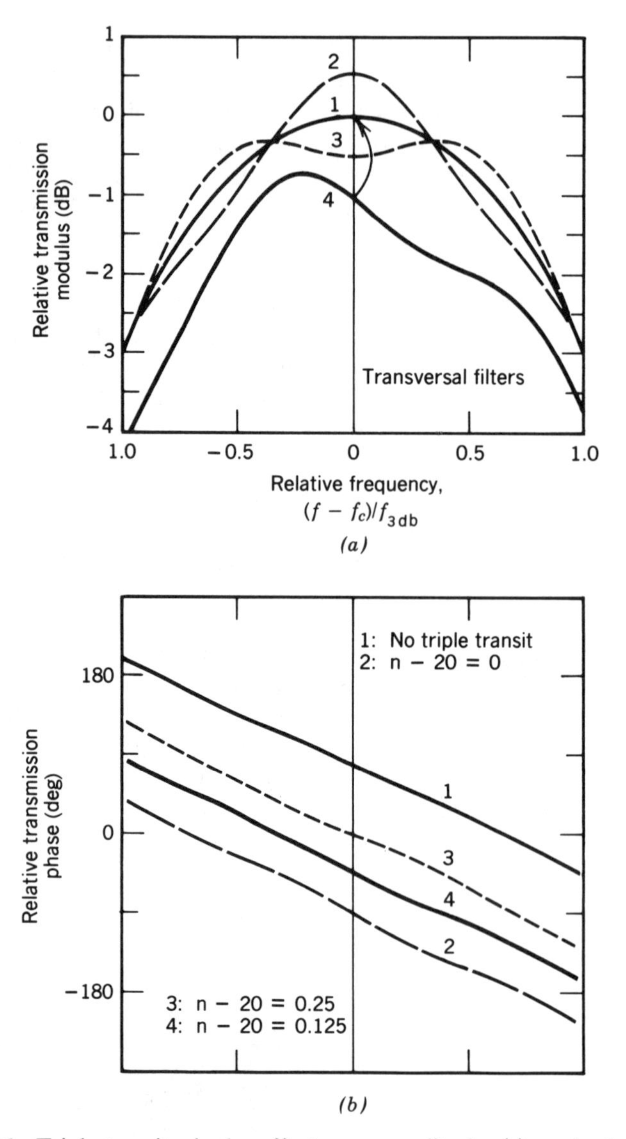

Figure 8.16 Triple-transit ripple effects on amplitude (a) and phase (b) in a crossed-field model of transversal filter transmission. Curves 1 omit triple-transit effects. Curves 2–4 correspond to transducer separations differing by $\lambda_{syn}/8$.

simple symmetrical three-port with no internal losses [47]. Except in a small region near $|S_{31}|^2 = 0.5$ (perfect impedance matching at port 3), the relative attenuation $|S_{11}|^2$ of the triple-transit signal falls off as $|S_{31}|^{2n}$, where $1 < n < 2$. This behavior is the basis of a simple strategy often used to control filter ripple: termination mismatch is increased to reduce $|S_{31}|^2$ to a point where the ripple magnitude is tolerable. The associated increase of insertion loss is compensated by electronic amplification as needed.

Other ripple control strategies have used more complex structures to either prevent or utilize bidirectionality. In such cases, an associated benefit is a dramatic reduction of the bidirectionality loss. Unidirectional transducer (UDT) designs ideally decouple one of the acoustic ports of a transducer from the other two ports, and use only the coupled ports in filter operation. The most straightforward UDT design is the three-phase transducer shown in Figure 8.17 [43]. Each transducer period has three fingers spaced at intervals of $\lambda_{syn}/3$. The three fingers are attached to three bus bars, and the voltages across the three pairs of bus bars are phased at intervals of 120°. The net effect at the synchronous frequency is in-phase coupling to the wave running in one direction, and completely destructive interference in coupling to the reverse wave. Two such transducers form a filter that transmits electrically in one direction only. As the operation depends on phasing, the effective band of unidirectionality tends to be rather narrow. Nevertheless, the passband ripple is well suppressed in many cases, and insertion loss at 100 MHz may be reduced to 1 dB [66] at the same time. (Residual insertion loss remains from sources such as ohmic and diffractive losses and acoustic radiation in unintended directions [16]). The use of three bus bars in the design of Figure 8.17 adds substantial manufacturing complexity, since finger contacts to one of the bus bars must cross over another bus bar, usually with an air gap to optimize electrical isolation. Although cross-over technology is improving gradually, other UDT schemes utilizing two phases driven in quadrature are also in use [66], and single-phase UDTs have recently been demonstrated [65] that utilize special acoustic reflection characteristics. Compared with bidirectional transducers, UDTs involve greater complexity in design and/or fabrication and/or matching, and show higher sensitivity to variations in processing or in the values of matching components and parasitics.

Another kind of transducer scattering modification is illustrated in Figure 8.18. The simplest version [31] of interleaved segmented transducers, where the output transducer has been divided into two segments placed at opposite ends of the input transducer, is shown in Figure 8.18(a). Since both generated acoustic waves are detected, there is no 3 dB acoustic loss associated with generation, and the minimum (single transit) array bidirec-

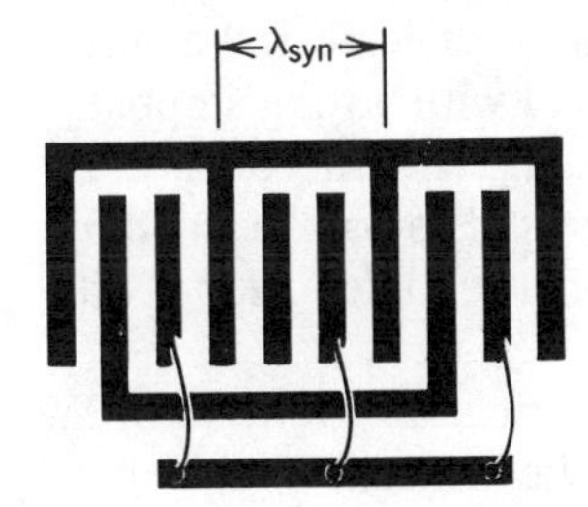

Figure 8.17 Electrode arrangement for a three-phase unidirectional transducer.

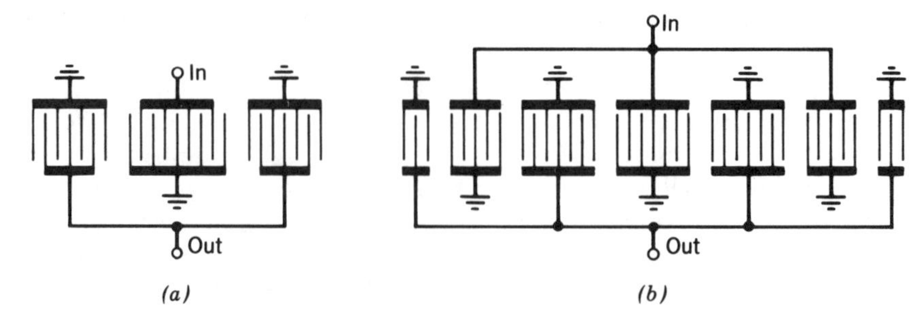

Figure 8.18 (a) Symmetrically divided output transducer to eliminate half of bidirectionality loss. (b) Arrangement for further reduction of bidirectionality loss.

tionality loss falls to 3 dB. The arrangement is in principle capable of suppressing the triple-transit ripple when the center transducer is perfectly matched, but ripple suppression can easily be spoiled by structural imperfections. Further insertion-loss reduction is available through further segmentation, illustrated in Figure 8.18(b). Apart from added design complexities, such designs show increased susceptibility to direct electromagnetic radiative feedthrough and ground loops in the package. Fairly recent work [23, 33] indicates that despite the difficulties, progress is being made with this approach.

Another scattering modification is achieved by trapping the acoustic energy in SAW resonators [2, 56]. As we shall see in §8.3, this method can be very effective in reducing insertion loss and eliminating ripple. Again, sensitivity to the values of matching components tends to be greater than in the case of simple transversal filters.

8.2.3.2. Filter Reflections. Before considering the special features of electrical reflection from SAW transversal filters, we should note that the basic circuit problems encountered are shared by all passive filter technologies. An ideal passive two-port filter would be perfectly transmitting in the passband and perfectly reflecting in the stopbands, with no internal losses in reactive designs. In real filters of any kind, this behavior is modified somewhat by internal losses and by a continuous variation of the transmission/reflection trade-off across the passband. Even with the usual internal losses, most of the incident energy in the stopband is reflected rather than absorbed. Circuits must therefore be designed to deal with strong stopband reflections and possibly significant passband reflections. If reflections must be held below a given level, resistive pads are often used as part of the filter design, even though filter transmission is thereby attenuated along with reflection.

In scattering terms, the coefficient of signal reflection from a SAW transversal filter with bidirectional transducers can be written

$$R(f) = S_{33}^t + S_{31}^t e^{-j\beta D} S_{11}^r e^{-j\beta D} S_{13}^t + \cdots \tag{8.8}$$

where superscripts t and r refer to the transmitting and receiving transducers, respectively. (The order of scattering symbols and indices follows the conventional right to left progression.) On the right-hand side of Eq. (8.8), S_{33}^t is the electrical reflection expected at the input port of any filter. The second term is a delayed double transit return signal generated by reflection of the first-transit acoustic wave at the receiving transducer. The acoustic propagation phase is represented by βD, where D is the effective transducer spacing and $\beta = 2\pi f/v_{ph}$. Higher-order reflections, represented by the ellipsis, are rarely important. Equation (8.8) exhibits a qualitative difference from conventional minimum-phase two-port filters. If the electrical input port is perfectly matched, the first term on the right will vanish, but the second term will not. In fact, if both electrical ports are matched, it can be shown [53] that the second term yields a reflection 12 dB below the input signal when internal losses are negligible. (Internal losses may be 2 dB or more in either transmission or reflection under electrically matched conditions.)

As electrical mismatch is increased, the second term in Eq. (8.8) decreases, but the first term becomes prominent very rapidly. The undelayed (direct) electrical reflection at the filter input port at center frequency is plotted as a function of filter insertion loss for three different filter types in Figure 8.19. In all cases, internal losses are assumed to vanish, so that the plotted curves represent upper bounds. The curve for conventional filters applies to any technology in which the filter electrical ports provide the only

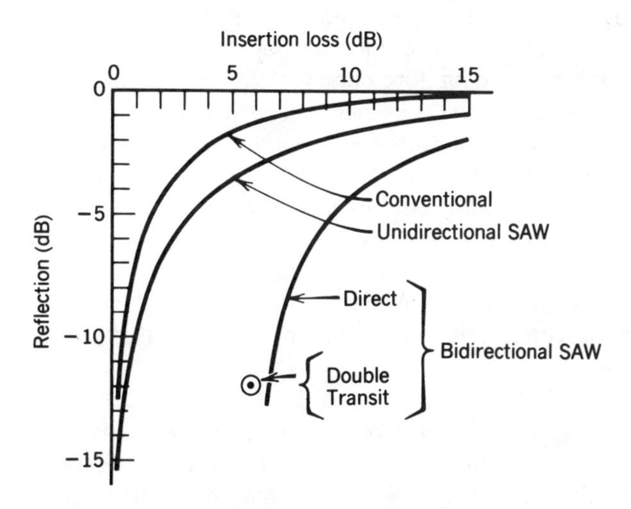

Figure 8.19 Direct electrical reflection versus insertion loss for conventional IIR filters and two types of SAW FIR filters (internal losses ignored).

channel for power flow. The point for SAW filters with bidirectional transducers represents the upper bound on double-transit reflections. The double-transit term falls rapidly with termination mismatch. The SAW curves show weaker reflection than conventional filters for a given insertion loss because the direct reflection $|S_{33}|$ occurs at just one transducer, which only accounts for half the dB insertion loss. The figure shows that a SAW filter with bidirectional transducers has by far the weakest direct electrical reflection at line center when insertion loss exceeds 6 dB. It is also clear that a low-loss SAW filter with unidirectional transducers can out-perform conventional filters with respect to reflections. (There is effectively no double-transit reflection from a unidirectional receiver at f_{syn}.) The inclusion of internal losses would tend to increase the differences between the curves in Figure 8.19.

8.2.3.3. Electrical Termination Compromises. At this point, it should be abundantly clear that the electrical terminations of a SAW transversal filter control some of the most important operating characteristics. The choice of terminations will usually be a compromise between insertion loss and triple-transit ripple requirements, and to some extent, reflections (VSWR, return loss). Another factor that may have to be considered, particularly in broader-band applications, is the modification of the array passband width by circuit elements in the termination. An excellent quantitative account of trade-offs between insertion loss, triple-transit suppression, operating bandwidth, and VSWR is given by Smith [54: 163–176], where the simplest and most often used termination circuits are considered. Cases evaluated explicitly are (a) a purely resistive termination, (b) shunt and series connected resistor-inductor combinations, where the inductor is used to tune out the dc capacitance C_{Tr} of the transducer, and (c) terminations as stated in (b) with a pad resistor added to assure adequate operating bandwidth or sufficient triple-transit suppression or low enough VSWR. As noted by Smith [54], other terminations are sometimes needed, for example, an "ell" section or a more complex circuit for broadband matching [42]. In addition, terminations for unidirectional transducers need special consideration [40]. Generally speaking, required matching components and values should be stated by the filter vendor, after agreement has been reached with the user on desired performance and the parameters of the adjacent circuits. Some of this complexity is avoided by a developing trend toward integration of SAW filters, matching components, and sometimes integrated circuits in a single package.

8.2.3.4. Insertion Loss/Bandwidth/Coupling Relationship. Before leaving the subject of terminations, we note a remarkably simple rule for the maximum bandwidth achievable with minimum insertion loss on a given substrate. For filters with like transducers at input and output ends, the largest fractional bandwidth at minimum loss is roughly $(4k^2/\pi)^{1/2}$ [21, 54].

The limit depends only on the electro-acoustic coupling coefficient k^2, and applies to both periodic and dispersive finger placements. For quartz and lithium niobate, the limits are about 4% and 25%, respectively. The limit derives from the relationship between the electrical bandwidth of a transducer ($(4k^2/\pi)N_{Tr}f_{syn}$ from Eqs. (8.4)–(8.7)) and its array bandwidth ($\sim N_{Tr}^{-1} f_{syn}$). When the array bandwidth is the narrower, the transducer can be impedance-matched across the array bandwidth without adding loss between the transducer and the load it serves. When the electrical bandwidth is the narrower, the array bandwidth can be recovered only by broadening the electrical bandwidth, for example, through resistive padding between the transducer and the load. The padding costs an additional 12 dB of filter insertion loss for each octave increase in the electrical bandwidth [21]. The cross-over between asymptotes leads to the minimum-loss filter bandwidth limit noted above.

8.2.4. Transversal Filter Size

The layout of a representative SAW transversal filter package is shown in Figure 8.20. Major regions include the finger patterns, the bus bars and contact pads, peripheral substrate regions of width a and length L_{end}, and shaded margins extending to the limits of the package enclosure. Available enclosures take a variety of forms, such as a metal cap and header with pins (TO or other), a dual in-line package (DIP), a ceramic flat pack with beam leads, or a vertical package with the plane of the fingers parallel to the signal pins. A metal header of the type shown is fairly typical. The minimum size of the package rests ultimately on the size of the finger patterns. Pattern sizes can vary greatly to meet the requirements of diverse applications, and may also vary somewhat according to the design preferences of the filter

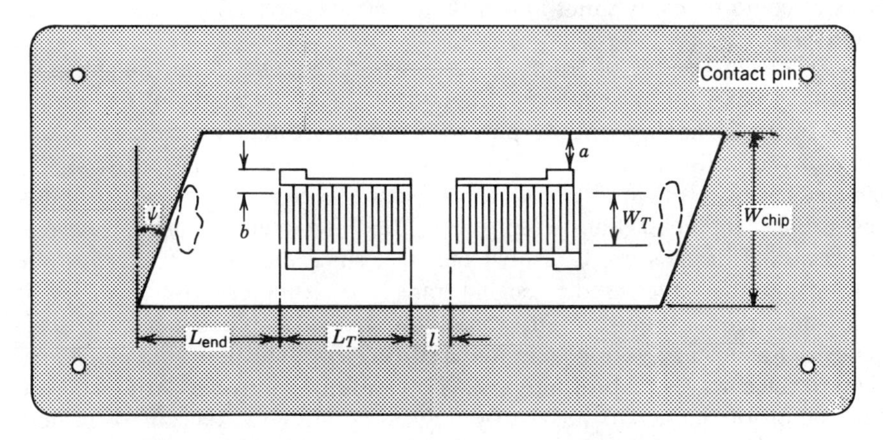

Figure 8.20 Sketch of common packaging arrangement.

supplier. To gain some idea of pattern size, we first consider a model with minimal dimensions, and then comment on enlargements required by applications.

The minimal model is the simplest transversal prototype, a filter with identical unweighted periodic transducers. The array 3 dB filter bandwidth Δf_a and the filter quality factor $Q_a \equiv f_{syn}/\Delta f_a$ are easily found from the $(\sin x/x)^4$ power transfer function implied by the array response given in Eq. (8.5). A little algebra then yields

$$N_{Tr} = 2Q_a/\pi \tag{8.9}$$

$$L_{Tr} \equiv N_{Tr}\lambda_{syn} = 2v_{ph}/(\pi\ \Delta f_a) \tag{8.10}$$

(Q_a, the Q of the array pair alone, may intentionally be made somewhat lower than the desired filter Q to allow for the band-narrowing effects of the transducer equivalent circuit and terminations [21].) Notice that the filter array bandwidth depends only on L_{Tr} and v_{ph}, regardless of the value of the synchronous frequency f_{syn} and the number of finger periods N_{Tr} spanned by L_{Tr}. (This behavior is consistent with a time-bandwidth product $\Delta f_a \cdot L_{Tr}/v_{ph}$ near unity in nondispersive devices.) The required number of transducer periods depends only on the corresponding Q_a value. Weighting a pattern of fixed length is equivalent to using fewer fingers, so that bandwidth is broadened and the Q_a is reduced. Hence, the unweighted prototype pattern length is a lower bound for a given filter array bandwidth. Apodization or other weighting methods, introduced to meet requirements on passband shape, sideband suppression, and so on, can easily increase the length of periodic designs several times over. The length multiplier tends to increase as specifications become more demanding. For the general run of filters with $Q \gtrsim 4$, a multiplier of 2.5 would be reasonable for a first estimate of weighted pattern length. Dispersive designs of the type shown in Figure 8.13(b) could be many times longer than a nondispersive design of the same bandwidth.

The angled regions of length L_{end} shown in Figure 8.20 at the substrate ends may comprise a significant fraction of the total length of the crystal chip. In the form shown, their purpose is to assure that waves propagating beyond the ends of the patterns will not be coherently reflected from the chip edges back into the transducer apertures, where they could seriously distort the intended filter response. (The unwanted acoustic waves in bidirectional designs are a distinct menace, since they tend to carry more power than the waves used for signal transport.) An alternative to angling is also indicated in Figure 8.20, where the irregular blobs near the ends represent acoustically absorbing material (e.g., epoxy) that attenuates incident surface waves. Absorbers must be carefully controlled to avoid contamination of the active surface by outgassing after the package is sealed. Angled ends are often preferred. A rough estimate of L_{end} can be obtained by assuming the length and angle to be adjusted so that no acoustic ray is specularly reflected back into the aperture W_{Tr}. For that simplified case, the minimum value of L_{end} is $(W_{Tr}W_{\text{chip}})^{1/2}$ and the angle ψ is $\tan^{-1}(W_{Tr}/W_{\text{chip}})$ [62].

For most applications, the active aperture W_{Tr} of the transducers tends to lie somewhere in the range 20 λ_{syn} to 100 λ_{syn}. Above 100 λ_{syn}, the finger resistance starts to produce a significant transverse variation in excitation voltage between fingers [25]. Small apertures are also avoided, especially with apodized transducer designs, since acoustic beam-spreading by diffraction becomes pronounced when the aperture of a source is only a few wavelengths across. The exact size of W_{Tr} may be chosen to match the impedance level of the terminations (e.g. through Eq. (8.7)), provided that W_{Tr} remains roughly between the limits mentioned. As shown in Figure 8.20, the total chip width includes not only W_{Tr}, but also margins a outside the pattern to guard against wafer dicing damage (perhaps 0.25–0.5 mm each) and the sum b of finger end-gap and bus bar and contact pad widths, amounting to a total b of perhaps 0.25 mm for GHz filters and 1.0 mm for filters near 50 MHz.

The separation l between transducers is usually made just large enough to obtain acceptable isolation of the input and output transducers against electromagnetic radiative coupling. A value of $l \simeq 20\ \lambda_{\text{syn}}$ is adequate in many cases when the filter housing provides suitable rf suppression.

The major dimensions of the illustrated chip have now been considered in a general way. The extent of the housing beyond the chip will depend partly on the chip mounting and contacting arrangements and partly on access requirements during package assembly. The dominant consideration may also turn out to be the housing sizes readily obtainable.

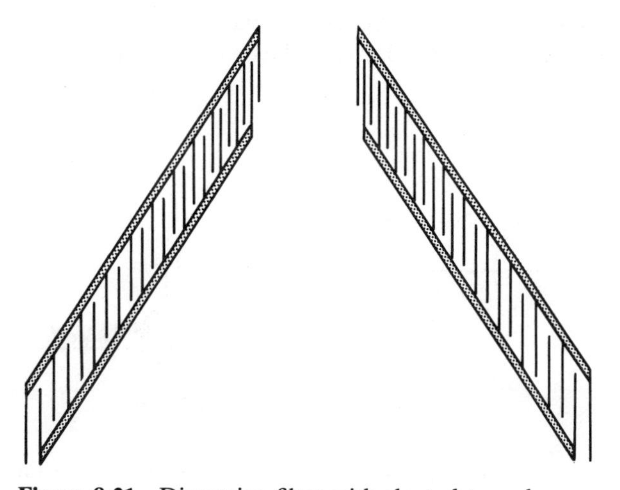

Figure 8.21 Dispersive filter with slanted transducers.

Package heights tend to be 3–10 mm or more, almost independently of the chip thickness, which is typically 20 to 30 λ_{syn} at lower frequencies and 0.5 mm at higher frequencies.

The size information presented here is necessarily incomplete, but may be useful for a preliminary estimate of required space. Accurate sizing can only be done in cooperation with the vendor. The sizes implied here could be greatly exceeded for special-purpose filters such as broadband dispersive designs, which are sometimes spread out laterally as shown in Figure 8.21 to avoid the perturbing effects of many fingers in the acoustic path.

8.2.5. Excitation of Bulk Modes

Solid materials are capable of supporting many qualitatively different propagating acoustic modes, most of which have little or no impact on SAW technology. We describe here the modes of possible importance in the SAW arena.

The impulse response of a Rayleigh-wave transversal filter characteristically exhibits unwanted bulk-wave responses to some degree, as illustrated in Figure 8.22(a). As indicated, the Rayleigh-wave response is the last to be received, since it always has the slowest phase velocity. The first signal is associated with a bulk longitudinal wave, which is always the fastest, and the intermediate-speed signal belongs to two bulk shear waves. One of the latter (quasi-horizontal shear) is often too weakly coupled to be seen. Due to the velocity differences, each mode has a distinct frequency at which the wave is synchronous with the transducer period. A frequency sweep typically produces a response like that in Figure 8.22(b). Near the Rayleigh-wave synchronous frequency, the transduction efficiency for bulk waves tends to be several tens of dBs lower.

At bulk-wave synchronous frequencies, rejection may be enhanced in a number of ways. In the case of narrower-band filters, inductive tuning for a conjugate reactive match to the transducer capacitance at f_{syn} can be very

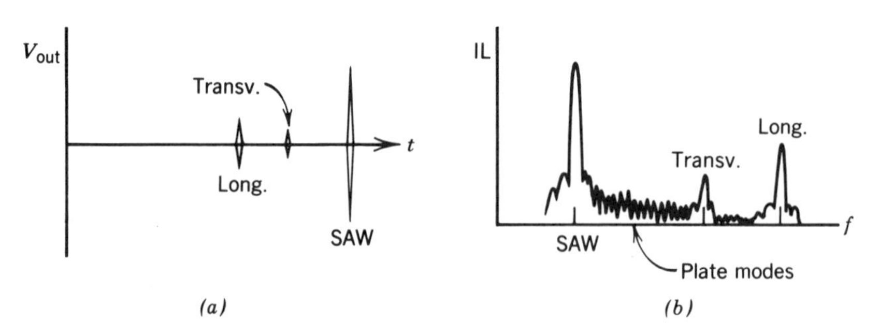

Figure 8.22 (a) Impulse response of a broadband SAW filter in the time-domain. (b) Corresponding cw response in the frequency-domain.

effective. Some crystal cuts give reduced transducer excitation of bulk waves. Further discrimination may occur through frequency cut-off by the circuit environment. A multistrip coupler can also enhance discrimination against bulk waves by transferring SAWs but not bulk waves from one acoustic track to another, as noted in §8.2.2. The band of low-level responses seen in Figure 8.22(b) starting near the Rayleigh mode comes from bulk plate modes guided by the two large surfaces of the substrate. These modes can be effectively eliminated by roughening the back face to scatter the waves randomly.

We should note that a bulk mode, usually horizontal shear, is sometimes used for signaling in place of the Rayleigh mode [6, 67]. Advantages include larger feature size in high-frequency transducer patterns and the availability of superior temperature stability. Crystal cuts for these surface-skimming bulk waves (SSBW), or shallow bulk acoustic waves (SBAW), are selected for good coupling to the signaling mode and an absence of coupling to Rayleigh waves. At frequencies above a GHz, excitation of SSBWs becomes less efficient than excitation of a Bleustein–Gulyaev (BG) wave [5], a variety of horizontal-shear surface wave with a much larger penetration depth than a Rayleigh wave. The BG wave is favored at high frequencies by a wave-trapping effect created by the transducer metallization [32].

8.2.6. Harmonic Behavior

The excitation of harmonics mf_{syn} of the synchronous SAW frequency of a periodic transducer depends crucially on the number of fingers per period and on their positions within the period [54]. Additional dependence on the relative widths of fingers and gaps is of secondary importance; the case of equal widths will suffice to describe the principal phenomena.

A simple transducer of the type shown in Figure 8.2 can only generate odd harmonics. The absence of even harmonics is purely a result of geometry. Even-harmonic standing waves would have one or more complete cycles between every pair of fingers; such waves cannot couple piezoelectrically to a voltage pattern that alternates in sign from one finger to the next. This behavior is not noticeably changed in the presence of weighting or of manufacturing inaccuracies. The odd harmonics generated by an unweighted periodic transducer separate into two groups, with the excitation strength of harmonics $M = 1, 5, 9, \ldots$ coupling roughly as $1/M$ and the group $M = 3, 7, 11, \ldots$ not excited at all. This difference is substantially modified in real, weighted designs, where the second group remains weaker than the first, but is by no means absent.

The relative weights of the odd harmonics are quite changed with double-finger transducers, illustrated in Figure 8.23. Double fingers were originally introduced as a means of suppressing acoustic reflections from finger edges [19], but their harmonic behavior made them important for another reason. The associated harmonic strengths exhibit pairing, with

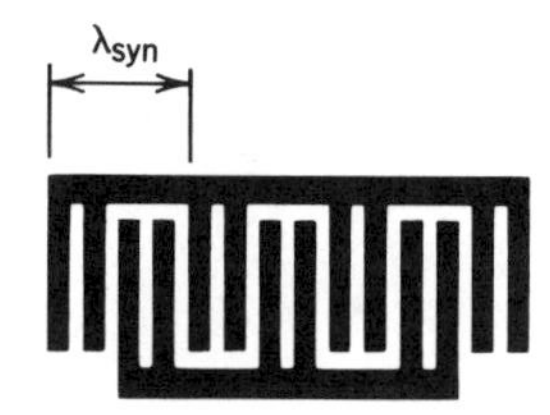

Figure 8.23 Prototype of a split-finger transducer.

stronger generation now at $M = 1, 3, 9, 11, \ldots$. In real filters, the third harmonic is often a bit stronger than the fundamental. A third harmonic transducer has a fabrication advantage in the GHz range, where its smallest feature size is 50% larger than the (sub-micrometer) feature size of a fundamental transducer.

A characteristic of harmonic operation is that all harmonics have the same 3 dB bandwidth, given by Eq. (8.10), in the unweighted periodic case. The third harmonic band thus has an array Q three times that of the fundamental. To recover the Q_a value of the fundamental, transducer length must be reduced by a factor of three. A third harmonic transducer then becomes the same length as a fundamental transducer designed for the same frequency and Q_a.

With certain exceptions, unwanted harmonic responses tend not to be an operational problem. The large frequency separation of odd-harmonic pass-bands usually allows electronic or passive cut-off of the unwanted bands. Harmonics could become a problem in the case of very broadband filters because of the possibility of aliasing. In the case of harmonic utilization, the fundamental could also be a problem, for example, in the timing recovery circuit of a digital regenerator.

8.3. SURFACE ACOUSTIC WAVE RESONATOR FILTERS

8.3.1. Resonator Structures and Behavior

The prototype of a SAW resonator filter [26] is shown in Figure 8.24. The structure is basically a prototype transversal filter enclosed between a pair of diffraction gratings that function as distributed reflectors for surface waves.

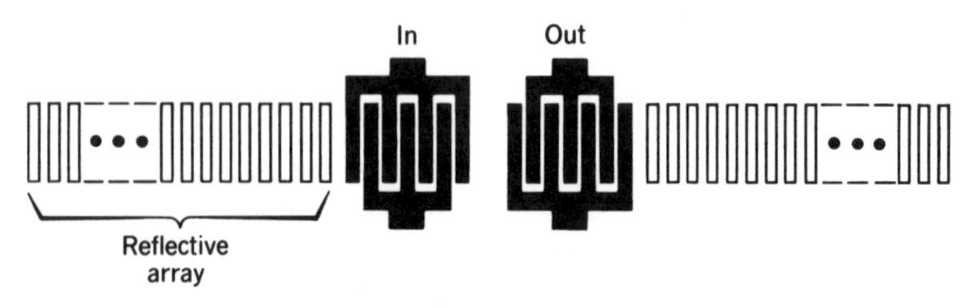

Figure 8.24 Prototype of a 2-port SAW resonator filter.

The gratings are periodic arrays of acoustic impedance discontinuities, typically quarter-wave grooves on half-wave centers at the frequency f_G of maximum grating reflectance, where the wavelength λ_G is v_G/f_G and v_G is the mean phase velocity at f_G in the grating region. The grooves are made shallow, roughly 1% of λ_G, to avoid excessive perturbation of the surface waves. Since the groove depth is only about 1% at the depth of the substrate layer excited by the SAWs, single grooves are weak reflectors. However, their cumulative effects add up to a very strong reflection at f_G, as illustrated in Figure 8.25(a). Roundtrip propagation-path increments of $\lambda_{G/2}$ between successive groove edges, combined with a reflectance phase difference of π between upsteps and downsteps, produce totally constructive interference in reflection at f_G.

The general frequency dependence of the grating power reflectance is shown in Figure 8.25(b). Reflection is strong across a transmission stop-band centered on f_G. As the number of grooves is increased, the stop-band narrows and its top flattens with the approach of reflectance toward unity. In the stop-band, the power transmission $P(f)$ tapers exponentially into the grating, as illustrated by the lower curve in Figure 8.26(a). In that case, the equivalent reflection plane of the grating lies at the $1/e$ power point L_p [13], the grating penetration depth, typically 20–50 λ_G from the entrance port at $z = 0$. Outside the stop-band limits $f_s^{\pm}$, there is a sequence of reflection nulls, separating subsidiary reflection bands that weaken as frequency deviates from f_G.

At frequencies far from f_G, the grating is nearly transparent to surface waves, and the taper in the transmitted power is effectively linear, as indicated by the upper curve in Figure 8.26(a). The effective reflection plane for the weak return wave then lies at the grating midline $L_G/2$. We note that the length of a grating in typical devices may be 200–400 λ_G (400–800 grooves) or more. Also, λ_G differs from the flat-surface wavelength by a very small amount, so that scaling for size can be done with the Rayleigh wavelength.

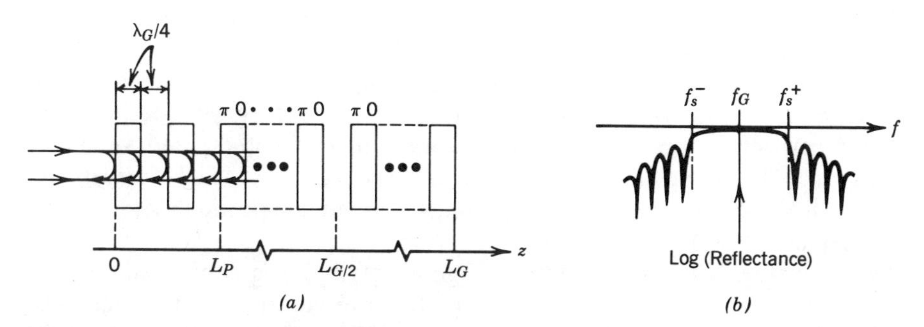

Figure 8.25 (a) Coherent interference of reflections from grating groove edges at synchronism. (b) Characteristic grating reflectance around synchronism. Edges of high-reflectance transmission stop-band indicated by $f_s^{\pm}$.

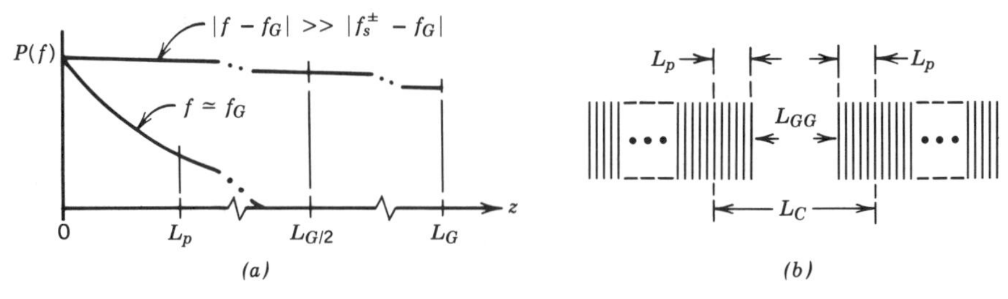

Figure 8.26 (a) Transmitted power versus distance from grating entrance port far from (upper curve) and near (lower curve) the synchronous reflection frequency. (b) Resonant cavity parameters.

When a pair of gratings is used for energy trapping, as shown in Figures 8.24 and 8.26(b), a cavity of length $L_C = L_{GG} + 2L_p$ is created, where L_{GG} is the distance between grating edges. Cavity resonances are possible at frequencies f_r, where the wavelength λ_r fits an integral number of times into $2L_C$. A resonance has a high Q only when f_r lies within the grating transmission stop-band between f_s^- and f_s^+. The actual location of resonances is complicated by a weak frequency dependence of the equivalent grating reflection plane across the transmission stop-band and also by the perturbing effects of internal transducers. At high-Q resonances, waves generated in both directions by the input transducers make many round trips in the cavity before they are effectively extracted by the output transducer. Since the transducers can be placed to obtain maximum coupling to waves traveling in both directions at resonance, the insertion loss is ideally zero. Representative transmission amplitude and phase responses [9] of a two-port SAW resonator filter with a Q of 10,000 are shown in Figure 8.27(a). The curves are effectively indistinguishable from those of a simple LC filter across the principal resonance band down to the sideband level. In the resonance band, SAW resonator filters behave like minimum-phase infinite-impulse-response (IIR) reactive filters with some internal loss. The first sidebands shown in the figure are vestiges of resonances adjacent to the main resonance. The peaks are low because their frequencies occur near the first grating reflection nulls on either side of the transmission stopband. The remaining sidebands are produced by the grating reflection sidebands, which cause weak energy trapping between the gratings, and by weak trapping in minor cavities formed by the transducers with each other, and with the gratings. The grating sideband ripple can be largely suppressed if necessary by weighting the outer ends of the gratings through apodization [28] or other techniques [55, 60].

The average off-resonance level (ignoring ripple) is controlled by the fractional power transmission $|S_{13}|^4$ of the transversal filter that remains when the gratings are effectively transparent. This behavior is apparent in the broader-band sweep illustrated in Figure 8.27(b), where the resonance

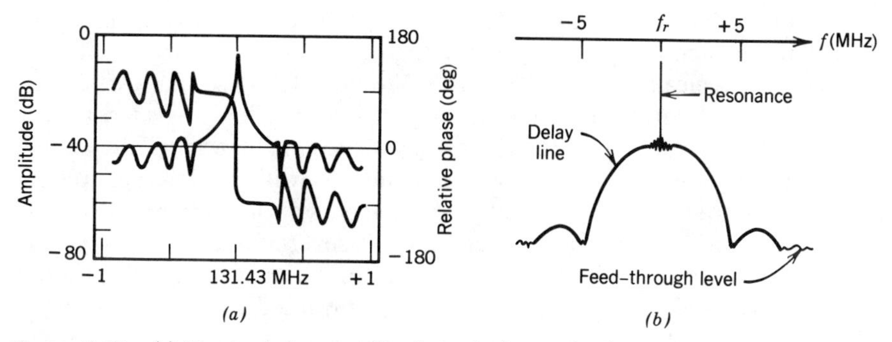

Figure 8.27 (a) Transmission amplitude and phase of a 2-port resonator filter near resonance. (b) Broadband amplitude transmission.

and its sidebands are seen to rise off the top of the broad central lobe of the transversal filter (delay line) response. An important point brought out by the figure is that a large value of relative sideband rejection can be achieved only when the transducers considered in isolation have severely mismatched terminations; only then will the internal transversal filter exhibit weak transmission in the sideband regions. A low insertion loss is nevertheless possible when a cavity resonance couples strongly to the transducers. The equivalent circuit seen by the terminations then looks like a good impedance match.

A quantitative account of the relationship between filter insertion loss, relative off-resonance rejection, filter Q, and the transducer conversion efficiency $|S_{13}|^2$ can be given through a few equations and a figure. The Q of a resonator filter like that shown in Figure 8.24 can be written.

$$Q \equiv \frac{f_r}{\Delta f_{3\,\mathrm{dB}}} = \frac{2\pi N_C\sqrt{m}}{(4/c)|S_{13}|^2 + m\alpha_u} \tag{8.12}$$

where $\Delta f_{3\,\mathrm{dB}}$ is the 3 dB bandwidth of the resonance, $N_C(\equiv L_C/\lambda_r)$ is the (integral) number of wavelengths across the resonant cavity, λ_r is the wavelength at f_r averaged over the various cavity regions (free surface, transducer regions, grating regions), and $2\alpha_u$ is the fractional unloaded (nonconversion) loss from a wave during a round trip inside the cavity. Unloaded losses include acoustic dissipation in the substrate, ohmic losses in the transducers, SAW diffraction losses, leakage transmission through the cavity gratings, acoustic scattering into bulk waves, and some other losses. Acoustic dissipation tends to become significant above a GHz and dominant at a few GHz. The coefficient $c(f)$ $(1 \le c \le 9/4)$ depends on both the conversion efficiency $|S_{13}|^2$ and the strength of grating reflection [44]. In the limit of weak conversion or weak reflection, c approaches unity. The symbol m has been introduced for later convenience; $m = 1$ for the two-port resonator filter of Figure 8.24.

The upper bound on achievable Q values is the unloaded Q, given by Eq. (8.12) in the limit $|S_{13}|^2 \to 0$. For a high-Q two-port resonator filter on quartz, α_u is typically about 0.05 and $2\pi N_C$ about 1,000, so that $Q \lesssim 2 \times 10^4$. A lower bound corresponds to almost perfectly matched transducers, with $|S_{13}|^2/c = 0.25$ in Eq. (8.12). Hence $Q \gtrsim 1{,}000$ on quartz. This value is impracticably low, however, since it coincides with an average relative rejection of about 8 dB. The central resonance will then be nearly indistinguishable from the sidebands, which may rise several dB above the average.

Between extremes, the insertion loss IL at resonance and the average value of sideband relative rejection $\overline{RR}$ are governed by

$$IL(\text{dB}) = -20 \log \left(1 - \sqrt{m}\alpha_u \frac{Q}{2\pi N_C}\right) \tag{8.13}$$

$$\overline{RR}(\text{dB}) = m\left[12 + 20 \log \left(\frac{Q}{2\pi N_C}\right)\right] - 20 \log c \tag{8.14}$$

These equations are plotted as the solid curves in Figure 8.28 for a simple two-port filter ($m = 1$) with $\alpha_u = 0.05$. Values of c have been calculated from a crossed-field transducer model [44]. At the low-Q end (larger fractional bandwidths), where the conversion efficiency term in Eq. (8.12) is stronger than unloaded loss, IL is low but $\overline{RR}$ is also low. At the high-Q end (smaller fractional bandwidths), where the conversion efficiency term is weaker than unloaded loss, $\overline{RR}$ is high but IL is also high. The $IL/\overline{RR}$ trade-off must be considered in applications.

SAW resonator filters are typically operated with inductor tuning to remove the degrading effects of the dc capacitance of the transducers. Since

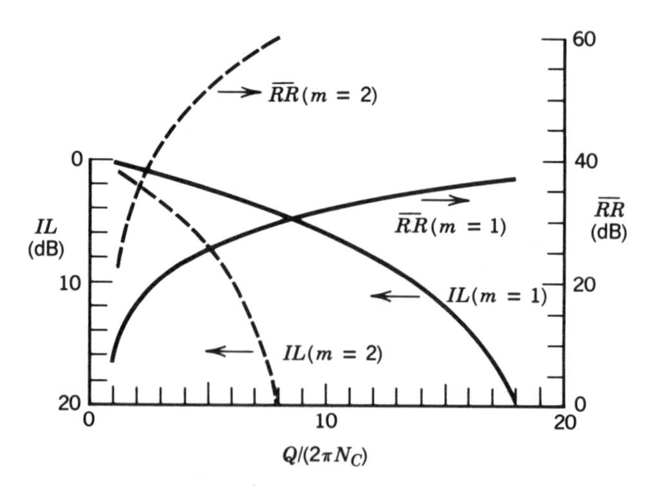

Figure 8.28 Insertion loss and mean relative rejection for one- and two-resonator SAW filters.

the required inductance is often quite small, of order 10 nH or less in the upper part of the useful frequency range, it is feasible to use spiral film inductors on the filter substrate itself. This practice is growing in popularity for applications where the additional substrate and package size are not important considerations.

8.3.2. Multiple-Resonator Filters

The use of coupled-resonator configurations provides the same benefits for SAW filters as for conventional resonator filters. A measure of control over passband shape is obtained at the same time that relative sideband rejection is improved. A number of different monolithic structures have been developed to couple energy from one cavity to another. Each structure has distinctive advantages and drawbacks [44]. One of the best coupling configurations is illustrated in Figure 8.29, which shows a pair of resonators placed side by side and coupled by direct electrical cross-connection of transducers in different cavities. The transducer coupler transfers power from the acoustic resonance on the input side to a second acoustic resonance on the output side by acoustic-to-electrical conversion in the first channel, followed by a sharing of electrical potentials with the second channel, followed by electrical-to-acoustic conversion in the second channel. The input and output transducers function just as they would in a two-port.

The structure can produce the same passband shapes as a classical multipole filter of second order. Figure 8.30(a) illustrates a nearly Butterworth response obtained with inductor tuning of the transducers [46]. The filter was designed with the help of a standard filter-synthesis handbook [68] to determine the intercavity coupling strength and the corresponding input and output conversion efficiencies needed to obtain a Butterworth response. The actual filter structure, a symmetrized version of Figure 8.29, is shown in

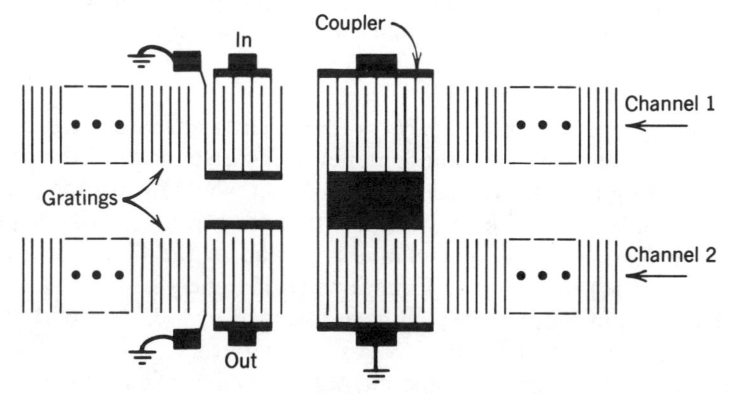

Figure 8.29 Basic layout of a transducer-coupled, two-resonator SAW filter.

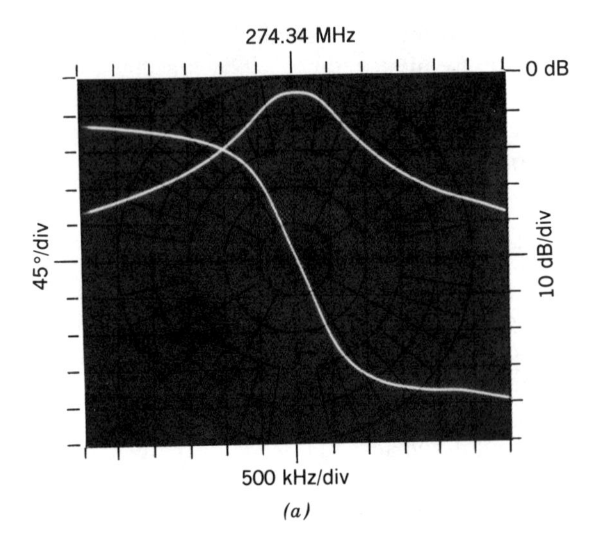

(a)

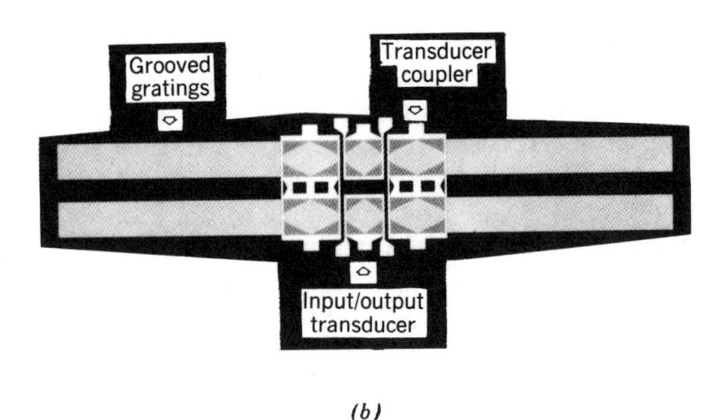

(b)

Figure 8.30 (a) Nearly Butterworth amplitude and phase transmission of the coupled-resonator filter structure (b) with transducer coupling.

Figure 8.30(b). For that filter, $2\pi N_C \simeq 1{,}395$. The transducers are apodized to match the amplitude of the fundamental transverse mode of the resonant cavities; coupling to other transverse modes that would distort the passband is thereby avoided. It can be shown [44] that for a Butterworth shape, the Q, insertion loss, and average relative rejection of the filter are given by Eqs. (8.12)–(8.14) with $m = 2$. Equations (8.13) and (8.14) are plotted as the dashed lines in Figure 8.28; we have assumed $\alpha_u = 0.08$, a fairly typical value for the structure of Figure 8.30(b) on quartz. For a fixed value of the abscissa, the single-resonator filter has the lower insertion loss, but also a much lower average relative rejection (by a factor of 2 or so in dB). Since

N_C may be 40% or more larger for transducer-coupled devices than for simple two-port resonators, the curves do not compare devices with the same Q. For that comparison, the ordinates on the $m = 2$ curves should be read at about 0.7 times the abscissa value at which ordinates are read on the $m = 1$ curves.

The average relative rejection given by Eq. (8.14) applies only to the transducer form of intercavity coupling. Other coupling structures show other types of behavior [11, 44]. The transducer coupling described here gives the best suppression of the nearest sidebands, however, and will often be the preferred choice. For practical reasons, however, transducer coupling on quartz cannot be made strong enough to reach the low-Q end of the range plotted in Figure 8.28. For that purpose, the crossed-resonator configuration [45] illustrated in Figure 8.31 provides the ultimate performance. There the cavities lie along two equivalent temperature-stable propagation axes, and are coupled by oblique acoustic reflections from a short array of grooves similar to the gratings that form the resonator cavities. Q values near 2,500 have been demonstrated with 35 dB relative rejection of sideband peaks. The oblique array coupler provides all the coupling strength necessary to reach the useful lower bound on relative rejection in resonator filters of second order. That bound is approximated in Figure 8.32, an unpublished result obtained by the writer with a 16-groove coupler. Relative rejection was about 14 dB at a Q of 1,480.

In principle, any number of SAW resonators can be coupled to make Chebyshev or other multipole filters of third or higher order. As a practical matter, however, fabrication in monolithic form becomes much more difficult beyond second order [10]. In third order and beyond, all intercavity

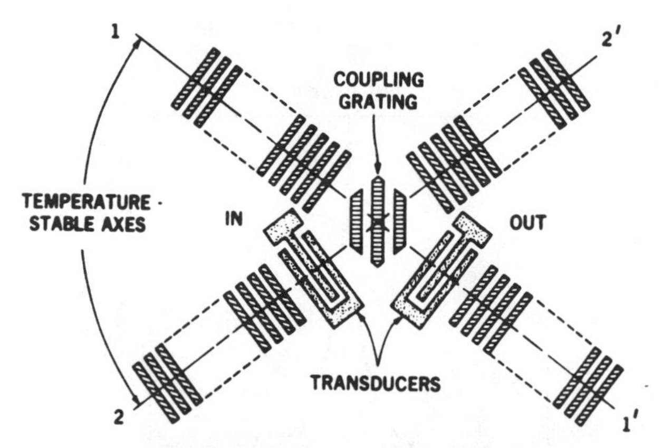

Figure 8.31 Crossed-resonator structure with coupling by grating reflection. Rosenberg and Coldren (1979) Crossed resonator SAW filter: a temperature-stable widerband filter on quartz. *Ultrasonics Symposium Proceedings*, pp. 836–840 © 1979 I.E.E.E.

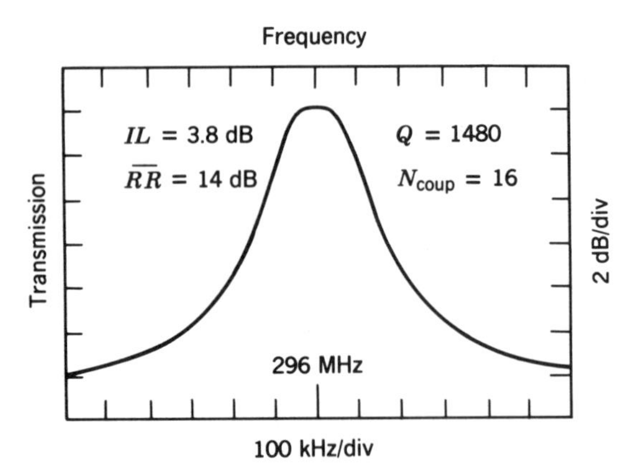

Figure 8.32 Approximately Butterworth response of a crossed-resonator filter near the useful lower bound on Q. The measured side-band relative rejection of 14 dB is a bit degraded by *rf* feedthrough.

coupling strengths and both conversion efficiencies must have unique values that depend on unloaded losses (i.e., on α_u) [68]. Present technology is not controlled finely enough to assure the achievement of required parameter values on the first try. Instead, considerable labor is required to trim resonance frequencies one at a time and adjust electrical coupling between sections [37, 61]. A sample response with electrical coupling is shown in Figure 8.33 [12], a fourth-order filter implemented by capacitive coupling of two second-order transducer-coupled filters. For this high-Q situation ($Q \simeq 10^4$), insertion loss is about 9 dB and the resonance peak is more than 75 dB above a sideband spike and 85 dB above the other sidebands.

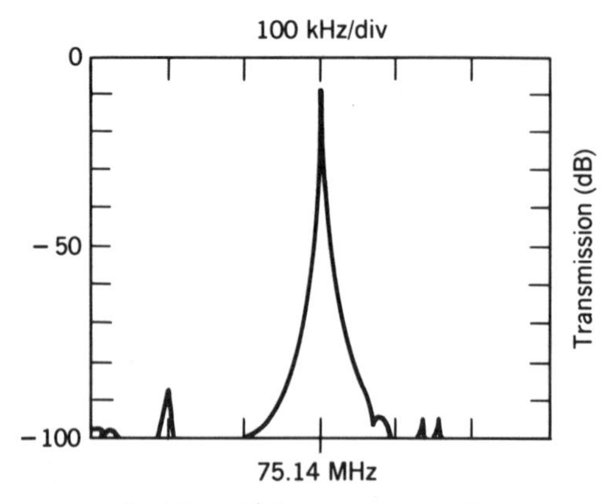

Figure 8.33 Response of a high-Q, four-resonator filter with capacitive coupling of transducer-coupled pairs.

The size of resonator filters on quartz can be estimated with far greater confidence than the size of transversal filters. A typical two-port resonator filter pattern will be about 900 λ_r in length and 48 λ_r + bus-bar + contact-pad requirements in width. Although designs do vary, the variations are unlikely to exceed 25%. The chip margins discussed in the transversal filter case must be added to the pattern dimensions, with certain modifications. Reflections from the chip ends are effectively ignorable in the resonance band, where the high grating reflectance prevents transmission of SAWs to the chip ends and back into the cavity. Nevertheless, slanted chip ends are used to control end reflections in the filter sidebands. A modest angle, for example, 10° from square, is usually sufficient even for coupled-resonator filters with parallel acoustic channels. Such filters have a pattern width about double that of the two-port, and a pattern length up to perhaps 25% greater to accommodate transducer coupling. In the special case of crossed-resonator filters, the channels themselves are slanted, the chips are rectangular as required by Figure 8.31, and the pattern diagonal is roughly 1,000 λ_r.

8.3.3. Other Resonator Structures

A number of other resonator structures have been demonstrated but not widely used to date. Only a few can be mentioned here. A one-port

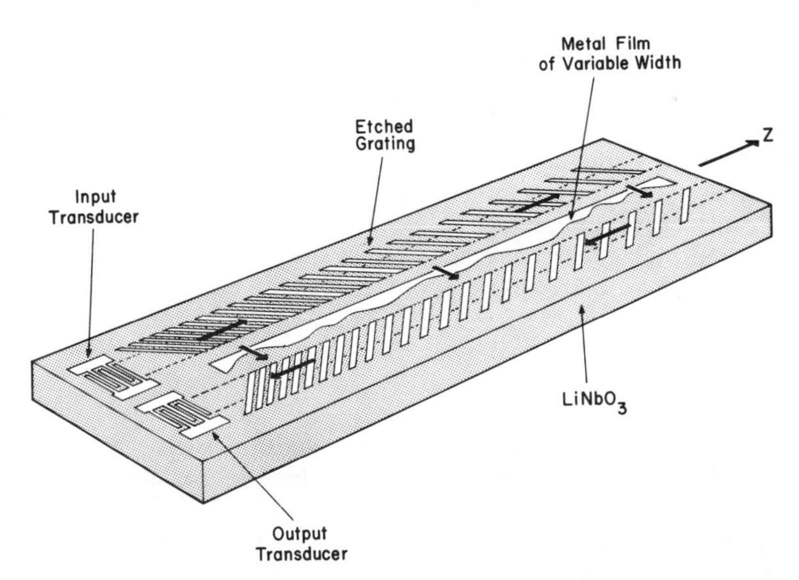

Figure 8.34 RAC structure for radar pulse compression with large time-bandwidth product. Williamson and Smith (1973) The use of surface-elastic-wave reflection gratings in large time-bandwidth pulse-compression filters. *I.E.E.E. Trans. Sonics Ultrasonics*, pp. 113–123 © 1973 I.E.E.E. (photo courtesy of MIT Lincoln Laboratory).

resonator tank circuit [2, 56] can be formed by omitting one of the transducers from the two-port filter shown in Figure 8.24. A single transducer with a great many periods can also be operated as a tank at the transducer antiresonance [27], where electrical reflectance rises effectively to 100%. Tanks usually require added circuit complexity to separate output from input. A two-port resonator structure of some potential utilizes horizontal-shear bulk waves bound to a surface by grooves [3]. Such a device makes possible high-Q SAW resonators on bulk-wave cuts with superior thermal stability, not only in quartz substrates, but also in a high-coupling material (lithium tantalate) for the first time.

A nonresonant device that often utilizes reflections from grooves is the reflective array compressor (RAC) [64], illustrated in Figure 8.34. The RAC uses two right-angle reflections by dispersive gratings to form a U-shaped transversal filter, having both large bandwidth and large variations of transmission time with frequency for radar ranging. RAC-based devices have been built with time-bandwidth products up to 33,000 [8].

8.3.4. Additional Comments on Resonator and Transversal Filters

It will be evident from the material already presented that the natural Q ranges of resonator filters and transversal filters are complementary. Transversal filters are most readily designed for low to intermediate Q values, up to 10^3 perhaps, where the number of transducer fingers becomes so large that higher order effects complicate the response. Resonator filters are most easily realized at higher Q values, from about 2×10^4 down to 2×10^3, where relative rejection starts to be inadequate for practical purposes. The Q-region between 10^3 and 2×10^3 is a challenge for either filter category. Candidate designs for that octave include harmonic transversal filters and resonator filters of first- or second-order. For reasons already discussed, neither choice is without problems. Resonator filters of third or higher order can solve the problem of inadequate relative rejection, but the technology does not yet lend itself to economical fabrication methods.

8.4. VENDOR INTERACTIONS

Vendors of SAW filters typically have a catalog of designs that have been developed for previous customers. For those designs, available information will be quite complete. Known properties will include much of the following list: center frequency, band shape within specified tolerances, delay versus frequency and/or phase versus frequency characteristics, out-of-band and spurious rejection, impulse response in the case of matched filtering, appropriate termination impedances, thermal characteristics, power-handling limits (generally in the 1–10 mW range in narrow-band filters and tending to diminish as the number of transducer fingers decreases), size, cost, and reliability. Catalog filters will usually be the simplest and lowest-cost answer to a specific need. Catalogs will vary from one vendor to another as a result of their market targets and design methods, so that user inquiries should cover the field.

TV IF filters, the largest commercial runner, are produced by the millions annually, and will be found in many catalogs. Prices are in the neighborhood of 1 US dollar as of this writing.

As manufactured quantities decrease, prices tend to increase in quantum leaps. Many filter varieties fall into a bracket between 50 and 500 US dollars. Filters requiring special design may also fall into this category after initial charges of roughly 5,000 dollars for engineering and perhaps an equal amount for photolithography masks. Since design iterations tend to be expensive, the prudent user will evaluate his requirements as thoroughly as possible beforehand, and will take full advantage of the vendor's expertise to arrive at a design suited to the purpose. A young trend in the industry is towards the production of functional modules incorporating the filters. This approach has significant advantages to both user and manufacturer, since only the overall functional requirements need to be specified by the user, and constraints on the manufacturer tend to be relaxed.

8.5. PERSPECTIVE ON THE PRESENT REVIEW

The present review of SAW filter technology is by no means exhaustive. The aim has been to acquaint prospective users with some fundamental concepts, a basic vocabulary, and an appreciation of practical matters such as thermal behavior, available electrical characteristics, the important role of electrical terminations, overall size, and some cost factors. In the limited space available, it has been necessary to gloss over or omit entirely many alternatives and variants to the specific structures described here. An effort has been made, however, to provide a sufficiently broad base so that a reader may quickly evaluate information coming from filter vendors, subsystem providers, and others. To assist in gaining a broader or deeper technical understanding, selected supplementary references are grouped by category after the cited references. The references in turn afford access to the extensive literature.

Although this review has focussed on passive filtering, the useful domain of SAW filter technology is much broader, as testified by the series of IEEE Ultrasonics Symposium Proceedings that go back to 1972. In those pages, the interested reader will find many papers on SAW-semiconductor hybrid modules capable of sophisticated signal processing functions. Examples are convolution, correlation, and Fourier and other transformations, with or without storage, programmable matched filtering, charge-transport processing, multiplexing, adaptive filtering, and spectrum analysis. One measure of effectiveness is the fact that SAW convolver modules presently provide the largest time-bandwidth products available in any technology [57]. SAW filters also have a growing presence in low-noise oscillators and voltage-controlled oscillators. The unique capabilities of SAW technology suggest that SAWs will be important for a long time to come.

REFERENCES

1 C. B. Armitage (1984) SAW filter retiming in the AT&T 432 Mb/s lightwave regenerator. *Proc. 10th European Conf. on Optical Communications*, September 1984.

2 E. A. Ash (1970) Surface wave grating reflectors and resonators. *I.E.E.E. Symp. on Microwave Theory and Techniques*, pp. 385–386.

3 B. A. Auld, J. J. Gagnepain and M. Tan (1976) Horizontal shear surface waves on corrugated surfaces. *Electron. Lett.* **12**, 650–651.

4 W. R. Bennett (1970) *Introduction to Signal Processing*. McGraw-Hill, New York.

5 J. L. Bleustein (1968) A new surface wave in piezoelectric crystals. *Appl. Phys. Lett.* **13**, 412–413.

6 T. I. Browning and M. F. Lewis (1977) A new class of quartz crystal oscillator controlled by surface-skimming bulk waves. *Proc. 31st Ann. Symp. on Frequency Control*, pp. 258–265.

7 I. Browning and M. Lewis (1978) A new cut of quartz giving improved temperature stability to SAW oscillators. *Proc. 32nd Ann. Symp. on Frequency Control*, pp. 87–94.

8 M. Chomiki and F. Genauzeau (1983) A SAW 330-MHz 100-μs chirp generator. *Ultrasonics Symposium Proceedings*, pp. 209–212.

9 L. A. Coldren and R. L. Rosenberg (1976) Scattering matrix approach to SAW resonators. *Ultrasonics Symposium Proceedings*, pp. 266–271.

10 L. A. Coldren, R. L. Rosenberg and J. A. Rentschler (1977) Monolithic transversely coupled SAW resonator filters. *Ultrasonics Symposium Proceedings*, pp. 888–893.

11 L. A. Coldren and R. L. Rosenberg (1978) SAW resonator filter overview: design and performance tradeoffs. *Ultrasonics Symposium Proceedings*, pp. 422–432.

12 L. A. Coldren and R. L. Rosenberg (1979) Surface-acoustic-wave resonator filters. *Proc. I.E.E.E.* **67**, 147–158.

13 P. S. Cross (1975) Reflective arrays for SAW resonators. *Ultrasonics Symposium Proceedings*, pp. 241–244.

14 A. J. DeVries and R. Adler (1976) Case history of a surface-wave TV IF filter for color television receivers. *Proc. I.E.E.E.* **64**, 671–676.

15 J. F. Dias, H. E. Karrer, J. A. Kusters, J. H. Matsinger and M. B. Schulz (1975) The temperature coefficient of delay-time for X-propagating acoustic surface-waves on rotated Y-cuts of α quartz. *I.E.E.E. Trans. Sonics Ultrasonics* **22**, 46–50.

16 G. Eberharter and H. P. Feurbaum (1980) Scanning electron microscope observations of propagating acoustic waves in surface acoustic wave devices. *Appl. Phys. Lett.* **37**, 698–699.

17 M. Feldmann and J. Henaff (1978) Design of SAW filter with minimum phase response. *Ultrasonics Symposium Proceedings*, pp. 720–723.

18 R. L. Filler, J. M. Frank, R. D. Peters and J. R. Vig (1978) Polyimid bonded resonators. *Proc. 32nd Ann. Symp. on Frequency Control*, pp. 290–298.

19 H. M. Gerard (1971) Experimental evaluation of non-ideal performance in interdigital surface-wave electrodes. *I.E.E.E. Ultrasonics Symposium*, paper J-1.

20 C. S. Hartmann (1973) Weighting interdigital surface-wave transducers by selective withdrawal of electrodes. *Ultrasonics Symposium Proceedings*, pp. 423–426.

21 C. S. Hartmann, D. T. Bell, Jr. and R. C. Rosenfeld (1973) Impulse model design of acoustic surface-wave filters. *I.E.E.E. Trans. Microwave Theory Tech.* **21**, 162–175.

22 M. Hikita, Y. Kinoshita, H. Kojima and T. Tabuchi (1980) Phase weighting for low loss filters. *Ultrasonics Symposium Proceedings*, pp. 308–312.

23 M. Hikita, T. Tabuchi, H. Kojima, A. Sumioka, A. Nakagoshi and Y. Kinoshita (1984) High performance SAW filters with several new technologies for cellular radio. *Ultrasonics Symposium Proceedings*, pp. 82–92.

24 R. C. Kinsman (1978) Temperature compensation of crystals with parabolic temperature coefficients. *Proc. 32nd Ann. Symp. on Frequency Control*, pp. 102–107.

25 K. M. Lakin (1974) Electrode resistance effects in interdigital transducers. *I.E.E.E. Trans. Microwave Theory Tech.* **22**, 418–424.

26 K. M. Lakin, T. Joseph and D. Penunuri (1974) Planar surface acoustic wave resonators. *Ultrasonics Symposium Proceedings*, pp. 263–267.

27 K. M. Lakin, T. Joseph and D. Penunuri (1974) A surface acoustic wave planar resonator employing an interdigital transducer. *Appl. Phys. Lett.* **25**, 363–365.

28 K. M. Lakin and T. R. Joseph (1975) Surface wave resonators. *Ultrasonics Symposium Proceedings*, pp. 269–278.

29 J. Lattanza, F. G. Herring, P. M. Krencik and A. F. Clerihew (1983) 240 MHz wideband programmable SAW matched filter. *Ultrasonics Symposium Proceedings*, pp. 143–150.

30 J. Ladd, C. Abdallah and T. O'Shea (1984) A temperature compensated L-band hybrid SAW oscillator and resonator filter. *Ultrasonics Symposium Proceedings*, pp. 191–196.

31 M. F. Lewis (1972) Triple transit suppression in surface-acoustic-wave devices. *Electron. Lett.* **8**, 553–554.

32 D. L. Lee (1981) Analysis of energy trapping effects for SH-type waves on rotated Y-cut quartz. *I.E.E.E. Trans. Sonics Ultrasonics* **28**, 330–341.

33 M. F. Lewis (1982) SAW filters employing interdigitated interdigital transducers. *Ultrasonics Symposium Proceedings*, pp. 12–17.

34 F. G. Marshall, C. O. Newton and E. G. S. Paige (1973) Surface acoustic wave multistrip components and their applications. *I.E.E.E. Trans. Microwave Theory Tech.* **21**, 216–224.

35 G. L. Matthaei, E. B. Savage and F. Barman (1978) Synthesis of acoustic surface wave resonator filters using any of various coupling mechanisms. *I.E.E.E. Trans. Sonics Ultrasonics* **25**, 72–84.

36 T. R. Meeker and W. R. Grise (1983) Packaging and reliability of SAW filters. *Ultrasonics Symposium Proceedings*, pp. 117–124.

37 P. C. Meyer and D. Gunes (1983) Design and fabrication of SAW multipole filters. *Ultrasonics Symposium Proceedings*, pp. 66–71.

38 J.-I. Minowa (1978) A method for accurately adjusting the center frequency of surface acoustic wave filters. *Rev. Electr. Commun. Lab.* (Fujitsu) **26**, 797–807.

39 C. H. Moor and C. F. Stolwyck (1973) Concatenated surface wave processors. *Ultrasonics Symposium Proceedings*, pp. 336–339.

40 B. R. Potter and C. S. Hartmann (1977) Low loss surface-acoustic-wave filters. *I.E.E.E. Trans. Parts, Hybrids, Packaging* **13**, pp. 348–353.

41 Lord Rayleigh (1885) On waves propagated along the plane surface of an elastic solid. *Proc. London Math. Soc.* **17**, 4–11.

42 T. M. Reeder, W. R. Shreve and P. L. Adams (1972) A new broadband coupling network for interdigital surface wave transducers. *I.E.E.E. Trans. Sonics Ultrasonics* **19**, 466–470.

43 R. C. Rosenfeld, C. S. Hartmann and R. B. Brown (1974) Low-loss unidirectional acoustic surface-wave filters. *Proc. 28th Ann. Symp. on Frequency Control*, pp. 299–303.

44 R. L. Rosenberg and L. A. Coldren (1979) Scattering analysis and design of SAW resonator filters. *I.E.E.E. Trans. Sonics Ultrasonics* **26**, 205–230.

45 R. L. Rosenberg and L. A. Coldren (1979) Crossed-resonator SAW filter: a temperature-stable wider-band filter on quartz. *Ultrasonics Symposium Proceedings*, pp. 836–840.

46 R. L. Rosenberg and L. A. Coldren (1980) Broader-band transducer-coupled SAW resonator filters with a single critical masking step. *Ultrasonics Symposium Proceedings*, pp. 164–168.

47 R. L. Rosenberg (1981) Wave scattering properties of interdigital SAW transducers. *I.E.E.E. Trans. Sonics Ultrasonics* **28**, 26–41.

48 R. L. Rosenberg, C. Chamzas and D. A. Fishman (1984) Timing recovery with SAW transversal filters in the regenerators of undersea long-haul fiber transmission systems. *J. Lightwave Technol.* **2**, 917–925.

49 M. B. Schulz, B. J. Matsinger and M. G. Holland (1970) Temperature dependence of surface acoustic wave velocity on α-quartz. *J. Appl. Phys.* **41**, 2755–2765.

50 M. Setrin, D. T. Bell, Jr., M. B. Schulz and M. G. Umkauf (1973) An IFF system using block programmable surface wave signal expander and compressor. *Ultrasonics Symposium Proceedings*, 1973, pp. 316–323.

51 A. J. Slobodnik, Jr. (1971) The temperature coefficients of acoustic surface wave velocity and delay on lithium niobate, lithium tantalate, quartz, and tellurium dioxide. AFCRL-72-1082, *Physical Sciences Research Papers*, No. 477, National Technical Information Service.

52 W. R. Smith, H. M. Gerard, J. H. Collins, T. M. Reeder and H. J. Shaw (1969) Analysis of interdigital surface wave transducers by use of an equivalent-circuit model. *I.E.E.E. Trans. Microwave Theory Tech.* **17**, 856–864.

53 W. R. Smith, H. M. Gerard, J. H. Collins, T. M. Reeder and H. J. Shaw (1969) Design of surface wave delay lines with interdigital transducers. *I.E.E.E. Trans. Microwave Theory Tech.* **17**, 865–873.

54 W. R. Smith (1981) Circuit model analysis and design of interdigital transducers for surface acoustic wave devices. In: *Physical Acoustics*, Vol. 15, edited by W. P. Mason and R. N. Thurston. Academic Press, New York, pp. 89–189.

55 L. P. Solie (1976) Surface acoustic wave reflective dot array, *Appl. Phys. Lett.* **28**, 420–422.

56 E. J. Staples, J. S. Schoenwald, R. C. Rosenfeld and C. S. Hartmann (1974) UHF surface acoustic wave resonators. *Ultrasonics Symposium Proceedings*, pp. 245–252.

57 E. Stern (1983) Comparison of new analog device technologies for signal processing. *Ultrasonics Symposium Proceedings*, pp. 129–136.

58 R. H. Tancrell and R. C. Williamson (1971) Wavefront distortion of acoustic surface waves from apodized interdigital transducers. *Appl. Phys. Lett.* **19**, 456–459.

59 R. H. Tancrell and M. G. Holland (1971) Acoustic surface wave filters. *Proc. I.E.E.E.* **59**, 393–409.

60 W. J. Tanski (1979) SAW resonators utilizing withdrawal weighted reflectors. *I.E.E.E. Trans. Sonics Ultrasonics* **26**, 404–410.

61 W. J. Tanski (1981) SAW frequency trimming of resonant and traveling-wave devices on quartz. *Appl. Phys. Lett.* **39**, 40–42.

62 W.-S. Tsay, private communication.

63 R. M. White and F. W. Voltmer (1965) Direct piezoelectric coupling to surface elastic waves. *Appl. Phys. Lett.* **7**, 314–316.

64 R. C. Williamson and H. I. Smith (1973) The use of surface-elastic-wave reflection gratings in large time-bandwidth pulse-compression filters. *I.E.E.E. Trans. Sonics Ultrasonics*, pp. 113–123.

65 P. V. Wright (1985) The natural single-phase unidirectional transducer: a new low-loss SAW transducer. *Ultrasonics Symposium Proceedings*, pp. 58–63.

66 K. Yamanouchi, F. M. Nyffeler and K. Shibayama (1975) Low insertion loss acoustic surface wave filter using group-type unidirectional interdigital transducer. *Ultrasonics Symposium Proceedings*, pp. 317–321.

67 K. H. Yen, K. L. Wang, R. S. Kagiwada and K. F. Lau (1977) Interdigital transducers — A means of efficient bulk wave excitation. *Proc. 31st Ann. Symp. on Frequency Control*, pp. 266–270.

68 A. I. Zverev (1967) *Handbook of Filter Synthesis*. Wiley, New York.

SUPPLEMENTARY READING

Device Review M. F. Lewis, C. L. West, J. M. Deacon and R. F. Humphryes (1984) Recent developments in SAW devices. *Proc. I.E.E.E.* PtA **131**, 186–215.

General Reviews R. M. White (1970) Surface elastic waves. *Proc. I.E.E.E.* **58**, 1238–1276.

M. G. Holland and L. T. Claiborne (1974) Practical surface acoustic wave devices. *Proc. I.E.E.E.* **62**, 582–611.

Design and Manufacturing G. Tobolka, W. Faber, G. Albrecht and D. Pilz (1984) High-volume TV-IF filter design, fabrication, and applications. *Ultrasonics Symposium Proceedings*, pp. 1–12.

Special Issues Microwave acoustics (1969) *I.E.E.E. Trans. Microwave Theory Tech.* **17**, 798–1052.

Microwave acoustic signal processing (1973) *I.E.E.E. Trans. Sonics Ultrasonics* **20**, 79–230.

Surface acoustic wave devices and applications (1976) *Proc. I.E.E.E.* **64**, 577–832.

Miniaturized Filters (1979) *Proc. I.E.E.E.* **67**, 1–192.

Surface-acoustic-wave device applications (1981) *I.E.E.E. Trans. Sonics Ultrasonics* **28**, 115–234.

SAW convolvers and correlators (1985) *I.E.E.E. Trans. Sonics Ultrasonics* **32**, 618–799.

Books B. A. Auld (1973) *Acoustic Fields and Waves in Solids*, Vols. I and II. Wiley, New York.

H. Matthews (Editor) (1977) *Surface Wave Filters*. Wiley, New York.

A. A. Oliner (Editor) (1978) *Acoustic Surface Waves*. Springer, Berlin.

E. Dieulesaint and D. Royer (1980) *Elastic Waves in Solids*. Wiley, Chichester.

V. M. Ristic (1983) *Principles of Acoustic Devices*. Wiley, New York.

E. A. Ash and E. G. S. Paige (Editors) (1985) *Rayleigh Wave Theory and Application*. Springer, Berlin.

D. P. Morgan (1985) *Surface Wave Devices for Signal Processing*. Elsevier, Amsterdam.

S. Datta (1986) *Surface Acoustic Wave Devices*. Prentice-Hall, Englewood Cliffs, NJ.

9 Digital Filters

KALYAN MONDAL AND JALIL FADAVI-ARDEKANI
AT&T Bell Laboratories, Allentown, PA

In recent years, there has been an abundance of digital filters used in many communication, speech, image, and radar processing systems. The segment of the applications where we focus our attention in this chapter, uses dedicated very large scale integration (VLSI) hardware (thereby miniaturized) to do the filtering. This is in contrast with general purpose computing systems (e.g., mainframes or minicomputers) which among other things (e.g., data processing) can be programmed to perform filtering. In general, there are two ways of implementing any desired digital filter in hardware form. One is programming a general purpose digital signal processor (DSP) chip, and the other is to have all of the steps of filtering algorithm implemented by special purpose hardware. One only needs multipliers, adders, and delay elements (in the form of storage devices) in order to implement any filter function.

A **digital filter** converts a sequence of numbers called the *input* to another sequence of numbers called the *output*. The output of an analog-to-digital (A/D) converter, which samples a continuous-time input signal and generates a sequence of finite-length binary numbers, is a typical input to a digital filter and is called a *digital signal*. An A/D converter essentially consists of a *sampler* (the output of which is a discrete-time signal) in cascade with a *quantizer-encoder* that truncates the amplitude of the signal to one of a finite number of levels. This causes an error in signal representation in any digital system. Thus one of the design considerations of a digital filter is the number of quantized levels (or *bits*) needed to represent the digital signal when binary encoding is complete. The number of bits used in the representation of a signal is also called its *wordlength*. A digital system will always process a signal in a finite wordlength representation. The larger the number of bits (i.e., larger the wordlength) used, the less error occurs in the filter implementation. Consequently, the price of the filter hardware goes up. Currently, 16- or 24-bit fixed-point and 22- or 32-bit floating-point representation [30] of digital signals are considered adequate for most audio, speech, radar, sonar, and video applications. Finally the filtered digital signal is passed through a digital-to-analog (D/A) converter to obtain the continuous-time output signal.

In this chapter, we concentrate on one-dimensional filtering (where signals that are functions of one independent variable such as time are filtered) although many concepts explained here extend readily to two-dimensional filtering [30]. An example of two-dimensional filtering is picture processing.

9.1. DESIGN OF DIGITAL FILTERS

As discussed in Chapter 1, the transfer function $H(z)$ of an Nth order **infinite-impulse response** (IIR) digital filter is given by a ratio of two polynomials in z^{-1} with real coefficients a_k and b_k

$$H(z) = \frac{Y(z)}{U(z)} = \frac{\sum_{k=0}^{N} b_k z^{-k}}{1 - \sum_{k=1}^{N} a_k z^{-k}} \tag{9.1}$$

Likewise, an Nth order **finite-impulse response** (FIR) digital filter is described by the z-domain transfer function

$$H(z) = \frac{Y(z)}{U(z)} = \sum_{k=0}^{N} b_k z^{-k} \tag{9.2}$$

It should be noted that in many applications, the nonrecursive realization of FIR filters (with no feedback) are preferred over IIR filters as they are always stable, can be designed with linear phase, and do not exhibit limit cycles [30] when implemented in digital systems with finite wordlength. On the other hand, in general, for a given frequency response, an IIR filter is computationally more efficient than an equivalent FIR filter [31].

There are several ways to implement the FIR digital filter transfer function of Eq. (9.2). The direct form realization of Eq. (9.2) corresponds to the implementation of the following linear difference equation

$$y[n] = \sum_{k=0}^{N} b_k u[n-k] \tag{9.3}$$

where $y[n]$ represents the output digital signal sequence of a given wordlength and $u[n-k]$ the corresponding input digital signal sequence delayed by k samples.

Similarly the direct form realization of the IIR digital filter of Eq. (9.1) is indeed the implementation of the following linear difference equation

$$y[n] = \sum_{k=0}^{N} b_k u[n-k] + \sum_{k=1}^{N} a_k y[n-k] \tag{9.4}$$

By introducing a new variable $d[n]$, one can rewrite the above equation in such a way that $d[n]$ only depends on the present value of $u[n]$ and the output $y[n]$ can be updated by using only the previous values of $d[n]$. Define

$$d[n] = 0\,, \quad \text{for } n < 0 \tag{9.5a}$$

$$d[n] = \sum_{k=1}^{N} a_k d[n-k] + u[n]\,, \quad \text{for } n \geq 0 \tag{9.5b}$$

Then the output $y[n]$ is given by

$$y[n] = \sum_{k=0}^{N} b_k d[n-k]\,, \quad \text{for } n \geq 0 \tag{9.5c}$$

Equations (9.5b) and (9.5c) represent the direct form II [30] realization of IIR filters. We will elaborate on such an implementation later in this section with an example.

Filter designers have advocated the realization of IIR filters with cascaded second-order sections due to superior filter properties, such as lower coefficient sensitivity due to finite wordlength implementation, greater design flexibility for reducing roundoff noise (to be defined later) by appropriate pole-zero pairing, and the potential savings in the number of multipliers [18]. In this format, an Nth order digital filter transfer function is written as

$$H(z) = \prod_{i=1}^{\lceil \frac{N}{2} \rceil} \frac{\beta_{0i} + \beta_{1i} z^{-1} + \beta_{2i} z^{-2}}{1 - \alpha_{1i} z^{-1} - \alpha_{2i} z^{-2}} \tag{9.6}$$

where $\lceil x \rceil$ denotes the closest integer greater than or equal to x. If N is odd, one of the second-order section degenerates to a first-order one.

Without any loss of generality, every second-order section can be implemented in direct form II with the difference equations for the ith section as shown in Eqs. (9.7a) and (9.7b).

$$d_i[n] = y_{i-1}[n] + \alpha_{1i} d_i[n-1] + \alpha_{2i} d_i[n-2] \tag{9.7a}$$

$$y_i[n] = \beta_{0i} d_i[n] + \beta_{1i} d_i[n-1] + \beta_{2i} d_i[n-2] \tag{9.7b}$$

where $i = 1, \ldots, \lceil \frac{N}{2} \rceil$. For the first stage of the filter, $i = 1$, and $y_0[n] = u[n]$.

At this point let us identify the basic steps involved in implementing a given transfer function. Consider the following simple transfer function as an example

$$H(z) = \frac{Y(z)}{U(z)} = \frac{b_0 + b_1 z^{-1}}{1 - a_1 z^{-1}} \tag{9.8}$$

To realize this transfer function, we first convert it into a difference equation (by cross-multiplication followed by inverse z transformation) as

$$y[n] = b_0 u[n] + b_1 u[n-1] + a_1 y[n-1] \tag{9.9}$$

Equation (9.9) denotes that the present output sample $y[n]$ is a weighted sum of the previous output sample $y[n-1]$, the present input sample $u[n]$, and the previous input sample $u[n-1]$. Next we implement the algorithm outlined in Eq. (9.9), which requires the following types of computing elements.

(i) *Delay units* implemented as shift registers or random access memory (RAM) addressed by a counter, to store past input and output sample values. These are denoted by boxes with z^{-1} [18] as in Figure 9.1.
(ii) *Multiplication units* to multiply signal samples by the coefficients that are usually stored in a read-only memory (ROM). These are denoted by a triangular symbol as in Figure 9.1.
(iii) *Summing units* to add up all the partial products on the right of Eq. (9.9). These are denoted by a circled plus sign as in Figure 9.1.

In terms of these building blocks, the direct form II [30] representation of the filter transfer function given in Eq. (9.8) is shown in Figure 9.1.

This figure clearly shows that in IIR filters we have both feedforward paths and feedback loops. On the contrary, FIR filter block diagrams will

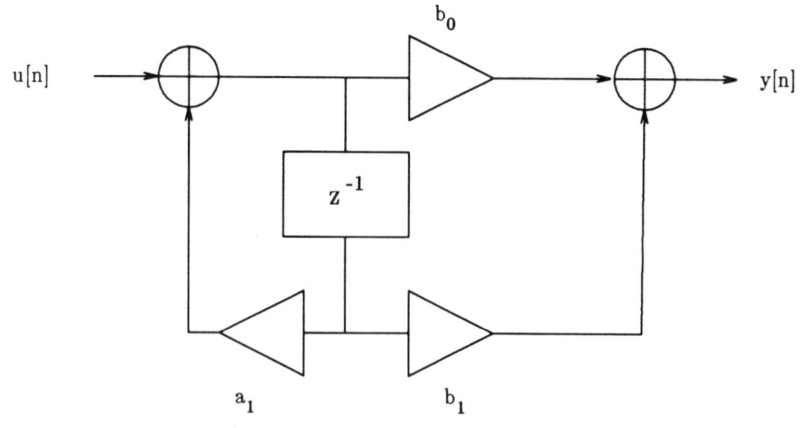

Figure 9.1 Block diagram of the IIR filter given by Eq. (9.8).

only have feedforward paths. The computation of an output sample using the above-mentioned building blocks will be feasible at any time only if the right side of the difference equation describing a filter such as Eq. (9.9), does not contain terms involving the present output. This requirement on a digital filter is expressed as the absence of **delay-free feedback loops** or the equivalent signal flow graph of the filter being **computable** [30].

Due to a finite wordlength of the filter coefficients, the realized filter would have an error in the computation of the output (in comparison to results computed with infinite precision coefficients), even though all the computations were done with infinite precision. In practice such infinite precision computation is not possible either, leading to truncation or rounding of all intermediate results. This leads to what is known as **roundoff noise** at the output of multipliers. Similarly, due to the finite word size of summation units, the outputs of the adders may overflow, which can sometimes lead to **overflow oscillations**. By increasing the number of bits in the signal representation and arithmetic elements, the occurrence of such problems can be reduced.

In the following two subsections we briefly discuss the design of digital filters using some examples. The first step in designing filters is to find a transfer function such that some aspects of the frequency response (like magnitude, phase, or group delay) or time response meet the desired specifications. This is basically a mathematical approximation problem. A vast body of literature [12, 39] exists showing the steps needed to design Butterworth, Chebyshev, Elliptic, Bessel, and other well-known analog filters. Thus, several digital filter design techniques have their roots in analog filter design. Other novel techniques (such as minimization of mean square error in the frequency-domain) that are independent of analog filter design have also been proposed.

Another extremely interesting class of filters is known as "wave digital filters". These filter structures are obtained by creating digital equivalent of analog ladder filters taking account of the wave variables (cf. transmission lines). An important feature of such filters is that their coefficients can be represented by a small number of bits and they possess a low sensitivity to parameter variation [15]. A detailed discussion on the design of such filters is beyond the scope of this book; interested readers can learn the synthesis of such filters from [34].

9.1.1. Finite-Impulse Response Filter Design

There are several techniques for designing FIR filters in both the time- and frequency-domain. The notable ones are: windowing, frequency sampling, and equiripple techniques (such as the Parks–McClellan method) [30]. The windowing method can be applied easily. Other FIR filter design methods (such as equiripple) are computer-aided iterative procedures. Several powerful design packages are available to aid in the design using windowing or

Parks–McClellan methods [13, 24]. In the following, we include a few illustrative design examples.

Example 1: Windowing Design of a Low-Pass FIR Filter

FIR filter design by windowing is based on truncating an infinite impulse response sequence (that characterizes a general digital filter described by Eq. (9.1)) to a finite length for implementation. To avoid drastic change in the resultant filter amplitude response, the original infinite impulse response coefficients are weighted by a windowing function as part of the truncation process. The designed FIR filters have symmetric impulse response and exhibit exact linear phase [30].

We use the program FWFIR [13] to design a 20-tap low-pass FIR digital filter using Hamming window [30]. The ideal normalized cut-off frequency (at which an equivalent ideal digital filter will have its magnitude response drop from 1 to 0) is specified to be 0.3. All the frequencies are normalized between 0 and 0.5 (assuming sampling frequency to be 1 rad/s). The program outputs 20 impulse response coefficients $H(i) = H(21 - i)$, where $i = 1, 2, \ldots, 20$ corresponding to the coefficients b_k in Eq. (9.2) with $k = i - 1$ and $N = 19$. The output from the program is listed below.

```
**LOWPASS FILTER DESIGN**
IDEAL LOWPASS CUTOFF=    0.3000000
        H(  1)=-0.21685762e-02 =H( 20)
        H(  2)=-0.12141996e-02 =H( 19)
        H(  3)= 0.75119208e-02 =H( 18)
        H(  4)=-0.43643550e-02 =H( 17)
        H(  5)=-0.19996367e-01 =H( 16)
        H(  6)= 0.33076014e-01 =H( 15)
        H(  7)= 0.20369060e-01 =H( 14)
        H(  8)=-0.10842266e+00 =H( 13)
        H(  9)= 0.61939765e-01 =H( 12)
        H( 10)= 0.51180500e+00 =H( 11)
PASSBAND CUTOFF 0.2103 RIPPLE   0.026 DB
STOPBAND CUTOFF 0.3856 RIPPLE -48.741 DB
```

It can be noted that the stop-band attenuation of the designed filter is −48.741 dB. In addition, the ideal cut-off frequency is between the pass-band and the stop-band cut-off frequencies. If this stop-band attenuation does not satisfy the required specification, the program can be re-run using a different number of taps to get a proper FIR filter. The coefficients obtained from the program are with double precision. They need to be converted to binary and truncated to a proper number of bits before implementation (on a DSP integrated circuit). This coefficient quantization will lead to a deterministic error in the characteristic of the filter.

Example 2: Equiripple Design of a Low-Pass FIR Filter

Here we use the program EQFIR [13] as adapted by Computer Audio Research Laboratory [11] to design a 20-tap low-pass FIR digital filter as in Example 1. The four normalized band edge frequencies are specified to be 0.0, 0.2103, 0.3856, and 0.5 to match exactly with the windowing design in Example 1. The output from the program is listed below.

```
*****************************************************************
                  finite impulse response (fir)
                linear phase digital filter design
                    remez exchange algorithm

                        bandpass filter

                       filter length = 20

                   ***** impulse response *****
               h( 1) = -0.19135618e-02 = h( 20)
               h( 2) =  0.22408522e-02 = h( 19)
               h( 3) =  0.83617112e-02 = h( 18)
               h( 4) = -0.11926175e-01 = h( 17)
               h( 5) = -0.14388399e-01 = h( 16)
               h( 6) =  0.41479409e-01 = h( 15)
               h( 7) =  0.51481165e-02 = h( 14)
               h( 8) = -0.10855851e+00 = h( 13)
               h( 9) =  0.81133753e-01 = h( 12)
               h(10) =  0.49943283e+00 = h( 11)

                       band 1           band 2
lower band edge      0.                0.3856000
upper band edge      0.2103000         0.5000000
desired value        1.0000000         0.
weighting            1.0000000        10.0000000
deviation            0.0020201         0.0002020
deviation in db      0.0175288       -73.8925323

extremal frequencies--maxima of the error curve
  0.          0.0656250   0.1187500   0.1624999   0.1968749
  0.2103000   0.3856000   0.3949750   0.4168501   0.4481002
  0.4824754

*****************************************************************
```

Note the tremendous improvement in stop-band attenuation compared to that obtained in Example 1 since the optimization technique used is superior.

9.1.2. Infinite-Impulse Response Filter Design

A standard technique for IIR digital filter design has been to first obtain an equivalent analog filter transfer function (with desired frequency response) and then convert it to a digital transfer function by applying a suitable transformation. This is an analytical technique and many filter design books carry the formulae and relevant coefficient tables. Other techniques (such as Steiglitz' minimum mean square error technique [30]) that use approximation of magnitude and/or phase response directly in the digital domain also exist.

Example 3: Bilinear Transformation Based IIR Filter Design

Design an equiripple low-pass IIR digital filter with a sampling frequency of 8 kHz, pass-band cut-off frequency of 2.1 kHz, stop-band cut-off frequency of 3.2 kHz, a pass-band ripple of 0.02 (0.1755 dB), and a stop-band tolerance of 0.001 (60 dB).

We use the program EQIIR [13] with no optimization for any given second-order section (the program allows optimization by proper pole-zero pairing, ordering, and using an optimal structure). An order five filter satisfies the given specification. A detailed listing of the output from the program is as follows.

```
REALIZED
NORM. CUTOFF FREQ.      1.649336     2.513274

CHOSEN DESIGN PAR.     CX = 0.6618671 (=) C =     20.0451107

UTILIZATION OF THE PASSBAND DELTA P = 0.0049186 = 24.59288 PERCENT
                   STOPBAND DELTA S = 0.0002478 = 24.77571 PERCENT

POLES AND ZEROS IN THE Z-DOMAIN

CONSTANT GAIN FACTOR =    0.7584426e-01

NUM.              POLES                NUM.             ZEROS

 2      -0.176175 +-J*  0.837236      2       -0.824683 +-J*  0.565595
 2       0.024753 +-J*  0.518851      2       -0.926671 +-J*  0.375875
 1       0.161381 +-J*  0.            1       -1.000000 +-J*  0.


BLOCKS OF SECOND ORDER

CONSTANT GAIN FACTOR =    0.7584426e-01

L      B2(L)         B1(L)         BO(L)          C1(L)         CO(L)

1   1.00000000   1.64936602   1.00000000    0.35235086   0.73200119
2   1.00000000   1.85334110   1.00000000   -0.04950522   0.26981941
3   1.00000000   1.00000000   0.           -0.16138065   0.
```

The coefficients obtained from the program correspond to those in Eq. (9.6) with the following equivalences

$$L = i, \quad B2(L) = \beta_{0i}, \quad B1(L) = \beta_{1i}, \quad B0(L) = \beta_{2i}$$

$$C1(L) = -\alpha_{1i}, \quad C0(L) = -\alpha_{2i}$$

9.2. REALIZATION OF DIGITAL FILTERS

9.2.1. Introduction

In a general purpose DSP chip, at least one fast multiplier, and an arithmetic and logic unit (ALU) are provided. We will refer to these as the computation blocks of a digital filter. There are no registers directly corresponding to the delay elements of the filter. Instead, the input signal samples and the intermediate signal values are held in a RAM to provide the delay action. The program running on the DSP chip is responsible for shifting the data out of the RAM and/or coefficients out of the ROM into the multiplier, or the ALU (as the case may be), and putting the results back into the RAM so that the desired filter algorithm is implemented.

The important issues in selecting an algorithm for filtering, and the means of implementing it, are: the signal throughput, the stability of the computations, and the accuracy of the results. The **throughput** is directly related to the architecture. By **architecture**, we mean the ensemble of choices for the representation of data, the mode of implementation of arithmetic functions (e.g., serial or parallel), the amount of resources in terms of the number of multipliers, adders, and registers, and the method of interconnection and control. The **stability and accuracy** of computations are related to the structure of the filter, that is, to the topology of its signal flow graph. We saw for instance the differences between FIR and IIR filters. Among the IIR filter structures, of course, there are some which are less prone to instability and computational errors. Next we will consider the main architectural issues in designing miniaturized digital filters.

9.2.2. General Purpose Digital Signal Processors

The single chip programmable DSP came about only after phenomenal advancement in integrated circuit (IC) design technology. After the introduction of the concepts of common control using microprogramming [36], it took more than two decades of activity in the areas of computers, sampled data processing, and IC design before manufacturers announced [7, 27, 29] their integrated DSPs [2]. In many applications, a single DSP is utilized to realize a collection of tasks including filtering [16]. On the other hand, several ICs are available that perform specific tasks of filtering [1, 26, 38]. We will examine the internals and the modes of operation of these ICs later. However, at this time we would like to explore the differences and the trade-offs of the two approaches to miniaturized filter design.

Consider a simple case first. Figure 9.2 shows the block diagram of a generic microprocessor. Here, at every clock pulse, an instruction is read from the memory (RAM or ROM), and is put into the instruction register of the control section. There, the instruction is decoded and, at the next clock, the corresponding control signals are activated. The executed instruc-

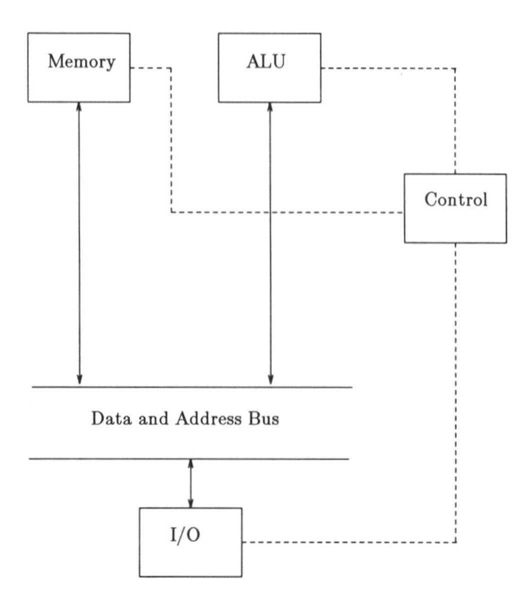

Figure 9.2 A generic microprocessor block diagram.

tion can in turn have requested transfer of one data item from the memory to the ALU, or to the input/output circuitry. Then the next instruction is read from the memory and is executed in a similar manner. Here, unless a (usually conditional) branch instruction is specified, the program counter is incremented, and the instructions are executed sequentially. This architecture has only one storage area (RAM) and a single shared bus for both instruction and data transfer, and the instructions are executed sequentially. It is known as the **von Neumann** architecture. The important feature of this type of architecture is that there is a maximum of two addresses mentioned in any move instruction, one that corresponds to the source of the data and the other corresponding to its destination. Instructions that need more than one data movement, manage this task by repeated single transfers. As there is only one memory space, instructions and data share the same storage space. Programming of such machines is tantamount to specifying transfer of data between the memory and the ALU in a repeated and orderly manner for executing the desired algorithm. This type of programming is simple in concept, and easily translatable from flow-charts to code. Most algorithms to date have been sequential in nature, thus the von Neumann architecture has been so prevalent for many years. The disadvantage of such an architecture for digital signal processing is also due to the sequential execution of instructions. For instance, in order to start a multiplication (assuming there is such a computation unit in a given processor), first two data items from the memory would have to be transferred to the multiplier. This means that the multiplier unit would remain idle for the duration of the two data transfers. Certainly, this architecture is not suitable for high throughput systems with large amounts of intermediate computations.

In order to alleviate this so-called von Neumann bottleneck, one can separate the instruction and the data memory spaces. This will allow simultaneous fetching of an instruction from the program memory, and execution of another instruction dealing with the data memory. The instruction and data usually travel on separate buses here. This kind of architecture is commonly known as the **Harvard** architecture. In fact, most of today's general purpose DSPs are based on variations of the Harvard architecture and allow simultaneous addressing of two operands in a single instruction. For instance, in the case of multiplication, two addresses are used for two required operands, and the third address indicates the destination for storing the result.

Figure 9.3 shows the block diagram of a generic programmable DSP. Two or more memory blocks (RAM and ROM) are provided in these DSPs. Data and instructions have their own separate memory spaces. Ordinarily, instructions reside in ROM. An adequate number of buses are built to ease communication between the memory blocks, ALU, multiplier-accumulator (MAC), and the I/O ports. In these DSPs, instructions can have three or more operands (addresses) specified. Presently, there are two types of arrangements for the data memories. Either there are two independent data memory spaces [17, 22], or there is only one data memory space [21, 28, 29]. Each memory space can itself be composed of several physical memory blocks. In the latter DSP [21] (Figure 9.10), while there are three physical

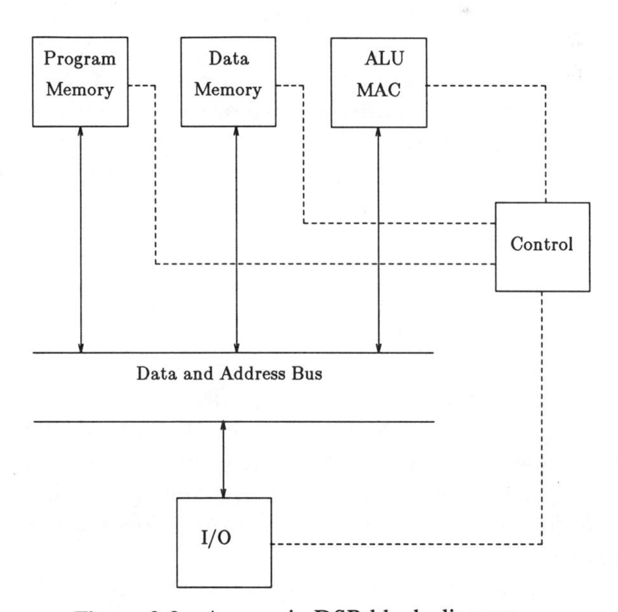

Figure 9.3 A generic DSP block diagram.

memory blocks, they are all in one memory space and accessible only through one set of address and data buses. In a true Harvard machine, the accesses to the ROM and the RAM are completely independent of each other. Here, these accesses have to be sequential because of the availability of only one set of buses. In order to improve the efficiency of transfers from/to the memories in this DSP, the two memory accesses are staggered in time with respect to each other. Thus the rate of memory reads and writes are doubled.

In any generic processor, the fetch, decode, and operate cycles of an instruction should all follow each other sequentially. However, one can add some facility for fetching a second instruction, after the first instruction is obtained, and is being decoded. Next, when the second instruction is being decoded and perhaps the first instruction is still completing its execution, a third instruction can be fetched. The resulting effect is called **pipelining**. When pipelining von Neumann machines containing only one physical memory block, the amount of clashes between different memory accesses (memory contention) is prohibitive. On the other hand, pipelined Harvard machines have higher throughput because of less memory contentions due to separation of data, and instruction memory spaces. In fact, instructions of these machines can target two items from the data memories to the computation units, and specify an address for the returning result, but they do not wait for the completion of the computations (e.g., multiplication) before they execute the next instruction. The result will be written back to the memory as scheduled by the controller (usually after one or more instructions depending on the computation pipeline). Pipelining in a computation unit refers to insertion of registers at convenient places in the unit to save intermediate values. In this manner, as long as intermediate data arrive at a safe place in the next set of registers, a new set of operands can enter the computation pipe without affecting the previous calculations. Therefore, instruction pipelining will be efficient only if lengthy computations are also pipelined. In this set-up, one tries to repeatedly issue instructions which target the computation unit in order to utilize the computation pipe to its fullest extent.

Thus by keeping the pipeline busy at all times, maximum throughput is achieved. The worst situation is when a computation unit, such as the multiplier, needs the result of its previous computations. Then, pipelining is not of any help and creates more sluggishness in the string of calculations. This is the so-called breaking of pipeline effect. Therefore, one has to be very careful in programming these machines. In order to facilitate writing such programs, some companies (third party support) provide development software that allows the design of filters and generation of assembly language code for the target DSP starting from frequency-domain specifications automatically [4].

Important metrics in selecting a general purpose DSP for filtering are: the instruction cycle time, accuracy of computations, amount and type of memory, ease of programming, and the I/O capabilities. All of the present day DSPs support multiply-accumulate instructions with an on-chip MAC. In almost all cases, a pipeline register separates the multiply and accumulate sections from each other. The instruction cycle time is thus very much related to the multiplication or accumulation time. This in itself depends on the wordlength and whether fixed or floating point is used for the representation of the data. The instruction cycle time of today's DSPs varies from 60 [17] to 200 ns, and above. The data widths are commonly 16 bits with fixed point arithmetic. However, 22-bit floating point (Oki MSM6992 chip), and 32-bit floating point [21] are also available. The amount of on-chip memory (RAM/ROM) varies from none (Analog Devices ADSP-2100), up to 128K bits (μPD77230 NEC) [20] and above. We should emphasize that the amount of available memory address space determines the size of application that can be fitted on a DSP, and not the amount of on-chip memory provided. In fact, in some applications, the arrangement provided by Analog Devices ADSP-2100 [3] is advantageous. The advantage of floating point in a DSP is the increase in dynamic range of operands. This reduces roundoff noise problems and frees the filter designers from the burden of scaling, a problem in programming fixed-point DSPs [8, 18]. The price paid for this capability is slower instruction cycle times, larger die size, and more power consumption. The number of I/O ports and their capabilities matter when one wants to communicate with the outside world. The serial ports are basically provided for PCM lines (e.g., codec applications) or for any serial mode of communication among several ICs. The parallel ports are mostly for sharing the global data memory spaces with other DSPs or for providing direct memory access (DMA) to other processors in a multiprocessing environment. When a master controller loads an application program into the on-chip memory of the DSPs, it usually uses the parallel port.

9.2.3. Case Studies

We will now look at two commonly used general purpose programmable DSPs in detail. We have chosen a fixed-point DSP, Texas Instruments (TI) TMS32020, and a floating-point DSP, AT&T WEDSP32 from two leading US DSP manufacturers.

In 1985, TI introduced its second-generation DSP, TMS32020 [27], and AT&T announced its 32-bit floating point DSP, WEDSP32 [21]. Both TMS32020 and WEDSP32 chips were first built in NMOS, and their CMOS versions are now available.

9.3. TEXAS INSTRUMENTS' TMS32020

A block diagram of the TMS32020 chip [14] is shown in Figure 9.4(a) and a photomicrograph of the chip is shown in Figure 9.4(b). The Harvard architecture of this chip is based on the use of two memory spaces (data and

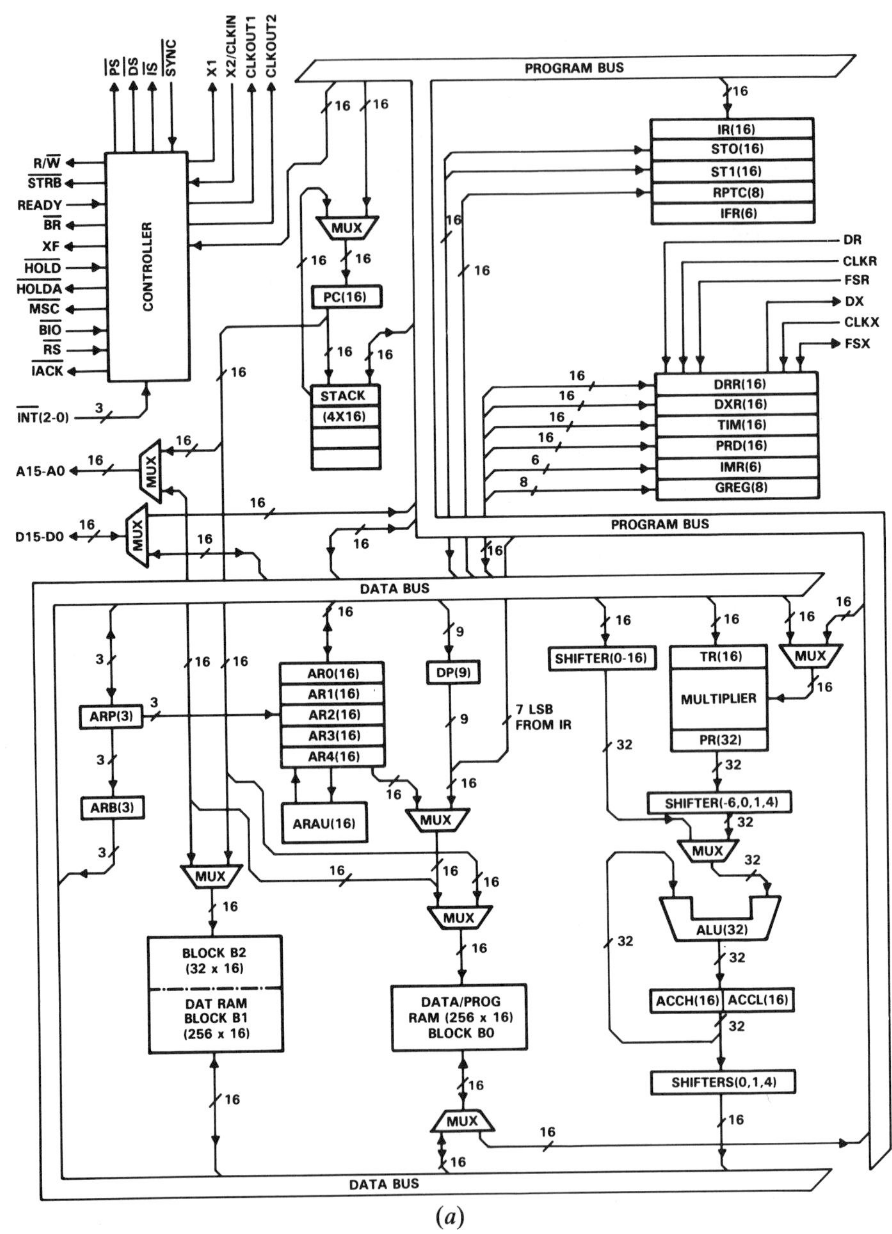

(*a*)

Figure 9.4 Functional block diagram and photomicrograph of the TMS32020. From TMS32020 User's Guide, © 1985, Texas Instruments, Inc.

(*b*)

Figure 9.4 (*Continued*)

instruction memory spaces), with each of these spaces containing on-chip and off-chip physical memory blocks. One physical memory block consists of an on-chip RAM of 256 16-bit words (block B0) and it can be allocated to either the data or program memory spaces. There are two other physical memory blocks, namely block B1 of 256 words and Block B2 of 32 words. These two RAMs are always used as data memory. The reason for assigning two RAM blocks to the second memory space is that they have noncontiguous address spaces. There is no (program or coefficients) ROM on chip, but a total of 128K words of off-chip memory address space is provided. External memory (RAM or ROM) can be supplied for both data (total data memory, 64K words) and program memory (total program memory, 64K words). Since the data and address buses for the corresponding memories are multiplexed, control signals are provided to select correct physical memory at any given time. There are two instructions, namely, configure block as data memory (CNFD) and configure block as program memory (CNFP) for specifying the desired configuration of the memories. In Figure 9.5(a), we see the address assigned to each block of memory. In particular, notice that block B0 starts from the hexadecimal address FF00 (of program memory) when configured as program memory and from the hexadecimal address 0200 (of data memory) otherwise. (Hexadecimal addresses will be distinguished by a greater than sign, >, from now on.) Also notice that block B1 always starts at address >0300.

There are two 16-bit wide buses inside the chip called data bus and program bus. The program bus carries the instruction code and immediate operands from the program memory. The data bus interconnects various elements, such as the central arithmetic logic unit (CALU) and the auxiliary register file to the data RAM. An important building block inside this chip is the 16×16 two's complement multiplier. Normally with one internal data bus, two instructions are needed for obtaining the results of multiplication of two numbers, one for bringing in the multiplier and the other for the multiplicand. By allowing transfers from the data and program memory spaces to the multiplier inputs, two operands can be fetched simultaneously. Thus an effective single instruction cycle multiplication-accumulation becomes possible. The multiplier can have its operands simultaneously from the two on chip data spaces, one via the data bus (block B1 or block B2), and the other (block B0) from the program bus. Therefore, when this mode of operation is desired, the block B0 of RAM should be configured as program memory so that the operand residing there can be addressed as a program item and can be transferred to the multiplier via the program bus. This configuration of the memory is shown in Figure 9.5(b). Other manufacturers [22, 29] have opted for two bypass buses directly between the data RAMs and the multiplier inputs. This, at the expense of more buses, simplifies the communication needed. The result of a multiplication is always captured in a register called product register (PR). The output of this register always feeds the ALU and therefore the result of the previous multiplication gets accumulated (in ALU) while a new multiplication is taking place (due to pipelining). The multiplication takes 150 ns (in the 2.4 μm NMOS version). The accumulation also takes 150 ns. The instruction cycle is also 150 ns. Because of pipelining, successive instructions of multiply–accumulate (along with proper data movements) can be executed to have an effective throughput of single instruction cycle time, that is, 150 ns. In fact an instruction exists (RPTK), that arranges for repeated execution of its following instruction. Therefore, for an FIR filter, every tap calculation (in a repeat instruction mode) can be done in 150 ns. One should be aware of the fact that for a small number of multiplication-accumulations the overhead of setting up the repeat instruction mode may defeat the advantages mentioned above, and therefore is not recommended [35]. In such a case, in-line coding (writing the multiplication-accumulation instruction as

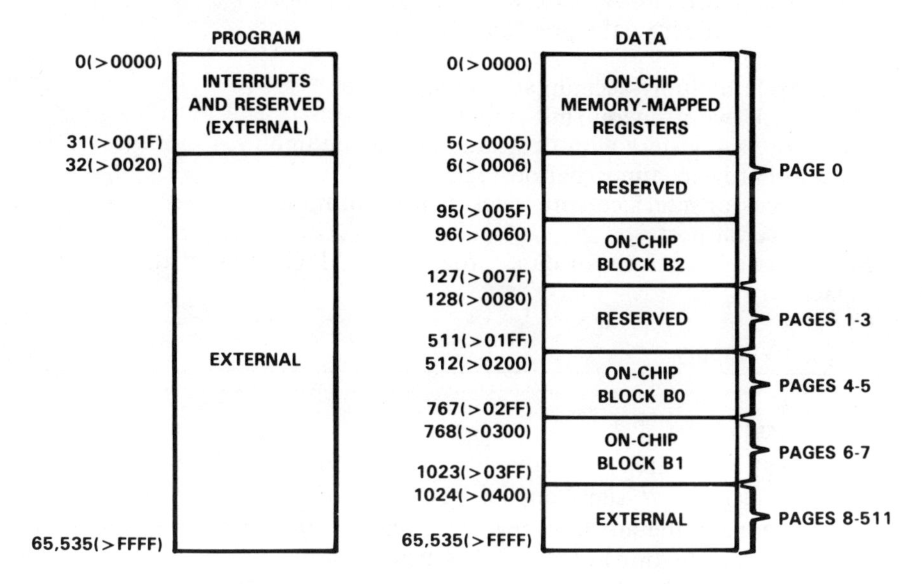

(a) ADDRESS MAPS AFTER A CNFD INSTRUCTION

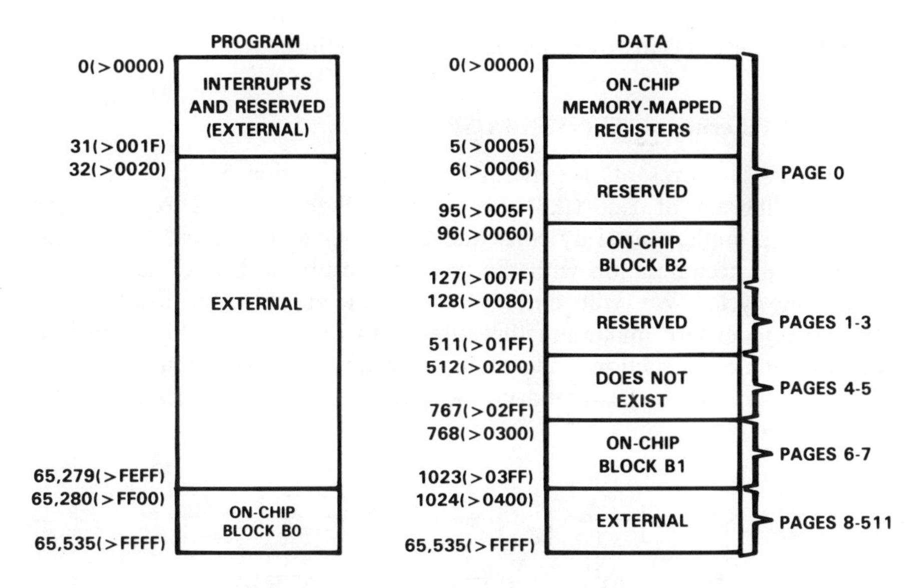

(b) ADDRESS MAPS AFTER A CNFP INSTRUCTION

Figure 9.5 TMS32020 memory maps. From TMS32020 User's Guide, © 1985, Texas Instruments, Inc.

Table 9.1 TMS32020 Key Features

- 64K word data memory, 64K word program memory spaces
- 544 × 16-bit on-chip data/program RAM
- 150 ns cycle time
- Single-cycle multiply-accumulate operation
- 32-bit ALU with several shift options
- Five address registers with dedicated arithmetic unit
- On-chip hardware timer/period registers
- Multiprocessor interface with clock synchronization
- 8/16-bit serial port
- 16-bit parallel interface for data, program, and I/O access with wait state capability
- 68-pin grid array package
- Single 5 V supply

many times as desired) is advised. A high degree of parallelism in TMS32020 also allows simultaneous operation of CALU and arithmetic/logic operations in the auxiliary register arithmetic unit (ARAU). Table 9.1 shows the salient features of the TMS32020.

Next, we are going to examine programming the TMS32020 DSP for filtering applications. There are more than 100 instructions for this chip. We will explain only the instructions we use for our purposes.

9.3.1. FIR Filtering on TMS32020 DSP [25]

Equation (9.3) represents a general difference equation for an FIR filter. The flow-diagram of Eq. (9.3) is shown in Figure 9.6. The z^{-1} delay elements are implemented by consequent data RAM locations. One multiplication and accumulation with the previous results will be done at every instruction cycle. We will employ all of the parallelism available in TMS32020 to ensure maximum throughput. In order to be able to benefit from the single instruction cycle multiply-accumulate operation (MACD), we have to configure the memory block B0 as program memory. We also

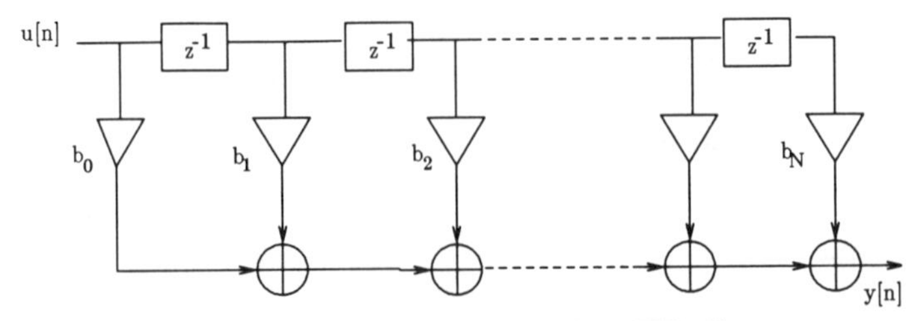

Figure 9.6 Signal flow graph of an FIR filter.

have to arrange the signal samples, and the coefficients in the memory in the manner shown in Figure 9.7.

The instruction MACD stands for multiply and accumulate with data move. It has several modes of operation. Among them, the following case is of interest. **MACD ⟨pma⟩**, {*−} will multiply the content of the program memory (block B0), addressed by the value of program memory address (i.e., ⟨pma⟩), by the content of the data memory (block B1), addressed by the current pointer (desired pointer register should be selected before this instruction). In the same instruction, after the operands have been fetched, the operand in the data memory will be copied to the next higher on-chip RAM location. This in effect realizes the z^{-1} delay action. Also, the current data memory pointer will be decremented (as requested by the *− convention in the MACD instruction above). This pointer decrement will prepare the machine to address the next coefficient. The program memory address ⟨pma⟩ which in this instruction is stored in the program counter (PC), is also automatically incremented. It can be recalled that the multiply-accumulate block is pipelined. Thus when the MACD instruction is completed (after one instruction cycle), the result of the latest multiplication has not yet accumulated to the previous result. This will be taken care of by an extra **APAC** (add p register to accumulator) instruction.

Another instruction that we can take advantage of is **RPTK ⟨constant⟩** (repeat following instruction as specified by immediate value). In this instruction, the 8-bit immediate value (constant) is loaded into a dedicated counter, called RPTC. Several other instructions (such as MACD) condition their operations on the content of this counter. That is, if the content of

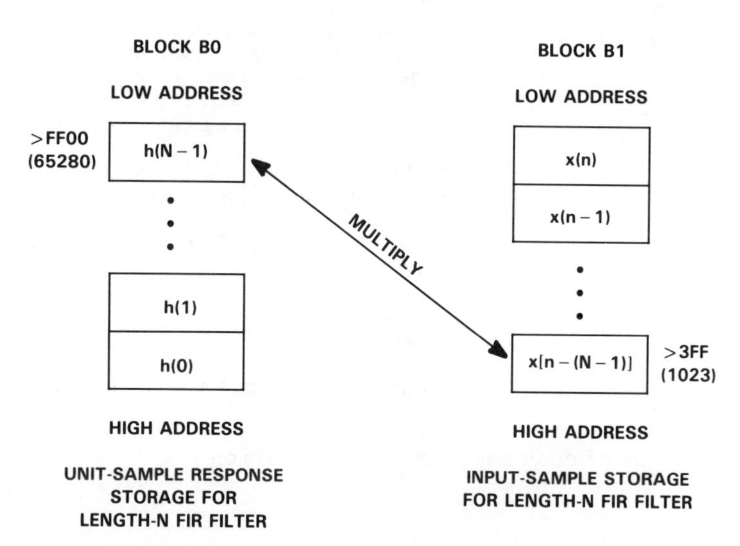

Figure 9.7 TMS32020 memory storage scheme. From TMS32020 User's Guide, © 1985, Texas Instruments, Inc.

RPTC is greater than zero, they will continue their normal operation and arrange for their re-execution while at the same time decrementing RPTC after each execution. However, if the content of RPTC is zero, these instructions will finish the execution of current instruction and arrange for the normal execution of the next instruction in the sequence. Notice that in order to repeat an instruction N times, one has to set the initial value of RPTC to $N-1$.

With the memory arrangement of the signal samples, and the coefficients as depicted in Figure 9.7, and the two instructions described above, we can now write the program for an Nth order FIR filter ($N \leq 255$).

```
*
      CNFP            * Use block B0 as program area.
*
* This section of code polls the input port to bring in the next
* sample.
*
WAIT BIOZ NXT         * ''BIO'' pin goes low when a new sample
                      * is available. This instruction looks
                      * to find out if ''BIO'' pin is zero. If
                      * so, a branch to NXT takes place. Else,
                      * the next instruction loops the control
                      * back to WAIT.
*
     B    WAIT        * Branch to WAIT.
*
NXT  IN   XN,PA0      * XN is a constant showing the desired place
                      * for storing the incoming data. This
                      * address, i.e. XN is specified as an offset to
                      * the beginning page of selected memory block.
                      * Its value depends on the order of the filter.
                      * PA0 refers to the I/O port address 0.
                      * IN brings the data (single sample) from
                      * peripheral on port address 0.
                      * NXT is the label of this instruction. It
                      * will be used whenever we like to jump to this
                      * instruction for bringing the next sample.
*
     LRLK AR1,>3FF    * The auxiliary register AR1, is loaded with
                      * the address of the highest point in block B1.
*
     LARP AR1         * By loading the ARP register with 1, we are
                      * selecting AR1, as our pointer register.
*
     MPYK 0           * Multiplication by constant 0, will clear the
                      * product register.
*
     ZAC              * Zero accumulator clears the accumulator.
*
*
* This section of code implements the equation
*
*
```

$$* \sum_{k=0}^{N-1} h[k]u[n-k]$$

```
*
*
```

```
        RPTK NM1          * NM1 is a constant equal to N-1.
                          * RPTK as explained above, will repeat the
                          * following instruction N times.
*
        MACD >FF00,*-     * This instruction was explained in detail
                          * above. It multiply-accumulates according to
                          * the above FIR equation.
*
        APAC              * Add Product Register to accumulator.
*
        SACH YN,1         * YN is a constant indicating the desired
                          * location in the data memory (B2)
                          * for storage of the output samples.
                          * We have assumed that all of the input samples
                          * and the coefficients have values ≤1.
                          * Therefore, the result of the all multiply-
                          * accumulations is 32 bits, with the decimal
                          * point between the 30th, and 31st bits.
                          * SACH copies the entire accumulator into a
                          * shifter. It then shifts this entire 32-bit
                          * number 1 bit (as we have asked) to the left,
                          * and stores the upper 16 bits of the shifted
                          * value into data memory at location YN.
*
        OUT   YN,PA1      * Outputs the filter response now stored at
                          * location YN, to peripheral on port address 1.
*
        B     WAIT        * Branch to get next input sample.
*
```


For an FIR filter with 80 coefficients, it takes 95 cycles to execute the above program. Therefore with a TMS32020 which has 150 ns instruction cycle time, we can filter signals at a rate as fast as $95 \times 150 = 14250$ ns intervals, or at 70 kHz sampling rate.

9.3.2. IIR Filtering on TMS32020 [25]

The flow graph of an IIR filter implemented with the cascade of second-order direct form II sections (Eq. (9.6)) is shown in Figure 9.8. One should

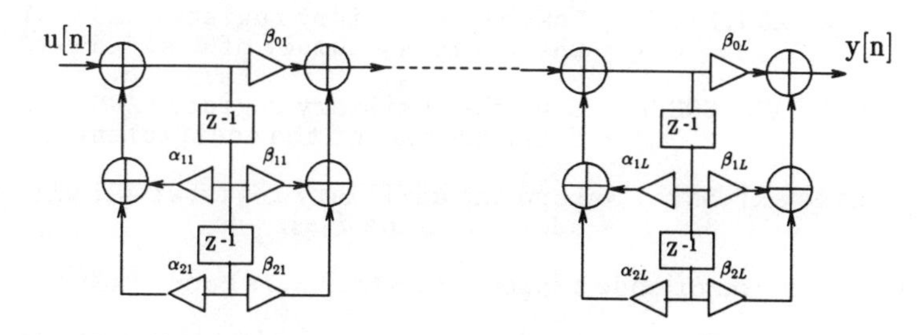

Figure 9.8 A cascade of direct form II second-order sections.

α_{11}
α_{21}
β_{21}
β_{11}
β_{01}
α_{12}
α_{22}
β_{22}
β_{12}
β_{02}
α_{13}
α_{23}
β_{23}
.
.
.
etc.

Figure 9.9 Coefficient storage sequence.

emphasize that because of the finite precision, and fixed point multiplier of TI TMS32020, one has to pair appropriately the poles and the zeroes. Also one should insert interstage multipliers for gain distribution [18]. However, in the implementation below these points are not explicitly shown.

Now we will show the TMS32020 code for the implementation of IIR filters in the form of cascaded second-order direct form II sections. We will assume that there will be S second-order sections. Also, we will store all the coefficients of the S sections as shown in Figure 9.9.

```
*
* This section of code polls the input port to bring in the next
* sample.
*
WAIT BIOZ NXZ         * The function of these two lines was
     B    WAIT        * explained in the FIR program.
*
INPT IN   XN,PA2      * Bring in the new sample u[n].
*
* This section of code sets up the counters.
*
     LARK AR3,SM1     * Load the auxiliary register AR3 with S-1,
                      * where S is the number of desired sections.
*
     LAR  AR0,COEFP   * Load the auxiliary register AR0 with the
                      * first address of the coefficients.
*
     LAR  AR1,DNP     * Load the auxiliary register AR1 with the
                      * address of the d_i[n].
*
* This section of code implements Eqs. (9.7a) and (9.7b).
*
LOOP LAC  XN,15       * Load (with correct bit alignment) the
                      * accumulator with the input to the filter.
```

```
*
      LARP AR1          * Choose AR1 as current pointer.
      MAR   *+          * Point at d_i[n - 1]
      LT    *+,AR0      * Load T register (one input to the multiplier)
                        * with the data from d_i[n - 1].
                        * Also point at the address of d_i[n - 2].
                        * Choose AR0 as next pointer
                        * (for addressing the coefficients.)
*
      MPY   *+,AR1      * Calculate α_1i d_i[n - 1].
                        * Point at α_2i.
                        * Choose AR1 as next pointer.
*
      LTA   *-,AR0      * Load T register with d_i[n - 2].
                        * Calculate u[n] + α_1i d_i[n - 1].
                        * Point at d_i[n - 1].
                        * Choose AR0 as next pointer.
*
      MPY   *+,AR1      * Calculate α_2i d_i[n - 2].
                        * Point at β_2i.
                        * Choose AR1 as next pointer.
*
      APAC              * By adding the product register to the
                        * accumulator,
                        * Calculate u[n] + α_1i d_i[n - 1] + α_2i d_i[n - 2].
*
      MAR   *-          * Point at d_i[n].
*     SACH  *+,1        * Store d_i[n] (with proper bit alignment).
                        * Point at d_i[n - 1].
*
      ZAC               * Clear accumulator. Also, notice that the
                        * multiplier T register is still loaded with
                        * d_i[n - 2].
*
      LARP AR0          * Choose AR0 (for coefficients) pointer.
      MPY   *+,AR1      * Calculate β_2i d_i[n - 2].
                        * Point at β_1i.
                        * Choose AR1 (for data) pointer.
*
      LTD   *-,AR0      * Load multiplier T register with d_i[n - 1].
                        * move β_2i d_i[n - 2] to the accumulator.
                        * Also, d_i[n - 2] = d_i[n - 1].
                        * Point at d_i[n].
                        * Choose AR0 (for coefficients) pointer.
*
      MPY   *+,AR1      * Calculate β_1i d_i[n - 1].
                        * Point at β_0i.
                        * Choose AR1 (for data) pointer.
*
      LTD   *+,AR0      * Load multiplier T register with d_i[n].
                        * Calculate β_2i d_i[n - 2] + β_1i d_i[n - 1].
                        * Also, d_i[n - 1] = d_i[n].
                        * Point at d_i[n - 1].
                        * Choose AR0 (for coefficients) pointer.
*
      MPY   *+,AR1      * Calculate β_0i d_i[n].
                        * Point at the α_1i of the next section.
                        * Choose AR1 (for data) pointer.
*
      MAR   *+          * Point at d_i[n - 2].
      MAR   *+          * Point at d_i[n] of the next section.
```

```
*
      APAC              * By adding the product register to the
                        * accumulator,
                        * Calculate  y_i[n] = β_0i d_i[n] + β_1i d_i[n-1]
                          + β_2i d_i[n-2].
*
      SACH XN,1         * Store the output of the ith section in the
                        * same place where the input of the i+1 section
                        * is expected.
*
      LARP AR3          * Use third auxiliary register.
      BANZ LOOP         * If all of the sections have not been
                        * calculated, go to process the next section.
*
OTPT OUT  XN,PA2        * Output the filter response y[n] (now in the
                        * same location as previous XN).
*
      B    WAIT         * Branch to process next input.
*
```

It takes 24 cycles to compute each second-order section. Therefore, signals sampled up to 90 kHz can conveniently be filtered with IIR filters composed of three second-order sections.

9.4. AT&T WEDSP32

The WEDSP32 is a high-speed, 32-bit **floating-point** digital signal processor with 32-bit instructions. It can be programmed to implement many signal processing applications.

A block diagram of WEDSP32 is shown in Figure 9.10(a) and a chip photomicrograph in Figure 9.10(b). This chip has one memory space that consists of 2,048 bytes of ROM (512 32-bit words), and 4,096 bytes of RAM (2×512 32-bit words). Data can be addressed as 8-, 16-, or 32-bit words. In the 100-pin pin-array package, memory can be expanded off-chip making 56K bytes of memory directly accessible.

The WEDSP32 has two execution units: the control arithmetic unit (CAU) and the data arithmetic unit (DAU). The CAU is a 16-bit, fixed-point integer unit that supports control, addressing, arithmetic, and logic functions. The DAU is a 32-bit floating-point unit for signal processing computations.

This chip has several on-chip I/O units. The serial I/O unit interfaces with most codecs and time-division multiplex (TDM) lines with few (if any) additional components. The parallel I/O interfaces the WEDSP32 device with a microprocessor. Data transfers can be made directly with the WEDSP32 memory (DMA) without program intervention.

The DAU is the main execution unit for filtering and other signal processing algorithms. It contains a floating-point multiplier and an adder, four static 40-bit accumulators, and the data arithmetic unit control register

(DAUC). The DAU performs multiply-accumulate operations on signal processing data and does data type conversions. The form of instruction executed in the DAU is: $a = b + c \times d$. The DAU multiplier inputs can be from memory, I/O registers, or from one of four accumulators. These inputs are 32-bit floating-point numbers with a 24-bit mantissa and an 8-bit exponent. The adder inputs can be from memory, I/O registers, or from an accumulator and are 8-, 16-, 32-, or 40-bits wide. An accumulator input of 40 bits (nominal 32 bits plus 8 mantissa guard bits) can only come from one of the four accumulators, a0 to a3, or from the multiplier.

The control arithmetic unit (CAU) is used for generating addresses to memory and executing instructions that operate on 16-bit integers. It is

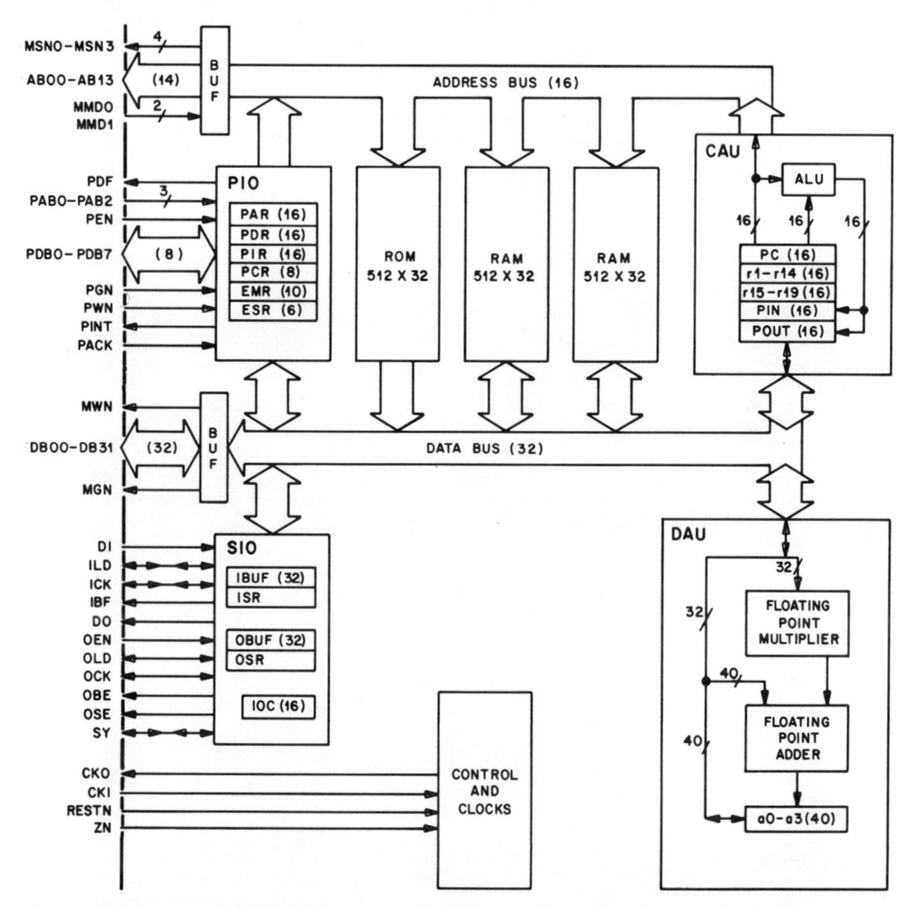

Figure 9.10 Functional block diagram and photomicrograph of the WEDSP32. From J. R. Boddie et al. (1986) The DSP32 digital signal processor and its application development tools. *AT&T Tech. J.* **65**, 89–104 © 1986, AT&T.

Figure 9.10 (*Continued*)

capable of executing a full set of such instructions. The CAU contains 21 16-bit general-purpose registers (r1–r21), a 16-bit program counter (PC), and an arithmetic logic unit (ALU).

Although in this chip there are two memory banks, the architecture is not truly Harvard. Here, the two memory banks share the same bus. Therefore transfer of the operands to the DAU and transfer of data to the CAU are time multiplexed on the internal bus. Every processor cycle (of 250 or 160 ns) is composed of four clock cycles. In three of the four clock cycles, read-from-the-memories operations are scheduled, while one clock cycle is reserved for writing back into the memory. Usually the three read operations in every processor cycle are devoted to instruction fetch and two operand fetches, respectively. The instructions of this processor are of a level higher than other DSPs. For example, a single instruction can request two operands from the memory, send them to the floating-point DAU to get them multiplied, the result to be added to one of the four available accumulators, then the new result to be registered in another accumulator, and finally a copy of the result to be written back in the memory at the address pointed to by one of the 21 registers of the CAU (the pointers themselves can also be post-modified inside the CAU). Interestingly enough, the processor is capable of executing other instructions while the data is being worked on in the DAU. Since the internal sections of the DAU itself are pipelined (two-stage multiply-accumulate), the next instruction can also send its operands to the DAU, without affecting the previous calculations. In this manner, at every given processor cycle, up to six instructions can be in different stages of fetch, decode, and execute. This also means that the result of an instruction may not be ready for use in the following instruction. The amount of **latency** for obtaining the result of each instruction varies and we will discuss them as we encounter them.

Although there is only one internal data and address bus, the time multiplexing of the bus has created a virtual Harvard architecture. On the other extreme, one could have provided a separate bus for each memory bank. Then the bottle-neck would have been the speed of floating-point multiplication. In that case, knowing that for a given technology, floating-point multiplication takes a certain amount of time, either the CAU had to sit idle, or more pipelining would have been necessary in the DAU section. The present scheme of time multiplexing the bus is a good compromise between the two extremes. Also, this level of pipelining makes WEDSP32 a very efficient machine [23]. Another important point here is that the floating-point multiplier and accumulator of the WEDSP32 produce normalized results. This is very convenient, and allows correct computations of the signal processing algorithms without the need for any special measures for normalization.

Table 9.2 shows some salient features of WEDSP32 device.

Because of the abundance of WEDSP32 instructions, only the instructions used in the example code fragments presented below will be explained.

Table 9.2 WEDSP32 Key Features

- On-chip 2048 bytes ROM and 4096 bytes RAM
- 16/25 MHz operation
- 32-bit floating-point (normalized) multiplication and addition
- 16-bit integer operations
- Four 40 bit accumulators
- Four user selectable memory arrangements
- 8/16/32-bit serial port with DMA
- 8-bit parallel port with DMA
- Off-chip memory interface (pin-array package only)
- 100-pin pin-grid array or 40 pin package
- Single 5 V supply

9.4.1. FIR Filtering on WEDSP32 [5, 8]

First we will examine programming the WEDSP32 for FIR filtering. The difference equation of the filter to be programmed is given in Eq. (9.3) for an Nth order FIR filter. The corresponding signal flow graph is shown in Figure 9.6. A direct form implementation of the above summation is available as "zfir" routine in AT&T DSP 32 Library Routine [5]. A single DSP32 instruction can perform all of the operations needed to perform one tap of FIR filter; fetch two operands, multiply, accumulate, and shift data samples. Optimum performance is achieved using in-line coding because the DSP32 lacks a low-overhead looping capability (the CMOS version of this chip contains this feature).

```
/* This section of code implements the equation
```

$$\sum_{i=0}^{N-1} h[i]u[n-i]$$

```
Arguments:
   N        Length of the FIR filter N(>2)
   r11      points to the previous signal samples
            where the samples are arranged from oldest to the
            newest and the pointer shows the oldest first.
            Memory bank 1 is used for storing the signal samples.
   r12      points to filter coefficients
            where the coefficients are arranged in reverse
            order and stored in consecutive memory locations.
            Memory bank 0 is used for storing the coefficients.
   r1       auxiliary pointer
   r2       holds the order of the filter
Performance
            The execution time for a FIR filter of length N (>2),
            and the above memory arrangement
            is (12+2*N) cycles. */

#define Nminus3 N-3     /* loop counter related to filter order */
```

```
.global sample, repeat, end, coef, data

        ioc = 0x0987   /* The serial I/O control register
                          is set to the following:
                        * internal frame synchronization -
                          (8 kHz)
                        * load and synchronization clocks from
                          the same source
                        * 8 samples per frame
                        * external input clock
                        * external loading of I/O buffer from
                          the input serial to parallel register
                        * input data length 16 bits
                        * internal output clock
                        * external loading of output serial to
                        * parallel register from the I/O buffer
                        * output data 16 bits
                        * clear sanity, no DMA */
sample: if (ibe) goto sample
                       /* If the input buffer is empty
                          loop on ''sample''. */
        r2 = Nminus3   /* Note that this
                          instruction is always executed. */

        r1 = r11       /* r1 will be used to shift samples */
        a0 = float(ibuf)
                       /* 16 bit input sample is converted to
                        * 32 bit floating-point data */
        a1 = *r11++ * *r12++
                       /* h[N-1] × u[N-1] is calculated, and
                          the pointers are advanced. */

repeat: if (r2-- >=0) goto repeat
                       /* N-2 times next instruction
                          will be executed. */

        a1 = a1 + (*r1++ =*r11++) * *r12++
                       /* sum of h[N-i] × u[N-i] is calculated
                          for i from 2 to N-1. At every step
                          u[i] is replacing u[i-1]. */

        a0 = a0 + (*r1=a0) * *r12
                       /* u[0] × h[0] is added to the
                        * total sum, i.e. the output.
                        * u[-1] is replaced by u[0]. */
        goto sample
end:    obuf= a0 = int(a0)
                       /* Output sample is sent to serial
                          I/O */

coef:   N*float        /* Specify N coefficients */

data:   N*float 0.0   /* Initialize signals to zero */
```

9.4.2. IIR Filtering on WEDSP32 [8]

We will now show how to program the WEDSP32 for implementing the transfer function in Eq. (9.1). The equivalent difference equation given in

Eq. (9.4) will be implemented as a cascade of second-order sections, as in Eqs. (9.7a) and (9.7b) (see Figure 9.8). In the following particular example, it is assumed that the WEDSP32 is connected to a 16-bit linear codec and it employs four-coefficient second-order IIR sections. With the WEDSP32 floating-point capability there is no need to consider pole-zero pairing, ordering of the sections, or gain distribution associated with a fixed point arithmetic implementation (which would also require five-multiplier sections to overcome these problems).

```
/* Cascaded 2nd order IIR filters
   r11 & r13 point to data (state variables)
   r12 points to filter coefficients */

/* The following define statements are added only for readability
   of the code. Each variable defined below will be replaced by
   its equivalent statement during assembly. */

#define Nminus2   7        /* Number of Sections - 2 */
#define Ai1  *r12++        /* Ai1, Ai2 are ith section
#define Ai2  *r12++           denominator coefficients */
#define Bi1  *r12++        /* Bi1, Bi2 are ith section
#define Bi2  *r12++           numerator coefficients */

.global sample, repeat, end, coeff, data

 ioc = 0x987               /* see the FIR program for
                              details */
sample:  if (ibe) goto sample
         r10=Nminus2       /* Nine-section IIR filter */

         a0 = float(ibuf)  /* Converts to 32-bit floating
                              point */

         r11=data          /* Initialize pointers */

         r13=data          /* Will be used for shifting
                              data */

         a0 = a0 * *r12++  /* (input) x (constant
                              multiplier) */

repeat:  a0 = a0 + *r11++ * Ai1
                           /* Eq. (9.7a) is being calculated */

         *r13++ = a0 = a0 + *r11-- * Ai2
                           /* Eq. (9.7a) is complete. The
                              pointer to the state variable d_i[n-2]
                              is set back to d_i[n-1].
                              The value of d_i[n] calculated
                              now, is being sent to memory to
                              replace d_i[n-1]. This will
                              happen after four instructions */

         a0 = a0 + (*r13++ = *r11++) * Bi1
                           /* Eq. (9.7b) is being calculated.
                              Also d_i[n-1] is going to replace
                              d_i[n-2] after four instructions. */
```

```
        if (r10-- >=0) goto repeat
                            /* Check to see if all sections are
                               done. Notice how the advantage of
                               pipelining is utilized. The next
                               instruction is always going to be
                               computed. That will complete
                               Eq. (9.7b) before computing for
                               the next section begins. */

        a0 = a0 + *r11++ * Bi2
                            /* One section output is calculated. */
        goto sample         /* Next instruction will be executed
                               before going back to sample. */
end:    obuf=a0=int(a0)     /* Converts back to 16-bit format and
                               sends the result to serial I/O. */

coef:   float 9.5124534e-7
                            /* Constant multiplier */
        float 1.3898448, -0.96402235, -1.5164367, 1.0
        float 1.3689732, -0.96678623, -1.2360012, 1.0
        float 1.4146495, -0.96782737, -1.5582712, 1.0
        .
        .
        .
        float 1.4683514, -0.99540254, 0.,

data:   30*float 0.0        /* Initializes data to zero */
```

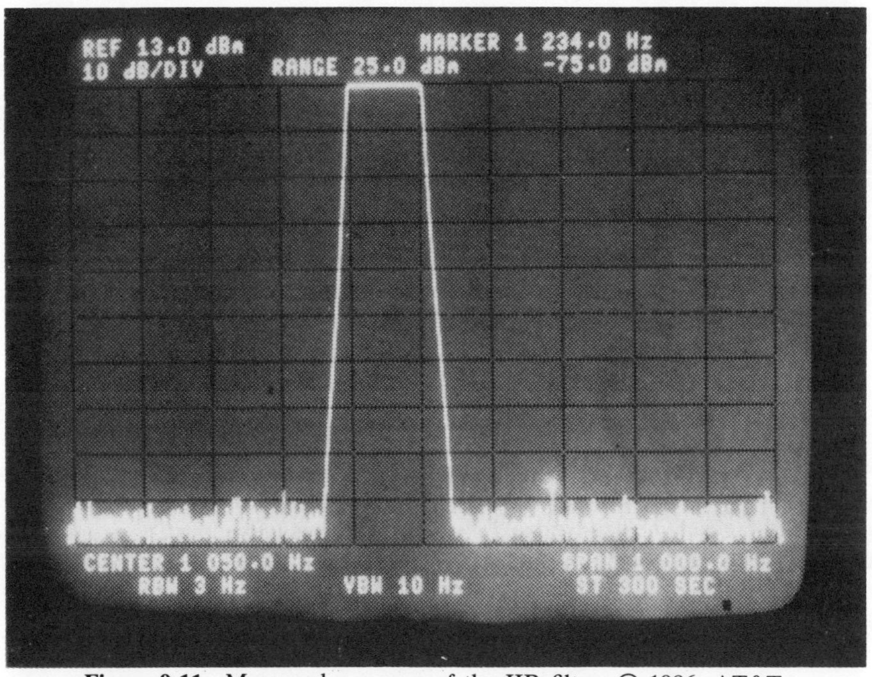

Figure 9.11 Measured response of the IIR filter. © 1986, AT&T.

In this example, the IIR filter coefficients were obtained directly from a computer synthesis program [33]. However, other programs for automatic synthesis and WEDSP32 implementation of digital filters are also available and run on personal computers [4]. The filter design shown above corresponds to a 0.01 dB ripple in the pass-band (950–1050 Hz) and 100 dB stop-band loss below 900 Hz and above 1100 Hz. Figure 9.11 shows the actual measured filter response on a HP3585A spectrum analyzer.

9.5. APPLICATION/ALGORITHM SPECIFIC FILTER INTEGRATED CIRCUITS

So far we have discussed how one can program a general purpose DSP chip to act as a specified digital filter. Another kind of monolithic digital filter processors is becoming available in the marketplace. Such chips incorporate sufficient hardware to implement specific filtering applications. These chips are not general purpose and are not software programmable. They implement, in hardware, low-order filtering algorithms. A sufficient number of them cascaded or connected in parallel over a common bus can implement a higher order filter. For instance, the IIR biquad sections [26], and the FIR filter building blocks [9, 38] are some examples of such products. The usefulness of these special purpose filter chips can be illustrated as follows. Consider the case of FIR filters working at video rates of 20 million samples/s. Most general-purpose DSPs can be programmed for FIR filtering. Since such general-purpose DSPs contain a single multiplier-accumulator (MAC), they can execute a single tap of an FIR filter in a single instruction cycle. For a 64-tap FIR filter with the DSP operating at 100 ns instruction cycle, it would take 6.4 μs to complete the processing of one sample on the DSP. The maximum sampling rate achievable in this technique is only 156 kHz which is much smaller than the video rate of 20 MHz.

Most vendors offering FIR filter chips implement the transpose of the direct form. By rearranging Eq. (9.2) in the following manner the transpose form results

$$H(z) = h[0] + z^{-1}(h[1] + z^{-1}(\cdots + z^{-1}h[N] \ldots) \tag{9.10}$$

A block diagram of the transpose form is shown in Figure 9.12. This is achieved in hardware by feeding the input sample to an array of multipliers, and multiple adder stages sum up the products from the current and previous stages. The transpose FIR filter can easily be partitioned into a cascade of lower order FIR sections. In Figure 9.12, a three taps partition is also shown. This allows multiple identical chips to be cascaded to form a high order FIR filter. Thus the trend is to fit in as many taps on a single chip as possible with the interface for cascade connection.

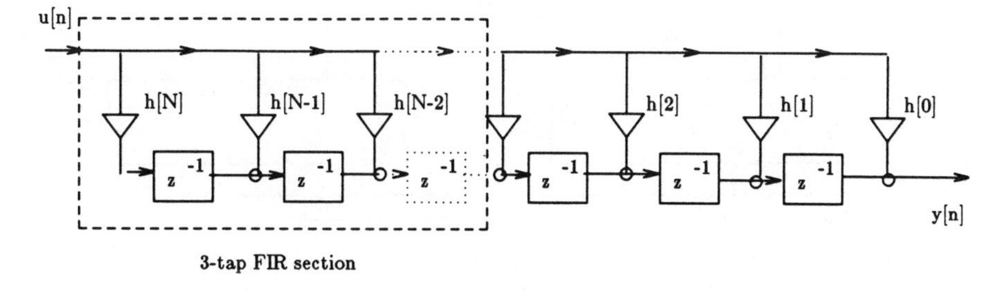

Figure 9.12 Transpose FIR filter.

9.5.1. Case Study

Several chips have been announced in the last few years mostly in the category of transversal filters. One such chip is IMS A100 which is a 16-bit, 32-stage cascadable digital transversal filter. It can be used in other digital signal processing applications, such as discrete Fourier transform (DFT), convolution, and correlation. A functional block diagram of the chip is shown in Figure 9.13. The input data wordlength is 16 bits, and coefficients

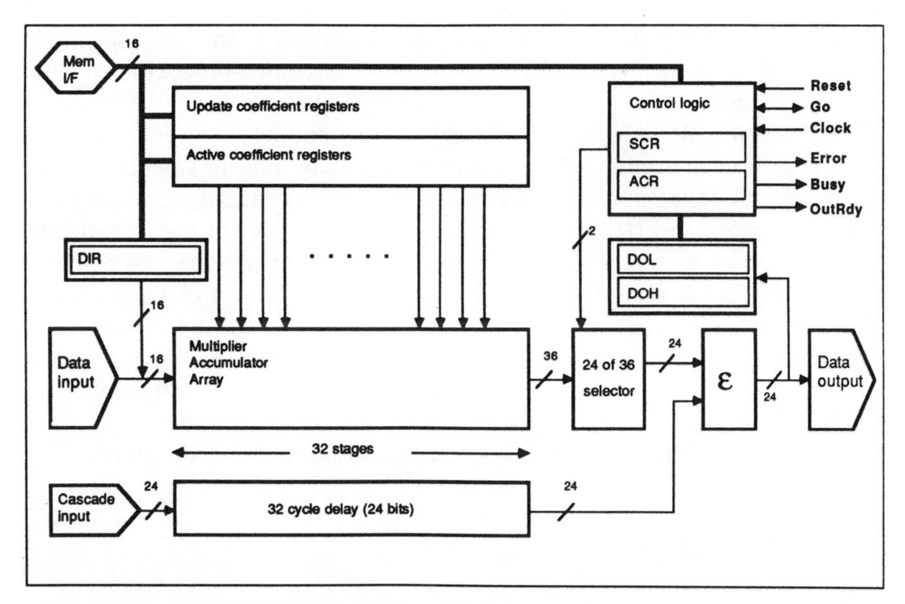

Figure 9.13 Functional block diagram of IMS A100. From H. Yassaie (1986) Discrete Fourier transform with the IMS A100. *IMS A100 Application Note 2*, © 1986, INMOS, Ltd.

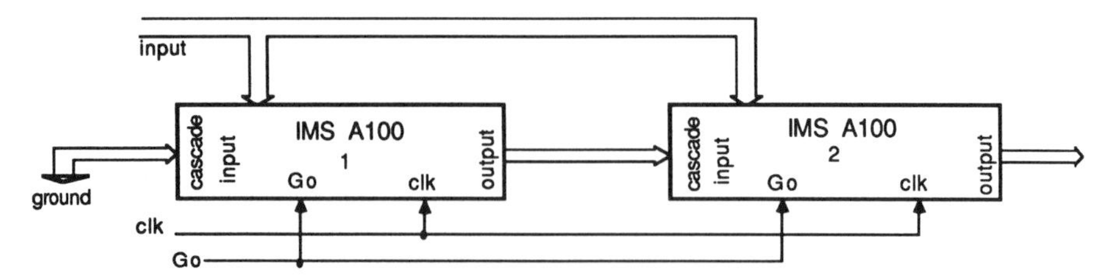

Figure 9.14 Cascading of IMS A100 chips. From H. Yassaie (1986) Digital filtering with the IMS A100. *IMS A100 Application Note 1*, © 1986, INMOS, Ltd.

are programmable to be 4, 8, 12, or 16 bits wide. Two's complement arithmetic is performed. The coefficients can be updated asynchronously, allowing the chip to perform adaptive algorithms as well.

Each data sample loaded into the IMS A100 is fed in parallel to all 32 multipliers on chip. At each multiplier the current input sample is multiplied by a coefficient stored in memory (from coefficient register), and added to the output of the previous stage delayed by one clock cycle.

The IMS A100 has four data interface ports. The memory interface port allows access to the coefficient registers, the configuration and status registers and the data input and output registers for the MAC array. Three ports are provided for high-speed data I/O and cascading of several devices. Typically an external microprocessor is needed to control a cascade of such devices over the memory interface.

There is a trade-off between the coefficient size and the speed of operation. For an IMS A100 operating at 20 MHz using 4-bit coefficients the maximum throughput is 10 MHz, similarly for 16-bit coefficients the throughput is proportionally degraded to 2.5 MHz. It should be mentioned that full 36-bit precision in accumulation is preserved in the interim calculations. A selected 24-bit result out of the 36-bit accumulator is output from the chip.

To implement an FIR filter of arbitrary order, a sufficient number of IMS A100 chips can be cascaded. The cascading does not involve any external components and is accomplished by connecting the output of the previous IC to the cascade input of the next chip and joining the data input ports together. In normal operation, the cascade input of the first device should be grounded. Figure 9.14 depicts the cascading of two IMS A100 devices. In this figure, we assume that the filter coefficients have been preloaded into the IMS A100 coefficient memories using external memory interface.

9.6. FUTURE TRENDS

In the arena of general purpose DSPs, finer line technologies will allow speedup, more parallelism in architecture and better memory management and I/O interfaces. Software support and peripheral devices will be enhanced. We notice that more microprocessor-like functions are being included into the DSPs, allowing them to be useful in nondigital-signal-processing applications as well. Many new application areas, where speed was a bottleneck, are becoming feasible with the ever increasing speed of commercially available DSPs. Features are provided to accommodate computational kernels other than multiply-add. One example of such a feature is the inclusion of a bit reversal instruction to facilitate fast Fourier transform (FFT) algorithms.

There is also a lack of standardization in floating-point operation, external interfaces, and so on, as has been true in the case of microprocessors. This manifests itself in a multiplicity of offerings from multiple vendors with no compatibility. In recent times high-level language compilers (notably in C) are being made available by various sources (such as AT&T, TI, etc.). Such compilers produce codes that may not be totally optimal, but do allow portability of applications to different DSPs.

On the other hand, with rapid advances in computer-aided (CAD) tools, versatile application or algorithm specific DSPs (ASDSPs) are being designed. This trend will continue to grow. There are some real-time applications that cannot be programmed into a general purpose DSP. In such cases special purpose ICs are the only solution. On the other hand, ASDSPs could be designed to make an application cost-effective. Some manufacturers of CAD tools provide tools for building your own VLSIs. In such situations, the system designer has to be aware of the architectural and technological possibilities and trade-offs. The level of flexibility of these systems in answering the design needs vary from capability of capturing the designs at the logic level [6] (standard cell library based design) to using prefabricated building blocks. Usually the software support for such semi-custom design is excellent and chip design turnaround time of a few weeks is quite feasible. These systems usually do not allow very high-performance circuit design and have to be used with good architectural choice to obtain a high-performance chip. This deficiency will be reduced in future with advances in CAD technology.

Some researchers are even providing CAD tools for automatically building the filters from high-level specifications [10, 19, 32, 37]. A problem with these tools is their limited flexibility in implementing any arbitrary architecture. More research is needed along these lines to make the tools attractive for automated design of complicated digital signal processing systems.

REFERENCES

1 A. Aliphas et al. (1985) High resolution digital filter chip. *I.E.E.E. Int. Conf. on Acoustics, Speech, and Signal Processing Record*, 224–227.

2 J. Allen (1985) Computer architecture for digital signal processing. *Proc. I.E.E.E.* **73**, 852–873.

3 ADSP-2100 Digital Signal Processor Data Sheet. Analog Devices, Inc., One Technology Way, Norwood, MA 02062, 1986.

4 Digital Filter Design Package (DFDP). Atlanta Signal Processors Inc. (ASPI), 770 Spring St. NW, Suite 208, Atlanta, GA 30308, 1984.

5 AT&T DSP Library Routines, AT&T, 555 Union Blvd., Allentown, PA 18103, 1986.

6 AT&T 1.25 micron CMOS Library: Standard Cells and Building Blocks. AT&T, 555 Union Blvd., Allentown, PA 18103, 1987.

7 J. R. Boddie et al. (1981) Digital signal processor: Architecture and performance. *Bell System Tech. J.* **60**, 1449–1462.

8 J. R. Boddie et al. (1986) The DSP32 digital signal processor and its application development tools. *AT&T Tech. J.* **65**, 89–104.

9 D. Bursky (1987) Three CMOS DSP chips rev up FIR filtering. *Electron. Design*, pp. 68–72.

10 P. R. Cappello and C.-W. Wu (1987) Computer-aided design of VLSI FIR filters. *Proc. I.E.E.E.* **75**, 1260–1271.

11 CARL Startup Kit (1985) Computer Audio Research Lab., Univ. of Calif., San Diego, La Jolla, CA 92093.

12 R. W. Daniels (1974) *Approximation Methods for Electronic Filter Design*. McGraw-Hill, New York.

13 Digital Signal Processing Committee (Editor) (1979) *Programs for Digital Signal Processing*. IEEE Press, New York.

14 D. Essig et al. (1986) A second-generation digital signal processor. *I.E.E.E. J. Solid-State Circuits* **21**, 86–91.

15 A. Fettweis (1971) Digital filter structures related to classical filter networks. *Arch. Elek. Ubertragung* **25**, 79–89.

16 H. Gambe et al. (1985) On the design of a high-performance LSI circuit digital signal processor for communication. *I.E.E.E. J. Selected Areas Commun.* **3**, 357–368.

17 C. J. Garen et al. (1987) A 60 ns CMOS DSP with on-chip instruction cache. *Int. Solid-State Circuits Conf. Digest of Papers*, pp. 156–157.

18 L. B. Jackson (1986) *Digital Filters and Signal Processing*. Kluwer, Boston, MA.

19 R. Jain et al. (1986) Custom design of a VLSI PCM-FDM transmultiplexer from system specifications to circuit layout using a computer-aided design system. *I.E.E.E. J. Solid State Circuits* **21**, 73–85.

20 Y. Kawakami et al. (1986) A 32 bit floating point CMOS digital signal processor. *Int. Solid-State Circuits Conf. Digest of Papers*, pp. 86–87.

21 R. N. Kershaw et al. (1985) A programmable digital signal processor with 32b floating point arithmetic, *Int. Solid-State Circuits Conf. Digest of Papers*, pp. 92–93.

22 H. Kikuchi et al. (1983) A 23K gate CMOS DSP with 100 ns multiplication. *Int. Solid-State Circuits Conf. Digest of Papers*, pp. 128–129.

23 P. M. Kogge (1981) *The Architecture of Pipelined Computers*. Hemisphere, Washington, DC.

24 D. A. Lager and S. G. Azevedo (1987) SIG — A general-purpose signal processing program. *Proc. I.E.E.E.* **75**, 1322–1332.

25 A. Lovrich and R. Simar, Jr. (1986) Implementation of FIR/IIR filters with the TMS32010/TMS32020. *Digital Signal Processing Applications with the TMS320 Family*. Texas Instruments, Inc., pp. 27–46.

26 P. Lowenstein (1985) Signal-processing chip whips through 1 million second order sections/sec. *Electron. Design*, pp. 141–152.

27 S. Magar et al. (1985) An NMOS digital signal processor with multiprocessing capability. *Int. Solid-State Circuits Conf. Digest of Papers*, pp. 90–91.

28 K. McDonough et al. (1982) Microcomputer with 32-bit arithmetic does high-precision number crunching. *Electronics*, pp. 105–110.

29 T. Nishitani et al. (1981) A single-chip digital signal processor for telecommunication applications. *I.E.E.E. J. Solid State Circuits* **16**, 372–376.

30 A. V. Oppenheim and R. W. Schafer (1975) *Digital Signal Processing*. Prentice-Hall, Englewood Cliffs, NJ.

31 L. R. Rabiner and B. Gold (1975) *Theory and Applications of Digital Signal Processing*. Prentice-Hall, Englewood Cliffs, NJ.

32 J. M. Rabaey, S. P. Pope and R. W. Brodersen (1985) An integrated automated layout generation system for DSP circuits. *I.E.E.E. Trans. Computer-Aided Design* **4**, 285–296.

33 G. Szentirmai (1977) FILSYN: a general purpose filter synthesis program. *Proc. I.E.E.E.* **65**, 1443–1458.

34 F. J. Taylor (1983) *Digital Filter Design Handbook*. Marcel Dekker, Inc., New York.

35 TMS32020 User's Guide (1985) Digital Signal Processor Products, Texas Instruments, Inc., P.O. Box 1443, MS 640, Houston, TX 77001, pp. 5.23–5.26.

36 M. V. Wilkes (1951) The best way to design an automated calculating machine. Manchester University Computer Inaugural Conference, Manchester, UK.

37 F. F. Yassa et al. (1987) A silicon compiler for digital signal processing: methodology, implementation, and applications. *Proc. I.E.E.E.* **75**, 1272–1282.

38 Zoran Digital Signal Processors Data Book, Zoran Corp., 3450 Central Expressway, Santa Clara, CA 95051, 1987.

39 A. I. Zverev (1967) *Handbook of Filter Design*. Wiley, New York.

Subject Index

M

N

O

V

W

Z